Third Edition

Data Structures and Algorithm Analysis in
Java™

Third Edition

Data Structures and Algorithm Analysis in

Java™

Mark Allen Weiss
Florida International University

PEARSON

Boston Columbus Indianapolis New York San Francisco Upper Saddle River
Amsterdam Cape Town Dubai London Madrid Milan Munich Paris Montreal Toronto
Delhi Mexico City Sao Paulo Sydney Hong Kong Seoul Singapore Taipei Tokyo

Editorial Director: Marcia Horton
Editor-in-Chief: Michael Hirsch
Editorial Assistant: Emma Snider
Director of Marketing: Patrice Jones
Marketing Manager: Yezan Alayan
Marketing Coordinator: Kathryn Ferranti
Director of Production: Vince O'Brien
Managing Editor: Jeff Holcomb
Production Project Manager: Kayla
   Smith-Tarbox

Project Manager: Pat Brown
Manufacturing Buyer: Pat Brown
Art Director: Jayne Conte
Cover Designer: Bruce Kenselaar
Cover Photo: © De-Kay Dreamstime.com
Media Editor: Daniel Sandin
Full-Service Project Management: Integra
Composition: Integra
Printer/Binder: LSC Communications
Cover Printer: LSC Communications
Text Font: Berkeley-Book

Many of the designations by manufacturers and sellers to distinguish their products are claimed as trademarks. Where those designations appear in this book, and the publisher was aware of a trademark claim, the designations have been printed in initial caps or all caps.

Library of Congress Cataloging-in-Publication Data

Weiss, Mark Allen.
   Data structures and algorithm analysis in Java / Mark Allen Weiss. – 3rd ed.
     p.   cm.
   ISBN-13: 978-0-13-257627-7 (alk. paper)
   ISBN-10: 0-13-257627-9 (alk. paper)
   1. Java (Computer program language)   2. Data structures (Computer science)
   3. Computer algorithms.   I. Title.
   QA76.73.J38W448 2012
   005.1–dc23   2011035536

13 16

ISBN 10: 0-13-257627-9
ISBN 13: 9780-13-257627-7

To the love of my life, Jill.

CONTENTS

Chapter 4   Trees   101

Chapter 7   Sorting   **271**

Chapter 10 Algorithm Design Techniques   429

Chapter 11   Amortized Analysis   513

Chapter 12   Advanced Data Structures
and Implementation   541

PREFACE

Purpose/Goals

This new Java edition describes *data structures*, methods of organizing large amounts of data, and *algorithm analysis*, the estimation of the running time of algorithms. As computers become faster and faster, the need for programs that can handle large amounts of input becomes more acute. Paradoxically, this requires more careful attention to efficiency, since inefficiencies in programs become most obvious when input sizes are large. By analyzing an algorithm before it is actually coded, students can decide if a particular solution will be feasible. For example, in this text students look at specific problems and see how careful implementations can reduce the time constraint for large amounts of data from centuries to less than a second. Therefore, no algorithm or data structure is presented without an explanation of its running time. In some cases, minute details that affect the running time of the implementation are explored.

Once a solution method is determined, a program must still be written. As computers have become more powerful, the problems they must solve have become larger and more complex, requiring development of more intricate programs. The goal of this text is to teach students good programming and algorithm analysis skills simultaneously so that they can develop such programs with the maximum amount of efficiency.

This book is suitable for either an advanced data structures (CS7) course or a first-year graduate course in algorithm analysis. Students should have some knowledge of intermediate programming, including such topics as object-based programming and recursion, and some background in discrete math.

Summary of the Most Significant Changes in the Third Edition

The third edition incorporates numerous bug fixes, and many parts of the book have undergone revision to increase the clarity of presentation. In addition,

- Chapter 4 includes implementation of the AVL tree deletion algorithm—a topic often requested by readers.

- Chapter 5 has been extensively revised and enlarged and now contains material on two newer algorithms: cuckoo hashing and hopscotch hashing. Additionally, a new section on universal hashing has been added.

- Chapter 7 now contains material on radix sort, and a new section on lower bound proofs has been added.

- Chapter 8 uses the new union/find analysis by Seidel and Sharir, and shows the $O(M\alpha(M, N))$ bound instead of the weaker $O(M \log^* N)$ bound in prior editions.

- Chapter 12 adds material on suffix trees and suffix arrays, including the linear-time suffix array construction algorithm by Karkkainen and Sanders (with implementation). The sections covering deterministic skip lists and AA-trees have been removed.

- Throughout the text, the code has been updated to use the diamond operator from Java 7.

Approach

Although the material in this text is largely language independent, programming requires the use of a specific language. As the title implies, we have chosen Java for this book.

Java is often examined in comparison with C++. Java offers many benefits, and programmers often view Java as a safer, more portable, and easier-to-use language than C++. As such, it makes a fine core language for discussing and implementing fundamental data structures. Other important parts of Java, such as threads and its GUI, although important, are not needed in this text and thus are not discussed.

Complete versions of the data structures, in both Java and C++, are available on the Internet. We use similar coding conventions to make the parallels between the two languages more evident.

Overview

Chapter 1 contains review material on discrete math and recursion. I believe the only way to be comfortable with recursion is to see good uses over and over. Therefore, recursion is prevalent in this text, with examples in every chapter except Chapter 5. Chapter 1 also presents material that serves as a review of inheritance in Java. Included is a discussion of Java generics.

Chapter 2 deals with algorithm analysis. This chapter explains asymptotic analysis and its major weaknesses. Many examples are provided, including an in-depth explanation of logarithmic running time. Simple recursive programs are analyzed by intuitively converting them into iterative programs. More complicated divide-and-conquer programs are introduced, but some of the analysis (solving recurrence relations) is implicitly delayed until Chapter 7, where it is performed in detail.

Chapter 3 covers lists, stacks, and queues. This chapter has been significantly revised from prior editions. It now includes a discussion of the Collections API `ArrayList` and `LinkedList` classes, and it provides implementations of a significant subset of the collections API `ArrayList` and `LinkedList` classes.

Chapter 4 covers trees, with an emphasis on search trees, including external search trees (B-trees). The UNIX file system and expression trees are used as examples. AVL trees and splay trees are introduced. More careful treatment of search tree implementation details is found in Chapter 12. Additional coverage of trees, such as file compression and game trees, is deferred until Chapter 10. Data structures for an external medium are considered as the final topic in several chapters. New to this edition is a discussion of the Collections API `TreeSet` and `TreeMap` classes, including a significant example that illustrates the use of three separate maps to efficiently solve a problem.

Chapter 5 discusses hash tables, including the classic algorithms such as separate chaining and linear and quadratic probing, as well as several newer algorithms, namely cuckoo hashing and hopscotch hashing. Universal hashing is also discussed, and extendible hashing is covered at the end of the chapter.

Chapter 6 is about priority queues. Binary heaps are covered, and there is additional material on some of the theoretically interesting implementations of priority queues. The Fibonacci heap is discussed in Chapter 11, and the pairing heap is discussed in Chapter 12.

Chapter 7 covers sorting. It is very specific with respect to coding details and analysis. All the important general-purpose sorting algorithms are covered and compared. Four algorithms are analyzed in detail: insertion sort, Shellsort, heapsort, and quicksort. New to this edition is radix sort and lower bound proofs for selection-related problems. External sorting is covered at the end of the chapter.

Chapter 8 discusses the disjoint set algorithm with proof of the running time. The analysis is new. This is a short and specific chapter that can be skipped if Kruskal's algorithm is not discussed.

Chapter 9 covers graph algorithms. Algorithms on graphs are interesting, not only because they frequently occur in practice, but also because their running time is so heavily dependent on the proper use of data structures. Virtually all the standard algorithms are presented along with appropriate data structures, pseudocode, and analysis of running time. To place these problems in a proper context, a short discussion on complexity theory (including *NP*-completeness and undecidability) is provided.

Chapter 10 covers algorithm design by examining common problem-solving techniques. This chapter is heavily fortified with examples. Pseudocode is used in these later chapters so that the student's appreciation of an example algorithm is not obscured by implementation details.

Chapter 11 deals with amortized analysis. Three data structures from Chapters 4 and 6 and the Fibonacci heap, introduced in this chapter, are analyzed.

Chapter 12 covers search tree algorithms, the suffix tree and array, the k-d tree, and the pairing heap. This chapter departs from the rest of the text by providing complete and careful implementations for the search trees and pairing heap. The material is structured so that the instructor can integrate sections into discussions from other chapters. For example, the top-down red-black tree in Chapter 12 can be discussed along with AVL trees (in Chapter 4).

Chapters 1–9 provide enough material for most one-semester data structures courses. If time permits, then Chapter 10 can be covered. A graduate course on algorithm analysis could cover Chapters 7–11. The advanced data structures analyzed in Chapter 11 can easily be referred to in the earlier chapters. The discussion of *NP*-completeness in Chapter 9 is far too brief to be used in such a course. You might find it useful to use an additional work on *NP*-completeness to augment this text.

Exercises

Exercises, provided at the end of each chapter, match the order in which material is presented. The last exercises may address the chapter as a whole rather than a specific section. Difficult exercises are marked with an asterisk, and more challenging exercises have two asterisks.

References

References are placed at the end of each chapter. Generally the references either are historical, representing the original source of the material, or they represent extensions and improvements to the results given in the text. Some references represent solutions to exercises.

Supplements

The following supplements are available to all readers at
www.pearsonhighered.com/cssupport:

- Source code for example programs

In addition, the following material is available only to qualified instructors at Pearson's Instructor Resource Center (www.pearsonhighered.com/irc). Visit the IRC or contact your campus Pearson representative for access.

- Solutions to selected exercises
- Figures from the book

Acknowledgments

Many, many people have helped me in the preparation of books in this series. Some are listed in other versions of the book; thanks to all.

As usual, the writing process was made easier by the professionals at Pearson. I'd like to thank my editor, Michael Hirsch, and production editor, Pat Brown. I'd also like to thank Abinaya Rajendran and her team in Integra Software Services for their fine work putting the final pieces together. My wonderful wife Jill deserves extra special thanks for everything she does.

Finally, I'd like to thank the numerous readers who have sent e-mail messages and pointed out errors or inconsistencies in earlier versions. My World Wide Web page www.cis.fiu.edu/~weiss contains updated source code (in Java and C++), an errata list, and a link to submit bug reports.

M.A.W.
Miami, Florida

CHAPTER 1

Introduction

In this chapter, we discuss the aims and goals of this text and briefly review programming concepts and discrete mathematics. We will

- See that how a program performs for reasonably large input is just as important as its performance on moderate amounts of input.
- Summarize the basic mathematical background needed for the rest of the book.
- Briefly review recursion.
- Summarize some important features of Java that are used throughout the text.

1.1 What's the Book About?

Suppose you have a group of N numbers and would like to determine the kth largest. This is known as the **selection problem**. Most students who have had a programming course or two would have no difficulty writing a program to solve this problem. There are quite a few "obvious" solutions.

One way to solve this problem would be to read the N numbers into an array, sort the array in decreasing order by some simple algorithm such as bubblesort, and then return the element in position k.

A somewhat better algorithm might be to read the first k elements into an array and sort them (in decreasing order). Next, each remaining element is read one by one. As a new element arrives, it is ignored if it is smaller than the kth element in the array. Otherwise, it is placed in its correct spot in the array, bumping one element out of the array. When the algorithm ends, the element in the kth position is returned as the answer.

Both algorithms are simple to code, and you are encouraged to do so. The natural questions, then, are which algorithm is better and, more important, is either algorithm good enough? A simulation using a random file of 30 million elements and $k = 15,000,000$ will show that neither algorithm finishes in a reasonable amount of time; each requires several days of computer processing to terminate (albeit eventually with a correct answer). An alternative method, discussed in Chapter 7, gives a solution in about a second. Thus, although our proposed algorithms work, they cannot be considered good algorithms, because they are entirely impractical for input sizes that a third algorithm can handle in a reasonable amount of time.

	1	2	3	4
1	t	h	i	s
2	w	a	t	s
3	o	a	h	g
4	f	g	d	t

Figure 1.1 Sample word puzzle

A second problem is to solve a popular word puzzle. The input consists of a two-dimensional array of letters and a list of words. The object is to find the words in the puzzle. These words may be horizontal, vertical, or diagonal in any direction. As an example, the puzzle shown in Figure 1.1 contains the words *this, two, fat,* and *that.* The word *this* begins at row 1, column 1, or (1,1), and extends to (1,4); *two* goes from (1,1) to (3,1); *fat* goes from (4,1) to (2,3); and *that* goes from (4,4) to (1,1).

Again, there are at least two straightforward algorithms that solve the problem. For each word in the word list, we check each ordered triple (*row, column, orientation*) for the presence of the word. This amounts to lots of nested for loops but is basically straightforward.

Alternatively, for each ordered quadruple (*row, column, orientation, number of characters*) that doesn't run off an end of the puzzle, we can test whether the word indicated is in the word list. Again, this amounts to lots of nested for loops. It is possible to save some time if the maximum number of characters in any word is known.

It is relatively easy to code up either method of solution and solve many of the real-life puzzles commonly published in magazines. These typically have 16 rows, 16 columns, and 40 or so words. Suppose, however, we consider the variation where only the puzzle board is given and the word list is essentially an English dictionary. Both of the solutions proposed require considerable time to solve this problem and therefore are not acceptable. However, it is possible, even with a large word list, to solve the problem in a matter of seconds.

An important concept is that, in many problems, writing a working program is not good enough. If the program is to be run on a large data set, then the running time becomes an issue. Throughout this book we will see how to estimate the running time of a program for large inputs and, more important, how to compare the running times of two programs without actually coding them. We will see techniques for drastically improving the speed of a program and for determining program bottlenecks. These techniques will enable us to find the section of the code on which to concentrate our optimization efforts.

1.2 Mathematics Review

This section lists some of the basic formulas you need to memorize or be able to derive and reviews basic proof techniques.

1.2.1 Exponents

$$X^A X^B = X^{A+B}$$

$$\frac{X^A}{X^B} = X^{A-B}$$

$$(X^A)^B = X^{AB}$$

$$X^N + X^N = 2X^N \neq X^{2N}$$

$$2^N + 2^N = 2^{N+1}$$

1.2.2 Logarithms

In computer science, all logarithms are to the base 2 unless specified otherwise.

Definition 1.1.
$X^A = B$ if and only if $\log_X B = A$
    Several convenient equalities follow from this definition.

Theorem 1.1.

$$\log_A B = \frac{\log_C B}{\log_C A}; \quad A, B, C > 0, \ A \neq 1$$

Proof.
Let $X = \log_C B$, $Y = \log_C A$, and $Z = \log_A B$. Then, by the definition of logarithms, $C^X = B$, $C^Y = A$, and $A^Z = B$. Combining these three equalities yields $C^X = B = (C^Y)^Z$. Therefore, $X = YZ$, which implies $Z = X/Y$, proving the theorem.

Theorem 1.2.

$$\log AB = \log A + \log B; \ A, B > 0$$

Proof.
Let $X = \log A$, $Y = \log B$, and $Z = \log AB$. Then, assuming the default base of 2, $2^X = A$, $2^Y = B$, and $2^Z = AB$. Combining the last three equalities yields $2^X 2^Y = AB = 2^Z$. Therefore, $X + Y = Z$, which proves the theorem.

Some other useful formulas, which can all be derived in a similar manner, follow.

$$\log A/B = \log A - \log B$$

$$\log(A^B) = B \log A$$

$$\log X < X \quad \text{for all } X > 0$$

$$\log 1 = 0, \quad \log 2 = 1, \quad \log 1{,}024 = 10, \quad \log 1{,}048{,}576 = 20$$

1.2.3 Series

The easiest formulas to remember are

$$\sum_{i=0}^{N} 2^i = 2^{N+1} - 1$$

and the companion,

$$\sum_{i=0}^{N} A^i = \frac{A^{N+1} - 1}{A - 1}$$

In the latter formula, if $0 < A < 1$, then

$$\sum_{i=0}^{N} A^i \leq \frac{1}{1 - A}$$

and as N tends to ∞, the sum approaches $1/(1 - A)$. These are the "geometric series" formulas.

We can derive the last formula for $\sum_{i=0}^{\infty} A^i$ $(0 < A < 1)$ in the following manner. Let S be the sum. Then

$$S = 1 + A + A^2 + A^3 + A^4 + A^5 + \cdots$$

Then

$$AS = A + A^2 + A^3 + A^4 + A^5 + \cdots$$

If we subtract these two equations (which is permissible only for a convergent series), virtually all the terms on the right side cancel, leaving

$$S - AS = 1$$

which implies that

$$S = \frac{1}{1 - A}$$

We can use this same technique to compute $\sum_{i=1}^{\infty} i/2^i$, a sum that occurs frequently. We write

$$S = \frac{1}{2} + \frac{2}{2^2} + \frac{3}{2^3} + \frac{4}{2^4} + \frac{5}{2^5} + \cdots$$

and multiply by 2, obtaining

$$2S = 1 + \frac{2}{2} + \frac{3}{2^2} + \frac{4}{2^3} + \frac{5}{2^4} + \frac{6}{2^5} + \cdots$$

Subtracting these two equations yields

$$S = 1 + \frac{1}{2} + \frac{1}{2^2} + \frac{1}{2^3} + \frac{1}{2^4} + \frac{1}{2^5} + \cdots$$

Thus, $S = 2$.

Another type of common series in analysis is the arithmetic series. Any such series can be evaluated from the basic formula.

$$\sum_{i=1}^{N} i = \frac{N(N+1)}{2} \approx \frac{N^2}{2}$$

For instance, to find the sum $2 + 5 + 8 + \cdots + (3k - 1)$, rewrite it as $3(1 + 2 + 3 + \cdots + k) - (1 + 1 + 1 + \cdots + 1)$, which is clearly $3k(k+1)/2 - k$. Another way to remember this is to add the first and last terms (total $3k + 1$), the second and next to last terms (total $3k + 1$), and so on. Since there are $k/2$ of these pairs, the total sum is $k(3k + 1)/2$, which is the same answer as before.

The next two formulas pop up now and then but are fairly uncommon.

$$\sum_{i=1}^{N} i^2 = \frac{N(N+1)(2N+1)}{6} \approx \frac{N^3}{3}$$

$$\sum_{i=1}^{N} i^k \approx \frac{N^{k+1}}{|k+1|} \quad k \neq -1$$

When $k = -1$, the latter formula is not valid. We then need the following formula, which is used far more in computer science than in other mathematical disciplines. The numbers H_N are known as the harmonic numbers, and the sum is known as a harmonic sum. The error in the following approximation tends to $\gamma \approx 0.57721566$, which is known as **Euler's constant**.

$$H_N = \sum_{i=1}^{N} \frac{1}{i} \approx \log_e N$$

These two formulas are just general algebraic manipulations.

$$\sum_{i=1}^{N} f(N) = N f(N)$$

$$\sum_{i=n_0}^{N} f(i) = \sum_{i=1}^{N} f(i) - \sum_{i=1}^{n_0-1} f(i)$$

1.2.4 Modular Arithmetic

We say that A is congruent to B modulo N, written $A \equiv B \pmod{N}$, if N divides $A - B$. Intuitively, this means that the remainder is the same when either A or B is divided by N. Thus, $81 \equiv 61 \equiv 1 \pmod{10}$. As with equality, if $A \equiv B \pmod{N}$, then $A + C \equiv B + C \pmod{N}$ and $AD \equiv BD \pmod{N}$.

Often, N is a prime number. In that case, there are three important theorems.

First, if N is prime, then $ab \equiv 0 \pmod{N}$ is true if and only if $a \equiv 0 \pmod{N}$ or $b \equiv 0 \pmod{N}$. In other words, if a prime number N divides a product of two numbers, it divides at least one of the two numbers.

Second, if N is prime, then the equation $ax \equiv 1 \pmod{N}$ has a unique solution $\pmod{N}$, for all $0 < a < N$. This solution $0 < x < N$, is the *multiplicative inverse*.

Third, if N is prime, then the equation $x^2 \equiv a \pmod{N}$ has either two solutions $\pmod{N}$, for all $0 < a < N$, or no solutions.

There are many theorems that apply to modular arithmetic, and some of them require extraordinary proofs in number theory. We will use modular arithmetic sparingly, and the preceding theorems will suffice.

1.2.5 The *P* Word

The two most common ways of proving statements in data structure analysis are proof by induction and proof by contradiction (and occasionally proof by intimidation, used by professors only). The best way of proving that a theorem is false is by exhibiting a counterexample.

Proof by Induction

A proof by induction has two standard parts. The first step is proving a *base case*, that is, establishing that a theorem is true for some small (usually degenerate) value(s); this step is almost always trivial. Next, an **inductive hypothesis** is assumed. Generally this means that the theorem is assumed to be true for all cases up to some limit k. Using this assumption, the theorem is then shown to be true for the next value, which is typically $k + 1$. This proves the theorem (as long as k is finite).

As an example, we prove that the Fibonacci numbers, $F_0 = 1, F_1 = 1, F_2 = 2, F_3 = 3, F_4 = 5, \ldots, F_i = F_{i-1} + F_{i-2}$, satisfy $F_i < (5/3)^i$, for $i \geq 1$. (Some definitions have $F_0 = 0$, which shifts the series.) To do this, we first verify that the theorem is true for the trivial cases. It is easy to verify that $F_1 = 1 < 5/3$ and $F_2 = 2 < 25/9$; this proves the basis. We assume that the theorem is true for $i = 1, 2, \ldots, k$; this is the inductive hypothesis. To prove the theorem, we need to show that $F_{k+1} < (5/3)^{k+1}$. We have

$$F_{k+1} = F_k + F_{k-1}$$

by the definition, and we can use the inductive hypothesis on the right-hand side, obtaining

$$F_{k+1} < (5/3)^k + (5/3)^{k-1}$$
$$< (3/5)(5/3)^{k+1} + (3/5)^2(5/3)^{k+1}$$
$$< (3/5)(5/3)^{k+1} + (9/25)(5/3)^{k+1}$$

which simplifies to

$$F_{k+1} < (3/5 + 9/25)(5/3)^{k+1}$$
$$< (24/25)(5/3)^{k+1}$$
$$< (5/3)^{k+1}$$

proving the theorem.

As a second example, we establish the following theorem.

Theorem 1.3.
If $N \geq 1$, then $\sum_{i=1}^{N} i^2 = \frac{N(N+1)(2N+1)}{6}$

Proof.
The proof is by induction. For the basis, it is readily seen that the theorem is true when $N = 1$. For the inductive hypothesis, assume that the theorem is true for $1 \leq k \leq N$. We will establish that, under this assumption, the theorem is true for $N + 1$. We have

$$\sum_{i=1}^{N+1} i^2 = \sum_{i=1}^{N} i^2 + (N+1)^2$$

Applying the inductive hypothesis, we obtain

$$\sum_{i=1}^{N+1} i^2 = \frac{N(N+1)(2N+1)}{6} + (N+1)^2$$
$$= (N+1)\left[\frac{N(2N+1)}{6} + (N+1)\right]$$
$$= (N+1)\frac{2N^2 + 7N + 6}{6}$$
$$= \frac{(N+1)(N+2)(2N+3)}{6}$$

Thus,

$$\sum_{i=1}^{N+1} i^2 = \frac{(N+1)[(N+1)+1][2(N+1)+1]}{6}$$

proving the theorem.

Proof by Counterexample

The statement $F_k \leq k^2$ is false. The easiest way to prove this is to compute $F_{11} = 144 > 11^2$.

Proof by Contradiction

Proof by contradiction proceeds by assuming that the theorem is false and showing that this assumption implies that some known property is false, and hence the original assumption was erroneous. A classic example is the proof that there is an infinite number of primes. To prove this, we assume that the theorem is false, so that there is some largest prime P_k. Let $P_1, P_2, \ldots, P_k$ be all the primes in order and consider

$$N = P_1 P_2 P_3 \cdots P_k + 1$$

Clearly, N is larger than P_k, so by assumption N is not prime. However, none of $P_1, P_2, \ldots, P_k$ divides N exactly, because there will always be a remainder of 1. This is a contradiction, because every number is either prime or a product of primes. Hence, the original assumption, that P_k is the largest prime, is false, which implies that the theorem is true.

1.3 A Brief Introduction to Recursion

Most mathematical functions that we are familiar with are described by a simple formula. For instance, we can convert temperatures from Fahrenheit to Celsius by applying the formula

$$C = 5(F - 32)/9$$

Given this formula, it is trivial to write a Java method; with declarations and braces removed, the one-line formula translates to one line of Java.

Mathematical functions are sometimes defined in a less standard form. As an example, we can define a function f, valid on nonnegative integers, that satisfies $f(0) = 0$ and $f(x) = 2f(x - 1) + x^2$. From this definition we see that $f(1) = 1, f(2) = 6, f(3) = 21$, and $f(4) = 58$. A function that is defined in terms of itself is called **recursive**. Java allows functions to be recursive.[1] It is important to remember that what Java provides is merely an attempt to follow the recursive spirit. Not all mathematically recursive functions are efficiently (or correctly) implemented by Java's simulation of recursion. The idea is that the recursive function f ought to be expressible in only a few lines, just like a nonrecursive function. Figure 1.2 shows the recursive implementation of f.

Lines 3 and 4 handle what is known as the **base case**, that is, the value for which the function is directly known without resorting to recursion. Just as declaring $f(x) = 2f(x-1) + x^2$ is meaningless, mathematically, without including the fact that $f(0) = 0$, the recursive Java method doesn't make sense without a base case. Line 6 makes the recursive call.

There are several important and possibly confusing points about recursion. A common question is: Isn't this just circular logic? The answer is that although we are defining a method in terms of itself, we are not defining a particular instance of the method in terms of itself. In other words, evaluating $f(5)$ by computing $f(5)$ would be circular. Evaluating $f(5)$ by computing $f(4)$ is not circular—unless, of course, $f(4)$ is evaluated by eventually computing $f(5)$. The two most important issues are probably the *how* and *why* questions.

---

[1] Using recursion for numerical calculations is usually a bad idea. We have done so to illustrate the basic points.

```
1        public static int f( int x )
2        {
3            if( x == 0 )
4                return 0;
5            else
6                return 2 * f( x - 1 ) + x * x;
7        }
```

Figure 1.2 A recursive method

In Chapter 3, the *how* and *why* issues are formally resolved. We will give an incomplete description here.

It turns out that recursive calls are handled no differently from any others. If f is called with the value of 4, then line 6 requires the computation of $2 * f(3) + 4 * 4$. Thus, a call is made to compute $f(3)$. This requires the computation of $2 * f(2) + 3 * 3$. Therefore, another call is made to compute $f(2)$. This means that $2 * f(1) + 2 * 2$ must be evaluated. To do so, $f(1)$ is computed as $2 * f(0) + 1 * 1$. Now, $f(0)$ must be evaluated. Since this is a base case, we know a priori that $f(0) = 0$. This enables the completion of the calculation for $f(1)$, which is now seen to be 1. Then $f(2)$, $f(3)$, and finally $f(4)$ can be determined. All the bookkeeping needed to keep track of pending calls (those started but waiting for a recursive call to complete), along with their variables, is done by the computer automatically. An important point, however, is that recursive calls will keep on being made until a base case is reached. For instance, an attempt to evaluate $f(-1)$ will result in calls to $f(-2)$, $f(-3)$, and so on. Since this will never get to a base case, the program won't be able to compute the answer (which is undefined anyway). Occasionally, a much more subtle error is made, which is exhibited in Figure 1.3. The error in Figure 1.3 is that bad(1) is defined, by line 6, to be bad(1). Obviously, this doesn't give any clue as to what bad(1) actually is. The computer will thus repeatedly make calls to bad(1) in an attempt to resolve its values. Eventually, its bookkeeping system will run out of space, and the program will terminate abnormally. Generally, we would say that this method doesn't work for one special case but is correct otherwise. This isn't true here, since bad(2) calls bad(1). Thus, bad(2) cannot be evaluated either. Furthermore, bad(3), bad(4), and bad(5) all make calls to bad(2). Since bad(2) is unevaluable, none of these values are either. In fact, this

```
1        public static int bad( int n )
2        {
3            if( n == 0 )
4                return 0;
5            else
6                return bad( n / 3 + 1 ) + n - 1;
7        }
```

Figure 1.3 A nonterminating recursive method

program doesn't work for any nonnegative value of n, except 0. With recursive programs, there is no such thing as a "special case."

These considerations lead to the first two fundamental rules of recursion:

1. *Base cases.* You must always have some base cases, which can be solved without recursion.
2. *Making progress.* For the cases that are to be solved recursively, the recursive call must always be to a case that makes progress toward a base case.

Throughout this book, we will use recursion to solve problems. As an example of a nonmathematical use, consider a large dictionary. Words in dictionaries are defined in terms of other words. When we look up a word, we might not always understand the definition, so we might have to look up words in the definition. Likewise, we might not understand some of those, so we might have to continue this search for a while. Because the dictionary is finite, eventually either (1) we will come to a point where we understand all of the words in some definition (and thus understand that definition and retrace our path through the other definitions) or (2) we will find that the definitions are circular and we are stuck, or that some word we need to understand for a definition is not in the dictionary.

Our recursive strategy to understand words is as follows: If we know the meaning of a word, then we are done; otherwise, we look the word up in the dictionary. If we understand all the words in the definition, we are done; otherwise, we figure out what the definition means by *recursively* looking up the words we don't know. This procedure will terminate if the dictionary is well defined but can loop indefinitely if a word is either not defined or circularly defined.

Printing Out Numbers

Suppose we have a positive integer, n, that we wish to print out. Our routine will have the heading printOut(n). Assume that the only I/O routines available will take a single-digit number and output it to the terminal. We will call this routine printDigit; for example, printDigit(4) will output a 4 to the terminal.

Recursion provides a very clean solution to this problem. To print out 76234, we need to first print out 7623 and then print out 4. The second step is easily accomplished with the statement printDigit(n%10), but the first doesn't seem any simpler than the original problem. Indeed it is virtually the same problem, so we can solve it recursively with the statement printOut(n/10).

This tells us how to solve the general problem, but we still need to make sure that the program doesn't loop indefinitely. Since we haven't defined a base case yet, it is clear that we still have something to do. Our base case will be printDigit(n) if $0 \leq n < 10$. Now printOut(n) is defined for every positive number from 0 to 9, and larger numbers are defined in terms of a smaller positive number. Thus, there is no cycle. The entire method is shown in Figure 1.4.

```
1        public static void printOut( int n )  /* Print nonnegative n */
2        {
3            if( n >= 10 )
4                printOut( n / 10 );
5            printDigit( n % 10 );
6        }
```

Figure 1.4　Recursive routine to print an integer

We have made no effort to do this efficiently. We could have avoided using the mod routine (which can be very expensive) because $n\%10 = n - \lfloor n/10 \rfloor * 10$.[2]

Recursion and Induction

Let us prove (somewhat) rigorously that the recursive number-printing program works. To do so, we'll use a proof by induction.

Theorem 1.4.

The recursive number-printing algorithm is correct for $n \geq 0$.

Proof (by induction on the number of digits in n).

First, if n has one digit, then the program is trivially correct, since it merely makes a call to printDigit. Assume then that printOut works for all numbers of k or fewer digits. A number of $k + 1$ digits is expressed by its first k digits followed by its least significant digit. But the number formed by the first k digits is exactly $\lfloor n/10 \rfloor$, which, by the inductive hypothesis, is correctly printed, and the last digit is n mod 10, so the program prints out any $(k + 1)$-digit number correctly. Thus, by induction, all numbers are correctly printed.

This proof probably seems a little strange in that it is virtually identical to the algorithm description. It illustrates that in designing a recursive program, all smaller instances of the same problem (which are on the path to a base case) may be *assumed* to work correctly. The recursive program needs only to combine solutions to smaller problems, which are "magically" obtained by recursion, into a solution for the current problem. The mathematical justification for this is proof by induction. This gives the third rule of recursion:

3. *Design rule.* Assume that all the recursive calls work.

This rule is important because it means that when designing recursive programs, you generally don't need to know the details of the bookkeeping arrangements, and you don't have to try to trace through the myriad of recursive calls. Frequently, it is extremely difficult to track down the actual sequence of recursive calls. Of course, in many cases this is an indication of a good use of recursion, since the computer is being allowed to work out the complicated details.

---

[2] $\lfloor x \rfloor$ is the largest integer that is less than or equal to x.

The main problem with recursion is the hidden bookkeeping costs. Although these costs are almost always justifiable, because recursive programs not only simplify the algorithm design but also tend to give cleaner code, recursion should never be used as a substitute for a simple for loop. We'll discuss the overhead involved in recursion in more detail in Section 3.6.

When writing recursive routines, it is crucial to keep in mind the four basic rules of recursion:

1. *Base cases.* You must always have some base cases, which can be solved without recursion.

2. *Making progress.* For the cases that are to be solved recursively, the recursive call must always be to a case that makes progress toward a base case.

3. *Design rule.* Assume that all the recursive calls work.

4. *Compound interest rule.* Never duplicate work by solving the same instance of a problem in separate recursive calls.

The fourth rule, which will be justified (along with its nickname) in later sections, is the reason that it is generally a bad idea to use recursion to evaluate simple mathematical functions, such as the Fibonacci numbers. As long as you keep these rules in mind, recursive programming should be straightforward.

1.4 Implementing Generic Components Pre-Java 5

An important goal of object-oriented programming is the support of code reuse. An important mechanism that supports this goal is the **generic** mechanism: If the implementation is identical except for the basic type of the object, a **generic implementation** can be used to describe the basic functionality. For instance, a method can be written to sort an array of items; the *logic* is independent of the types of objects being sorted, so a generic method could be used.

Unlike many of the newer languages (such as C++, which uses templates to implement generic programming), before version 1.5, Java did not support generic implementations directly. Instead, generic programming was implemented using the basic concepts of inheritance. This section describes how generic methods and classes can be implemented in Java using the basic principles of inheritance.

Direct support for generic methods and classes was announced by Sun in June 2001 as a future language addition. Finally, in late 2004, Java 5 was released and provided support for generic methods and classes. However, using generic classes requires an understanding of the pre-Java 5 idioms for generic programming. As a result, an understanding of how inheritance is used to implement generic programs is essential, even in Java 5.

1.4.1 Using `Object` for Genericity

The basic idea in Java is that we can implement a generic class by using an appropriate superclass, such as `Object`. An example is the `MemoryCell` class shown in Figure 1.5.

There are two details that must be considered when we use this strategy. The first is illustrated in Figure 1.6, which depicts a `main` that writes a `"37"` to a `MemoryCell` object and then reads from the `MemoryCell` object. To access a specific method of the object, we must downcast to the correct type. (Of course, in this example, we do not need the downcast, since we are simply invoking the `toString` method at line 9, and this can be done for any object.)

A second important detail is that primitive types cannot be used. Only reference types are compatible with `Object`. A standard workaround to this problem is discussed momentarily.

```
1    // MemoryCell class
2    //   Object read( )          --> Returns the stored value
3    //   void write( Object x ) --> x is stored
4
5    public class MemoryCell
6    {
7            // Public methods
8        public Object read( )          { return storedValue; }
9        public void write( Object x ) { storedValue = x; }
10
11           // Private internal data representation
12        private Object storedValue;
13   }
```

Figure 1.5 A generic `MemoryCell` class (pre-Java 5)

```
1    public class TestMemoryCell
2    {
3        public static void main( String [ ] args )
4        {
5            MemoryCell m = new MemoryCell( );
6
7            m.write( "37" );
8            String val = (String) m.read( );
9            System.out.println( "Contents are: " + val );
10       }
11   }
```

Figure 1.6 Using the generic `MemoryCell` class (pre-Java 5)

1.4.2 Wrappers for Primitive Types

When we implement algorithms, often we run into a language typing problem: We have an object of one type, but the language syntax requires an object of a different type.

This technique illustrates the basic theme of a **wrapper class**. One typical use is to store a primitive type, and add operations that the primitive type either does not support or does not support correctly.

In Java, we have already seen that although every reference type is compatible with Object, the eight primitive types are not. As a result, Java provides a wrapper class for each of the eight primitive types. For instance, the wrapper for the int type is Integer. Each wrapper object is **immutable** (meaning its state can never change), stores one primitive value that is set when the object is constructed, and provides a method to retrieve the value. The wrapper classes also contain a host of static utility methods.

As an example, Figure 1.7 shows how we can use the MemoryCell to store integers.

1.4.3 Using Interface Types for Genericity

Using Object as a generic type works only if the operations that are being performed can be expressed using only methods available in the Object class.

Consider, for example, the problem of finding the maximum item in an array of items. The basic code is type-independent, but it does require the ability to compare any two objects and decide which is larger and which is smaller. Thus we cannot simply find the maximum of an array of Object—we need more information. The simplest idea would be to find the maximum of an array of Comparable. To determine order, we can use the compareTo method that we know must be available for all Comparables. The code to do this is shown in Figure 1.8, which provides a main that finds the maximum in an array of String or Shape.

It is important to mention a few caveats. First, only objects that implement the Comparable interface can be passed as elements of the Comparable array. Objects that have a compareTo method but do not declare that they implement Comparable are not Comparable, and do not have the requisite *IS-A* relationship. Thus, it is presumed that Shape implements

```
1   public class WrapperDemo
2   {
3       public static void main( String [ ] args )
4       {
5           MemoryCell m = new MemoryCell( );
6
7           m.write( new Integer( 37 ) );
8           Integer wrapperVal = (Integer) m.read( );
9           int val = wrapperVal.intValue( );
10          System.out.println( "Contents are: " + val );
11      }
12  }
```

Figure 1.7 An illustration of the Integer wrapper class

```
1    class FindMaxDemo
2    {
3        /**
4         * Return max item in arr.
5         * Precondition: arr.length > 0
6         */
7        public static Comparable findMax( Comparable [ ] arr )
8        {
9            int maxIndex = 0;
10
11           for( int i = 1; i < arr.length; i++ )
12               if( arr[ i ].compareTo( arr[ maxIndex ] ) > 0 )
13                   maxIndex = i;
14
15           return arr[ maxIndex ];
16       }
17
18       /**
19        * Test findMax on Shape and String objects.
20        */
21       public static void main( String [ ] args )
22       {
23           Shape [ ] sh1 = { new Circle( 2.0 ),
24                             new Square( 3.0 ),
25                             new Rectangle( 3.0, 4.0 ) };
26
27           String [ ] st1 = { "Joe", "Bob", "Bill", "Zeke" };
28
29           System.out.println( findMax( sh1 ) );
30           System.out.println( findMax( st1 ) );
31       }
32   }
```

Figure 1.8 A generic findMax routine, with demo using shapes and strings (pre-Java 5)

the Comparable interface, perhaps comparing areas of Shapes. It is also implicit in the test program that Circle, Square, and Rectangle are subclasses of Shape.

Second, if the Comparable array were to have two objects that are incompatible (e.g., a String and a Shape), the compareTo method would throw a ClassCastException. This is the expected (indeed, required) behavior.

Third, as before, primitives cannot be passed as Comparables, but the wrappers work because they implement the Comparable interface.

Fourth, it is not required that the interface be a standard library interface.

Finally, this solution does not always work, because it might be impossible to declare that a class implements a needed interface. For instance, the class might be a library class, while the interface is a user-defined interface. And if the class is final, we can't extend it to create a new class. Section 1.6 offers another solution for this problem, which is the **function object**. The function object uses interfaces also and is perhaps one of the central themes encountered in the Java library.

1.4.4 Compatibility of Array Types

One of the difficulties in language design is how to handle inheritance for aggregate types. Suppose that Employee *IS-A* Person. Does this imply that Employee[] *IS-A* Person[]? In other words, if a routine is written to accept Person[] as a parameter, can we pass an Employee[] as an argument?

At first glance, this seems like a no-brainer, and Employee[] should be type-compatible with Person[]. However, this issue is trickier than it seems. Suppose that in addition to Employee, Student *IS-A* Person. Suppose the Employee[] is type-compatible with Person[]. Then consider this sequence of assignments:

```
Person[] arr = new Employee[ 5 ]; // compiles: arrays are compatible
arr[ 0 ] = new Student( ... );     // compiles: Student IS-A Person
```

Both assignments compile, yet arr[0] is actually referencing an Employee, and Student *IS-NOT-A* Employee. Thus we have type confusion. The runtime system cannot throw a ClassCastException since there is no cast.

The easiest way to avoid this problem is to specify that the arrays are not type-compatible. However, in Java the arrays *are* type-compatible. This is known as a **covariant array type**. Each array keeps track of the type of object it is allowed to store. If an incompatible type is inserted into the array, the Virtual Machine will throw an ArrayStoreException.

The covariance of arrays was needed in earlier versions of Java because otherwise the calls on lines 29 and 30 in Figure 1.8 would not compile.

1.5 Implementing Generic Components Using Java 5 Generics

Java 5 supports generic classes that are very easy to use. However, writing generic classes requires a little more work. In this section, we illustrate the basics of how generic classes and methods are written. We do not attempt to cover all the constructs of the language, which are quite complex and sometimes tricky. Instead, we show the syntax and idioms that are used throughout this book.

1.5.1 Simple Generic Classes and Interfaces

Figure 1.9 shows a generic version of the MemoryCell class previously depicted in Figure 1.5. Here, we have changed the name to GenericMemoryCell because neither class is in a package and thus the names cannot be the same.

When a generic class is specified, the class declaration includes one or more *type parameters* enclosed in angle brackets <> after the class name. Line 1 shows that the GenericMemoryCell takes one type parameter. In this instance, there are no explicit restrictions on the type parameter, so the user can create types such as GenericMemoryCell<String> and GenericMemoryCell<Integer> but not GenericMemoryCell<int>. Inside the GenericMemoryCell class declaration, we can declare fields of the generic type and methods that use the generic type as a parameter or return type. For example, in line 5 of Figure 1.9, the write method for GenericMemoryCell<String> requires a parameter of type String. Passing anything else will generate a compiler error.

Interfaces can also be declared as generic. For example, prior to Java 5 the Comparable interface was not generic, and its compareTo method took an Object as the parameter. As a result, any reference variable passed to the compareTo method would compile, even if the variable was not a sensible type, and only at runtime would the error be reported as a ClassCastException. In Java 5, the Comparable class is generic, as shown in Figure 1.10. The String class, for instance, now implements Comparable<String> and has a compareTo method that takes a String as a parameter. By making the class generic, many of the errors that were previously only reported at runtime become compile-time errors.

```
1    public class GenericMemoryCell<AnyType>
2    {
3        public AnyType read( )
4          { return storedValue; }
5        public void write( AnyType x )
6          { storedValue = x; }
7
8        private AnyType storedValue;
9    }
```

Figure 1.9 Generic implementation of the MemoryCell class

```
1    package java.lang;
2
3    public interface Comparable<AnyType>
4    {
5        public int compareTo( AnyType other );
6    }
```

Figure 1.10 Comparable interface, Java 5 version which is generic

1.5.2 Autoboxing/Unboxing

The code in Figure 1.7 is annoying to write because using the wrapper class requires creation of an `Integer` object prior to the call to `write`, and then the extraction of the `int` value from the `Integer`, using the `intValue` method. Prior to Java 5, this is required because if an `int` is passed in a place where an `Integer` object is required, the compiler will generate an error message, and if the result of an `Integer` object is assigned to an `int`, the compiler will generate an error message. This resulting code in Figure 1.7 accurately reflects the distinction between primitive types and reference types, yet it does not cleanly express the programmer's intent of storing `int`s in the collection.

Java 5 rectifies this situation. If an `int` is passed in a place where an `Integer` is required, the compiler will insert a call to the `Integer` constructor behind the scenes. This is known as autoboxing. And if an `Integer` is passed in a place where an `int` is required, the compiler will insert a call to the `intValue` method behind the scenes. This is known as auto-unboxing. Similar behavior occurs for the seven other primitive/wrapper pairs. Figure 1.11a illustrates the use of autoboxing and unboxing in Java 5. Note that the entities referenced in the `GenericMemoryCell` are still `Integer` objects; `int` cannot be substituted for `Integer` in the `GenericMemoryCell` instantiations.

1.5.3 The Diamond Operator

In Figure 1.11a, line 5 is annoying because since m is of type `GenericMemoryCell<Integer>`, it is obvious that object being created must also be `GenericMemoryCell<Integer>`; any other type parameter would generate a compiler error. Java 7 adds a new language feature, known as the diamond operator, that allows line 5 to be rewritten as

```
GenericMemoryCell<Integer> m = new GenericMemoryCell<>( );
```

The diamond operator simplifies the code, with no cost to the developer, and we use it throughout the text. Figure 1.11b shows the Java 7 version, incorporating the diamond operator.

```
1   class BoxingDemo
2   {
3       public static void main( String [ ] args )
4       {
5           GenericMemoryCell<Integer> m = new GenericMemoryCell<Integer>( );
6
7           m.write( 37 );
8           int val = m.read( );
9           System.out.println( "Contents are: " + val );
10      }
11  }
```

Figure 1.11a Autoboxing and unboxing (Java 5)

```
1    class BoxingDemo
2    {
3        public static void main( String [ ] args )
4        {
5            GenericMemoryCell<Integer> m = new GenericMemoryCell<>( );
6
7            m.write( 5 );
8            int val = m.read( );
9            System.out.println( "Contents are: " + val );
10       }
11   }
```

Figure 1.11b Autoboxing and unboxing (Java 7, using diamond operator)

1.5.4 Wildcards with Bounds

Figure 1.12 shows a static method that computes the total area in an array of Shapes (we assume Shape is a class with an area method; Circle and Square extend Shape). Suppose we want to rewrite the method so that it works with a parameter that is Collection<Shape>. Collection is described in Chapter 3; for now, the only important thing about it is that it stores a collection of items that can be accessed with an enhanced for loop. Because of the enhanced for loop, the code should be identical, and the resulting code is shown in Figure 1.13. If we pass a Collection<Shape>, the code works. However, what happens if we pass a Collection<Square>? The answer depends on whether a Collection<Square> IS-A Collection<Shape>. Recall from Section 1.4.4 that the technical term for this is whether we have covariance.

In Java, as we mentioned in Section 1.4.4, arrays are covariant. So Square[] IS-A Shape[]. On the one hand, consistency would suggest that if arrays are covariant, then collections should be covariant too. On the other hand, as we saw in Section 1.4.4, the covariance of arrays leads to code that compiles but then generates a runtime exception (an ArrayStoreException). Because the entire reason to have generics is to generate compiler

```
1        public static double totalArea( Shape [ ] arr )
2        {
3            double total = 0;
4
5            for( Shape s : arr )
6                if( s != null )
7                    total += s.area( );
8
9            return total;
10       }
```

Figure 1.12 totalArea method for Shape[]

```
1    public static double totalArea( Collection<Shape> arr )
2    {
3        double total = 0;
4
5        for( Shape s : arr )
6            if( s != null )
7                total += s.area( );
8
9        return total;
10   }
```

Figure 1.13 totalArea method that does not work if passed a Collection<Square>

```
1    public static double totalArea( Collection<? extends Shape> arr )
2    {
3        double total = 0;
4
5        for( Shape s : arr )
6            if( s != null )
7                total += s.area( );
8
9        return total;
10   }
```

Figure 1.14 totalArea method revised with wildcards that works if passed a Collection<Square>

errors rather than runtime exceptions for type mismatches, generic collections are not covariant. As a result, we cannot pass a Collection<Square> as a parameter to the method in Figure 1.13.

What we are left with is that generics (and the generic collections) are not covariant (which makes sense), but arrays are. Without additional syntax, users would tend to avoid collections because the lack of covariance makes the code less flexible.

Java 5 makes up for this with **wildcards**. Wildcards are used to express subclasses (or superclasses) of parameter types. Figure 1.14 illustrates the use of wildcards with a bound to write a totalArea method that takes as parameter a Collection<T>, where T *IS-A* Shape. Thus, Collection<Shape> and Collection<Square> are both acceptable parameters. Wildcards can also be used without a bound (in which case extends Object is presumed) or with super instead of extends (to express superclass rather than subclass); there are also some other syntax uses that we do not discuss here.

1.5.5 Generic Static Methods

In some sense, the totalArea method in Figure 1.14 is generic, since it works for different types. But there is no specific type parameter list, as was done in the GenericMemoryCell

```
1   public static <AnyType> boolean contains( AnyType [ ] arr, AnyType x )
2   {
3       for( AnyType val : arr )
4           if( x.equals( val ) )
5               return true;
6
7       return false;
8   }
```

Figure 1.15 Generic static method to search an array

class declaration. Sometimes the specific type is important perhaps because one of the following reasons apply:

1. The type is used as the return type.
2. The type is used in more than one parameter type.
3. The type is used to declare a local variable.

If so, then an explicit generic method with type parameters must be declared.

For instance, Figure 1.15 illustrates a generic static method that performs a sequential search for value x in array arr. By using a generic method instead of a nongeneric method that uses Object as the parameter types, we can get compile-time errors if searching for an Apple in an array of Shapes.

The generic method looks much like the generic class in that the type parameter list uses the same syntax. The type parameters in a generic method precede the return type.

1.5.6 Type Bounds

Suppose we want to write a findMax routine. Consider the code in Figure 1.16. This code cannot work because the compiler cannot prove that the call to compareTo at line 6 is valid; compareTo is guaranteed to exist only if AnyType is Comparable. We can solve this problem

```
1   public static <AnyType> AnyType findMax( AnyType [ ] arr )
2   {
3       int maxIndex = 0;
4
5       for( int i = 1; i < arr.length; i++ )
6           if( arr[ i ].compareTo( arr[ maxIndex ] ) > 0 )
7               maxIndex = i;
8
9       return arr[ maxIndex ];
10  }
```

Figure 1.16 Generic static method to find largest element in an array that does not work

```
1    public static <AnyType extends Comparable<? super AnyType>>
2    AnyType findMax( AnyType [ ] arr )
3    {
4        int maxIndex = 0;
5
6        for( int i = 1; i < arr.length; i++ )
7            if( arr[ i ].compareTo( arr[ maxIndex ] ) > 0 )
8                maxIndex = i;
9
10       return arr[ maxIndex ];
11   }
```

Figure 1.17 Generic static method to find largest element in an array. Illustrates a bounds on the type parameter

by using a **type bound**. The type bound is specified inside the angle brackets <>, and it specifies properties that the parameter types must have. A naïve attempt is to rewrite the signature as

```
public static <AnyType extends Comparable> ...
```

This is naïve because, as we know, the Comparable interface is now generic. Although this code would compile, a better attempt would be

```
public static <AnyType extends Comparable<AnyType>> ...
```

However, this attempt is not satisfactory. To see the problem, suppose Shape implements Comparable<Shape>. Suppose Square extends Shape. Then all we know is that Square implements Comparable<Shape>. Thus, a Square *IS-A* Comparable<Shape>, but it *IS-NOT-A* Comparable<Square>!

As a result, what we need to say is that AnyType *IS-A* Comparable<T> where T is a superclass of AnyType. Since we do not need to know the exact type T, we can use a wildcard. The resulting signature is

```
public static <AnyType extends Comparable<? super AnyType>>
```

Figure 1.17 shows the implementation of findMax. The compiler will accept arrays of types T only such that T implements the Comparable<S> interface, where T *IS-A* S. Certainly the bounds declaration looks like a mess. Fortunately, we won't see anything more complicated than this idiom.

1.5.7 Type Erasure

Generic types, for the most part, are constructs in the Java language but not in the Virtual Machine. Generic classes are converted by the compiler to nongeneric classes by a process known as **type erasure**. The simplified version of what happens is that the compiler generates a **raw class** with the same name as the generic class with the type parameters removed. The type variables are replaced with their bounds, and when calls are made to generic methods that have an erased return type, casts are inserted automatically. If a generic class is used without a type parameter, the raw class is used.

One important consequence of type erasure is that the generated code is not much different than the code that programmers have been writing before generics and in fact is not any faster. The significant benefit is that the programmer does not have to place casts in the code, and the compiler will do significant type checking.

1.5.8 Restrictions on Generics

There are numerous restrictions on generic types. Every one of the restrictions listed here is required because of type erasure.

Primitive Types

Primitive types cannot be used for a type parameter. Thus GenericMemoryCell<int> is illegal. You must use wrapper classes.

instanceof *tests*

instanceof tests and typecasts work only with raw type. In the following code

```
GenericMemoryCell<Integer> cell1 = new GenericMemoryCell<>( );
cell1.write( 4 );
Object cell = cell1;
GenericMemoryCell<String> cell2 = (GenericMemoryCell<String>) cell;
String s = cell2.read( );
```

the typecast succeeds at runtime since all types are GenericMemoryCell. Eventually, a run-time error results at the last line because the call to read tries to return a String but cannot. As a result, the typecast will generate a warning, and a corresponding instanceof test is illegal.

Static Contexts

In a generic class, static methods and fields cannot refer to the class's type variables since, after erasure, there are no type variables. Further, since there is really only one raw class, static fields are shared among the class's generic instantiations.

Instantiation of Generic Types

It is illegal to create an instance of a generic type. If T is a type variable, the statement

```
T obj = new T( );        // Right-hand side is illegal
```

is illegal. T is replaced by its bounds, which could be Object (or even an abstract class), so the call to new cannot make sense.

Generic Array Objects

It is illegal to create an array of a generic type. If T is a type variable, the statement

```
T [ ] arr = new T[ 10 ];    // Right-hand side is illegal
```

is illegal. T would be replaced by its bounds, which would probably be Object, and then the cast (generated by type erasure) to T[] would fail because Object[] *IS-NOT-A* T[]. Because we cannot create arrays of generic objects, generally we must create an array of the erased type and then use a typecast. This typecast will generate a compiler warning about an unchecked type conversion.

Arrays of Parameterized Types

Instantiation of arrays of parameterized types is illegal. Consider the following code:

```
1   GenericMemoryCell<String> [ ] arr1 = new GenericMemoryCell<>[ 10 ];
2   GenericMemoryCell<Double> cell = new GenericMemoryCell<>( ); cell.write( 4.5 );
3   Object [ ] arr2 = arr1;
4   arr2[ 0 ] = cell;
5   String s = arr1[ 0 ].read( );
```

Normally, we would expect that the assignment at line 4, which has the wrong type, would generate an ArrayStoreException. However, after type erasure, the array type is GenericMemoryCell[], and the object added to the array is GenericMemoryCell, so there is no ArrayStoreException. Thus, this code has no casts, yet it will eventually generate a ClassCastException at line 5, which is exactly the situation that generics are supposed to avoid.

1.6 Function Objects

In Section 1.5, we showed how to write generic algorithms. As an example, the generic method in Figure 1.16 can be used to find the maximum item in an array.

However, that generic method has an important limitation: It works only for objects that implement the Comparable interface, using compareTo as the basis for all comparison decisions. In many situations, this approach is not feasible. For instance, it is a stretch to presume that a Rectangle class will implement Comparable, and even if it does, the compareTo method that it has might not be the one we want. For instance, given a 2-by-10 rectangle and a 5-by-5 rectangle, which is the larger rectangle? The answer would depend on whether we are using area or width to decide. Or perhaps if we are trying to fit the rectangle through an opening, the larger rectangle is the rectangle with the larger minimum dimension. As a second example, if we wanted to find the maximum string (alphabetically last) in an array of strings, the default compareTo does not ignore case distinctions, so "ZEBRA" would be considered to precede "alligator" alphabetically, which is probably not what we want.

The solution in these situations is to rewrite findMax to accept two parameters: an array of objects and a comparison function that explains how to decide which of two objects is the larger and which is the smaller. In effect, the objects no longer know how to compare themselves; instead, this information is completely decoupled from the objects in the array.

An ingenious way to pass functions as parameters is to notice that an object contains both data and methods, so we can define a class with no data and one method and pass

```
1       // Generic findMax, with a function object.
2       // Precondition: a.size( ) > 0.
3       public static <AnyType>
4       AnyType findMax( AnyType [ ] arr, Comparator<? super AnyType> cmp )
5       {
6           int maxIndex = 0;
7
8           for( int i = 1; i < arr.size( ); i++ )
9               if( cmp.compare( arr[ i ], arr[ maxIndex ] ) > 0 )
10                  maxIndex = i;
11
12          return arr[ maxIndex ];
13      }
14
15   class CaseInsensitiveCompare implements Comparator<String>
16   {
17       public int compare( String lhs, String rhs )
18          { return lhs.compareToIgnoreCase( rhs ); }
19   }
20
21   class TestProgram
22   {
23       public static void main( String [ ] args )
24       {
25           String [ ] arr = { "ZEBRA", "alligator", "crocodile" };
26           System.out.println( findMax( arr, new CaseInsensitiveCompare( ) ) )
27       }
28   }
```

Figure 1.18   Using a function object as a second parameter to findMax; output is ZEBRA

an instance of the class. In effect, a function is being passed by placing it inside an object. This object is commonly known as a **function object**.

Figure 1.18 shows the simplest implementation of the function object idea. findMax takes a second parameter, which is an object of type Comparator. The Comparator interface is specified in java.util and contains a compare method. This interface is shown in Figure 1.19.

Any class that implements the Comparator<AnyType> interface type must have a method named compare that takes two parameters of the generic type (AnyType) and returns an int, following the same general contract as compareTo. Thus, in Figure 1.18, the call to compare at line 9 can be used to compare array items. The bounded wildcard at line 4 is used to signal that if we are finding the maximum in an array of items, the comparator must know how to compare items, or objects of the items' supertype. To use this version of findMax, at line 26, we can see that findMax is called by passing an array of String and an object that

```
1    package java.util;
2
3    public interface Comparator<AnyType>
4    {
5        int compare( AnyType lhs, AnyType rhs );
6    }
```

Figure 1.19   The Comparator interface

implements Comparator<String>. This object is of type CaseInsensitiveCompare, which is a class we write.

In Chapter 4, we will give an example of a class that needs to order the items it stores. We will write most of the code using Comparable and show the adjustments needed to use the function objects. Elsewhere in the book, we will avoid the detail of function objects to keep the code as simple as possible, knowing that it is not difficult to add function objects later.

Summary

This chapter sets the stage for the rest of the book. The time taken by an algorithm confronted with large amounts of input will be an important criterion for deciding if it is a good algorithm. (Of course, correctness is most important.) Speed is relative. What is fast for one problem on one machine might be slow for another problem or a different machine. We will begin to address these issues in the next chapter and will use the mathematics discussed here to establish a formal model.

Exercises

1.1   Write a program to solve the selection problem. Let $k = N/2$. Draw a table showing the running time of your program for various values of N.

1.2   Write a program to solve the word puzzle problem.

1.3   Write a method to output an arbitrary double number (which might be negative) using only printDigit for I/O.

1.4   C allows statements of the form

         #include *filename*

which reads *filename* and inserts its contents in place of the *include* statement. *Include* statements may be nested; in other words, the file *filename* may itself contain an *include* statement, but, obviously, a file can't include itself in any chain. Write a program that reads in a file and outputs the file as modified by the *include* statements.

1.5 Write a recursive method that returns the number of 1's in the binary representation of N. Use the fact that this is equal to the number of 1's in the representation of $N/2$, plus 1, if N is odd.

1.6 Write the routines with the following declarations:

```
public  void permute( String str );
private void permute( char [ ] str, int low, int high );
```

The first routine is a driver that calls the second and prints all the permutations of the characters in String str. If str is "abc", then the strings that are output are abc, acb, bac, bca, cab, and cba. Use recursion for the second routine.

1.7 Prove the following formulas:
a. $\log X < X$ for all $X > 0$
b. $\log(A^B) = B \log A$

1.8 Evaluate the following sums:
a. $\sum_{i=0}^{\infty} \frac{1}{4^i}$
b. $\sum_{i=0}^{\infty} \frac{i}{4^i}$
*c. $\sum_{i=0}^{\infty} \frac{i^2}{4^i}$
**d. $\sum_{i=0}^{\infty} \frac{i^N}{4^i}$

1.9 Estimate

$$\sum_{i=\lfloor N/2 \rfloor}^{N} \frac{1}{i}$$

***1.10** What is 2^{100} (mod 5)?

1.11 Let F_i be the Fibonacci numbers as defined in Section 1.2. Prove the following:
a. $\sum_{i=1}^{N-2} F_i = F_N - 2$
b. $F_N < \phi^N$, with $\phi = (1 + \sqrt{5})/2$
**c. Give a precise closed-form expression for F_N.

1.12 Prove the following formulas:
a. $\sum_{i=1}^{N}(2i - 1) = N^2$
b. $\sum_{i=1}^{N} i^3 = \left(\sum_{i=1}^{N} i\right)^2$

1.13 Design a generic class, Collection, that stores a collection of Objects (in an array), along with the current size of the collection. Provide public methods isEmpty, makeEmpty, insert, remove, and isPresent. isPresent(x) returns true if and only if an Object that is equal to x (as defined by equals) is present in the collection.

1.14 Design a generic class, OrderedCollection, that stores a collection of Comparables (in an array), along with the current size of the collection. Provide public methods isEmpty, makeEmpty, insert, remove, findMin, and findMax. findMin and findMax return references to the smallest and largest, respectively, Comparable in the collection (or null if the collection is empty).

1.15 Define a `Rectangle` class that provides `getLength` and `getWidth` methods. Using the `findMax` routines in Figure 1.18, write a `main` that creates an array of `Rectangle` and finds the largest `Rectangle` first on the basis of area, and then on the basis of perimeter.

References

There are many good textbooks covering the mathematics reviewed in this chapter. A small subset is [1], [2], [3], [11], [13], and [14]. Reference [11] is specifically geared toward the analysis of algorithms. It is the first volume of a three-volume series that will be cited throughout this text. More advanced material is covered in [8].

Throughout this book we will assume a knowledge of Java [4], [6], [7]. The material in this chapter is meant to serve as an overview of the features that we will use in this text. We also assume familiarity with recursion (the recursion summary in this chapter is meant to be a quick review). We will attempt to provide hints on its use where appropriate throughout the textbook. Readers not familiar with recursion should consult [14] or any good intermediate programming textbook.

General programming style is discussed in several books. Some of the classics are [5], [9], and [10].

1. M. O. Albertson and J. P. Hutchinson, *Discrete Mathematics with Algorithms,* John Wiley & Sons, New York, 1988.

2. Z. Bavel, *Math Companion for Computer Science,* Reston Publishing Co., Reston, Va., 1982.

3. R. A. Brualdi, *Introductory Combinatorics,* North-Holland, New York, 1977.

4. G. Cornell and C. S. Horstmann, *Core Java,* Vol. I, 8th ed., Prentice Hall, Upper Saddle River, N.J., 2009.

5. E. W. Dijkstra, *A Discipline of Programming,* Prentice Hall, Englewood Cliffs, N.J., 1976.

6. D. Flanagan, *Java in a Nutshell,* 5th ed., O'Reilly and Associates, Sebastopol, Calif., 2005.

7. J. Gosling, B. Joy, G. Steele, and G. Bracha, *The Java Language Specification,* 3d ed., Addison-Wesley, Reading, Mass., 2005.

8. R. L. Graham, D. E. Knuth, and O. Patashnik, *Concrete Mathematics,* Addison-Wesley, Reading, Mass., 1989.

9. D. Gries, *The Science of Programming,* Springer-Verlag, New York, 1981.

10. B. W. Kernighan and P. J. Plauger, *The Elements of Programming Style,* 2d ed., McGraw-Hill, New York, 1978.

11. D. E. Knuth, *The Art of Computer Programming, Vol. 1: Fundamental Algorithms,* 3d ed., Addison-Wesley, Reading, Mass., 1997.

12. F. S. Roberts, *Applied Combinatorics,* Prentice Hall, Englewood Cliffs, N.J., 1984.

13. A. Tucker, *Applied Combinatorics,* 2d ed., John Wiley & Sons, New York, 1984.

14. M. A. Weiss, *Data Structures and Problem Solving Using Java,* 4th ed., Addison-Wesley, Boston, Mass., 2010.

CHAPTER 2

Algorithm Analysis

An **algorithm** is a clearly specified set of simple instructions to be followed to solve a problem. Once an algorithm is given for a problem and decided (somehow) to be correct, an important step is to determine how much in the way of resources, such as time or space, the algorithm will require. An algorithm that solves a problem but requires a year is hardly of any use. Likewise, an algorithm that requires hundreds of gigabytes of main memory is not (currently) useful on most machines.

In this chapter, we shall discuss

- How to estimate the time required for a program.
- How to reduce the running time of a program from days or years to fractions of a second.
- The results of careless use of recursion.
- Very efficient algorithms to raise a number to a power and to compute the greatest common divisor of two numbers.

2.1 Mathematical Background

The analysis required to estimate the resource use of an algorithm is generally a theoretical issue, and therefore a formal framework is required. We begin with some mathematical definitions.

Throughout the book we will use the following four definitions:

Definition 2.1.
$T(N) = O(f(N))$ if there are positive *constants* c and n_0 such that $T(N) \leq cf(N)$ when $N \geq n_0$.

Definition 2.2.
$T(N) = \Omega(g(N))$ if there are positive *constants* c and n_0 such that $T(N) \geq cg(N)$ when $N \geq n_0$.

Definition 2.3.
$T(N) = \Theta(h(N))$ if and only if $T(N) = O(h(N))$ and $T(N) = \Omega(h(N))$.

Definition 2.4.

$T(N) = o(p(N))$ if for all positive constants c there exists an n_0 such that $T(N) < cp(N)$ when $N > n_0$. Less formally, $T(N) = o(p(N))$ if $T(N) = O(p(N))$ and $T(N) \neq \Theta(p(N))$.

The idea of these definitions is to establish a relative order among functions. Given two functions, there are usually points where one function is smaller than the other function, so it does not make sense to claim, for instance, $f(N) < g(N)$. Thus, we compare their **relative rates of growth.** When we apply this to the analysis of algorithms, we shall see why this is the important measure.

Although $1{,}000N$ is larger than N^2 for small values of N, N^2 grows at a faster rate, and thus N^2 will eventually be the larger function. The turning point is $N = 1{,}000$ in this case. The first definition says that eventually there is some point n_0 past which $c \cdot f(N)$ is always at least as large as $T(N)$, so that if constant factors are ignored, $f(N)$ is at least as big as $T(N)$. In our case, we have $T(N) = 1{,}000N$, $f(N) = N^2$, $n_0 = 1{,}000$, and $c = 1$. We could also use $n_0 = 10$ and $c = 100$. Thus, we can say that $1{,}000N = O(N^2)$ (order N-squared). This notation is known as **Big-Oh notation**. Frequently, instead of saying "order . . . ," one says "Big-Oh . . . ."

If we use the traditional inequality operators to compare growth rates, then the first definition says that the growth rate of $T(N)$ is less than or equal to ($\leq$) that of $f(N)$. The second definition, $T(N) = \Omega(g(N))$ (pronounced "omega"), says that the growth rate of $T(N)$ is greater than or equal to ($\geq$) that of $g(N)$. The third definition, $T(N) = \Theta(h(N))$ (pronounced "theta"), says that the growth rate of $T(N)$ equals ($=$) the growth rate of $h(N)$. The last definition, $T(N) = o(p(N))$ (pronounced "little-oh"), says that the growth rate of $T(N)$ is less than ($<$) the growth rate of $p(N)$. This is different from Big-Oh, because Big-Oh allows the possibility that the growth rates are the same.

To prove that some function $T(N) = O(f(N))$, we usually do not apply these definitions formally but instead use a repertoire of known results. In general, this means that a proof (or determination that the assumption is incorrect) is a very simple calculation and should not involve calculus, except in extraordinary circumstances (not likely to occur in an algorithm analysis).

When we say that $T(N) = O(f(N))$, we are guaranteeing that the function $T(N)$ grows at a rate no faster than $f(N)$; thus $f(N)$ is an **upper bound** on $T(N)$. Since this implies that $f(N) = \Omega(T(N))$, we say that $T(N)$ is a **lower bound** on $f(N)$.

As an example, N^3 grows faster than N^2, so we can say that $N^2 = O(N^3)$ or $N^3 = \Omega(N^2)$. $f(N) = N^2$ and $g(N) = 2N^2$ grow at the same rate, so both $f(N) = O(g(N))$ and $f(N) = \Omega(g(N))$ are true. When two functions grow at the same rate, then the decision of whether or not to signify this with $\Theta()$ can depend on the particular context. Intuitively, if $g(N) = 2N^2$, then $g(N) = O(N^4)$, $g(N) = O(N^3)$, and $g(N) = O(N^2)$ are all technically correct, but the last option is the best answer. Writing $g(N) = \Theta(N^2)$ says not only that $g(N) = O(N^2)$, but also that the result is as good (tight) as possible.

Function	Name
c	Constant
$\log N$	Logarithmic
$\log^2 N$	Log-squared
N	Linear
$N \log N$	
N^2	Quadratic
N^3	Cubic
2^N	Exponential

Figure 2.1 Typical growth rates

The important things to know are

Rule 1.
If $T_1(N) = O(f(N))$ and $T_2(N) = O(g(N))$, then
    (a) $T_1(N) + T_2(N) = O(f(N) + g(N))$ (intuitively and less formally it is
        $O(\max(f(N), g(N))))$,
    (b) $T_1(N) * T_2(N) = O(f(N) * g(N))$.

Rule 2.
If $T(N)$ is a polynomial of degree k, then $T(N) = \Theta(N^k)$.

Rule 3.
$\log^k N = O(N)$ for any constant k. This tells us that logarithms grow very slowly.

This information is sufficient to arrange most of the common functions by growth rate (see Figure 2.1).

Several points are in order. First, it is very bad style to include constants or low-order terms inside a Big-Oh. Do not say $T(N) = O(2N^2)$ or $T(N) = O(N^2 + N)$. In both cases, the correct form is $T(N) = O(N^2)$. This means that in any analysis that will require a Big-Oh answer, all sorts of shortcuts are possible. Lower-order terms can generally be ignored, and constants can be thrown away. Considerably less precision is required in these cases.

Second, we can always determine the relative growth rates of two functions $f(N)$ and $g(N)$ by computing $\lim_{N \to \infty} f(N)/g(N)$, using L'Hôpital's rule if necessary.[1] The limit can have four possible values:

- The limit is 0: This means that $f(N) = o(g(N))$.

- The limit is $c \neq 0$: This means that $f(N) = \Theta(g(N))$.

---

[1] L'Hôpital's rule states that if $\lim_{N \to \infty} f(N) = \infty$ and $\lim_{N \to \infty} g(N) = \infty$, then $\lim_{N \to \infty} f(N)/g(N) = \lim_{N \to \infty} f'(N)/g'(N)$, where $f'(N)$ and $g'(N)$ are the derivatives of $f(N)$ and $g(N)$, respectively.

- The limit is ∞: This means that $g(N) = o(f(N))$.

- The limit does not exist: There is no relation (this will not happen in our context).

Using this method almost always amounts to overkill. Usually the relation between $f(N)$ and $g(N)$ can be derived by simple algebra. For instance, if $f(N) = N \log N$ and $g(N) = N^{1.5}$, then to decide which of $f(N)$ and $g(N)$ grows faster, one really needs to determine which of $\log N$ and $N^{0.5}$ grows faster. This is like determining which of $\log^2 N$ or N grows faster. This is a simple problem, because it is already known that N grows faster than any power of a log. Thus, $g(N)$ grows faster than $f(N)$.

One stylistic note: It is bad to say $f(N) \leq O(g(N))$, because the inequality is implied by the definition. It is wrong to write $f(N) \geq O(g(N))$, which does not make sense.

As an example of the typical kinds of analysis that are performed, consider the problem of downloading a file over the Internet. Suppose there is an initial 3-sec delay (to set up a connection), after which the download proceeds at 1.5 M(bytes)/sec. Then it follows that if the file is N megabytes, the time to download is described by the formula $T(N) = N/1.5 + 3$. This is a **linear function**. Notice that the time to download a 1,500M file (1,003 sec) is approximately (but not exactly) twice the time to download a 750M file (503 sec). This is typical of a linear function. Notice, also, that if the speed of the connection doubles, both times decrease, but the 1,500M file still takes approximately twice the time to download as a 750M file. This is the typical characteristic of linear-time algorithms, and it is why we write $T(N) = O(N)$, ignoring constant factors. (Although using Big-Theta would be more precise, Big-Oh answers are typically given.)

Observe, too, that this behavior is not true of all algorithms. For the first selection algorithm described in Section 1.1, the running time is controlled by the time it takes to perform a sort. For a simple sorting algorithm, such as the suggested bubble sort, when the amount of input doubles, the running time increases by a factor of four for large amounts of input. This is because those algorithms are not linear. Instead, as we will see when we discuss sorting, trivial sorting algorithms are $O(N^2)$, or quadratic.

2.2 Model

In order to analyze algorithms in a formal framework, we need a model of computation. Our model is basically a normal computer, in which instructions are executed sequentially. Our model has the standard repertoire of simple instructions, such as addition, multiplication, comparison, and assignment, but, unlike the case with real computers, it takes exactly one time unit to do anything (simple). To be reasonable, we will assume that, like a modern computer, our model has fixed-size (say, 32-bit) integers and that there are no fancy operations, such as matrix inversion or sorting, that clearly cannot be done in one time unit. We also assume infinite memory.

This model clearly has some weaknesses. Obviously, in real life, not all operations take exactly the same time. In particular, in our model one disk read counts the same as an addition, even though the addition is typically several orders of magnitude faster. Also, by assuming infinite memory, we ignore the fact that the cost of a memory access can increase when slower memory is used due to larger memory requirements.

2.3 What to Analyze

The most important resource to analyze is generally the running time. Several factors affect the running time of a program. Some, such as the compiler and computer used, are obviously beyond the scope of any theoretical model, so, although they are important, we cannot deal with them here. The other main factors are the algorithm used and the input to the algorithm.

Typically, the size of the input is the main consideration. We define two functions, $T_{avg}(N)$ and $T_{worst}(N)$, as the average and worst-case running time, respectively, used by an algorithm on input of size N. Clearly, $T_{avg}(N) \leq T_{worst}(N)$. If there is more than one input, these functions may have more than one argument.

Occasionally the best-case performance of an algorithm is analyzed. However, this is often of little interest, because it does not represent typical behavior. Average-case performance often reflects typical behavior, while worst-case performance represents a guarantee for performance on any possible input. Notice, also, that, although in this chapter we analyze Java code, these bounds are really bounds for the algorithms rather than programs. Programs are an implementation of the algorithm in a particular programming language, and almost always the details of the programming language do not affect a Big-Oh answer. If a program is running much more slowly than the algorithm analysis suggests, there may be an implementation inefficiency. This is more common in languages (like C++) where arrays can be inadvertently copied in their entirety, instead of passed with references. However, this can occur in Java, too. Thus in future chapters we will analyze the algorithms rather than the programs.

Generally, the quantity required is the worst-case time, unless otherwise specified. One reason for this is that it provides a bound for all input, including particularly bad input, which an average-case analysis does not provide. The other reason is that average-case bounds are usually much more difficult to compute. In some instances, the definition of "average" can affect the result. (For instance, what is average input for the following problem?)

As an example, in the next section, we shall consider the following problem:

Maximum Subsequence Sum Problem.
Given (possibly negative) integers $A_1, A_2, \ldots, A_N$, find the maximum value of $\sum_{k=i}^{j} A_k$. (For convenience, the maximum subsequence sum is 0 if all the integers are negative.) Example:
   For input $-2, 11, -4, 13, -5, -2$, the answer is 20 (A_2 through A_4).

This problem is interesting mainly because there are so many algorithms to solve it, and the performance of these algorithms varies drastically. We will discuss four algorithms to solve this problem. The running time on some computer (the exact computer is unimportant) for these algorithms is given in Figure 2.2.

There are several important things worth noting in this table. For a small amount of input, the algorithms all run in a blink of the eye, so if only a small amount of input is expected, it might be silly to expend a great deal of effort to design a clever algorithm. On the other hand, there is a large market these days for rewriting programs that were written five years ago based on a no-longer-valid assumption of small input size. These

| Input | Algorithm Time | | | |
Size	1 $O(N^3)$	2 $O(N^2)$	3 $O(N \log N)$	4 $O(N)$
$N = 100$	0.000159	0.000006	0.000005	0.000002
$N = 1{,}000$	0.095857	0.000371	0.000060	0.000022
$N = 10{,}000$	86.67	0.033322	0.000619	0.000222
$N = 100{,}000$	NA	3.33	0.006700	0.002205
$N = 1{,}000{,}000$	NA	NA	0.074870	0.022711

Figure 2.2 Running times of several algorithms for maximum subsequence sum (in seconds)

programs are now too slow, because they used poor algorithms. For large amounts of input, algorithm 4 is clearly the best choice (although algorithm 3 is still usable).

Second, the times given do not include the time required to read the input. For algorithm 4, the time merely to read in the input from a disk is likely to be an order of magnitude larger than the time required to solve the problem. This is typical of many efficient algorithms. Reading the data is generally the bottleneck; once the data are read, the problem can be solved quickly. For inefficient algorithms this is not true, and significant computer resources must be used. Thus it is important that, whenever possible, algorithms be efficient enough not to be the bottleneck of a problem.

Notice that algorithm 4, which is linear, exhibits the nice behavior that as the problem size increases by a factor of ten, the running time also increases by a factor of ten.

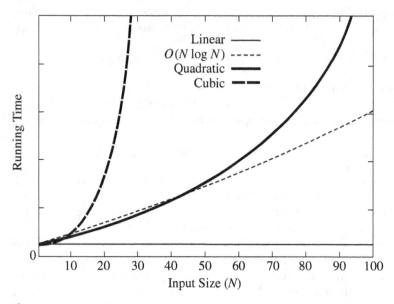

Figure 2.3 Plot (N vs. time) of various algorithms

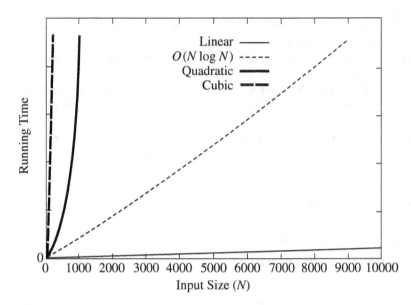

Figure 2.4 Plot (N vs. time) of various algorithms

Algorithm 2, which is quadratic, does not have this behavior; a tenfold increase in input size yields roughly a hundredfold (10^2) increase in running time. And algorithm 1, which is cubic, yields a thousandfold (10^3) increase in running time. We would expect algorithm 1 to take nearly 9,000 seconds (or two and half hours) to complete for $N = 100,000$. Similarly, we would expect algorithm 2 to take roughly 333 seconds to complete for $N = 1,000,000$. However, it is possible that Algorithm 2 could take somewhat longer to complete due to the fact that $N = 1,000,000$ could also yield slower memory accesses than $N = 100,000$ on modern computers, depending on the size of the memory cache.

Figure 2.3 shows the growth rates of the running times of the four algorithms. Even though this graph encompasses only values of N ranging from 10 to 100, the relative growth rates are still evident. Although the graph for the $O(N \log N)$ algorithm seems linear, it is easy to verify that it is not by using a straight-edge (or piece of paper). Although the graph for the $O(N)$ algorithm seems constant, this is only because for small values of N, the constant term is larger than the linear term. Figure 2.4 shows the performance for larger values. It dramatically illustrates how useless inefficient algorithms are for even moderately large amounts of input.

2.4 Running Time Calculations

There are several ways to estimate the running time of a program. The previous table was obtained empirically. If two programs are expected to take similar times, probably the best way to decide which is faster is to code them both up and run them!

Generally, there are several algorithmic ideas, and we would like to eliminate the bad ones early, so an analysis is usually required. Furthermore, the ability to do an analysis usually provides insight into designing efficient algorithms. The analysis also generally pinpoints the bottlenecks, which are worth coding carefully.

To simplify the analysis, we will adopt the convention that there are no particular units of time. Thus, we throw away leading constants. We will also throw away low-order terms, so what we are essentially doing is computing a Big-Oh running time. Since Big-Oh is an upper bound, we must be careful never to underestimate the running time of the program. In effect, the answer provided is a guarantee that the program will terminate within a certain time period. The program may stop earlier than this, but never later.

2.4.1 A Simple Example

Here is a simple program fragment to calculate $\sum_{i=1}^{N} i^3$:

```
    public static int sum( int n )
    {
        int partialSum;

1       partialSum = 0;
2       for( int i = 1; i <= n; i++ )
3           partialSum += i * i * i;
4       return partialSum;
    }
```

The analysis of this fragment is simple. The declarations count for no time. Lines 1 and 4 count for one unit each. Line 3 counts for four units per time executed (two multiplications, one addition, and one assignment) and is executed N times, for a total of $4N$ units. Line 2 has the hidden costs of initializing i, testing $i \leq N$, and incrementing i. The total cost of all these is 1 to initialize, $N + 1$ for all the tests, and N for all the increments, which is $2N + 2$. We ignore the costs of calling the method and returning, for a total of $6N + 4$. Thus, we say that this method is $O(N)$.

If we had to perform all this work every time we needed to analyze a program, the task would quickly become infeasible. Fortunately, since we are giving the answer in terms of Big-Oh, there are lots of shortcuts that can be taken without affecting the final answer. For instance, line 3 is obviously an $O(1)$ statement (per execution), so it is silly to count precisely whether it is two, three, or four units; it does not matter. Line 1 is obviously insignificant compared with the for loop, so it is silly to waste time here. This leads to several general rules.

2.4.2 General Rules

Rule 1—*for* loops.
The running time of a for loop is at most the running time of the statements inside the for loop (including tests) times the number of iterations.

Rule 2—Nested loops.
Analyze these inside out. The total running time of a statement inside a group of nested loops is the running time of the statement multiplied by the product of the sizes of all the loops.

As an example, the following program fragment is $O(N^2)$:

```
for( i = 0; i < n; i++ )
    for( j = 0; j < n; j++ )
        k++;
```

Rule 3—Consecutive Statements.
These just add (which means that the maximum is the one that counts; see rule 1(a) on page 31).

As an example, the following program fragment, which has $O(N)$ work followed by $O(N^2)$ work, is also $O(N^2)$:

```
for( i = 0; i < n; i++ )
    a[ i ] = 0;
for( i = 0; i < n; i++ )
    for( j = 0; j < n; j++ )
        a[ i ] += a[ j ] + i + j;
```

Rule 4—*if/else*.
For the fragment

```
if( condition )
    S1
else
    S2
```

the running time of an `if/else` statement is never more than the running time of the test plus the larger of the running times of S1 and S2.

Clearly, this can be an overestimate in some cases, but it is never an underestimate.

Other rules are obvious, but a basic strategy of analyzing from the inside (or deepest part) out works. If there are method calls, these must be analyzed first. If there are recursive methods, there are several options. If the recursion is really just a thinly veiled `for` loop, the analysis is usually trivial. For instance, the following method is really just a simple loop and is $O(N)$:

```
public static long factorial( int n )
{
    if( n <= 1 )
        return 1;
    else
        return n * factorial( n - 1 );
}
```

This example is really a poor use of recursion. When recursion is properly used, it is difficult to convert the recursion into a simple loop structure. In this case, the analysis will involve a recurrence relation that needs to be solved. To see what might happen, consider the following program, which turns out to be a horrible use of recursion:

```
        public static long fib( int n )
        {
1           if( n <= 1 )
2               return 1;
            else
3               return fib( n - 1 ) + fib( n - 2 );
        }
```

At first glance, this seems like a very clever use of recursion. However, if the program is coded up and run for values of N around 40, it becomes apparent that this program is terribly inefficient. The analysis is fairly simple. Let $T(N)$ be the running time for the method call fib(n). If $N = 0$ or $N = 1$, then the running time is some constant value, which is the time to do the test at line 1 and return. We can say that $T(0) = T(1) = 1$ because constants do not matter. The running time for other values of N is then measured relative to the running time of the base case. For $N > 2$, the time to execute the method is the constant work at line 1 plus the work at line 3. Line 3 consists of an addition and two method calls. Since the method calls are not simple operations, they must be analyzed by themselves. The first method call is fib(n - 1) and hence, by the definition of T, requires $T(N - 1)$ units of time. A similar argument shows that the second method call requires $T(N - 2)$ units of time. The total time required is then $T(N - 1) + T(N - 2) + 2$, where the 2 accounts for the work at line 1 plus the addition at line 3. Thus, for $N \geq 2$, we have the following formula for the running time of fib(n):

$$T(N) = T(N - 1) + T(N - 2) + 2$$

Since $fib(N) = fib(N - 1) + fib(N - 2)$, it is easy to show by induction that $T(N) \geq fib(N)$. In Section 1.2.5, we showed that $fib(N) < (5/3)^N$. A similar calculation shows that (for $N > 4$) $fib(N) \geq (3/2)^N$, and so the running time of this program grows *exponentially*. This is about as bad as possible. By keeping a simple array and using a **for** loop, the running time can be reduced substantially.

This program is slow because there is a huge amount of redundant work being performed, violating the fourth major rule of recursion (the compound interest rule), which was presented in Section 1.3. Notice that the first call on line 3, fib(n - 1), actually computes fib(n - 2) at some point. This information is thrown away and recomputed by the second call on line 3. The amount of information thrown away compounds recursively and results in the huge running time. This is perhaps the finest example of the maxim "Don't compute anything more than once" and should not scare you away from using recursion. Throughout this book, we shall see outstanding uses of recursion.

2.4.3 Solutions for the Maximum Subsequence Sum Problem

We will now present four algorithms to solve the maximum subsequence sum problem posed earlier. The first algorithm, which merely exhaustively tries all possibilities, is depicted in Figure 2.5. The indices in the for loop reflect the fact that in Java, arrays begin at 0, instead of 1. Also, the algorithm does not compute the actual subsequences; additional code is required to do this.

Convince yourself that this algorithm works (this should not take much convincing). The running time is $O(N^3)$ and is entirely due to lines 13 and 14, which consist of an $O(1)$ statement buried inside three nested for loops. The loop at line 8 is of size N.

The second loop has size $N - i$ which could be small but could also be of size N. We must assume the worst, with the knowledge that this could make the final bound a bit high. The third loop has size $j - i + 1$, which, again, we must assume is of size N. The total is $O(1 \cdot N \cdot N \cdot N) = O(N^3)$. Line 6 takes only $O(1)$ total, and lines 16 and 17 take only $O(N^2)$ total, since they are easy expressions inside only two loops.

It turns out that a more precise analysis, taking into account the actual size of these loops, shows that the answer is $\Theta(N^3)$ and that our estimate above was a factor of 6 too high (which is all right, because constants do not matter). This is generally true in these kinds of problems. The precise analysis is obtained from the sum $\sum_{i=0}^{N-1} \sum_{j=i}^{N-1} \sum_{k=i}^{j} 1$,

```
1      /**
2       * Cubic maximum contiguous subsequence sum algorithm.
3       */
4      public static int maxSubSum1( int [ ] a )
5      {
6          int maxSum = 0;
7
8          for( int i = 0; i < a.length; i++ )
9              for( int j = i; j < a.length; j++ )
10             {
11                 int thisSum = 0;
12
13                 for( int k = i; k <= j; k++ )
14                     thisSum += a[ k ];
15
16                 if( thisSum > maxSum )
17                     maxSum = thisSum;
18             }
19
20          return maxSum;
21     }
```

Figure 2.5 Algorithm 1

which tells how many times line 14 is executed. The sum can be evaluated inside out, using formulas from Section 1.2.3. In particular, we will use the formulas for the sum of the first N integers and first N squares. First we have

$$\sum_{k=i}^{j} 1 = j - i + 1$$

Next we evaluate

$$\sum_{j=i}^{N-1}(j - i + 1) = \frac{(N - i + 1)(N - i)}{2}$$

This sum is computed by observing that it is just the sum of the first $N - i$ integers. To complete the calculation, we evaluate

$$\sum_{i=0}^{N-1} \frac{(N - i + 1)(N - i)}{2} = \sum_{i=1}^{N} \frac{(N - i + 1)(N - i + 2)}{2}$$

$$= \frac{1}{2}\sum_{i=1}^{N} i^2 - \left(N + \frac{3}{2}\right)\sum_{i=1}^{N} i + \frac{1}{2}(N^2 + 3N + 2)\sum_{i=1}^{N} 1$$

$$= \frac{1}{2}\frac{N(N + 1)(2N + 1)}{6} - \left(N + \frac{3}{2}\right)\frac{N(N + 1)}{2} + \frac{N^2 + 3N + 2}{2}N$$

$$= \frac{N^3 + 3N^2 + 2N}{6}$$

We can avoid the cubic running time by removing a for loop. This is not always possible, but in this case there are an awful lot of unnecessary computations present in the algorithm. The inefficiency that the improved algorithm corrects can be seen by noticing that $\sum_{k=i}^{j} A_k = A_j + \sum_{k=i}^{j-1} A_k$, so the computation at lines 13 and 14 in algorithm 1 is unduly expensive. Figure 2.6 shows an improved algorithm. Algorithm 2 is clearly $O(N^2)$; the analysis is even simpler than before.

There is a recursive and relatively complicated $O(N \log N)$ solution to this problem, which we now describe. If there didn't happen to be an $O(N)$ (linear) solution, this would be an excellent example of the power of recursion. The algorithm uses a "divide-and-conquer" strategy. The idea is to split the problem into two roughly equal subproblems, which are then solved recursively. This is the "divide" part. The "conquer" stage consists of patching together the two solutions of the subproblems, and possibly doing a small amount of additional work, to arrive at a solution for the whole problem.

In our case, the maximum subsequence sum can be in one of three places. Either it occurs entirely in the left half of the input, or entirely in the right half, or it crosses the middle and is in both halves. The first two cases can be solved recursively. The last case can be obtained by finding the largest sum in the first half that includes the last element

```
1        /**
2         * Quadratic maximum contiguous subsequence sum algorithm.
3         */
4        public static int maxSubSum2( int [ ] a )
5        {
6            int maxSum = 0;
7
8            for( int i = 0; i < a.length; i++ )
9            {
10               int thisSum = 0;
11               for( int j = i; j < a.length; j++ )
12               {
13                   thisSum += a[ j ];
14
15                   if( thisSum > maxSum )
16                       maxSum = thisSum;
17               }
18           }
19
20           return maxSum;
21       }
```

Figure 2.6 Algorithm 2

in the first half, and the largest sum in the second half that includes the first element in the second half. These two sums can then be added together. As an example, consider the following input:

First Half	Second Half

4	−3	5	−2	−1	2	6	−2

The maximum subsequence sum for the first half is 6 (elements A_1 through A_3) and for the second half is 8 (elements A_6 through A_7).

The maximum sum in the first half that includes the last element in the first half is 4 (elements A_1 through A_4), and the maximum sum in the second half that includes the first element in the second half is 7 (elements A_5 through A_7). Thus, the maximum sum that spans both halves and goes through the middle is $4 + 7 = 11$ (elements A_1 through A_7).

We see, then, that among the three ways to form a large maximum subsequence, for our example, the best way is to include elements from both halves. Thus, the answer is 11. Figure 2.7 shows an implementation of this strategy.

The code for algorithm 3 deserves some comment. The general form of the call for the recursive method is to pass the input array along with the left and right borders, which

```
 1      /**
 2       * Recursive maximum contiguous subsequence sum algorithm.
 3       * Finds maximum sum in subarray spanning a[left..right].
 4       * Does not attempt to maintain actual best sequence.
 5       */
 6      private static int maxSumRec( int [ ] a, int left, int right )
 7      {
 8          if( left == right )  // Base case
 9              if( a[ left ] > 0 )
10                  return a[ left ];
11              else
12                  return 0;
13
14          int center = ( left + right ) / 2;
15          int maxLeftSum  = maxSumRec( a, left, center );
16          int maxRightSum = maxSumRec( a, center + 1, right );
17
18          int maxLeftBorderSum = 0, leftBorderSum = 0;
19          for( int i = center; i >= left; i-- )
20          {
21              leftBorderSum += a[ i ];
22              if( leftBorderSum > maxLeftBorderSum )
23                  maxLeftBorderSum = leftBorderSum;
24          }
25
26          int maxRightBorderSum = 0, rightBorderSum = 0;
27          for( int i = center + 1; i <= right; i++ )
28          {
29              rightBorderSum += a[ i ];
30              if( rightBorderSum > maxRightBorderSum )
31                  maxRightBorderSum = rightBorderSum;
32          }
33
34          return max3( maxLeftSum, maxRightSum,
35                      maxLeftBorderSum + maxRightBorderSum );
36      }
37
38      /**
39       * Driver for divide-and-conquer maximum contiguous
40       * subsequence sum algorithm.
41       */
42      public static int maxSubSum3( int [ ] a )
43      {
44          return maxSumRec( a, 0, a.length - 1 );
45      }
```

Figure 2.7 Algorithm 3

delimit the portion of the array that is operated upon. A one-line driver program sets this up by passing the borders 0 and $N - 1$ along with the array.

Lines 8 to 12 handle the base case. If `left == right`, there is one element, and it is the maximum subsequence if the element is nonnegative. The case `left > right` is not possible unless N is negative (although minor perturbations in the code could mess this up). Lines 15 and 16 perform the two recursive calls. We can see that the recursive calls are always on a smaller problem than the original, although minor perturbations in the code could destroy this property. Lines 18 to 24 and 26 to 32 calculate the two maximum sums that touch the center divider. The sum of these two values is the maximum sum that spans both halves. The routine `max3` (not shown) returns the largest of the three possibilities.

Algorithm 3 clearly requires more effort to code than either of the two previous algorithms. However, shorter code does not always mean better code. As we have seen in the earlier table showing the running times of the algorithms, this algorithm is considerably faster than the other two for all but the smallest of input sizes.

The running time is analyzed in much the same way as for the program that computes the Fibonacci numbers. Let $T(N)$ be the time it takes to solve a maximum subsequence sum problem of size N. If $N = 1$, then the program takes some constant amount of time to execute lines 8 to 12, which we shall call one unit. Thus, $T(1) = 1$. Otherwise, the program must perform two recursive calls, the two `for` loops between lines 19 and 32, and some small amount of bookkeeping, such as lines 14 and 18. The two `for` loops combine to touch every element in the subarray, and there is constant work inside the loops, so the time expended in lines 19 to 32 is $O(N)$. The code in lines 8 to 14, 18, 26, and 34 is all a constant amount of work and can thus be ignored compared with $O(N)$. The remainder of the work is performed in lines 15 and 16. These lines solve two subsequence problems of size $N/2$ (assuming N is even). Thus, these lines take $T(N/2)$ units of time each, for a total of $2T(N/2)$. The total time for the algorithm then is $2T(N/2) + O(N)$. This gives the equations

$$T(1) = 1$$
$$T(N) = 2T(N/2) + O(N)$$

To simplify the calculations, we can replace the $O(N)$ term in the equation above with N; since $T(N)$ will be expressed in Big-Oh notation anyway, this will not affect the answer. In Chapter 7, we shall see how to solve this equation rigorously. For now, if $T(N) = 2T(N/2) + N$, and $T(1) = 1$, then $T(2) = 4 = 2*2$, $T(4) = 12 = 4*3$, $T(8) = 32 = 8*4$, and $T(16) = 80 = 16*5$. The pattern that is evident, and can be derived, is that if $N = 2^k$, then $T(N) = N * (k + 1) = N \log N + N = O(N \log N)$.

This analysis assumes N is even, since otherwise $N/2$ is not defined. By the recursive nature of the analysis, it is really valid only when N is a power of 2, since otherwise we eventually get a subproblem that is not an even size, and the equation is invalid. When N is not a power of 2, a somewhat more complicated analysis is required, but the Big-Oh result remains unchanged.

In future chapters, we will see several clever applications of recursion. Here, we present a fourth algorithm to find the maximum subsequence sum. This algorithm is simpler to implement than the recursive algorithm and also is more efficient. It is shown in Figure 2.8.

```
1      /**
2       * Linear-time maximum contiguous subsequence sum algorithm.
3       */
4      public static int maxSubSum4( int [ ] a )
5      {
6          int maxSum = 0, thisSum = 0;
7
8          for( int j = 0; j < a.length; j++ )
9          {
10             thisSum += a[ j ];
11
12             if( thisSum > maxSum )
13                 maxSum = thisSum;
14             else if( thisSum < 0 )
15                 thisSum = 0;
16         }
17
18         return maxSum;
19     }
```

Figure 2.8 Algorithm 4

It should be clear why the time bound is correct, but it takes a little thought to see why the algorithm actually works. To sketch the logic, note that, like algorithms 1 and 2, j is representing the end of the current sequence, while i is representing the start of the current sequence. It happens that the use of i can be optimized out of the program if we do not need to know where the actual best subsequence is, so in designing the algorithm, let's pretend that i is needed, and that we are trying to improve algorithm 2. One observation is that if a[i] is negative, then it cannot possibly represent the start of the optimal sequence, since any subsequence that begins by including a[i] would be improved by beginning with a[i+1]. Similarly, any negative subsequence cannot possibly be a prefix of the optimal subsequence (same logic). If, in the inner loop, we detect that the subsequence from a[i] to a[j] is negative, then we can advance i. The crucial observation is that not only can we advance i to i+1, but we can also actually advance it all the way to j+1. To see this, let p be any index between i+1 and j. Any subsequence that starts at index p is not larger than the corresponding subsequence that starts at index i and includes the subsequence from a[i] to a[p-1], since the latter subsequence is not negative (j is the first index that causes the subsequence starting at index i to become negative). Thus advancing i to j+1 is risk free: we cannot miss an optimal solution.

This algorithm is typical of many clever algorithms: The running time is obvious, but the correctness is not. For these algorithms, formal correctness proofs (more formal than the sketch above) are almost always required; even then, however, many people still are not convinced. In addition, many of these algorithms require trickier programming, leading to longer development. But when these algorithms work, they run quickly, and we can test much of the code logic by comparing it with an inefficient (but easily implemented) brute-force algorithm using small input sizes.

An extra advantage of this algorithm is that it makes only one pass through the data, and once a[i] is read and processed, it does not need to be remembered. Thus, if the array is on a disk or is being transmitted over the Internet, it can be read sequentially, and there is no need to store any part of it in main memory. Furthermore, at any point in time, the algorithm can correctly give an answer to the subsequence problem for the data it has already read (the other algorithms do not share this property). Algorithms that can do this are called **online algorithms**. An online algorithm that requires only constant space and runs in linear time is just about as good as possible.

2.4.4 Logarithms in the Running Time

The most confusing aspect of analyzing algorithms probably centers around the logarithm. We have already seen that some divide-and-conquer algorithms will run in $O(N \log N)$ time. Besides divide-and-conquer algorithms, the most frequent appearance of logarithms centers around the following general rule: *An algorithm is $O(\log N)$ if it takes constant ($O(1)$) time to cut the problem size by a fraction (which is usually $\frac{1}{2}$).* On the other hand, if constant time is required to merely reduce the problem by a constant *amount* (such as to make the problem smaller by 1), then the algorithm is $O(N)$.

It should be obvious that only special kinds of problems can be $O(\log N)$. For instance, if the input is a list of N numbers, an algorithm must take $\Omega(N)$ merely to read the input in. Thus, when we talk about $O(\log N)$ algorithms for these kinds of problems, we usually presume that the input is preread. We provide three examples of logarithmic behavior.

Binary Search
The first example is usually referred to as binary search.

Binary Search.
Given an integer X and integers $A_0, A_1, \ldots, A_{N-1}$, which are presorted and already in memory, find i such that $A_i = X$, or return $i = -1$ if X is not in the input.

The obvious solution consists of scanning through the list from left to right and runs in linear time. However, this algorithm does not take advantage of the fact that the list is sorted and is thus not likely to be best. A better strategy is to check if X is the middle element. If so, the answer is at hand. If X is smaller than the middle element, we can apply the same strategy to the sorted subarray to the left of the middle element; likewise, if X is larger than the middle element, we look to the right half. (There is also the case of when to stop.) Figure 2.9 shows the code for binary search (the answer is mid). As usual, the code reflects Java's convention that arrays begin with index 0.

Clearly, all the work done inside the loop takes $O(1)$ per iteration, so the analysis requires determining the number of times around the loop. The loop starts with high - low $= N - 1$ and finishes with high - low ≥ -1. Every time through the loop the value high - low must be at least halved from its previous value; thus, the number of times around the loop is at most $\lceil \log(N - 1) \rceil + 2$. (As an example, if high - low $= 128$, then

```
1     /**
2      * Performs the standard binary search.
3      * @return index where item is found, or -1 if not found.
4      */
5     public static <AnyType extends Comparable<? super AnyType>>
6     int binarySearch( AnyType [ ] a, AnyType x )
7     {
8         int low = 0, high = a.length - 1;
9
10        while( low <= high )
11        {
12            int mid = ( low + high ) / 2;
13
14            if( a[ mid ].compareTo( x ) < 0 )
15                low = mid + 1;
16            else if( a[ mid ].compareTo( x ) > 0 )
17                high = mid - 1;
18            else
19                return mid;    // Found
20        }
21        return NOT_FOUND;      // NOT_FOUND is defined as -1
22    }
```

Figure 2.9 Binary search

the maximum values of high - low after each iteration are 64, 32, 16, 8, 4, 2, 1, 0, −1.)
Thus, the running time is $O(\log N)$. Equivalently, we could write a recursive formula for
the running time, but this kind of brute-force approach is usually unnecessary when you
understand what is really going on and why.

Binary search can be viewed as our first data structure implementation. It supports the
contains operation in $O(\log N)$ time, but all other operations (in particular insert) require
$O(N)$ time. In applications where the data are static (that is, insertions and deletions are
not allowed), this could be very useful. The input would then need to be sorted once,
but afterward accesses would be fast. An example is a program that needs to maintain
information about the periodic table of elements (which arises in chemistry and physics).
This table is relatively stable, as new elements are added infrequently. The element names
could be kept sorted. Since there are only about 118 elements, at most eight accesses would
be required to find an element. Performing a sequential search would require many more
accesses.

Euclid's Algorithm

A second example is Euclid's algorithm for computing the greatest common divisor. The
greatest common divisor (gcd) of two integers is the largest integer that divides both. Thus,
$gcd(50, 15) = 5$. The algorithm in Figure 2.10 computes $gcd(M, N)$, assuming $M \geq N$. (If
$N > M$, the first iteration of the loop swaps them.)

```
1        public static long gcd( long m, long n )
2        {
3            while( n != 0 )
4            {
5                long rem = m % n;
6                m = n;
7                n = rem;
8            }
9            return m;
10       }
```

Figure 2.10   Euclid's algorithm

The algorithm works by continually computing remainders until 0 is reached. The last nonzero remainder is the answer. Thus, if $M = 1,989$ and $N = 1,590$, then the sequence of remainders is 399, 393, 6, 3, 0. Therefore, $gcd(1989, 1590) = 3$. As the example shows, this is a fast algorithm.

As before, estimating the entire running time of the algorithm depends on determining how long the sequence of remainders is. Although $\log N$ seems like a good answer, it is not at all obvious that the value of the remainder has to decrease by a constant factor, since we see that the remainder went from 399 to only 393 in the example. Indeed, the remainder *does not* decrease by a constant factor in one iteration. However, we can prove that after two iterations, the remainder is at most half of its original value. This would show that the number of iterations is at most $2 \log N = O(\log N)$ and establish the running time. This proof is easy, so we include it here. It follows directly from the following theorem.

Theorem 2.1.
If $M > N$, then $M \bmod N < M/2$.

Proof.
There are two cases. If $N \leq M/2$, then since the remainder is smaller than N, the theorem is true for this case. The other case is $N > M/2$. But then N goes into M once with a remainder $M - N < M/2$, proving the theorem.

One might wonder if this is the best bound possible, since $2 \log N$ is about 20 for our example, and only seven operations were performed. It turns out that the constant can be improved slightly, to roughly $1.44 \log N$, in the worst case (which is achievable if M and N are consecutive Fibonacci numbers). The average-case performance of Euclid's algorithm requires pages and pages of highly sophisticated mathematical analysis, and it turns out that the average number of iterations is about $(12 \ln 2 \ln N)/\pi^2 + 1.47$.

Exponentiation

Our last example in this section deals with raising an integer to a power (which is also an integer). Numbers that result from exponentiation are generally quite large, so an analysis works only if we can assume that we have a machine that can store such large integers

```
1        public static long pow( long x, int n )
2        {
3            if( n == 0 )
4                return 1;
5            if( n == 1 )
6                return x;
7            if( isEven( n ) )
8                return pow( x * x, n / 2 );
9            else
10               return pow( x * x, n / 2 ) * x;
11       }
```

Figure 2.11   Efficient exponentiation

(or a compiler that can simulate this). We will count the number of multiplications as the measurement of running time.

The obvious algorithm to compute X^N uses $N-1$ multiplications. A recursive algorithm can do better. $N \le 1$ is the base case of the recursion. Otherwise, if N is even, we have $X^N = X^{N/2} \cdot X^{N/2}$, and if N is odd, $X^N = X^{(N-1)/2} \cdot X^{(N-1)/2} \cdot X$.

For instance, to compute X^{62}, the algorithm does the following calculations, which involve only nine multiplications:

$$X^3 = (X^2)X, X^7 = (X^3)^2 X, X^{15} = (X^7)^2 X, X^{31} = (X^{15})^2 X, X^{62} = (X^{31})^2$$

The number of multiplications required is clearly at most $2 \log N$, because at most two multiplications (if N is odd) are required to halve the problem. Again, a recurrence formula can be written and solved. Simple intuition obviates the need for a brute-force approach.

Figure 2.11 implements this idea.[2] It is sometimes interesting to see how much the code can be tweaked without affecting correctness. In Figure 2.11, lines 5 to 6 are actually unnecessary, because if N is 1, then line 10 does the right thing. Line 10 can also be rewritten as

```
10                   return pow( x, n - 1 ) * x;
```

without affecting the correctness of the program. Indeed, the program will still run in $O(\log N)$, because the sequence of multiplications is the same as before. However, all of the following alternatives for line 8 are bad, even though they look correct:

```
8a                   return pow( pow( x, 2 ), n / 2 );
8b                   return pow( pow( x, n / 2 ), 2 );
8c                   return pow( x, n / 2 ) * pow( x, n / 2 );
```

---

[2] Java provides a BigInteger class that can be used to manipulate arbitrarily large integers. Translating Figure 2.11 to use BigInteger instead of long is straightforward.

Both lines 8a and 8b are incorrect because when N is 2, one of the recursive calls to pow has 2 as the second argument. Thus no progress is made, and an infinite loop results (in an eventual abnormal termination).

Using line 8c affects the efficiency, because there are now two recursive calls of size $N/2$ instead of only one. An analysis will show that the running time is no longer $O(\log N)$. We leave it as an exercise to the reader to determine the new running time.

2.4.5 A Grain of Salt

Sometimes the analysis is shown empirically to be an overestimate. If this is the case, then either the analysis needs to be tightened (usually by a clever observation), or it may be that the *average* running time is significantly less than the worst-case running time and no improvement in the bound is possible. For many complicated algorithms the worst-case bound is achievable by some bad input but is usually an overestimate in practice. Unfortunately, for most of these problems, an average-case analysis is extremely complex (in many cases still unsolved), and a worst-case bound, even though overly pessimistic, is the best analytical result known.

Summary

This chapter gives some hints on how to analyze the complexity of programs. Unfortunately, it is not a complete guide. Simple programs usually have simple analyses, but this is not always the case. As an example, later in the text we shall see a sorting algorithm (Shellsort, Chapter 7) and an algorithm for maintaining disjoint sets (Chapter 8), each of which requires about 20 lines of code. The analysis of Shellsort is still not complete, and the disjoint set algorithm has an analysis that is extremely difficult and requires pages and pages of intricate calculations. Most of the analyses that we will encounter here will be simple and involve counting through loops.

An interesting kind of analysis, which we have not touched upon, is lower-bound analysis. We will see an example of this in Chapter 7, where it is proved that any algorithm that sorts by using only comparisons requires $\Omega(N \log N)$ comparisons in the worst case. Lower-bound proofs are generally the most difficult, because they apply not to an algorithm but to a class of algorithms that solve a problem.

We close by mentioning that some of the algorithms described here have real-life application. The *gcd* algorithm and the exponentiation algorithm are both used in cryptography. Specifically, a 600-digit number is raised to a large power (usually another 600-digit number), with only the low 600 or so digits retained after each multiplication. Since the calculations require dealing with 600-digit numbers, efficiency is obviously important. The straightforward algorithm for exponentiation would require about 10^{600} multiplications, whereas the algorithm presented requires only about 4,000, in the worst case.

Exercises

2.1   Order the following functions by growth rate: N, $\sqrt{N}$, $N^{1.5}$, N^2, $N \log N$, $N \log \log N$, $N \log^2 N$, $N \log(N^2)$, $2/N$, 2^N, $2^{N/2}$, 37, $N^2 \log N$, N^3. Indicate which functions grow at the same rate.

2.2   Suppose $T_1(N) = O(f(N))$ and $T_2(N) = O(f(N))$. Which of the following are true?
a. $T_1(N) + T_2(N) = O(f(N))$
b. $T_1(N) - T_2(N) = o(f(N))$
c. $\dfrac{T_1(N)}{T_2(N)} = O(1)$
d. $T_1(N) = O(T_2(N))$

2.3   Which function grows faster: $N \log N$ or $N^{1+\epsilon/\sqrt{\log N}}, \epsilon > 0$?

2.4   Prove that for any constant, k, $\log^k N = o(N)$.

2.5   Find two functions $f(N)$ and $g(N)$ such that neither $f(N) = O(g(N))$ nor $g(N) = O(f(N))$.

2.6   In a recent court case, a judge cited a city for contempt and ordered a fine of \$2 for the first day. Each subsequent day, until the city followed the judge's order, the fine was squared (that is, the fine progressed as follows: \$2, \$4, \$16, \$256, \$65,536, ...).
a. What would be the fine on day N?
b. How many days would it take the fine to reach D dollars? (A Big-Oh answer will do.)

2.7   For each of the following six program fragments:
a. Give an analysis of the running time (Big-Oh will do).
b. Implement the code in Java, and give the running time for several values of N.
c. Compare your analysis with the actual running times.

```
(1)  sum = 0;
     for( i = 0; i < n; i++ )
         sum++;

(2)  sum = 0;
     for( i = 0; i < n; i++ )
         for( j = 0; j < n; j++ )
             sum++;

(3)  sum = 0;
     for( i = 0; i < n; i++ )
         for( j = 0; j < n * n; j++ )
             sum++;

(4)  sum = 0;
     for( i = 0; i < n; i++ )
         for( j = 0; j < i; j++ )
             sum++;
```

```
(5)  sum = 0;
         for( i = 0; i < n; i++ )
             for( j = 0; j < i * i; j++ )
                 for( k = 0; k < j; k++ )
                     sum++;

(6)  sum = 0;
         for( i = 1; i < n; i++ )
             for( j = 1; j < i * i; j++ )
                 if( j % i == 0 )
                     for( k = 0; k < j; k++ )
                         sum++;
```

2.8   Suppose you need to generate a *random* permutation of the first N integers. For example, {4, 3, 1, 5, 2} and {3, 1, 4, 2, 5} are legal permutations, but {5, 4, 1, 2, 1} is not, because one number (1) is duplicated and another (3) is missing. This routine is often used in simulation of algorithms. We assume the existence of a random number generator, r, with method randInt(i, j), that generates integers between i and j with equal probability. Here are three algorithms:

1. Fill the array a from a[0] to a[n-1] as follows: To fill a[i], generate random numbers until you get one that is not already in a[0], a[1], ..., a[i-1].
2. Same as algorithm (1), but keep an extra array called the used array. When a random number, ran, is first put in the array a, set used[ran] = true. This means that when filling a[i] with a random number, you can test in one step to see whether the random number has been used, instead of the (possibly) i steps in the first algorithm.
3. Fill the array such that a[i] = i + 1. Then

    ```
    for( i = 1; i < n; i++ )
        swapReferences( a[ i ], a[ randInt( 0, i ) ] );
    ```

    a. Prove that all three algorithms generate only legal permutations and that all permutations are equally likely.
    b. Give as accurate (Big-Oh) an analysis as you can of the *expected* running time of each algorithm.
    c. Write (separate) programs to execute each algorithm 10 times, to get a good average. Run program (1) for $N = 250, 500, 1,000, 2,000$; program (2) for $N = 25,000, 50,000, 100,000, 200,000, 400,000, 800,000$; and program (3) for $N = 100,000, 200,000, 400,000, 800,000, 1,600,000, 3,200,000, 6,400,000$.
    d. Compare your analysis with the actual running times.
    e. What is the worst-case running time of each algorithm?

2.9   Complete the table in Figure 2.2 with estimates for the running times that were too long to simulate. Interpolate the running times for these algorithms and estimate the time required to compute the maximum subsequence sum of 1 million numbers. What assumptions have you made?

2.10   Determine, for the typical algorithms that you use to perform calculations by hand, the running time to do the following:
a. Add two N-digit integers.
b. Multiply two N-digit integers.
c. Divide two N-digit integers.

2.11   An algorithm takes 0.5 ms for input size 100. How long will it take for input size 500 if the running time is the following (assume low-order terms are negligible):
a. linear
b. $O(N \log N)$
c. quadratic
d. cubic

2.12   An algorithm takes 0.5 ms for input size 100. How large a problem can be solved in 1 min if the running time is the following (assume low-order terms are negligible):
a. linear
b. $O(N \log N)$
c. quadratic
d. cubic

2.13   How much time is required to compute $f(x) = \sum_{i=0}^{N} a_i x^i$:
a. Using a simple routine to perform exponentiation?
b. Using the routine in Section 2.4.4?

2.14   Consider the following algorithm (known as *Horner's rule*) to evaluate $f(x) = \sum_{i=0}^{N} a_i x^i$:

```
poly = 0;
for( i = n; i >= 0; i-- )
    poly = x * poly + a[i];
```

a. Show how the steps are performed by this algorithm for $x = 3$, $f(x) = 4x^4 + 8x^3 + x + 2$.
b. Explain why this algorithm works.
c. What is the running time of this algorithm?

2.15   Give an efficient algorithm to determine if there exists an integer i such that $A_i = i$ in an array of integers $A_1 < A_2 < A_3 < \cdots < A_N$. What is the running time of your algorithm?

2.16   Write an alternative gcd algorithm based on the following observations (arrange so that $a > b$):

- $gcd(a, b) = 2gcd(a/2, b/2)$ if a and b are both even.
- $gcd(a, b) = gcd(a/2, b)$ if a is even and b is odd.
- $gcd(a, b) = gcd(a, b/2)$ if a is odd and b is even.
- $gcd(a, b) = gcd((a + b)/2, (a - b)/2)$ if a and b are both odd.

2.17 Give efficient algorithms (along with running time analyses) to:
  a. Find the minimum subsequence sum.
  *b. Find the minimum *positive* subsequence sum.
  *c. Find the maximum subsequence *product*.

2.18 An important problem in numerical analysis is to find a solution to the equation $f(X) = 0$ for some arbitrary f. If the function is continuous and has two points *low* and *high* such that $f(low)$ and $f(high)$ have opposite signs, then a root must exist between *low* and *high* and can be found by a binary search. Write a function that takes as parameters f, *low*, and *high* and solves for a zero. (To implement a generic function as a parameter, pass a function object that implements the Function interface, which you can define to contain a single method f.) What must you do to ensure termination?

2.19 The maximum contiguous subsequence sum algorithms in the text do not give any indication of the actual sequence. Modify them so that they return in a single object the value of the maximum subsequence and the indices of the actual sequence.

2.20 a. Write a program to determine if a positive integer, N, is prime.
  b. In terms of N, what is the worst-case running time of your program? (You should be able to do this in $O(\sqrt{N})$.)
  c. Let B equal the number of bits in the binary representation of N. What is the value of B?
  d. In terms of B, what is the worst-case running time of your program?
  e. Compare the running times to determine if a 20-bit number and a 40-bit number are prime.
  f. Is it more reasonable to give the running time in terms of N or B? Why?

*2.21 The Sieve of Eratosthenes is a method used to compute all primes less than N. We begin by making a table of integers 2 to N. We find the smallest integer, i, that is not crossed out, print i, and cross out i, $2i$, $3i$, ... When $i > \sqrt{N}$, the algorithm terminates. What is the running time of this algorithm?

2.22 Show that X^{62} can be computed with only eight multiplications.

2.23 Write the fast exponentiation routine without recursion.

2.24 Give a precise count on the number of multiplications used by the fast exponentiation routine. (*Hint:* Consider the binary representation of N.)

2.25 Programs A and B are analyzed and found to have worst-case running times no greater than $150N \log_2 N$ and N^2, respectively. Answer the following questions, if possible:
  a. Which program has the better guarantee on the running time, for large values of N ($N > 10{,}000$)?
  b. Which program has the better guarantee on the running time, for small values of N ($N < 100$)?
  c. Which program will run faster *on average* for $N = 1{,}000$?
  d. Is it possible that program B will run faster than program A on *all* possible inputs?

2.26   A majority element in an array, A, of size N is an element that appears more than $N/2$ times (thus, there is at most one). For example, the array

$$3, 3, 4, 2, 4, 4, 2, 4, 4$$

has a majority element (4), whereas the array

$$3, 3, 4, 2, 4, 4, 2, 4$$

does not. If there is no majority element, your program should indicate this. Here is a sketch of an algorithm to solve the problem:

First, a candidate majority element is found (this is the harder part). This candidate is the only element that could possibly be the majority element. The second step determines if this candidate is actually the majority. This is just a sequential search through the array. To find a candidate in the array, A, form a second array, B. Then compare A_1 and A_2. If they are equal, add one of these to B; otherwise do nothing. Then compare A_3 and A_4. Again if they are equal, add one of these to B; otherwise do nothing. Continue in this fashion until the entire array is read. Then recursively find a candidate for B; this is the candidate for A (why?).

   a. How does the recursion terminate?
  *b. How is the case where N is odd handled?
  *c. What is the running time of the algorithm?
   d. How can we avoid using an extra array B?
  *e. Write a program to compute the majority element.

2.27   The input is an N by N matrix of numbers that is already in memory. Each individual row is increasing from left to right. Each individual column is increasing from top to bottom. Give an $O(N)$ worst-case algorithm that decides if a number X is in the matrix.

2.28   Design efficient algorithms that take an array of positive numbers a, and determine:
   a. the maximum value of a[j]+a[i], with j ≥ i.
   b. the maximum value of a[j]-a[i], with j ≥ i.
   c. the maximum value of a[j]*a[i], with j ≥ i.
   d. the maximum value of a[j]/a[i], with j ≥ i.

 *2.29   Why is it important to assume that integers in our computer model have a fixed size?

2.30   Consider the word puzzle problem described in Chapter 1. Suppose we fix the size of the longest word to be 10 characters.
   a. In terms of R and C, which are the number of rows and columns in the puzzle, and W, which is the number of words, what are the running times of the algorithms described in Chapter 1?
   b. Suppose the word list is presorted. Show how to use binary search to obtain an algorithm with significantly better running time.

2.31 Suppose that line 15 in the binary search routine had the statement `low = mid` instead of `low = mid + 1`. Would the routine still work?

2.32 Implement the binary search so that only one two-way comparison is performed in each iteration. (The text implementation uses three-way comparisons. Assume that only a `lessThan` method is available.)

2.33 Suppose that lines 15 and 16 in algorithm 3 (Fig. 2.7) are replaced by

```
15          int maxLeftSum  = maxSubSum( a, left, center - 1 );
16          int maxRightSum = maxSubSum( a, center, right );
```

Would the routine still work?

*2.34 The inner loop of the cubic maximum subsequence sum algorithm performs $N(N+1)(N+2)/6$ iterations of the innermost code. The quadratic version performs $N(N + 1)/2$ iterations. The linear version performs N iterations. What pattern is evident? Can you give a combinatoric explanation of this phenomenon?

References

Analysis of the running time of algorithms was first made popular by Knuth in the three-part series [5], [6], and [7]. Analysis of the *gcd* algorithm appears in [6]. Another early text on the subject is [1].

Big-Oh, Big-Omega, Big-Theta, and little-oh notation were advocated by Knuth in [8]. There is still not uniform agreement on the matter, especially when it comes to using $\Theta()$. Many people prefer to use $O()$, even though it is less expressive. Additionally, $O()$ is still used in some corners to express a lower bound, when $\Omega()$ is called for.

The maximum subsequence sum problem is from [3]. The series of books [2], [3], and [4] show how to optimize programs for speed.

1. A. V. Aho, J. E. Hopcroft, and J. D. Ullman, *The Design and Analysis of Computer Algorithms,* Addison-Wesley, Reading, Mass., 1974.
2. J. L. Bentley, *Writing Efficient Programs,* Prentice Hall, Englewood Cliffs, N.J., 1982.
3. J. L. Bentley, *Programming Pearls,* Addison-Wesley, Reading, Mass., 1986.
4. J. L. Bentley, *More Programming Pearls,* Addison-Wesley, Reading, Mass., 1988.
5. D. E. Knuth, *The Art of Computer Programming, Vol 1: Fundamental Algorithms,* 3d ed., Addison-Wesley, Reading, Mass., 1997.
6. D. E. Knuth, *The Art of Computer Programming, Vol 2: Seminumerical Algorithms,* 3d ed., Addison-Wesley, Reading, Mass., 1998.
7. D. E. Knuth, *The Art of Computer Programming, Vol 3: Sorting and Searching,* 2d ed., Addison-Wesley, Reading, Mass., 1998.
8. D. E. Knuth, "Big Omicron and Big Omega and Big Theta," *ACM SIGACT News,* 8 (1976), 18–23.

CHAPTER 3

Lists, Stacks, and Queues

This chapter discusses three of the most simple and basic data structures. Virtually every significant program will use at least one of these structures explicitly, and a stack is always implicitly used in a program, whether or not you declare one. Among the highlights of this chapter, we will

- Introduce the concept of Abstract Data Types (ADTs).
- Show how to efficiently perform operations on lists.
- Introduce the stack ADT and its use in implementing recursion.
- Introduce the queue ADT and its use in operating systems and algorithm design.

In this chapter, we provide code that implements a significant subset of two library classes: `ArrayList` and `LinkedList`.

3.1 Abstract Data Types (ADTs)

An **abstract data type** (ADT) is a set of objects together with a set of operations. Abstract data types are mathematical abstractions; nowhere in an ADT's definition is there any mention of *how* the set of operations is implemented. Objects such as lists, sets, and graphs, along with their operations, can be viewed as abstract data types, just as integers, reals, and booleans are data types. Integers, reals, and booleans have operations associated with them, and so do abstract data types. For the set ADT, we might have such operations as *add, remove,* and *contains.* Alternatively, we might only want the two operations *union* and *find,* which would define a different ADT on the set.

The Java class allows for the implementation of ADTs, with appropriate hiding of implementation details. Thus any other part of the program that needs to perform an operation on the ADT can do so by calling the appropriate method. If for some reason implementation details need to be changed, it should be easy to do so by merely changing the routines that perform the ADT operations. This change, in a perfect world, would be completely transparent to the rest of the program.

There is no rule telling us which operations must be supported for each ADT; this is a design decision. Error handling and tie breaking (where appropriate) are also generally up to the program designer. The three data structures that we will study in this chapter are primary examples of ADTs. We will see how each can be implemented in several ways, but if they are done correctly, the programs that use them will not necessarily need to know which implementation was used.

3.2 The List ADT

We will deal with a general list of the form $A_0, A_1, A_2, \ldots, A_{N-1}$. We say that the size of this list is N. We will call the special list of size 0 an **empty list**.

For any list except the empty list, we say that A_i follows (or succeeds) A_{i-1} ($i < N$) and that A_{i-1} precedes A_i ($i > 0$). The first element of the list is A_0, and the last element is A_{N-1}. We will not define the predecessor of A_0 or the successor of A_{N-1}. The **position** of element A_i in a list is i. Throughout this discussion, we will assume, to simplify matters, that the elements in the list are integers, but in general, arbitrarily complex elements are allowed (and easily handled by a generic Java class).

Associated with these "definitions" is a set of operations that we would like to perform on the list ADT. Some popular operations are printList and makeEmpty, which do the obvious things; find, which returns the position of the first occurrence of an item; insert and remove, which generally insert and remove some element from some position in the list; and findKth, which returns the element in some position (specified as an argument). If the list is 34, 12, 52, 16, 12, then find(52) might return 2; insert(x,2) might make the list into 34, 12, x, 52, 16, 12 (if we insert into the position given); and remove(52) might turn that list into 34, 12, x, 16, 12.

Of course, the interpretation of what is appropriate for a method is entirely up to the programmer, as is the handling of special cases (for example, what does find(1) return above?). We could also add operations such as next and previous, which would take a position as argument and return the position of the successor and predecessor, respectively.

3.2.1 Simple Array Implementation of Lists

All these instructions can be implemented just by using an array. Although arrays are created with a fixed capacity, we can create a different array with double the capacity when needed. This solves the most serious problem with using an array, namely that historically, to use an array, an estimate of the maximum size of the list was required. This estimate is not needed in Java, or any modern programming language. The following code fragment illustrates how an array, arr, which initially has length 10, can be expanded as needed:

```
int [ ] arr = new int[ 10 ];
    ...
// Later on we decide arr needs to be larger.
int [ ] newArr = new int[ arr.length * 2 ];
for( int i = 0; i < arr.length; i++ )
    newArr[ i ] = arr[ i ];
arr = newArr;
```

An array implementation allows `printList` to be carried out in linear time, and the `findKth` operation takes constant time, which is as good as can be expected. However, insertion and deletion are potentially expensive, depending on where the insertions and deletions occur. In the worst case, inserting into position 0 (in other words, at the front of the list) requires pushing the entire array down one spot to make room, and deleting the first element requires shifting all the elements in the list up one spot, so the worst case for these operations is $O(N)$. On average, half of the list needs to be moved for either operation, so linear time is still required. On the other hand, if all the operations occur at the high end of the list, then no elements need to be shifted, and then adding and deleting take $O(1)$ time.

There are many situations where the list is built up by insertions at the high end, and then only array accesses (i.e., `findKth` operations) occur. In such a case, the array is a suitable implementation. However, if insertions and deletions occur throughout the list, and in particular, at the front of the list, then the array is not a good option. The next subsection deals with the alternative: the *linked list*.

3.2.2 Simple Linked Lists

In order to avoid the linear cost of insertion and deletion, we need to ensure that the list is not stored contiguously, since otherwise entire parts of the list will need to be moved. Figure 3.1 shows the general idea of a **linked list**.

The linked list consists of a series of nodes, which are not necessarily adjacent in memory. Each node contains the element and a link to a node containing its successor. We call this the `next` link. The last cell's `next` link references `null`.

To execute `printList` or `find(x)` we merely start at the first node in the list and then traverse the list by following the `next` links. This operation is clearly linear-time, as in the array implementation, although the constant is likely to be larger than if an array implementation were used. The `findKth` operation is no longer quite as efficient as an array implementation; `findKth(i)` takes $O(i)$ time and works by traversing down the list in the obvious manner. In practice, this bound is pessimistic, because frequently the calls to `findKth` are in sorted order (by i). As an example, `findKth(2)`, `findKth(3)`, `findKth(4)`, and `findKth(6)` can all be executed in one scan down the list.

The `remove` method can be executed in one `next` reference change. Figure 3.2 shows the result of deleting the third element in the original list.

The `insert` method requires obtaining a new node from the system by using a `new` call and then executing two reference maneuvers. The general idea is shown in Figure 3.3. The dashed line represents the old `next` reference.

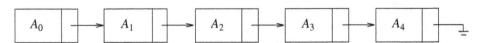

Figure 3.1 A linked list

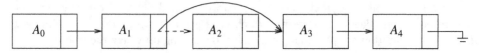

Figure 3.2 Deletion from a linked list

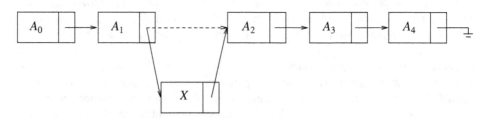

Figure 3.3 Insertion into a linked list

As we can see, in principle, if we know where a change is to be made, inserting or removing an item from a linked list does not require moving lots of items and instead involves only a constant number of changes to node links.

The special case of adding to the front or removing the first item is thus a constant-time operation, presuming of course that a link to the front of the linked list is maintained. The special case of adding at the end (i.e., making the new item as the last item) can be constant-time, as long as we maintain a link to the last node. Thus, a typical linked list keeps links to both ends of the list. Removing the last item is trickier, because we have to find the next-to-last item, change its *next* link to null, and then update the link that maintains the last node. In the classic linked list, where each node stores a link to its next node, having a link to the last node provides no information about the next-to-last node.

The obvious idea of maintaining a third link to the next-to-last node doesn't work, because it too would need to be updated during a remove. Instead, we have every node maintain a link to its previous node in the list. This is shown in Figure 3.4 and is known as a **doubly linked list.**

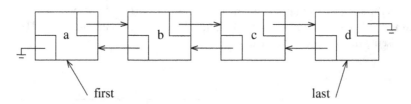

Figure 3.4 A doubly linked list

3.3 Lists in the Java Collections API

The Java language includes, in its library, an implementation of common data structures. This part of the language is popularly known as the **Collections API**. The List ADT is one of the data structures implemented in the Collections API. We will see some others in Chapters 4 and 5.

3.3.1 Collection **Interface**

The Collections API resides in package java.util. The notion of a **collection**, which stores a collection of identically typed objects, is abstracted in the Collection interface. Figure 3.5 shows the most important parts of this interface (some methods are not shown).

Many of the methods in the Collection interface do the obvious things that their names suggest. So size returns the number of items in the collection; isEmpty returns true if and only if the size of the collection is zero. contains returns true if x is in the collection. Note that the interface doesn't specify how the collection decides if x is in the collection—this is determined by the actual classes that implement the Collection interface. add and remove add and remove item x from the collection, returning true if the operation succeeds and false if it fails for a plausible (nonexceptional) reason. For instance, a remove can fail if the item is not present in the collection, and if the particular collection does not allow duplicates, then add can fail when an attempt is made to insert a duplicate.

The Collection interface extends the Iterable interface. Classes that implement the Iterable interface can have the enhanced for loop used on them to view all their items. For instance, the routine in Figure 3.6 can be used to print all the items in any collection. The implementation of this version of print is identical, character-for-character, with a corresponding implementation that could be used if coll had type AnyType[].

3.3.2 Iterator s

Collections that implement the Iterable interface must provide a method named iterator that returns an object of type Iterator. The Iterator is an interface defined in package java.util and is shown in Figure 3.7.

```
1    public interface Collection<AnyType> extends Iterable<AnyType>
2    {
3        int size( );
4        boolean isEmpty( );
5        void clear( );
6        boolean contains( AnyType x );
7        boolean add( AnyType x );
8        boolean remove( AnyType x );
9        java.util.Iterator<AnyType> iterator( );
10   }
```

Figure 3.5 Subset of the Collection interface in package java.util

```
1       public static <AnyType> void print( Collection<AnyType> coll )
2       {
3           for( AnyType item : coll )
4               System.out.println( item );
5       }
```

Figure 3.6   Using the enhanced for loop on an Iterable type

```
1    public interface Iterator<AnyType>
2    {
3        boolean hasNext( );
4        AnyType next( );
5        void remove( );
6    }
```

Figure 3.7   The Iterator interface in package java.util

The idea of the Iterator is that via the iterator method, each collection can create, and return to the client, an object that implements the Iterator interface and stores internally its notion of a current position.

Each call to next gives the next item in the collection (that has not yet been seen). Thus the first call to next gives the first item, the second call gives the second item, and so forth. hasNext can be used to tell you if there is a next item. When the compiler sees an enhanced for loop being used on an object that is Iterable, it mechanically replaces the enhanced for loop with calls to the iterator method to obtain an Iterator and then calls to next and hasNext. Thus the previously seen print routine is rewritten by the compiler as shown in Figure 3.8.

Because of the limited set of methods available in the Iterator interface, it is hard to use the Iterator for anything more than a simple traversal through the Collection. The Iterator interface also contains a method called remove. With this method you can remove the last item returned by next (after which you cannot call remove again until after another

```
1       public static <AnyType> void print( Collection<AnyType> coll )
2       {
3           Iterator<AnyType> itr = coll.iterator( );
4           while( itr.hasNext( ) )
5           {
6               AnyType item = itr.next( );
7               System.out.println( item );
8           }
9       }
```

Figure 3.8   The enhanced for loop on an Iterable type rewritten by the compiler to use an iterator

call to **next**). Although the **Collection** interface also contains a **remove** method, there are presumably advantages to using the **Iterator**'s **remove** method instead.

The main advantage of the **Iterator**'s **remove** method is that the **Collection**'s **remove** method must first find the item to remove. Presumably it is much less expensive to remove an item if you know exactly where it is. An example that we will see in the next section removes every other item in the collection. This code is easy to write with an iterator, and potentially more efficient than using the **Collection**'s **remove** method.

When using the iterator directly (rather than indirectly via an enhanced **for** loop) it is important to keep in mind a fundamental rule: If you make a structural change to the collection being iterated (i.e., an **add**, **remove**, or **clear** method is applied on the collection), then the iterator is no longer valid (and a **ConcurrentModificationException** is thrown on subsequent attempts to use the iterator). This is necessary to avoid ugly situations in which the iterator is prepared to give a certain item as the next item, and then that item is either removed, or perhaps a new item is inserted just prior to the next item. This means that you shouldn't obtain an iterator until immediately prior to the need to use it. However, if the iterator invokes its **remove** method, then the iterator is still valid. This is a second reason to prefer the iterator's **remove** method sometimes.

3.3.3 The **List** Interface, **ArrayList**, and **LinkedList**

The collection that concerns us the most in this section is the list, which is specified by the **List** interface in package **java.util**. The **List** interface extends **Collection**, so it contains all the methods in the **Collection** interface, plus a few others. Figure 3.9 illustrates the most important of these methods.

get and **set** allow the client to access or change an item at the specified position in the list, given by its index, **idx**. Index 0 is the front of the list, index **size()-1** represents the last item in the list, and index **size()** represents the position where a newly added item can be placed. **add** allows the placement of a new item in position **idx** (pushing subsequent items one position higher). Thus, an **add** at position 0 is adding at the front, whereas an **add** at position **size()** is adding an item as the new last item. In addition to the standard **remove** that takes **AnyType** as a parameter, **remove** is overloaded to remove an item at a specified position. Finally, the **List** interface specifies the **listIterator** method that produces a more

```
1    public interface List<AnyType> extends Collection<AnyType>
2    {
3        AnyType get( int idx );
4        AnyType set( int idx, AnyType newVal );
5        void add( int idx, AnyType x );
6        void remove( int idx );
7
8        ListIterator<AnyType> listIterator( int pos );
9    }
```

Figure 3.9 Subset of the **List** interface in package **java.util**

complicated iterator than normally expected. The `ListIterator` interface is discussed in Section 3.3.5.

There are two popular implementations of the List ADT. The `ArrayList` provides a growable array implementation of the List ADT. The advantage of using the `ArrayList` is that calls to `get` and `set` take constant time. The disadvantage is that insertion of new items and removal of existing items is expensive, unless the changes are made at the end of the `ArrayList`. The `LinkedList` provides a doubly linked list implementation of the List ADT. The advantage of using the `LinkedList` is that insertion of new items and removal of existing items is cheap, provided that the position of the changes is known. This means that adds and removes from the front of the list are constant-time operations, so much so that the `LinkedList` provides methods `addFirst` and `removeFirst`, `addLast` and `removeLast`, and `getFirst` and `getLast` to efficiently add, remove, and access the items at both ends of the list. The disadvantage is that the `LinkedList` is not easily indexable, so calls to `get` are expensive unless they are very close to one of the ends of the list (if the call to `get` is for an item near the back of the list, the search can proceed from the back of the list). To see the differences, we look at some methods that operate on a `List`. First, suppose we construct a `List` by adding items at the end.

```
public static void makeList1( List<Integer> lst, int N )
{
    lst.clear( );
    for( int i = 0; i < N; i++ )
        lst.add( i );
}
```

Regardless of whether an `ArrayList` or `LinkedList` is passed as a parameter, the running time of `makeList1` is $O(N)$ because each call to `add`, being at the end of the list, takes constant time (the occasional expansion of the `ArrayList` is safe to ignore). On the other hand, if we construct a `List` by adding items at the front,

```
public static void makeList2( List<Integer> lst, int N )
{
    lst.clear( );
    for( int i = 0; i < N; i++ )
        lst.add( 0, i );
}
```

the running time is $O(N)$ for a `LinkedList`, but $O(N^2)$ for an `ArrayList`, because in an `ArrayList`, adding at the front is an $O(N)$ operation.

The next routine attempts to compute the sum of the numbers in a `List`:

```
public static int sum( List<Integer> lst )
{
    int total = 0;
    for( int i = 0; i < N; i++ )
        total += lst.get( i );
    return total;
}
```

Here, the running time is $O(N)$ for an ArrayList, but $O(N^2)$ for a LinkedList, because in a LinkedList, calls to get are $O(N)$ operations. Instead, use an enhanced for loop, which will make the running time $O(N)$ for any List, because the iterator will efficiently advance from one item to the next.

Both ArrayList and LinkedList are inefficient for searches, so calls to the Collection contains and remove methods (that take an AnyType as parameter) take linear time.

In an ArrayList, there is a notion of a capacity, which represents the size of the underlying array. The ArrayList automatically increases the capacity as needed to ensure that it is at least as large as the size of the list. If an early estimate of the size is available, ensureCapacity can set the capacity to a sufficiently large amount to avoid a later expansion of the array capacity. Also, trimToSize can be used after all ArrayList adds are completed to avoid wasted space.

3.3.4 Example: Using remove on a LinkedList

As an example, we provide a routine that removes all even-valued items in a list. Thus, if the list contains 6, 5, 1, 4, 2, then after the method is invoked it will contain 5, 1.

There are several possible ideas for an algorithm that deletes items from the list as they are encountered. Of course, one idea is to construct a new list containing all the odd numbers, and then clear the original list and copy the odd numbers back into it. But we are more interested in writing a clean version that avoids making a copy and instead removes items from the list as they are encountered.

This is almost certainly a losing strategy for an ArrayList, since removing from almost anywhere in an ArrayList is expensive. In a LinkedList, there is some hope, as we know that removing from a known position can be done efficiently by rearranging some links.

Figure 3.10 shows the first attempt. On an ArrayList, as expected, the remove is not efficient, so the routine takes quadratic time. A LinkedList exposes two problems. First, the call to get is not efficient, so the routine takes quadratic time. Additionally, the call to remove is equally inefficient, because it is expensive to get to position i.

Figure 3.11 shows one attempt to rectify the problem. Instead of using get, we use an iterator to step through the list. This is efficient. But then we use the Collection's remove

```
1       public static void removeEvensVer1( List<Integer> lst )
2       {
3           int i = 0;
4           while( i < lst.size( ) )
5               if( lst.get( i ) % 2 == 0 )
6                   lst.remove( i );
7               else
8                   i++;
9       }
```

Figure 3.10 Removes the even numbers in a list; quadratic on all types of lists

```
1       public static void removeEvensVer2( List<Integer> lst )
2       {
3           for( Integer x : lst )
4               if( x % 2 == 0 )
5                   lst.remove( x );
6       }
```

Figure 3.11   Removes the even numbers in a list; doesn't work because of ConcurrentModificationException

```
1       public static void removeEvensVer3( List<Integer> lst )
2       {
3           Iterator<Integer> itr = lst.iterator( );
4
5           while( itr.hasNext( ) )
6               if( itr.next( ) % 2 == 0 )
7                   itr.remove( );
8       }
```

Figure 3.12   Removes the even numbers in a list; quadratic on ArrayList, but linear time for LinkedList

method to remove an even-valued item. This is not an efficient operation because the remove method has to search for the item again, which takes linear time. But if we run the code, we find out that the situation is even worse: The program generates an exception because when an item is removed, the underlying iterator used by the enhanced for loop is invalidated. (The code in Figure 3.10 explains why: we cannot expect the enhanced for loop to understand that it must advance only if an item is not removed.)

Figure 3.12 shows an idea that works: After the iterator finds an even-valued item, we can use the iterator to remove the value it has just seen. For a LinkedList, the call to the iterator's remove method is only constant time, because the iterator is at (or near) the node that needs to be removed. Thus, for a LinkedList, the entire routine takes linear time, rather than quadratic time. For an ArrayList, even though the iterator is at the point that needs to be removed, the remove is still expensive, because array items must be shifted, so as expected, the entire routine still takes quadratic time for an ArrayList.

If we run the code in Figure 3.12, passing a LinkedList<Integer>, it takes 0.039 seconds for an 800,000-item list, and 0.073 seconds for a 1,600,000 item LinkedList, and is clearly a linear-time routine, because the running time increases by the same factor as the input size. When we pass an ArrayList<Integer>, the routine takes almost five minutes for an 800,000-item ArrayList, and about twenty minutes for a 1,600,000-item ArrayList; the fourfold increase in running time when the input increases by only a factor of two is consistent with quadratic behavior.

```
1    public interface ListIterator<AnyType> extends Iterator<AnyType>
2    {
3        boolean hasPrevious( );
4        AnyType previous( );
5
6        void add( AnyType x );
7        void set( AnyType newVal );
8    }
```

Figure 3.13   Subset of the ListIterator interface in package java.util

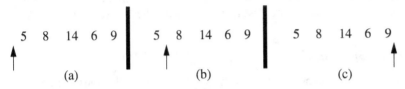

(a)   (b)   (c)

Figure 3.14   (a) Normal starting point: next returns 5, previous is illegal, add places item before 5; (b) next returns 8, previous returns 5, add places item between 5 and 8; (c) next is illegal, previous returns 9, add places item after 9

3.3.5   ListIterators

Figure 3.13 shows that a ListIterator extends the functionality of an Iterator for Lists. previous and hasPrevious allow traversal of the list from the back to the front. add places a new item into the list in the current position. The notion of the current position is abstracted by viewing the iterator as being between the item that would be given by a call to next and the item that would be given by a call to previous, an abstraction that is illustrated in Figure 3.14. add is a constant-time operation for a LinkedList but is expensive for an ArrayList. set changes the last value seen by the iterator and is convenient for LinkedLists. As an example, it can be used to subtract 1 from all the even numbers in a List, which would be hard to do on a LinkedList without using the ListIterator's set method.

3.4   Implementation of ArrayList

In this section, we provide the implementation of a usable ArrayList generic class. To avoid ambiguities with the library class, we will name our class MyArrayList. We do not provide a MyCollection or MyList interface; rather, MyArrayList is standalone. Before examining the (nearly one hundred lines of) MyArrayList code, we outline the main details.

1. The MyArrayList will maintain the underlying array, the array capacity, and the current number of items stored in the MyArrayList.

2. The MyArrayList will provide a mechanism to change the capacity of the underlying array. The capacity is changed by obtaining a new array, copying the old array into the new array, and allowing the Virtual Machine to reclaim the old array.

3. The MyArrayList will provide an implementation of get and set.

4. The MyArrayList will provide basic routines, such as size, isEmpty, and clear, which are typically one-liners; a version of remove; and also two versions of add. The add routines will increase capacity if the size and capacity are the same.

5. The MyArrayList will provide a class that implements the Iterator interface. This class will store the index of the next item in the iteration sequence and provide implementations of next, hasNext, and remove. The MyArrayList's iterator method simply returns a newly constructed instance of the class that implements the Iterator interface.

3.4.1 The Basic Class

Figure 3.15 and Figure 3.16 show the MyArrayList class. Like its Collections API counterpart, there is some error checking to ensure valid bounds; however, in order to concentrate on the basics of writing the iterator class, we do not check for a structural modification that could invalidate an iterator, nor do we check for an illegal iterator remove. These checks are shown in the subsequent implementation of MyLinkedList in Section 3.5 and are exactly the same for both list implementations.

As shown on lines 5–6, the MyArrayList stores the size and array as its data members.

A host of short routines, namely clear, doClear (used to avoid having the constructor invoke an overridable method), size, trimToSize, isEmpty, get, and set, are implemented in lines 11 to 38.

The ensureCapacity routine is shown at lines 40 to 49. Expanding capacity is done with the same logic outlined earlier: saving a reference to the original array at line 45, allocation of a new array at line 46, and copying of the old contents at lines 47 to 48. As shown at lines 42 to 43, the ensureCapacity routine can also be used to shrink the underlying array, but only if the specified new capacity is at least as large as the size. If it isn't, the ensureCapacity request is ignored. At line 46, we see an idiom that is required because generic array creation is illegal. Instead, we create an array of the generic type's bound and then use an array cast. This will generate a compiler warning but is unavoidable in the implementation of generic collections.

Two versions of add are shown. The first adds at the end of the list and is trivially implemented by calling the more general version that adds at the specified position. That version is computationally expensive because it requires shifting elements that are at or after the specified position an additional position higher. add may require increasing capacity. Expanding capacity is very expensive, so if the capacity is expanded, it is made twice as large as the size to avoid having to change the capacity again unless the size increases dramatically (the +1 is used in case the size is 0).

The remove method is similar to add, in that elements that are at or after the specified position must be shifted to one position lower.

The remaining routine deals with the iterator method and the implementation of the associated iterator class. In Figure 3.16, this is shown at lines 77 to 96. The iterator

```java
1   public class MyArrayList<AnyType> implements Iterable<AnyType>
2   {
3       private static final int DEFAULT_CAPACITY = 10;
4
5       private int theSize;
6       private AnyType [ ] theItems;
7
8       public MyArrayList( )
9         { doClear( ); }
10
11      public void clear( )
12        { doClear( ); }
13
14      private void doClear( )
15        { theSize = 0; ensureCapacity( DEFAULT_CAPACITY ); }
16
17      public int size( )
18        { return theSize; }
19      public boolean isEmpty( )
20        { return size( ) == 0; }
21      public void trimToSize( )
22        { ensureCapacity( size( ) ); }
23
24      public AnyType get( int idx )
25      {
26          if( idx < 0 || idx >= size( ) )
27              throw new ArrayIndexOutOfBoundsException( );
28          return theItems[ idx ];
29      }
30
31      public AnyType set( int idx, AnyType newVal )
32      {
33          if( idx < 0 || idx >= size( ) )
34              throw new ArrayIndexOutOfBoundsException( );
35          AnyType old = theItems[ idx ];
36          theItems[ idx ] = newVal;
37          return old;
38      }
39
40      public void ensureCapacity( int newCapacity )
41      {
42          if( newCapacity < theSize )
43              return;
44
45          AnyType [ ] old = theItems;
46          theItems = (AnyType []) new Object[ newCapacity ];
47          for( int i = 0; i < size( ); i++ )
48              theItems[ i ] = old[ i ];
49      }
```

Figure 3.15 MyArrayList class (Part 1 of 2)

```java
50      public boolean add( AnyType x )
51      {
52          add( size( ), x );
53          return true;
54      }
55
56      public void add( int idx, AnyType x )
57      {
58          if( theItems.length == size( ) )
59              ensureCapacity( size( ) * 2 + 1 );
60          for( int i = theSize; i > idx; i-- )
61              theItems[ i ] = theItems[ i - 1 ];
62          theItems[ idx ] = x;
63
64          theSize++;
65      }
66
67      public AnyType remove( int idx )
68      {
69          AnyType removedItem = theItems[ idx ];
70          for( int i = idx; i < size( ) - 1; i++ )
71              theItems[ i ] = theItems[ i + 1 ];
72
73          theSize--;
74          return removedItem;
75      }
76
77      public java.util.Iterator<AnyType> iterator( )
78        { return new ArrayListIterator( ); }
79
80      private class ArrayListIterator implements java.util.Iterator<AnyType>
81      {
82          private int current = 0;
83
84          public boolean hasNext( )
85            { return current < size( ); }
86
87          public AnyType next( )
88          {
89              if( !hasNext( ) )
90                  throw new java.util.NoSuchElementException( );
91              return theItems[ current++ ];
92          }
93
94          public void remove( )
95            { MyArrayList.this.remove( --current ); }
96      }
97  }
```

Figure 3.16 MyArrayList class (Part 2 of 2)

method simply returns an instance of `ArrayListIterator`, which is a class that implements the `Iterator` interface. The `ArrayListIterator` stores the notion of a current position, and provides implementations of `hasNext`, `next`, and `remove`. The current position represents the (array index of the) next element that is to be viewed, so initially the current position is 0.

3.4.2 The Iterator and Java Nested and Inner Classes

The `ArrayListIterator` class uses a tricky Java construct known as the **inner class**. Clearly the class is declared inside of the `MyArrayList` class, a feature that is supported by many languages. However, an inner class in Java has a more subtle property.

To see how an inner class works, Figure 3.17 sketches the iterator idea (however, the code is flawed), making `ArrayListIterator` a top-level class. We focus only on the data fields of `MyArrayList`, the iterator method in `MyArrayList`, and the `ArrayListIterator` (but not its `remove` method).

In Figure 3.17, `ArrayListIterator` is generic, it stores a current position, and the code attempts to use the current position in `next` to index the array and then advance. Note that if `arr` is an array, `arr[idx++]` uses `idx` to the array, and then advances `idx`. The positioning of the `++` matters. The form we used is called the **postfix ++ operator**, in which the `++` is after `idx`. But in the **prefix ++ operator**, `arr[++idx]` advances `idx` and then uses the new `idx` to index the array. The problem with Figure 3.17 is that `theItems[current++]` is illegal, because `theItems` is not part of the `ArrayListIterator` class; it is part of the `MyArrayList`. Thus the code doesn't make sense at all.

```
1    public class MyArrayList<AnyType> implements Iterable<AnyType>
2    {
3        private int theSize;
4        private AnyType [ ] theItems;
5            ...
6        public java.util.Iterator<AnyType> iterator( )
7          { return new ArrayListIterator<AnyType>( ); }
8    }
9    class ArrayListIterator<AnyType> implements java.util.Iterator<AnyType>
10   {
11       private int current = 0;
12           ...
13       public boolean hasNext( )
14         { return current < size( ); }
15       public AnyType next( )
16         { return theItems[ current++ ]; }
17   }
```

Figure 3.17 Iterator Version #1 (doesn't work): The iterator is a top-level class and stores the current position. It doesn't work because `theItems` and `size()` are not part of the `ArrayListIterator` class

```
1    public class MyArrayList<AnyType> implements Iterable<AnyType>
2    {
3        private int theSize;
4        private AnyType [ ] theItems;
5            ...
6        public java.util.Iterator<AnyType> iterator( )
7          { return new ArrayListIterator<AnyType>( this ); }
8    }
9    class ArrayListIterator<AnyType> implements java.util.Iterator<AnyType>
10   {
11       private int current = 0;
12       private MyArrayList<AnyType> theList;
13           ...
14       public ArrayListIterator( MyArrayList<AnyType> list )
15         { theList = list; }
16
17       public boolean hasNext( )
18         { return current < theList.size( ); }
19       public AnyType next( )
20         { return theList.theItems[ current++ ]; }
21   }
```

Figure 3.18 Iterator Version #2 (almost works): The iterator is a top-level class and stores the current position and a link to the MyArrayList. It doesn't work because theItems is private in the MyArrayList class

The simplest solution is shown in Figure 3.18, which is unfortunately also flawed, but in a more minor way. In Figure 3.18, we solve the problem of not having the array in the iterator by having the iterator store a reference to the MyArrayList that it is iterating over. This reference is a second data field and is initialized by a new one-parameter constructor for ArrayListIterator. Now that we have a reference to MyArrayList, we can access the array field that is contained in MyArrayList (and also get the size of the MyArrayList, which is needed in hasNext).

The flaw in Figure 3.18 is that theItems is a private field in MyArrayList, and since ArrayListIterator is a different class, it is illegal to access theItems in the next method. The simplest fix would be to change the visibility of theItems in MyArrayList from private to something less restrictive (such as public, or the default which is known as **package visibility**). But this violates basic principles of good object-oriented programming, which requires data to be as hidden as possible.

Instead, Figure 3.19 shows a solution that works: Make the ArrayListIterator class a **nested class**. When we make ArrayListIterator a nested class, it is placed inside of another class (in this case MyArrayList) which is the **outer class**. We must use the word static to signify that it is nested; without static we will get an inner class, which is sometimes good and sometimes bad. The nested class is the type of class that is typical of many programming languages. Observe that the nested class can be made private, which is nice

```
1    public class MyArrayList<AnyType> implements Iterable<AnyType>
2    {
3        private int theSize;
4        private AnyType [ ] theItems;
5            ...
6        public java.util.Iterator<AnyType> iterator( )
7          { return new ArrayListIterator<AnyType>( this ); }
8
9        private static class ArrayListIterator<AnyType>
10                                        implements java.util.Iterator<AnyType>
11        {
12            private int current = 0;
13            private MyArrayList<AnyType> theList;
14                ...
15            public ArrayListIterator( MyArrayList<AnyType> list )
16              { theList = list; }
17
18            public boolean hasNext( )
19              { return current < theList.size( ); }
20            public AnyType next( )
21              { return theList.theItems[ current++ ]; }
22        }
23    }
```

Figure 3.19 Iterator Version #3 (works): The iterator is a nested class and stores the current position and a link to the MyArrayList. It works because the nested class is considered part of the MyArrayList class

because then it is inaccessible except by the outer class MyArrayList. More importantly, because the nested class is considered to be part of the outer class, there are no visibility issues that arise: theItems is a visible member of class MyArrayList, because next is part of MyArrayList.

Now that we have a nested class, we can discuss the inner class. The problem with the nested class is that in our original design, when we wrote theItems without referring to MyArrayList that it was contained in, the code looked nice, and kind of made sense, but was illegal because it was impossible for the compiler to deduce which MyArrayList was being referred to. It would be nice not to have to keep track of this ourselves. This is exactly what an inner class does for you.

When you declare an inner class, the compiler adds an implicit reference to the outer class object that caused the inner class object's construction. If the name of the outer class is Outer, then the implicit reference is Outer.this. Thus if ArrayListIterator is declared as an inner class, without the static, then MyArrayList.this and theList would both be referencing the same MyArrayList. Thus theList would be redundant and could be removed.

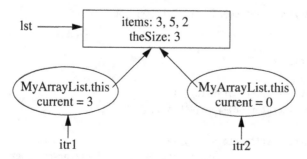

Figure 3.20 Iterator/container with inner classes

The inner class is useful in a situation in which each inner class object is associated with exactly one instance of an outer class object. In such a case, the inner class object can never exist without having an outer class object with which to be associated. In the case of the MyArrayList and its iterator, Figure 3.20 shows the relationship between the iterator class and MyArrayList class, when inner classes are used to implement the iterator.

The use of theList.theItems could be replaced with MyArrayList.this.theItems. This is hardly an improvement, but a further simplification is possible. Just as this.data can be written simply as data (provided there is no other variable named data that could clash), MyArrayList.this.theItems can be written simply as theItems. Figure 3.21 shows the simplification of the ArrayListIterator.

```
1    public class MyArrayList<AnyType> implements Iterable<AnyType>
2    {
3        private int theSize;
4        private AnyType [ ] theItems;
5            ...
6        public java.util.Iterator<AnyType> iterator( )
7          { return new ArrayListIterator( ); }
8
9        private class ArrayListIterator implements java.util.Iterator<AnyType>
10        {
11            private int current = 0;
12
13            public boolean hasNext( )
14              { return current < size( ); }
15            public AnyType next( )
16              { return theItems[ current++ ]; }
17            public void remove( )
18              { MyArrayList.this.remove( --current ); }
19        }
20    }
```

Figure 3.21 Iterator Version #4 (works): The iterator is an inner class and stores the current position and an implicit link to the MyArrayList

First, the `ArrayListIterator` is implicitly generic, since it is now tied to `MyArrayList`, which is generic; we don't have to say so.

Second, `theList` is gone, and we use `size()` and `theItems[current++]` as shorthands for `MyArrayList.this.size()` and `MyArrayList.this.theItems[current++]`. The removal of `the-List` as a data member also removes the associated constructor, so the code reverts back to the style in Version #1.

We can implement the iterator's `remove` by calling `MyArrayList`'s `remove`. Since `MyArrayList`'s `remove` would conflict with `ArrayListIterator`'s `remove`, we have to use `MyArrayList.this.remove`. Note that after the item is removed, elements shift, so for `current` to be viewing the same element, it must also shift. Hence the use of `--`, rather than `-1`.

Inner classes are a syntactical convenience for Java programmers. They are not needed to write any Java code, but their presence in the language allows the Java programmer to write code in the style that was natural (like Version #1), with the compiler writing the extra code required to associate the inner class object with the outer class object.

3.5 Implementation of LinkedList

In this section, we provide the implementation of a usable `LinkedList` generic class. As in the case of the `ArrayList` class, our list class will be named `MyLinkedList` to avoid ambiguities with the library class.

Recall that the `LinkedList` class will be implemented as a doubly linked list, and that we will need to maintain references to both ends of the list. Doing so allows us to maintain constant time cost per operation, so long as the operation occurs at a known position. The known position can be either end, or at a position specified by an iterator (however, we do not implement a `ListIterator`, thus leaving some code for the reader).

In considering the design, we will need to provide three classes:

1. The `MyLinkedList` class itself, which contains links to both ends, the size of the list, and a host of methods.
2. The `Node` class, which is likely to be a private nested class. A node contains the data and links to the previous and next nodes, along with appropriate constructors.
3. The `LinkedListIterator` class, which abstracts the notion of a position and is a private inner class, implementing the `Iterator` interface. It provides implementations of `next`, `hasNext`, and `remove`.

Because the iterator classes store a reference to the "current node," and the end marker is a valid position, it makes sense to create an extra node at the end of the list to represent the end marker. Further, we can create an extra node at the front of the list, logically representing the beginning marker. These extra nodes are sometimes known as **sentinel nodes**; specifically, the node at the front is sometimes known as a **header node**, and the node at the end is sometimes known as a **tail node**.

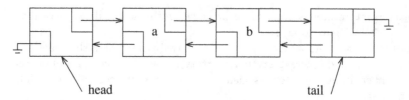

Figure 3.22   A doubly linked list with header and tail nodes

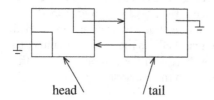

Figure 3.23   An empty doubly linked list with header and tail nodes

The advantage of using these extra nodes is that they greatly simplify the coding by removing a host of special cases. For instance, if we do not use a header node, then removing the first node becomes a special case, because we must reset the list's link to the first node during the remove, and also because the remove algorithm in general needs to access the node prior to the node being removed (and without a header node, the first node does not have a node prior to it). Figure 3.22 shows a doubly linked list with header and tail nodes. Figure 3.23 shows an empty list. Figure 3.24 shows the outline and partial implementation of the MyLinkedList class.

We can see at line 3 the beginning of the declaration of the private nested Node class. Figure 3.25 shows the Node class, consisting of the stored item, links to the previous and next Node, and a constructor. All the data members are public. Recall that in a class, the data members are normally private. However, members in a nested class are visible even in the outer class. Since the Node class is private, the visibility of the data members in the Node class is irrelevant; the MyLinkedList methods can see all Node data members, and classes outside of MyLinkedList cannot see the Node class at all.

Back in Figure 3.24, lines 46 to 49 contain the data members for MyLinkedList, namely the reference to the header and tail nodes. We also keep track of the size in a data member, so that the size method can be implemented in constant time. At line 47, we have one additional data field that is used to help the iterator detect changes in the collection. modCount represents the number of changes to the linked list since construction. Each call to add or remove will update modCount. The idea is that when an iterator is created, it will store the modCount of the collection. Each call to an iterator method (next or remove) will check the stored modCount in the iterator with the current modCount in the linked list and will throw a ConcurrentModificationException if these two counts don't match.

The rest of the MyLinkedList class consists of the constructor, the implementation of the iterator, and a host of methods. Many of the methods are one-liners.

```
 1    public class MyLinkedList<AnyType> implements Iterable<AnyType>
 2    {
 3        private static class Node<AnyType>
 4          { /* Figure 3.25 */ }
 5
 6        public MyLinkedList( )
 7          { doClear( ); }
 8
 9        public void clear( )
10          { /* Figure 3.26 */ }
11        public int size( )
12          { return theSize; }
13        public boolean isEmpty( )
14          { return size( ) == 0; }
15
16        public boolean add( AnyType x )
17          { add( size( ), x );  return true; }
18        public void add( int idx, AnyType x )
19          { addBefore( getNode( idx, 0, size( ) ), x ); }
20        public AnyType get( int idx )
21          { return getNode( idx ).data; }
22        public AnyType set( int idx, AnyType newVal )
23        {
24            Node<AnyType> p = getNode( idx );
25            AnyType oldVal = p.data;
26            p.data = newVal;
27            return oldVal;
28        }
29        public AnyType remove( int idx )
30          { return remove( getNode( idx ) ); }
31
32        private void addBefore( Node<AnyType> p, AnyType x )
33          { /* Figure 3.28 */ }
34        private AnyType remove( Node<AnyType> p )
35          { /* Figure 3.30 */ }
36        private Node<AnyType> getNode( int idx )
37          { /* Figure 3.31 */ }
38        private Node<AnyType> getNode( int idx, int lower, int upper )
39          { /* Figure 3.31 */ }
40
41        public java.util.Iterator<AnyType> iterator( )
42           { return new LinkedListIterator( ); }
43        private class LinkedListIterator implements java.util.Iterator<AnyType>
44          { /* Figure 3.32 */ }
45
46        private int theSize;
47        private int modCount = 0;
48        private Node<AnyType> beginMarker;
49        private Node<AnyType> endMarker;
50    }
```

Figure 3.24 MyLinkedList class

```
1        private static class Node<AnyType>
2        {
3            public Node( AnyType d, Node<AnyType> p, Node<AnyType> n )
4              { data = d; prev = p; next = n; }
5
6            public AnyType data;
7            public Node<AnyType>   prev;
8            public Node<AnyType>   next;
9        }
```

Figure 3.25 Nested Node class for MyLinkedList class

```
1        public void clear( )
2          { doClear( ); }
3
4        private void doClear( )
5        {
6            beginMarker = new Node<AnyType>( null, null, null );
7            endMarker = new Node<AnyType>( null, beginMarker, null );
8            beginMarker.next = endMarker;
9
10           theSize = 0;
11           modCount++;
12       }
```

Figure 3.26 clear routine for MyLinkedList class, which invokes private doClear

The doClear method in Figure 3.26 is invoked by the constructor. It creates and connects the header and tail nodes and then sets the size to 0.

In Figure 3.24, at line 43 we see the beginning of the declaration of the private inner LinkedListIterator class. We'll discuss those details when we see the actual implementations later.

Figure 3.27 illustrates how a new node containing x is spliced in between a node referenced by p and p.prev. The assignment to the node links can be described as follows:

```
Node newNode = new Node( x, p.prev, p );  // Steps 1 and 2
p.prev.next = newNode;                    // Step 3
p.prev = newNode;                         // Step 4
```

Steps 3 and 4 can be combined, yielding only two lines:

```
Node newNode = new Node( x, p.prev, p );  // Steps 1 and 2
p.prev = p.prev.next = newNode;           // Steps 3 and 4
```

But then these two lines can also be combined, yielding:

```
p.prev = p.prev.next = new Node( x, p.prev, p );
```

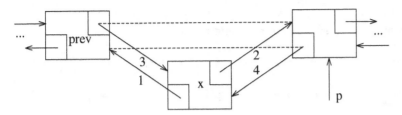

Figure 3.27 Insertion in a doubly linked list by getting new node and then changing pointers in the order indicated

This makes the `addBefore` routine in Figure 3.28 short.

Figure 3.29 shows the logic of removing a node. If p references the node being removed, only two links change before the node is disconnected and eligible to be reclaimed by the Virtual Machine:

```
p.prev.next = p.next;
p.next.prev = p.prev;
```

```
1     /**
2      * Adds an item to this collection, at specified position p.
3      * Items at or after that position are slid one position higher.
4      * @param p Node to add before.
5      * @param x any object.
6      * @throws IndexOutOfBoundsException if idx is not between 0 and size(),.
7      */
8     private void addBefore( Node<AnyType> p, AnyType x )
9     {
10        Node<AnyType> newNode = new Node<>( x, p.prev, p );
11        newNode.prev.next = newNode;
12        p.prev = newNode;
13        theSize++;
14        modCount++;
15    }
```

Figure 3.28 add routine for `MyLinkedList` class

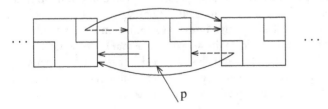

Figure 3.29 Removing node specified by p from a doubly linked list

```
1       /**
2        * Removes the object contained in Node p.
3        * @param p the Node containing the object.
4        * @return the item was removed from the collection.
5        */
6       private AnyType remove( Node<AnyType> p )
7       {
8           p.next.prev = p.prev;
9           p.prev.next = p.next;
10          theSize--;
11          modCount++;
12
13          return p.data;
14      }
```

Figure 3.30   remove routine for MyLinkedList class

Figure 3.30 shows the basic private remove routine that contains the two lines of code shown above.

Figure 3.31 has the previously mentioned private getNode methods. If the index represents a node in the first half of the list, then at lines 29 to 34 we step through the linked list, in the forward direction. Otherwise, we go backward, starting at the end, as shown in lines 37 to 39.

The LinkedListIterator, shown in Figure 3.32, has logic that is similar to the ArrayListIterator but incorporates significant error checking. The iterator maintains a current position, shown at line 3. current represents the node containing the item that is to be returned by a call to next. Observe that when current is positioned at the endMarker, a call to next is illegal.

In order to detect a situation in which the collection has been modified during the iteration, at line 4 the iterator stores in the data field expectedModCount the modCount of the linked list at the time the iterator is constructed. At line 5, the Boolean data field okToRemove is true if a next has been performed, without a subsequent remove. Thus okToRemove is initially false, set to true in next, and set to false in remove.

hasNext is fairly routine. As in java.util.LinkedList's iterator, it does not check for modification of the linked list.

The next method advances current (line 18) after getting the value in the node (line 17) that is to be returned (line 20). okToRemove is updated at line 19.

Finally, the iterator's remove method is shown at lines 23 to 33. It is mostly error checking (which is why we avoided the error checks in the ArrayListIterator). The actual remove at line 30 mimics the logic in the ArrayListIterator. But here, current remains unchanged, because the node that current is viewing is unaffected by the removal of the prior node (in the ArrayListIterator, items shifted, requiring an update of current).

```
1      /**
2       * Gets the Node at position idx, which must range from 0 to size( ) - 1.
3       * @param idx index to search at.
4       * @return internal node corresponding to idx.
5       * @throws IndexOutOfBoundsException if idx is not
6       *          between 0 and size( ) - 1, inclusive.
7       */
8      private Node<AnyType> getNode( int idx )
9      {
10          return getNode( idx, 0, size( ) - 1 );
11     }
12
13     /**
14      * Gets the Node at position idx, which must range from lower to upper.
15      * @param idx index to search at.
16      * @param lower lowest valid index.
17      * @param upper highest valid index.
18      * @return internal node corresponding to idx.
19      * @throws IndexOutOfBoundsException if idx is not
20      *          between lower and upper, inclusive.
21      */
22     private Node<AnyType> getNode( int idx, int lower, int upper )
23     {
24         Node<AnyType> p;
25
26         if( idx < lower || idx > upper )
27             throw new IndexOutOfBoundsException( );
28
29         if( idx < size( ) / 2 )
30         {
31             p = beginMarker.next;
32             for( int i = 0; i < idx; i++ )
33                 p = p.next;
34         }
35         else
36         {
37             p = endMarker;
38             for( int i = size( ); i > idx; i-- )
39                 p = p.prev;
40         }
41
42         return p;
43     }
```

Figure 3.31   Private getNode routine for MyLinkedList class

```
1       private class LinkedListIterator implements java.util.Iterator<AnyType>
2       {
3           private Node<AnyType> current = beginMarker.next;
4           private int expectedModCount = modCount;
5           private boolean okToRemove = false;
6
7           public boolean hasNext( )
8             { return current != endMarker; }
9
10          public AnyType next( )
11          {
12              if( modCount != expectedModCount )
13                  throw new java.util.ConcurrentModificationException( );
14              if( !hasNext( ) )
15                  throw new java.util.NoSuchElementException( );
16
17              AnyType nextItem = current.data;
18              current = current.next;
19              okToRemove = true;
20              return nextItem;
21          }
22
23          public void remove( )
24          {
25              if( modCount != expectedModCount )
26                  throw new java.util.ConcurrentModificationException( );
27              if( !okToRemove )
28                  throw new IllegalStateException( );
29
30              MyLinkedList.this.remove( current.prev );
31              expectedModCount++;
32              okToRemove = false;
33          }
34      }
```

Figure 3.32   Inner Iterator class for MyList class

3.6 The Stack ADT

3.6.1 Stack Model

A **stack** is a list with the restriction that insertions and deletions can be performed in only one position, namely, the end of the list, called the **top**. The fundamental operations on a stack are push, which is equivalent to an insert, and pop, which deletes the most recently

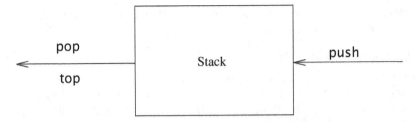

Figure 3.33 Stack model: input to a stack is by push, output is by pop and top

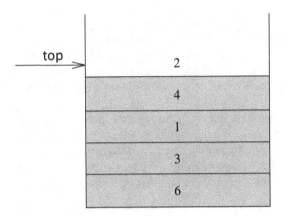

Figure 3.34 Stack model: Only the top element is accessible

inserted element. The most recently inserted element can be examined prior to performing a pop by use of the top routine. A pop or top on an empty stack is generally considered an error in the stack ADT. On the other hand, running out of space when performing a push is an implementation limit but not an ADT error.

Stacks are sometimes known as LIFO (last in, first out) lists. The model depicted in Figure 3.33 signifies only that pushes are input operations and pops and tops are output. The usual operations to make empty stacks and test for emptiness are part of the repertoire, but essentially all that you can do to a stack is push and pop.

Figure 3.34 shows an abstract stack after several operations. The general model is that there is some element that is at the top of the stack, and it is the only element that is visible.

3.6.2 Implementation of Stacks

Since a stack is a list, any list implementation will do. Clearly ArrayList and LinkedList support stack operations; 99% of the time they are the most reasonable choice. Occasionally it can be faster to design a special-purpose implementation (for instance, if the items being placed on the stack are a primitive type). Because stack operations are constant-time operations, this is unlikely to yield any discernable improvement except under very unique circumstances. For these special times, we will give two popular implementations. One uses a linked structure and the other uses an array, and both simplify the logic in `ArrayList` and `LinkedList`, so we do not provide code.

Linked List Implementation of Stacks

The first implementation of a stack uses a singly linked list. We perform a `push` by inserting at the front of the list. We perform a `pop` by deleting the element at the front of the list. A `top` operation merely examines the element at the front of the list, returning its value. Sometimes the `pop` and `top` operations are combined into one.

Array Implementation of Stacks

An alternative implementation avoids links and is probably the more popular solution. Mimicking the `ArrayList` add operation, the implementation is trivial. Associated with each stack is `theArray` and `topOfStack`, which is -1 for an empty stack (this is how an empty stack is initialized). To push some element x onto the stack, we increment `topOfStack` and then set `theArray[topOfStack]` = x. To pop, we set the return value to `theArray[topOfStack]` and then decrement `topOfStack`.

Notice that these operations are performed in not only constant time, but very fast constant time. On some machines, `pushes` and `pops` (of integers) can be written in one machine instruction, operating on a register with auto-increment and auto-decrement addressing. The fact that most modern machines have stack operations as part of the instruction set enforces the idea that the stack is probably the most fundamental data structure in computer science, after the array.

3.6.3 Applications

It should come as no surprise that if we restrict the operations allowed on a list, those operations can be performed very quickly. The big surprise, however, is that the small number of operations left are so powerful and important. We give three of the many applications of stacks. The third application gives a deep insight into how programs are organized.

Balancing Symbols

Compilers check your programs for syntax errors, but frequently a lack of one symbol (such as a missing brace or comment starter) will cause the compiler to spill out a hundred lines of diagnostics without identifying the real error. (Fortunately, most Java compilers are pretty good about this. But not all languages and compilers are as responsible.)

A useful tool in this situation is a program that checks whether everything is balanced. Thus, every right brace, bracket, and parenthesis must correspond to its left counterpart. The sequence [()] is legal, but [(]) is wrong. Obviously, it is not worthwhile writing a huge program for this, but it turns out that it is easy to check these things. For simplicity, we will just check for balancing of parentheses, brackets, and braces and ignore any other character that appears.

The simple algorithm uses a stack and is as follows:

Make an empty stack. Read characters until end of file. If the character is an opening symbol, push it onto the stack. If it is a closing symbol, then if the stack is empty report an error. Otherwise, pop the stack. If the symbol popped is not the corresponding opening symbol, then report an error. At end of file, if the stack is not empty report an error.

You should be able to convince yourself that this algorithm works. It is clearly linear and actually makes only one pass through the input. It is thus online and quite fast. Extra work can be done to attempt to decide what to do when an error is reported—such as identifying the likely cause.

Postfix Expressions

Suppose we have a pocket calculator and would like to compute the cost of a shopping trip. To do so, we add a list of numbers and multiply the result by 1.06; this computes the purchase price of some items with local sales tax added. If the items are 4.99, 5.99, and 6.99, then a natural way to enter this would be the sequence

$$4.99 + 5.99 + 6.99 * 1.06 =$$

Depending on the calculator, this produces either the intended answer, 19.05, or the scientific answer, 18.39. Most simple four-function calculators will give the first answer, but many advanced calculators know that multiplication has higher precedence than addition.

On the other hand, some items are taxable and some are not, so if only the first and last items were actually taxable, then the sequence

$$4.99 * 1.06 + 5.99 + 6.99 * 1.06 =$$

would give the correct answer (18.69) on a scientific calculator and the wrong answer (19.37) on a simple calculator. A scientific calculator generally comes with parentheses, so we can always get the right answer by parenthesizing, but with a simple calculator we need to remember intermediate results.

A typical evaluation sequence for this example might be to multiply 4.99 and 1.06, saving this answer as A_1. We then add 5.99 and A_1, saving the result in A_1. We multiply 6.99 and 1.06, saving the answer in A_2, and finish by adding A_1 and A_2, leaving the final answer in A_1. We can write this sequence of operations as follows:

$$4.99 \ 1.06 * 5.99 + 6.99 \ 1.06 \ * +$$

This notation is known as **postfix** or **reverse Polish notation** and is evaluated exactly as we have described above. The easiest way to do this is to use a stack. When a number is seen, it is pushed onto the stack; when an operator is seen, the operator is applied to the two numbers (symbols) that are popped from the stack, and the result is pushed onto the stack. For instance, the postfix expression

$$6 \ 5 \ 2 \ 3 + 8 * + 3 + *$$

is evaluated as follows: The first four symbols are placed on the stack. The resulting stack is

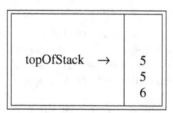

Next a '+' is read, so 3 and 2 are popped from the stack and their sum, 5, is pushed.

Next 8 is pushed.

Now a '*' is seen, so 8 and 5 are popped and $5 * 8 = 40$ is pushed.

Next a '+' is seen, so 40 and 5 are popped and $5 + 40 = 45$ is pushed.

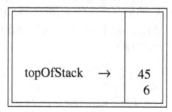

Now, 3 is pushed.

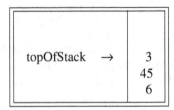

Next '+' pops 3 and 45 and pushes $45 + 3 = 48$.

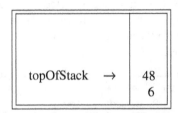

Finally, a '*' is seen and 48 and 6 are popped; the result, $6 * 48 = 288$, is pushed.

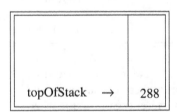

The time to evaluate a postfix expression is $O(N)$, because processing each element in the input consists of stack operations and thus takes constant time. The algorithm to do so is very simple. Notice that when an expression is given in postfix notation, there is no need to know any precedence rules; this is an obvious advantage.

Infix to Postfix Conversion

Not only can a stack be used to evaluate a postfix expression, but we can also use a stack to convert an expression in standard form (otherwise known as **infix**) into postfix. We will concentrate on a small version of the general problem by allowing only the operators +, *, (,), and insisting on the usual precedence rules. We will further assume that the expression is legal. Suppose we want to convert the infix expression

    a + b * c + (d * e + f) * g

into postfix. A correct answer is a b c * + d e * f + g * +.

When an operand is read, it is immediately placed onto the output. Operators are not immediately output, so they must be saved somewhere. The correct thing to do is to place operators that have been seen, but not placed on the output, onto the stack. We will also stack left parentheses when they are encountered. We start with an initially empty stack.

If we see a right parenthesis, then we pop the stack, writing symbols until we encounter a (corresponding) left parenthesis, which is popped but not output.

If we see any other symbol (+, *, ()), then we pop entries from the stack until we find an entry of lower priority. One exception is that we never remove a (from the stack except when processing a). For the purposes of this operation, + has lowest priority and (highest. When the popping is done, we push the operator onto the stack.

Finally, if we read the end of input, we pop the stack until it is empty, writing symbols onto the output.

The idea of this algorithm is that when an operator is seen, it is placed on the stack. The stack represents pending operators. However, some of the operators on the stack that have high precedence are now known to be completed and should be popped, as they will no longer be pending. Thus prior to placing the operator on the stack, operators that are on the stack and are to be completed prior to the current operator, are popped. This is illustrated in the following table:

Expression	Stack When Third Operator Is Processed	Action
a*b-c+d	-	- is completed; + is pushed
a/b+c*d	+	Nothing is completed; * is pushed
a-b*c/d	- *	* is completed; / is pushed
a-b*c+d	- *	* and - are completed; + is pushed

Parentheses simply add an additional complication. We can view a left parenthesis as a high-precedence operator when it is an input symbol (so that pending operators remain pending), and a low-precedence operator when it is on the stack (so that it is not accidentally removed by an operator). Right parentheses are treated as the special case.

To see how this algorithm performs, we will convert the long infix expression above into its postfix form. First, the symbol a is read, so it is passed through to the output. Then + is read and pushed onto the stack. Next b is read and passed through to the output. The state of affairs at this juncture is as follows:

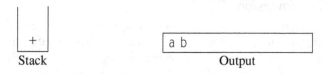

Stack   Output

Next a * is read. The top entry on the operator stack has lower precedence than *, so nothing is output and * is put on the stack. Next, c is read and output. Thus far, we have

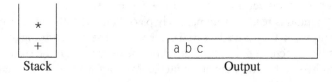

Stack   Output

The next symbol is a +. Checking the stack, we find that we will pop a * and place it on the output; pop the other +, which is not of *lower* but equal priority, on the stack; and then push the +.

Stack   Output

The next symbol read is a (, which, being of highest precedence, is placed on the stack. Then d is read and output.

Stack   Output

We continue by reading a *. Since open parentheses do not get removed except when a closed parenthesis is being processed, there is no output. Next, e is read and output.

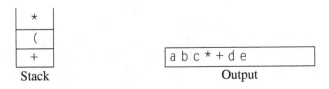

Stack   Output

The next symbol read is a +. We pop and output * and then push +. Then we read and output f.

Stack   Output

Now we read a), so the stack is emptied back to the (. We output a +.

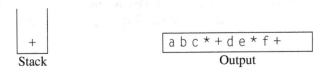

Stack   Output

We read a * next; it is pushed onto the stack. Then g is read and output.

Stack   Output

The input is now empty, so we pop and output symbols from the stack until it is empty.

Stack   Output

As before, this conversion requires only $O(N)$ time and works in one pass through the input. We can add subtraction and division to this repertoire by assigning subtraction and addition equal priority and multiplication and division equal priority. A subtle point is that the expression a-b-c will be converted to ab-c- and not abc--. Our algorithm does the right thing, because these operators associate from left to right. This is not necessarily the case in general, since exponentiation associates right to left: $2^{2^3} = 2^8 = 256$, not $4^3 = 64$. We leave as an exercise the problem of adding exponentiation to the repertoire of operators.

Method Calls

The algorithm to check balanced symbols suggests a way to implement method calls in compiled procedural and object-oriented languages.[1] The problem here is that when a call is made to a new method, all the variables local to the calling routine need to be saved by the system, since otherwise the new method will overwrite the memory used by the calling routine's variables. Furthermore, the current location in the routine must be saved so that the new method knows where to go after it is done. The variables have generally been assigned by the compiler to machine registers, and there are certain to be conflicts (usually all methods get some variables assigned to register #1), especially if recursion is involved. The reason that this problem is similar to balancing symbols is that a method call and method return are essentially the same as an open parenthesis and closed parenthesis, so the same ideas should work.

When there is a method call, all the important information that needs to be saved, such as register values (corresponding to variable names) and the return address (which can be obtained from the program counter, which is typically in a register), is saved "on a piece of paper" in an abstract way and put at the top of a pile. Then the control is transferred to the new method, which is free to replace the registers with its values. If it makes other method calls, it follows the same procedure. When the method wants to return, it looks at the "paper" at the top of the pile and restores all the registers. It then makes the return jump.

---

[1] Since Java is interpreted, rather than compiled, some details in this section may not apply to Java, but the general concepts still do in Java and many other languages.

```
 1      /**
 2       * Print container from itr.
 3       */
 4      public static <AnyType> void printList( Iterator<AnyType> itr )
 5      {
 6          if( !itr.hasNext( ) )
 7              return;
 8
 9          System.out.println( itr.next( ) );
10          printList( itr );
11      }
```

Figure 3.35 A bad use of recursion: printing a linked list

Clearly, all of this work can be done using a stack, and that is exactly what happens in virtually every programming language that implements recursion. The information saved is called either an **activation record** or **stack frame**. Typically, a slight adjustment is made: The current environment is represented at the top of the stack. Thus, a return gives the previous environment (without copying). The stack in a real computer frequently grows from the high end of your memory partition downward, and on many non-Java systems there is no checking for overflow. There is always the possibility that you will run out of stack space by having too many simultaneously active methods. Needless to say, running out of stack space is always a fatal error.

In languages and systems that do not check for stack overflow, programs crash without an explicit explanation. In Java, an exception is thrown.

In normal events, you should not run out of stack space; doing so is usually an indication of runaway recursion (forgetting a base case). On the other hand, some perfectly legal and seemingly innocuous programs can cause you to run out of stack space. The routine in Figure 3.35, which prints out a collection, is perfectly legal and actually correct. It properly handles the base case of an empty collection, and the recursion is fine. This program can be *proven* correct. Unfortunately, if the collection contains 20,000 elements to print, there will be a stack of 20,000 activation records representing the nested calls of line 10. Activation records are typically large because of all the information they contain, so this program is likely to run out of stack space. (If 20,000 elements are not enough to make the program crash, replace the number with a larger one.)

This program is an example of an extremely bad use of recursion known as **tail recursion.** Tail recursion refers to a recursive call at the last line. Tail recursion can be mechanically eliminated by enclosing the body in a `while` loop and replacing the recursive call with one assignment per method argument. This simulates the recursive call because nothing needs to be saved; after the recursive call finishes, there is really no need to know the saved values. Because of this, we can just go to the top of the method with the values that would have been used in a recursive call. The method in Figure 3.36 shows the mechanically improved version. Removal of tail recursion is so simple that some compilers do it automatically. Even so, it is best not to find out that yours does not.

```
1      /**
2       * Print container from itr.
3       */
4      public static <AnyType> void printList( Iterator<AnyType> itr )
5      {
6          while( true )
7          {
8              if( !itr.hasNext( ) )
9                  return;
10
11             System.out.println( itr.next( ) );
12         }
13     }
```

Figure 3.36   Printing a list without recursion; a compiler might do this

Recursion can always be completely removed (compilers do so in converting to assembly language), but doing so can be quite tedious. The general strategy requires using a stack and is worthwhile only if you can manage to put the bare minimum on the stack. We will not dwell on this further, except to point out that although nonrecursive programs are certainly generally faster than equivalent recursive programs, the speed advantage rarely justifies the lack of clarity that results from removing the recursion.

3.7 The Queue ADT

Like stacks, **queues** are lists. With a queue, however, insertion is done at one end, whereas deletion is performed at the other end.

3.7.1 Queue Model

The basic operations on a queue are enqueue, which inserts an element at the end of the list (called the rear), and dequeue, which deletes (and returns) the element at the start of the list (known as the front). Figure 3.37 shows the abstract model of a queue.

3.7.2 Array Implementation of Queues

As with stacks, any list implementation is legal for queues. Like stacks, both the linked list and array implementations give fast $O(1)$ running times for every operation. The linked list implementation is straightforward and left as an exercise. We will now discuss an array implementation of queues.

For each queue data structure, we keep an array, theArray, and the positions front and back, which represent the ends of the queue. We also keep track of the number of elements

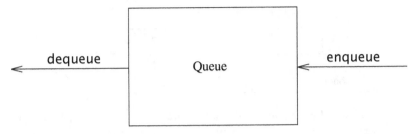

Figure 3.37 Model of a queue

that are actually in the queue, currentSize. The following figure shows a queue in some intermediate state.

		5	2	7	1			

front ↑ is at 5, back ↑ is at 1

The operations should be clear. To enqueue an element x, we increment currentSize and back, then set theArray[back]=x. To dequeue an element, we set the return value to theArray[front], decrement currentSize, and then increment front. Other strategies are possible (this is discussed later). We will comment on checking for errors presently.

There is one potential problem with this implementation. After 10 enqueues, the queue appears to be full, since back is now at the last array index, and the next enqueue would be in a nonexistent position. However, there might only be a few elements in the queue, because several elements may have already been dequeued. Queues, like stacks, frequently stay small even in the presence of a lot of operations.

The simple solution is that whenever front or back gets to the end of the array, it is wrapped around to the beginning. The following figures show the queue during some operations. This is known as a **circular array** implementation.

Initial State

							2	4

front ↑ at 2, back ↑ at 4

After enqueue(1)

1							2	4

back ↑ at 1, front ↑ at 2

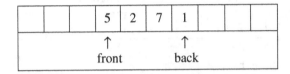
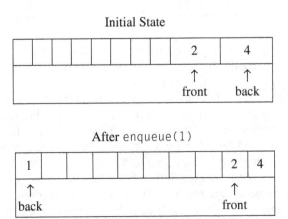

After enqueue(3)

1	3							2	4
	↑							↑	
	back							front	

After dequeue, Which Returns 2

1	3							2	4
	↑								↑
	back								front

After dequeue, Which Returns 4

1	3							2	4
↑	↑								
front	back								

After dequeue, Which Returns 1

1	3							2	4
	↑								
	back								
	front								

After dequeue, Which Returns 3
and Makes the Queue Empty

1	3							2	4
	↑	↑							
	back	front							

The extra code required to implement the wraparound is minimal (although it probably doubles the running time). If incrementing either back or front causes it to go past the array, the value is reset to the first position in the array.

Some programmers use different ways of representing the front and back of a queue. For instance, some do not use an entry to keep track of the size, because they rely on the base case that when the queue is empty, back = front-1. The size is computed implicitly by comparing back and front. This is a very tricky way to go, because there are some special cases, so be very careful if you need to modify code written this way. If the currentSize is not maintained as an explicit data field, then the queue is full when there are theArray.length-1 elements, since only theArray.length different sizes can be differentiated, and one of these is 0. Pick any style you like and make sure that all your routines are consistent. Since there are a few options for implementation, it is probably worth a comment or two in the code, if you don't use the currentSize field.

In applications where you are sure that the number of enqueues is not larger than the capacity of the queue, the wraparound is not necessary. As with stacks, dequeues are rarely performed unless the calling routines are certain that the queue is not empty. Thus error checks are frequently skipped for this operation, except in critical code. This is generally not justifiable, because the time savings that you are likely to achieve are minimal.

3.7.3 Applications of Queues

There are many algorithms that use queues to give efficient running times. Several of these are found in graph theory, and we will discuss them in Chapter 9. For now, we will give some simple examples of queue usage.

When jobs are submitted to a printer, they are arranged in order of arrival. Thus, essentially, jobs sent to a line printer are placed on a queue.[2]

Virtually every real-life line is (supposed to be) a queue. For instance, lines at ticket counters are queues, because service is first-come first-served.

Another example concerns computer networks. There are many network setups of personal computers in which the disk is attached to one machine, known as the **file server**. Users on other machines are given access to files on a first-come first-served basis, so the data structure is a queue.

Further examples include the following:

- Calls to large companies are generally placed on a queue when all operators are busy.

- In large universities, where resources are limited, students must sign a waiting list if all terminals are occupied. The student who has been at a terminal the longest is forced off first, and the student who has been waiting the longest is the next user to be allowed on.

A whole branch of mathematics, known as **queuing theory**, deals with computing, probabilistically, how long users expect to wait on a line, how long the line gets, and other such questions. The answer depends on how frequently users arrive to the line and how long it takes to process a user once the user is served. Both of these parameters are given as probability distribution functions. In simple cases, an answer can be computed analytically. An example of an easy case would be a phone line with one operator. If the operator is busy, callers are placed on a waiting line (up to some maximum limit). This problem is important for businesses, because studies have shown that people are quick to hang up the phone.

If there are k operators, then this problem is much more difficult to solve. Problems that are difficult to solve analytically are often solved by a simulation. In our case, we would need to use a queue to perform the simulation. If k is large, we also need other data structures to do this efficiently. We shall see how to do this simulation in Chapter 6. We

---

[2] We say *essentially* because jobs can be killed. This amounts to a deletion from the middle of the queue, which is a violation of the strict definition.

could then run the simulation for several values of k and choose the minimum k that gives a reasonable waiting time.

Additional uses for queues abound, and as with stacks, it is staggering that such a simple data structure can be so important.

Summary

This chapter describes the concept of ADTs and illustrates the concept with three of the most common abstract data types. The primary objective is to separate the implementation of the abstract data types from their function. The program must know what the operations do, but it is actually better off not knowing how it is done.

Lists, stacks, and queues are perhaps the three fundamental data structures in all of computer science, and their use is documented through a host of examples. In particular, we saw how stacks are used to keep track of method calls and how recursion is actually implemented. This is important to understand, not just because it makes procedural languages possible, but because knowing how recursion is implemented removes a good deal of the mystery that surrounds its use. Although recursion is very powerful, it is not an entirely free operation; misuse and abuse of recursion can result in programs crashing.

Exercises

3.1 You are given a list, L, and another list, P, containing integers sorted in ascending order. The operation `printLots(L,P)` will print the elements in L that are in positions specified by P. For instance, if $P = 1, 3, 4, 6$, the elements in positions $1, 3, 4$, and 6 in L are printed. Write the procedure `printLots(L,P)`. You may use only the public Collections API container operations. What is the running time of your procedure?

3.2 Swap two adjacent elements by adjusting only the links (and not the data) using:
a. Singly linked lists.
b. Doubly linked lists.

3.3 Implement the `contains` routine for `MyLinkedList`.

3.4 Given two sorted lists, L_1 and L_2, write a procedure to compute $L_1 \cap L_2$ using only the basic list operations.

3.5 Given two sorted lists, L_1 and L_2, write a procedure to compute $L_1 \cup L_2$ using only the basic list operations.

3.6 The *Josephus problem* is the following game: N people, numbered 1 to N, are sitting in a circle. Starting at person 1, a hot potato is passed. After M passes, the person holding the hot potato is eliminated, the circle closes ranks, and the game continues with the person who was sitting after the eliminated person picking up the hot potato. The last remaining person wins. Thus, if $M = 0$ and $N = 5$, players are eliminated in order, and player 5 wins. If $M = 1$ and $N = 5$, the order of elimination is 2, 4, 1, 5.

    a. Write a program to solve the Josephus problem for general values of *M* and *N*. Try to make your program as efficient as possible. Make sure you dispose of cells.

    b. What is the running time of your program?

3.7   What is the running time of the following code?

```
public static List<Integer> makeList( int N )
{
    ArrayList<Integer> lst = new ArrayList<>( );

    for( int i = 0; i < N; i++ )
    {
        lst.add( i );
        lst.trimToSize( );
    }
}
```

3.8   The following routine removes the first half of the list passed as a parameter:

```
public static void removeFirstHalf( List<?> lst )
{
    int theSize = lst.size( ) / 2;

    for( int i = 0; i < theSize; i++ )
        lst.remove( 0 );
}
```

    a. Why is `theSize` saved prior to entering the `for` loop?
    b. What is the running time of `removeFirstHalf` if `lst` is an `ArrayList`?
    c. What is the running time of `removeFirstHalf` if `lst` is a `LinkedList`?
    d. Does using an iterator make `removeHalf` faster for either type of `List`?

3.9   Provide an implementation of an `addAll` method for the `MyArrayList` class. Method `addAll` adds all items in the specified collection given by `items` to the end of the `MyArrayList`. Also provide the running time of your implementation. The method signature for you to use is slightly different than the one in the Java Collections API, and is as follows:

```
public void addAll( Iterable<? extends AnyType> items )
```

3.10   Provide an implementation of a `removeAll` method for the `MyLinkedList` class. Method `removeAll` removes all items in the specified collection given by `items` from the `MyLinkedList`. Also provide the running time of your implementation. The method signature for you to use is slightly different than the one in the Java Collections API, and is as follows:

```
public void removeAll( Iterable<? extends AnyType> items )
```

3.11   Assume that a singly linked list is implemented with a header node, but no tail node, and that it maintains only a reference to the header node. Write a class that includes methods to
   a. return the size of the linked list
   b. print the linked list
   c. test if a value x is contained in the linked list
   d. add a value x if it is not already contained in the linked list
   e. remove a value x if it is contained in the linked list

3.12   Repeat Exercise 3.11, maintaining the singly linked list in sorted order.

3.13   Add support for a `ListIterator` to the `MyArrayList` class. The `ListIterator` interface in `java.util` has more methods than are shown in Section 3.3.5. Notice that you will write a `listIterator` method to return a newly constructed `ListIterator`, and further, that the existing iterator method can return a newly constructed `ListIterator`. Thus you will change `ArrayListIterator` so that it implements `ListIterator` instead of `Iterator`. Throw an `UnsupportedOperationException` for methods not listed in Section 3.3.5.

3.14   Add support for a `ListIterator` to the `MyLinkedList` class, as was done in Exercise 3.13.

3.15   Add a `splice` operation to the `LinkedList` class. The method declaration

```
public void splice(Iterator<T> itr, MyLinkedList<? extends T> lst )
```

removes all the items from `lst` (making `lst` empty), placing them prior to `itr` in `MyLinkedList this`. `lst` and `this` must be different lists. Your routine must run in constant time.

3.16   An alternative to providing a `ListIterator` is to provide a method with signature

```
Iterator<AnyType> reverseIterator( )
```

that returns an `Iterator`, initialized to the last item, and for which `next` and `hasNext` are implemented to be consistent with the iterator advancing toward the front of the list, rather than the back. Then you could print a `MyArrayList L` in reverse by using the code

```
Iterator<AnyType> ritr = L.reverseIterator( );
while( ritr.hasNext( ) )
    System.out.println( ritr.next( ) );
```

Implement an `ArrayListReverseIterator` class, with this logic, and have `reverseIterator` return a newly constructed `ArrayListReverseIterator`.

3.17   Modify the `MyArrayList` class to provide stringent iterator checking by using the techniques seen in Section 3.5 for `MyLinkedList`.

3.18   For `MyLinkedList`, implement `addFirst`, `addLast`, `removeFirst`, `removeLast`, `getFirst`, and `getLast` by making calls to the private `add`, `remove`, and `getNode` routines, respectively.

3.19   Rewrite the `MyLinkedList` class without using header and tail nodes and describe the differences between the class and the class provided in Section 3.5.

3.20   An alternative to the deletion strategy we have given is to use **lazy deletion.** To delete an element, we merely mark it deleted (using an extra bit field). The number of deleted and nondeleted elements in the list is kept as part of the data structure. If there are as many deleted elements as nondeleted elements, we traverse the entire list, performing the standard deletion algorithm on all marked nodes.
a. List the advantages and disadvantages of lazy deletion.
b. Write routines to implement the standard linked list operations using lazy deletion.

3.21   Write a program to check for balancing symbols in the following languages:
a. Pascal (`begin`/`end`, (), [], {}).
b. Java (`/* */`, (), [], {}).
*c. Explain how to print out an error message that is likely to reflect the probable cause.

3.22   Write a program to evaluate a postfix expression.

3.23   a. Write a program to convert an infix expression that includes (,), +, -, *, and / to postfix.
b. Add the exponentiation operator to your repertoire.
c. Write a program to convert a postfix expression to infix.

3.24   Write routines to implement two stacks using only one array. Your stack routines should not declare an overflow unless every slot in the array is used.

3.25   *a. Propose a data structure that supports the stack `push` and `pop` operations and a third operation `findMin`, which returns the smallest element in the data structure, all in $O(1)$ worst-case time.
*b. Prove that if we add the fourth operation `deleteMin` which finds and removes the smallest element, then at least one of the operations must take $\Omega(\log N)$ time. (This requires reading Chapter 7.)

3.26   Show how to implement three stacks in one array.

3.27   If the recursive routine in Section 2.4 used to compute Fibonacci numbers is run for $N = 50$, is stack space likely to run out? Why or why not?

3.28   A **deque** is a data structure consisting of a list of items, on which the following operations are possible:
`push(x)`: Insert item x on the front end of the deque.
`pop()`: Remove the front item from the deque and return it.
`inject(x)`: Insert item x on the rear end of the deque.
`eject()`: Remove the rear item from the deque and return it.
Write routines to support the deque that take $O(1)$ time per operation.

3.29   Write an algorithm for printing a singly linked list in reverse, using only constant extra space. This instruction implies that you cannot use recursion, but you may assume that your algorithm is a list member function.

3.30 a. Write an array implementation of self-adjusting lists. In a **self-adjusting list**, all insertions are performed at the front. A self-adjusting list adds a find operation, and when an element is accessed by a find, it is moved to the front of the list without changing the relative order of the other items.

  b. Write a linked list implementation of self-adjusting lists.

  *c. Suppose each element has a fixed probability, p_i, of being accessed. Show that the elements with highest access probability are expected to be close to the front.

3.31 Efficiently implement a stack class using a singly linked list, with no header or tail nodes.

3.32 Efficiently implement a queue class using a singly linked list, with no header or tail nodes.

3.33 Efficiently implement a queue class using a circular array.

3.34 A linked list contains a cycle if, starting from some node p, following a sufficient number of next links brings us back to node p. p does not have to be the first node in the list. Assume that you are given a linked list that contains N nodes. However, the value of N is unknown.

  a. Design an $O(N)$ algorithm to determine if the list contains a cycle. You may use $O(N)$ extra space.

  *b. Repeat part (a), but use only $O(1)$ extra space. (*Hint:* Use two iterators that are initially at the start of the list, but advance at different speeds.)

3.35 One way to implement a queue is to use a circular linked list. In a **circular linked list**, the last node's next link links to the first node. Assume the list does not contain a header and that we can maintain, at most, one iterator corresponding to a node in the list. For which of the following representations can all basic queue operations be performed in constant worst-case time? Justify your answers.

  a. Maintain an iterator that corresponds to the first item in the list.

  b. Maintain an iterator that corresponds to the last item in the list.

3.36 Suppose we have a reference to a node in a singly linked list that is guaranteed *not to be the last node* in the list. We do not have references to any other nodes (except by following links). Describe an $O(1)$ algorithm that logically removes the value stored in such a node from the linked list, maintaining the integrity of the linked list. (*Hint:* Involve the next node.)

3.37 Suppose that a singly linked list is implemented with both a header and a tail node. Describe constant-time algorithms to

  a. Insert item x before position p (given by an iterator).

  b. Remove the item stored at position p (given by an iterator).

CHAPTER 4

Trees

For large amounts of input, the linear access time of linked lists is prohibitive. In this chapter we look at a simple data structure for which the running time of most operations is $O(\log N)$ on average. We also sketch a conceptually simple modification to this data structure that guarantees the above time bound in the worst case and discuss a second modification that essentially gives an $O(\log N)$ running time per operation for a long sequence of instructions.

The data structure that we are referring to is known as a **binary search tree**. The binary search tree is the basis for the implementation of two library collections classes, `TreeSet` and `TreeMap`, which are used in many applications. *Trees* in general are very useful abstractions in computer science, so we will discuss their use in other, more general applications. In this chapter, we will

- See how trees are used to implement the file system of several popular operating systems.

- See how trees can be used to evaluate arithmetic expressions.

- Show how to use trees to support searching operations in $O(\log N)$ average time, and how to refine these ideas to obtain $O(\log N)$ worst-case bounds. We will also see how to implement these operations when the data are stored on a disk.

- Discuss and use the `TreeSet` and `TreeMap` classes.

4.1 Preliminaries

A **tree** can be defined in several ways. One natural way to define a tree is recursively. A tree is a collection of nodes. The collection can be empty; otherwise, a tree consists of a distinguished node r, called the **root**, and zero or more nonempty (sub)trees $T_1, T_2, \ldots, T_k$, each of whose roots are connected by a directed **edge** from r.

The root of each subtree is said to be a **child** of r, and r is the **parent** of each subtree root. Figure 4.1 shows a typical tree using the recursive definition.

From the recursive definition, we find that a tree is a collection of N nodes, one of which is the root, and $N - 1$ edges. That there are $N - 1$ edges follows from the fact that each edge connects some node to its parent, and every node except the root has one parent (see Figure 4.2).

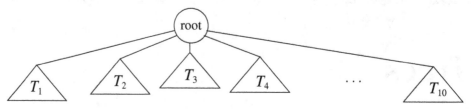

Figure 4.1   Generic tree

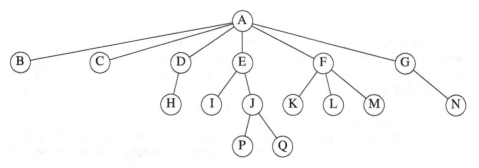

Figure 4.2   A tree

In the tree of Figure 4.2, the root is A. Node F has A as a parent and K, L, and M as children. Each node may have an arbitrary number of children, possibly zero. Nodes with no children are known as **leaves**; the leaves in the tree above are $B, C, H, I, P, Q, K, L, M$, and N. Nodes with the same parent are **siblings**; thus K, L, and M are all siblings. **Grandparent** and **grandchild** relations can be defined in a similar manner.

A **path** from node n_1 to n_k is defined as a sequence of nodes $n_1, n_2, \ldots, n_k$ such that n_i is the parent of n_{i+1} for $1 \leq i < k$. The **length** of this path is the number of edges on the path, namely $k - 1$. There is a path of length zero from every node to itself. Notice that in a tree there is exactly one path from the root to each node.

For any node n_i, the **depth** of n_i is the length of the unique path from the root to n_i. Thus, the root is at depth 0. The **height** of n_i is the length of the longest path from n_i to a leaf. Thus all leaves are at height 0. The height of a tree is equal to the height of the root. For the tree in Figure 4.2, E is at depth 1 and height 2; F is at depth 1 and height 1; the height of the tree is 3. The depth of a tree is equal to the depth of the deepest leaf; this is always equal to the height of the tree.

If there is a path from n_1 to n_2, then n_1 is an **ancestor** of n_2 and n_2 is a **descendant** of n_1. If $n_1 \neq n_2$, then n_1 is a **proper ancestor** of n_2 and n_2 is a **proper descendant** of n_1.

4.1.1 Implementation of Trees

One way to implement a tree would be to have in each node, besides its data, a link to each child of the node. However, since the number of children per node can vary so greatly and is not known in advance, it might be infeasible to make the children direct links in the data

```
class TreeNode
{
    Object   element;
    TreeNode firstChild;
    TreeNode nextSibling;
}
```

Figure 4.3 Node declarations for trees

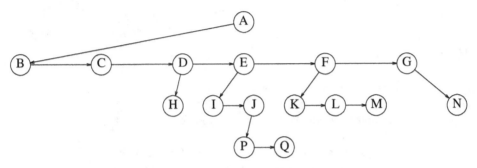

Figure 4.4 First child/next sibling representation of the tree shown in Figure 4.2

structure, because there would be too much wasted space. The solution is simple: Keep the children of each node in a linked list of tree nodes. The declaration in Figure 4.3 is typical.

Figure 4.4 shows how a tree might be represented in this implementation. Arrows that point downward are firstChild links. Horizontal arrows are nextSibling links. Null links are not drawn, because there are too many.

In the tree of Figure 4.4, node *E* has both a link to a sibling (*F*) and a link to a child (*I*), while some nodes have neither.

4.1.2 Tree Traversals with an Application

There are many applications for trees. One of the popular uses is the directory structure in many common operating systems, including UNIX and DOS. Figure 4.5 is a typical directory in the UNIX file system.

The root of this directory is */usr*. (The asterisk next to the name indicates that */usr* is itself a directory.) */usr* has three children, *mark, alex,* and *bill,* which are themselves directories. Thus, */usr* contains three directories and no regular files. The filename */usr/mark/book/ch1.r* is obtained by following the leftmost child three times. Each / after the first indicates an edge; the result is the full **pathname**. This hierarchical file system is very popular, because it allows users to organize their data logically. Furthermore, two files in different directories can share the same name, because they must have different paths from the root and thus have different pathnames. A directory in the UNIX file system is just a file with a list of all its children, so the directories are structured almost exactly in accordance

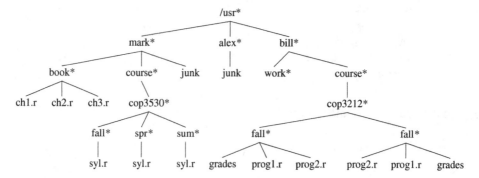

Figure 4.5 UNIX directory

```
        private void listAll( int depth )
        {
1           printName( depth ); // Print the name of the object
2           if( isDirectory( ) )
3               for each file c in this directory (for each child)
4                   c.listAll( depth + 1 );
        }

        public void listAll( )
        {
            listAll( 0 );
        }
```

Figure 4.6 Pseudocode to list a directory in a hierarchical file system

with the type declaration above.[1] Indeed, on some versions of UNIX, if the normal command to print a file is applied to a directory, then the names of the files in the directory can be seen in the output (along with other non-ASCII information).

Suppose we would like to list the names of all of the files in the directory. Our output format will be that files that are depth d_i will have their names indented by d_i tabs. Our algorithm is given in Figure 4.6 as pseudocode.[2]

The heart of the algorithm is the recursive method listAll. This routine needs to be started with a depth of 0, to signify no indenting for the root. This depth is an internal bookkeeping variable and is hardly a parameter that a calling routine should be expected to know about. Thus the driver routine is used to interface the recursive routine to the outside world.

---

[1] Each directory in the UNIX file system also has one entry that points to itself and another entry that points to the parent of the directory. Thus, technically, the UNIX file system is not a tree, but is treelike.

[2] The Java code to implement this is provided in the file *FileSystem.java* online. It uses Java features that have not been discussed in the text.

```
/usr
    mark
        book
            ch1.r
            ch2.r
            ch3.r
        course
            cop3530
                fall
                    syl.r
                spr
                    syl.r
                sum
                    syl.r
        junk
    alex
        junk
    bill
        work
        course
            cop3212
                fall
                    grades
                    prog1.r
                    prog2.r
                fall
                    prog2.r
                    prog1.r
                    grades
```

Figure 4.7 The (preorder) directory listing

The logic of the algorithm is simple to follow. The name of the file object is printed out with the appropriate number of tabs. If the entry is a directory, then we process all children recursively, one by one. These children are one level deeper and thus need to be indented an extra space. The output is in Figure 4.7.

This traversal strategy is known as a **preorder traversal**. In a preorder traversal, work at a node is performed before (*pre*) its children are processed. When this program is run, it is clear that line 1 is executed exactly once per node, since each name is output once. Since line 1 is executed at most once per node, line 2 must also be executed once per node. Furthermore, line 4 can be executed at most once for each child of each node. But the number of children is exactly one less than the number of nodes. Finally, the for loop iterates once per execution of line 4, plus once each time the loop ends. Thus, the total amount of work is constant per node. If there are N file names to be output, then the running time is $O(N)$.

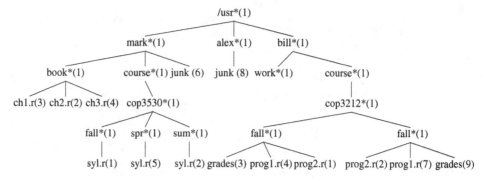

Figure 4.8 UNIX directory with file sizes obtained via postorder traversal

```
        public int size( )
        {
1           int totalSize = sizeOfThisFile( );

2           if( isDirectory( ) )
3               for each file c in this directory (for each child)
4                   totalSize += c.size( );

5           return totalSize;
        }
```

Figure 4.9 Pseudocode to calculate the size of a directory

Another common method of traversing a tree is the **postorder traversal**. In a postorder traversal, the work at a node is performed after (*post*) its children are evaluated. As an example, Figure 4.8 represents the same directory structure as before, with the numbers in parentheses representing the number of disk blocks taken up by each file.

Since the directories are themselves files, they have sizes too. Suppose we would like to calculate the total number of blocks used by all the files in the tree. The most natural way to do this would be to find the number of blocks contained in the subdirectories */usr/mark* (30), */usr/alex* (9), and */usr/bill* (32). The total number of blocks is then the total in the subdirectories (71) plus the one block used by */usr*, for a total of 72. The pseudocode method size in Figure 4.9 implements this strategy.

If the current object is not a directory, then size merely returns the number of blocks it uses. Otherwise, the number of blocks used by the directory is added to the number of blocks (recursively) found in all the children. To see the difference between the postorder traversal strategy and the preorder traversal strategy, Figure 4.10 shows how the size of each directory or file is produced by the algorithm.

```
       ch1.r              3
       ch2.r              2
       ch3.r              4
    book                 10
                syl.r     1
          fall            2
                syl.r     5
          spr             6
                syl.r     2
            sum           3
        cop3530          12
      course            13
      junk               6
  mark                  30
      junk               8
  alex                   9
      work               1
                grades    3
                prog1.r   4
                prog2.r   1
          fall            9
                prog2.r   2
                prog1.r   7
                grades    9
            fall         19
        cop3212         29
      course            30
    bill                32
  /usr                  72
```

Figure 4.10   Trace of the size function

4.2 Binary Trees

A binary tree is a tree in which no node can have more than two children.

Figure 4.11 shows that a binary tree consists of a root and two subtrees, T_L and T_R, both of which could possibly be empty.

A property of a binary tree that is sometimes important is that the depth of an average binary tree is considerably smaller than N. An analysis shows that the average depth is $O(\sqrt{N})$, and that for a special type of binary tree, namely the *binary search tree*, the average value of the depth is $O(\log N)$. Unfortunately, the depth can be as large as $N - 1$, as the example in Figure 4.12 shows.

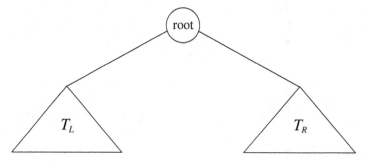

Figure 4.11   Generic binary tree

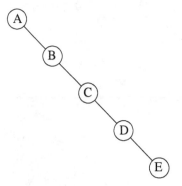

Figure 4.12   Worst-case binary tree

4.2.1 Implementation

Because a binary tree node has at most two children, we can keep direct links to them. The declaration of tree nodes is similar in structure to that for doubly linked lists in that a node is a structure consisting of the `element` information plus two references (`left` and `right`) to other nodes (see Figure 4.13).

We could draw the binary trees using the rectangular boxes that are customary for linked lists, but trees are generally drawn as circles connected by lines, because they are

```
class BinaryNode
{
        // Friendly data; accessible by other package routines
    Object      element;        // The data in the node
    BinaryNode left;            // Left child
    BinaryNode right;           // Right child
}
```

Figure 4.13   Binary tree node class

actually graphs. We also do not explicitly draw null links when referring to trees, because every binary tree with N nodes would require $N + 1$ null links.

Binary trees have many important uses not associated with searching. One of the principal uses of binary trees is in the area of compiler design, which we will now explore.

4.2.2 An Example: Expression Trees

Figure 4.14 shows an example of an **expression tree**. The leaves of an expression tree are **operands**, such as constants or variable names, and the other nodes contain **operators**. This particular tree happens to be binary, because all the operators are binary, and although this is the simplest case, it is possible for nodes to have more than two children. It is also possible for a node to have only one child, as is the case with the **unary minus** operator. We can evaluate an expression tree, T, by applying the operator at the root to the values obtained by recursively evaluating the left and right subtrees. In our example, the left subtree evaluates to a + (b * c) and the right subtree evaluates to ((d * e) + f) * g. The entire tree therefore represents (a + (b * c)) + (((d * e) + f) * g).

We can produce an (overly parenthesized) infix expression by recursively producing a parenthesized left expression, then printing out the operator at the root, and finally recursively producing a parenthesized right expression. This general strategy (left, node, right) is known as an **inorder traversal**; it is easy to remember because of the type of expression it produces.

An alternate traversal strategy is to recursively print out the left subtree, the right subtree, and then the operator. If we apply this strategy to our tree above, the output is a b c * + d e * f + g * +, which is easily seen to be the postfix representation of Section 3.6.3. This traversal strategy is generally known as a *postorder* traversal. We have seen this traversal strategy earlier in Section 4.1.

A third traversal strategy is to print out the operator first and then recursively print out the left and right subtrees. The resulting expression, + + a * b c * + * d e f g, is the less useful *prefix* notation and the traversal strategy is a *preorder* traversal, which we have also seen earlier in Section 4.1. We will return to these traversal strategies later in the chapter.

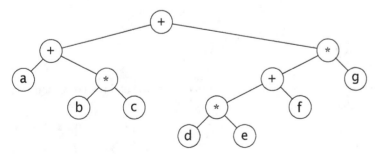

Figure 4.14   Expression tree for (a + b * c) + ((d * e + f) * g)

Constructing an Expression Tree

We now give an algorithm to convert a postfix expression into an expression tree. Since we already have an algorithm to convert infix to postfix, we can generate expression trees from the two common types of input. The method we describe strongly resembles the postfix evaluation algorithm of Section 3.6.3. We read our expression one symbol at a time. If the symbol is an operand, we create a one-node tree and push it onto a stack. If the symbol is an operator, we pop two trees T_1 and T_2 from the stack (T_1 is popped first) and form a new tree whose root is the operator and whose left and right children are T_2 and T_1, respectively. This new tree is then pushed onto the stack.

As an example, suppose the input is

    a b + c d e + * *

The first two symbols are operands, so we create one-node trees and push them onto a stack.[3]

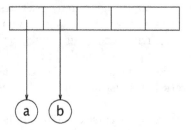

Next, a + is read, so two trees are popped, a new tree is formed, and it is pushed onto the stack.

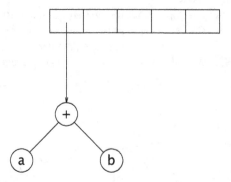

Next, c, d, and e are read, and for each a one-node tree is created and the corresponding tree is pushed onto the stack.

---

[3] For convenience, we will have the stack grow from left to right in the diagrams.

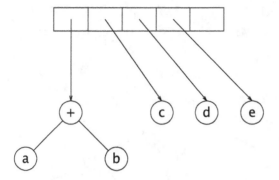

Now a + is read, so two trees are merged.

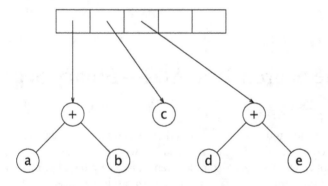

Continuing, a * is read, so we pop two trees and form a new tree with a * as root.

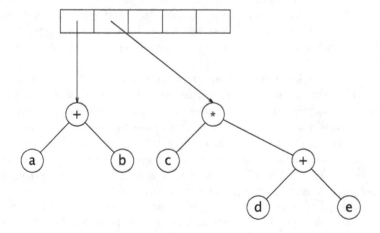

Finally, the last symbol is read, two trees are merged, and the final tree is left on the stack.

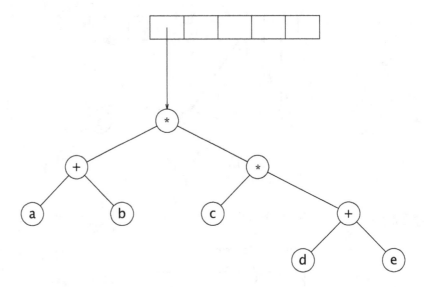

4.3 The Search Tree ADT—Binary Search Trees

An important application of binary trees is their use in searching. Let us assume that each node in the tree stores an item. In our examples, we will assume for simplicity that these are integers, although arbitrarily complex items are easily handled in Java. We will also assume that all the items are distinct and deal with duplicates later.

The property that makes a binary tree into a binary search tree is that for every node, X, in the tree, the values of all the items in its left subtree are smaller than the item in X, and the values of all the items in its right subtree are larger than the item in X. Notice that this implies that all the elements in the tree can be ordered in some consistent manner. In Figure 4.15, the tree on the left is a binary search tree, but the tree on the right is not. The tree on the right has a node with item 7 in the left subtree of a node with item 6 (which happens to be the root).

We now give brief descriptions of the operations that are usually performed on binary search trees. Note that because of the recursive definition of trees, it is common to write these routines recursively. Because the average depth of a binary search tree turns out to be $O(\log N)$, we generally do not need to worry about running out of stack space.

The binary search tree requires that all the items can be ordered. To write a generic class, we need to provide an `interface` type that represents this property. This interface is `Comparable`, as described in Chapter 1. The interface tells us that two items in the tree can always be compared using a `compareTo` method. From this, we can determine all other possible relationships. Specifically, we do not use the `equals` method. Instead, two items are equal if and only if the `compareTo` method returns 0. An alternative, described in Section 4.3.1, is to allow a function object. Figure 4.16 also shows the `BinaryNode` class that, like the node class in the linked list class, is a nested class.

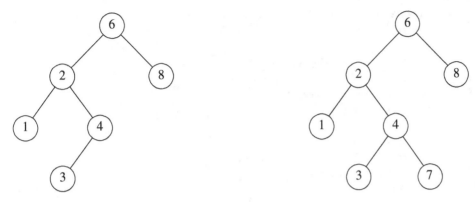

Figure 4.15 Two binary trees (only the left tree is a search tree)

```
1       private static class BinaryNode<AnyType>
2       {
3               // Constructors
4           BinaryNode( AnyType theElement )
5               { this( theElement, null, null ); }
6
7           BinaryNode( AnyType theElement, BinaryNode<AnyType> lt, BinaryNode<AnyType> rt )
8               { element  = theElement; left = lt; right = rt; }
9
10          AnyType element;              // The data in the node
11          BinaryNode<AnyType> left;     // Left child
12          BinaryNode<AnyType> right;    // Right child
13      }
```

Figure 4.16 The BinaryNode class

Figure 4.17 shows the BinarySearchTree class skeleton. The single data field is a reference to the root node; this reference is null for empty trees. The public methods use the general technique of calling private recursive methods.

We can now describe some of the private methods.

4.3.1 contains

This operation requires returning **true** if there is a node in tree T that has item X, or **false** if there is no such node. The structure of the tree makes this simple. If T is empty, then we can just return **false**. Otherwise, if the item stored at T is X, we can return **true**. Otherwise, we make a recursive call on a subtree of T, either left or right, depending on the relationship of X to the item stored in T. The code in Figure 4.18 is an implementation of this strategy.

Notice the order of the tests. It is crucial that the test for an empty tree be performed first, since otherwise, we would generate a NullPointerException attempting to access a data field through a null reference. The remaining tests are arranged with the least likely case

```java
 1    public class BinarySearchTree<AnyType extends Comparable<? super AnyType>>
 2    {
 3        private static class BinaryNode<AnyType>
 4          { /* Figure 4.16 */ }
 5
 6        private BinaryNode<AnyType> root;
 7
 8        public BinarySearchTree( )
 9          { root = null; }
10
11        public void makeEmpty( )
12          { root = null; }
13        public boolean isEmpty( )
14          { return root == null; }
15
16        public boolean contains( AnyType x )
17          { return contains( x, root ); }
18        public AnyType findMin( )
19          { if( isEmpty( ) ) throw new UnderflowException( );
20            return findMin( root ).element;
21          }
22        public AnyType findMax( )
23          { if( isEmpty( ) ) throw new UnderflowException( );
24            return findMax( root ).element;
25          }
26        public void insert( AnyType x )
27          { root = insert( x, root ); }
28        public void remove( AnyType x )
29          { root = remove( x, root ); }
30        public void printTree( )
31          { /* Figure 4.56 */ }
32
33        private boolean contains( AnyType x, BinaryNode<AnyType> t )
34          { /* Figure 4.18 */ }
35        private BinaryNode<AnyType> findMin( BinaryNode<AnyType> t )
36          { /* Figure 4.20 */ }
37        private BinaryNode<AnyType> findMax( BinaryNode<AnyType> t )
38          { /* Figure 4.20 */ }
39
40        private BinaryNode<AnyType> insert( AnyType x, BinaryNode<AnyType> t )
41          { /* Figure 4.22 */ }
42        private BinaryNode<AnyType> remove( AnyType x, BinaryNode<AnyType> t )
43          { /* Figure 4.25 */ }
44        private void printTree( BinaryNode<AnyType> t )
45          { /* Figure 4.56 */ }
46    }
```

Figure 4.17 Binary search tree class skeleton

```
1       /**
2        * Internal method to find an item in a subtree.
3        * @param x is item to search for.
4        * @param t the node that roots the subtree.
5        * @return true if the item is found; false otherwise.
6        */
7       private boolean contains( AnyType x, BinaryNode<AnyType> t )
8       {
9           if( t == null )
10              return false;
11
12          int compareResult = x.compareTo( t.element );
13
14          if( compareResult < 0 )
15              return contains( x, t.left );
16          else if( compareResult > 0 )
17              return contains( x, t.right );
18          else
19              return true;     // Match
20      }
```

Figure 4.18   contains operation for binary search trees

last. Also note that both recursive calls are actually tail recursions and can be easily removed with a `while` loop. The use of tail recursion is justifiable here because the simplicity of algorithmic expression compensates for the decrease in speed, and the amount of stack space used is expected to be only $O(\log N)$. Figure 4.19 shows the trivial changes required to use a function object rather than requiring that the items be `Comparable`. This mimics the idioms in Section 1.6.

4.3.2 `findMin` and `findMax`

These `private` routines return a reference to the node containing the smallest and largest elements in the tree, respectively. To perform a `findMin`, start at the root and go left as long as there is a left child. The stopping point is the smallest element. The `findMax` routine is the same, except that branching is to the right child.

This is so easy that many programmers do not bother using recursion. We will code the routines both ways by doing `findMin` recursively and `findMax` nonrecursively (see Figure 4.20).

Notice how we carefully handle the degenerate case of an empty tree. Although this is always important to do, it is especially crucial in recursive programs. Also notice that it is safe to change t in `findMax`, since we are only working with a copy of a reference. Always be extremely careful, however, because a statement such as `t.right = t.right.right` will make changes.

```
1    public class BinarySearchTree<AnyType>
2    {
3        private BinaryNode<AnyType> root;
4        private Comparator<? super AnyType> cmp;
5
6        public BinarySearchTree( )
7          { this( null ); }
8
9        public BinarySearchTree( Comparator<? super AnyType> c )
10         { root = null; cmp = c; }
11
12        private int myCompare( AnyType lhs, AnyType rhs )
13        {
14            if( cmp != null )
15                return cmp.compare( lhs, rhs );
16            else
17                return ((Comparable)lhs).compareTo( rhs );
18        }
19
20        private boolean contains( AnyType x, BinaryNode<AnyType> t )
21        {
22            if( t == null )
23                return false;
24
25            int compareResult = myCompare( x, t.element );
26
27            if( compareResult < 0 )
28                return contains( x, t.left );
29            else if( compareResult > 0 )
30                return contains( x, t.right );
31            else
32                return true;      // Match
33        }
34
35        // Remainder of class is similar with calls to compareTo replaced by myCompare
36   }
```

Figure 4.19   Illustrates use of a function object to implement binary search tree

4.3.3 insert

The insertion routine is conceptually simple. To insert X into tree T, proceed down the tree as you would with a contains. If X is found, do nothing (or "update" something). Otherwise, insert X at the last spot on the path traversed. Figure 4.21 shows what happens.

```
1        /**
2         * Internal method to find the smallest item in a subtree.
3         * @param t the node that roots the subtree.
4         * @return node containing the smallest item.
5         */
6        private BinaryNode<AnyType> findMin( BinaryNode<AnyType> t )
7        {
8            if( t == null )
9                return null;
10           else if( t.left == null )
11               return t;
12           return findMin( t.left );
13       }
14
15       /**
16        * Internal method to find the largest item in a subtree.
17        * @param t the node that roots the subtree.
18        * @return node containing the largest item.
19        */
20       private BinaryNode<AnyType> findMax( BinaryNode<AnyType> t )
21       {
22           if( t != null )
23               while( t.right != null )
24                   t = t.right;
25
26           return t;
27       }
```

Figure 4.20 Recursive implementation of `findMin` and nonrecursive implementation of `findMax` for binary search trees

To insert 5, we traverse the tree as though a `contains` were occurring. At the node with item 4, we need to go right, but there is no subtree, so 5 is not in the tree, and this is the correct spot.

Duplicates can be handled by keeping an extra field in the node record indicating the frequency of occurrence. This adds some extra space to the entire tree but is better than putting duplicates in the tree (which tends to make the tree very deep). Of course, this strategy does not work if the key that guides the `compareTo` method is only part of a larger structure. If that is the case, then we can keep all of the structures that have the same key in an auxiliary data structure, such as a list or another search tree.

Figure 4.22 shows the code for the insertion routine. Since t references the root of the tree, and the root changes on the first insertion, `insert` is written as a method that returns a reference to the root of the new tree. Lines 15 and 17 recursively insert and attach x into the appropriate subtree.

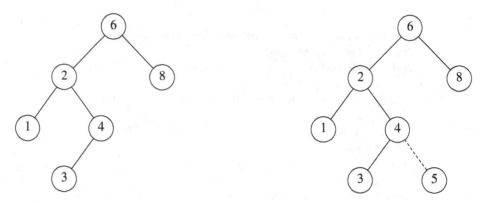

Figure 4.21   Binary search trees before and after inserting 5

```
1     /**
2      * Internal method to insert into a subtree.
3      * @param x the item to insert.
4      * @param t the node that roots the subtree.
5      * @return the new root of the subtree.
6      */
7     private BinaryNode<AnyType> insert( AnyType x, BinaryNode<AnyType> t )
8     {
9         if( t == null )
10            return new BinaryNode<>( x, null, null );
11
12        int compareResult = x.compareTo( t.element );
13
14        if( compareResult < 0 )
15            t.left = insert( x, t.left );
16        else if( compareResult > 0 )
17            t.right = insert( x, t.right );
18        else
19            ;  // Duplicate; do nothing
20        return t;
21    }
```

Figure 4.22   Insertion into a binary search tree

4.3.4 remove

As is common with many data structures, the hardest operation is deletion. Once we have found the node to be deleted, we need to consider several possibilities.

If the node is a leaf, it can be deleted immediately. If the node has one child, the node can be deleted after its parent adjusts a link to bypass the node (we will draw the link directions explicitly for clarity). See Figure 4.23.

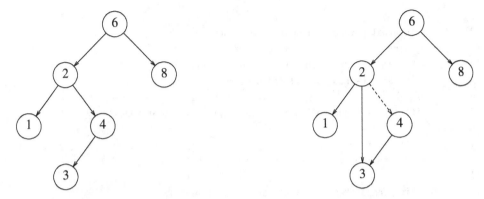

Figure 4.23 Deletion of a node (4) with one child, before and after

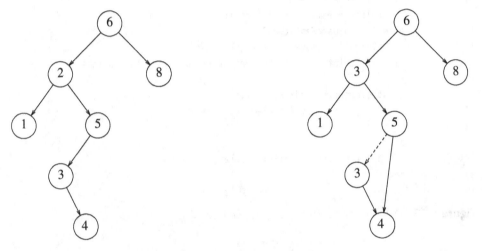

Figure 4.24 Deletion of a node (2) with two children, before and after

The complicated case deals with a node with two children. The general strategy is to replace the data of this node with the smallest data of the right subtree (which is easily found) and recursively delete that node (which is now empty). Because the smallest node in the right subtree cannot have a left child, the second `remove` is an easy one. Figure 4.24 shows an initial tree and the result of a deletion. The node to be deleted is the left child of the root; the key value is 2. It is replaced with the smallest data in its right subtree (3), and then that node is deleted as before.

The code in Figure 4.25 performs deletion. It is inefficient, because it makes two passes down the tree to find and delete the smallest node in the right subtree when this is appropriate. It is easy to remove this inefficiency by writing a special `removeMin` method, and we have left it in only for simplicity.

If the number of deletions is expected to be small, then a popular strategy to use is **lazy deletion**: When an element is to be deleted, it is left in the tree and merely *marked*

```
1     /**
2      * Internal method to remove from a subtree.
3      * @param x the item to remove.
4      * @param t the node that roots the subtree.
5      * @return the new root of the subtree.
6      */
7     private BinaryNode<AnyType> remove( AnyType x, BinaryNode<AnyType> t )
8     {
9         if( t == null )
10            return t;    // Item not found; do nothing
11
12        int compareResult = x.compareTo( t.element );
13
14        if( compareResult < 0 )
15            t.left = remove( x, t.left );
16        else if( compareResult > 0 )
17            t.right = remove( x, t.right );
18        else if( t.left != null && t.right != null ) // Two children
19        {
20            t.element = findMin( t.right ).element;
21            t.right = remove( t.element, t.right );
22        }
23        else
24            t = ( t.left != null ) ? t.left : t.right;
25        return t;
26    }
```

Figure 4.25   Deletion routine for binary search trees

as being deleted. This is especially popular if duplicate items are present, because then the field that keeps count of the frequency of appearance can be decremented. If the number of real nodes in the tree is the same as the number of "deleted" nodes, then the depth of the tree is only expected to go up by a small constant (why?), so there is a very small time penalty associated with lazy deletion. Also, if a deleted item is reinserted, the overhead of allocating a new cell is avoided.

4.3.5 Average-Case Analysis

Intuitively, we expect that all of the operations of the previous section should take $O(\log N)$ time, because in constant time we descend a level in the tree, thus operating on a tree that is now roughly half as large. Indeed, the running time of all the operations is $O(d)$, where d is the depth of the node containing the accessed item (in the case of remove this may be the replacement node in the two-child case).

We prove in this section that the average depth over all nodes in a tree is $O(\log N)$ on the assumption that all insertion sequences are equally likely.

The sum of the depths of all nodes in a tree is known as the **internal path length**. We will now calculate the average internal path length of a binary search tree, where the average is taken over all possible insertion sequences into binary search trees.

Let $D(N)$ be the internal path length for some tree T of N nodes. $D(1) = 0$. An N-node tree consists of an i-node left subtree and an $(N - i - 1)$-node right subtree, plus a root at depth zero for $0 \le i < N$. $D(i)$ is the internal path length of the left subtree with respect to its root. In the main tree, all these nodes are one level deeper. The same holds for the right subtree. Thus, we get the recurrence

$$D(N) = D(i) + D(N - i - 1) + N - 1$$

If all subtree sizes are equally likely, which is true for binary search trees (since the subtree size depends only on the relative rank of the first element inserted into the tree), but not binary trees, then the average value of both $D(i)$ and $D(N - i - 1)$ is $(1/N) \sum_{j=0}^{N-1} D(j)$. This yields

$$D(N) = \frac{2}{N} \left[\sum_{j=0}^{N-1} D(j) \right] + N - 1$$

This recurrence will be encountered and solved in Chapter 7, obtaining an average value of $D(N) = O(N \log N)$. Thus, the expected depth of any node is $O(\log N)$. As an example, the randomly generated 500-node tree shown in Figure 4.26 has nodes at expected depth 9.98.

It is tempting to say immediately that this result implies that the average running time of all the operations discussed in the previous section is $O(\log N)$, but this is not entirely

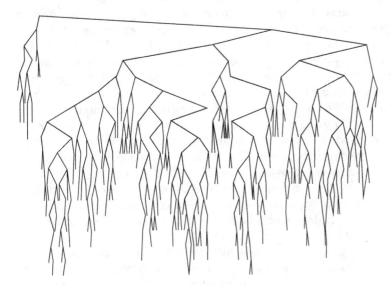

Figure 4.26 A randomly generated binary search tree

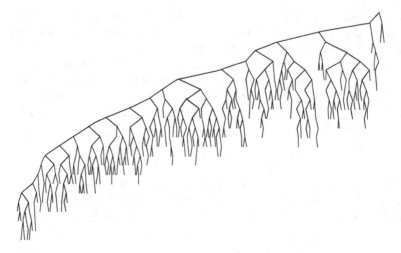

Figure 4.27   Binary search tree after $\Theta(N^2)$ insert/remove pairs

true. The reason for this is that because of deletions, it is not clear that all binary search trees are equally likely. In particular, the deletion algorithm described above favors making the left subtrees deeper than the right, because we are always replacing a deleted node with a node from the right subtree. The exact effect of this strategy is still unknown, but it seems only to be a theoretical novelty. It has been shown that if we alternate insertions and deletions $\Theta(N^2)$ times, then the trees will have an expected depth of $\Theta(\sqrt{N})$. After a quarter-million random insert/remove pairs, the tree that was somewhat right-heavy in Figure 4.26 looks decidedly unbalanced (average depth equals 12.51). See Figure 4.27.

We could try to eliminate the problem by randomly choosing between the smallest element in the right subtree and the largest in the left when replacing the deleted element. This apparently eliminates the bias and should keep the trees balanced, but nobody has actually proved this. In any event, this phenomenon appears to be mostly a theoretical novelty, because the effect does not show up at all for small trees, and stranger still, if $o(N^2)$ insert/remove pairs are used, then the tree seems to gain balance!

The main point of this discussion is that deciding what "average" means is generally extremely difficult and can require assumptions that may or may not be valid. In the absence of deletions, or when lazy deletion is used, we can conclude that the average running times of the operations above are $O(\log N)$. Except for strange cases like the one discussed above, this result is very consistent with observed behavior.

If the input comes into a tree presorted, then a series of inserts will take quadratic time and give a very expensive implementation of a linked list, since the tree will consist only of nodes with no left children. One solution to the problem is to insist on an extra structural condition called *balance:* No node is allowed to get too deep.

There are quite a few general algorithms to implement balanced trees. Most are quite a bit more complicated than a standard binary search tree, and all take longer on average for updates. They do, however, provide protection against the embarrassingly simple cases. Below, we will sketch one of the oldest forms of balanced search trees, the AVL tree.

A second, newer method is to forgo the balance condition and allow the tree to be arbitrarily deep, but after every operation, a restructuring rule is applied that tends to make future operations efficient. These types of data structures are generally classified as **self-adjusting**. In the case of a binary search tree, we can no longer guarantee an $O(\log N)$ bound on any single operation but can show that any *sequence* of M operations takes total time $O(M \log N)$ in the worst case. This is generally sufficient protection against a bad worst case. The data structure we will discuss is known as a *splay tree;* its analysis is fairly intricate and is discussed in Chapter 11.

4.4 AVL Trees

An AVL (Adelson-Velskii and Landis) tree is a binary search tree with a **balance condition**. The balance condition must be easy to maintain, and it ensures that the depth of the tree is $O(\log N)$. The simplest idea is to require that the left and right subtrees have the same height. As Figure 4.28 shows, this idea does not force the tree to be shallow.

Another balance condition would insist that every node must have left and right subtrees of the same height. If the height of an empty subtree is defined to be -1 (as is usual), then only perfectly balanced trees of $2^k - 1$ nodes would satisfy this criterion. Thus, although this guarantees trees of small depth, the balance condition is too rigid to be useful and needs to be relaxed.

An AVL **tree** is identical to a binary search tree, except that for every node in the tree, the height of the left and right subtrees can differ by at most 1. (The height of an empty tree is defined to be -1.) In Figure 4.29 the tree on the left is an AVL tree, but the tree on the right is not. Height information is kept for each node (in the node structure). It can be shown that the height of an AVL tree is at most roughly $1.44 \log(N + 2) - 1.328$, but in practice it is only slightly more than $\log N$. As an example, the AVL tree of height 9 with the fewest nodes (143) is shown in Figure 4.30. This tree has as a left subtree an AVL tree of height 7 of minimum size. The right subtree is an AVL tree of height 8 of minimum size. This tells us that the minimum number of nodes, $S(h)$, in an AVL tree of height h is given by $S(h) = S(h - 1) + S(h - 2) + 1$. For $h = 0$, $S(h) = 1$. For $h = 1$, $S(h) = 2$. The function

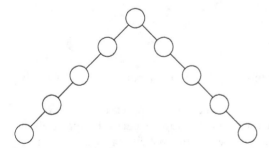

Figure 4.28   A bad binary tree. Requiring balance at the root is not enough

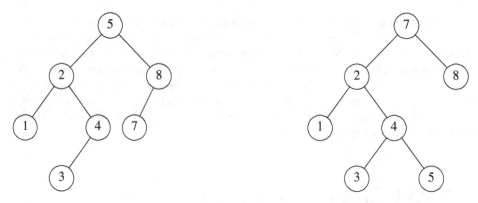

Figure 4.29 Two binary search trees. Only the left tree is AVL

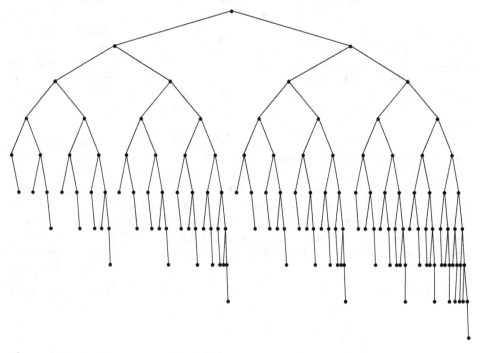

Figure 4.30 Smallest AVL tree of height 9

$S(h)$ is closely related to the Fibonacci numbers, from which the bound claimed above on the height of an AVL tree follows.

Thus, all the tree operations can be performed in $O(\log N)$ time, except possibly insertion (we will assume lazy deletion). When we do an insertion, we need to update all the balancing information for the nodes on the path back to the root, but the reason that insertion is potentially difficult is that inserting a node could violate the AVL tree property. (For instance, inserting 6 into the AVL tree in Figure 4.29 would destroy the balance condition at the node with key 8.) If this is the case, then the property has to be restored before the insertion step is considered over. It turns out that this can always be done with a simple modification to the tree, known as a **rotation**.

After an insertion, only nodes that are on the path from the insertion point to the root might have their balance altered because only those nodes have their subtrees altered. As we follow the path up to the root and update the balancing information, we may find a node whose new balance violates the AVL condition. We will show how to rebalance the tree at the first (i.e., deepest) such node, and we will prove that this rebalancing guarantees that the entire tree satisfies the AVL property.

Let us call the node that must be rebalanced α. Since any node has at most two children, and a height imbalance requires that α's two subtrees' height differ by two, it is easy to see that a violation might occur in four cases:

1. An insertion into the left subtree of the left child of α.
2. An insertion into the right subtree of the left child of α.
3. An insertion into the left subtree of the right child of α.
4. An insertion into the right subtree of the right child of α.

Cases 1 and 4 are mirror image symmetries with respect to α, as are cases 2 and 3. Consequently, as a matter of theory, there are two basic cases. From a programming perspective, of course, there are still four cases.

The first case, in which the insertion occurs on the "outside" (i.e., left–left or right–right), is fixed by a **single rotation** of the tree. The second case, in which the insertion occurs on the "inside" (i.e., left–right or right–left) is handled by the slightly more complex **double rotation**. These are fundamental operations on the tree that we'll see used several times in balanced-tree algorithms. The remainder of this section describes these rotations, proves that they suffice to maintain balance, and gives a casual implementation of the AVL tree. Chapter 12 describes other balanced-tree methods with an eye toward a more careful implementation.

4.4.1 Single Rotation

Figure 4.31 shows the *single rotation* that fixes case 1. The before picture is on the left, and the after is on the right. Let us analyze carefully what is going on. Node k_2 violates the AVL balance property because its left subtree is two levels deeper than its right subtree (the dashed lines in the middle of the diagram mark the levels). The situation depicted is the only possible case 1 scenario that allows k_2 to satisfy the AVL property before an insertion but violate it afterwards. Subtree X has grown to an extra level, causing it to be exactly two levels deeper than Z. Y cannot be at the same level as the new X because then k_2 would have been out of balance *before* the insertion, and Y cannot be at the same level as Z because then k_1 would be the first node on the path toward the root that was in violation of the AVL balancing condition.

To ideally rebalance the tree, we would like to move X up a level and Z down a level. Note that this is actually more than the AVL property would require. To do this, we rearrange

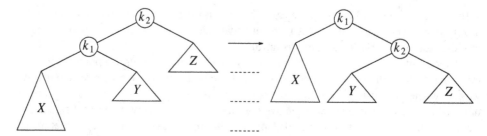

Figure 4.31 Single rotation to fix case 1

nodes into an equivalent tree as shown in the second part of Figure 4.31. Here is an abstract scenario: Visualize the tree as being flexible, grab the child node k_1, close your eyes, and shake it, letting gravity take hold. The result is that k_1 will be the new root. The binary search tree property tells us that in the original tree $k_2 > k_1$, so k_2 becomes the right child of k_1 in the new tree. X and Z remain as the left child of k_1 and right child of k_2, respectively. Subtree Y, which holds items that are between k_1 and k_2 in the original tree, can be placed as k_2's left child in the new tree and satisfy all the ordering requirements.

As a result of this work, which requires only a few link changes, we have another binary search tree that is an AVL tree. This happens because X moves up one level, Y stays at the same level, and Z moves down one level. k_2 and k_1 not only satisfy the AVL requirements, but they also have subtrees that are exactly the same height. Furthermore, the new height of the entire subtree is *exactly the same* as the height of the original subtree prior to the insertion that caused X to grow. Thus no further updating of heights on the path to the root is needed, and consequently *no further rotations are needed*. Figure 4.32 shows that after the insertion of 6 into the original AVL tree on the left, node 8 becomes unbalanced. Thus, we do a single rotation between 7 and 8, obtaining the tree on the right.

As we mentioned earlier, case 4 represents a symmetric case. Figure 4.33 shows how a single rotation is applied. Let us work through a rather long example. Suppose we start with an initially empty AVL tree and insert the items 3, 2, 1, and then 4 through 7 in sequential

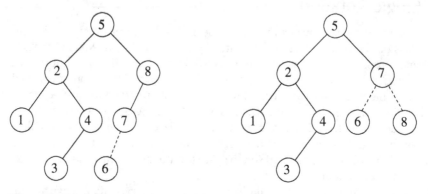

Figure 4.32 AVL property destroyed by insertion of 6, then fixed by a single rotation

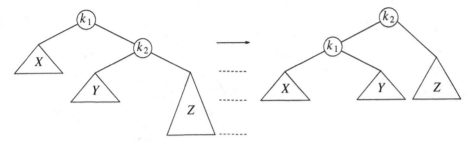

Figure 4.33 Single rotation fixes case 4

order. The first problem occurs when it is time to insert 1 because the AVL property is violated at the root. We perform a single rotation between the root and its left child to fix the problem. Here are the before and after trees:

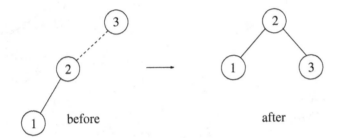

A dashed line joins the two nodes that are the subject of the rotation. Next we insert 4, which causes no problems, but the insertion of 5 creates a violation at node 3 that is fixed by a single rotation. Besides the local change caused by the rotation, the programmer must remember that the rest of the tree has to be informed of this change. Here this means that 2's right child must be reset to link to 4 instead of 3. Forgetting to do so is easy and would destroy the tree (4 would be inaccessible).

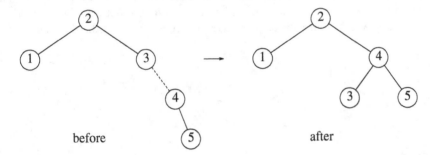

Next we insert 6. This causes a balance problem at the root, since its left subtree is of height 0 and its right subtree would be height 2. Therefore, we perform a single rotation at the root between 2 and 4.

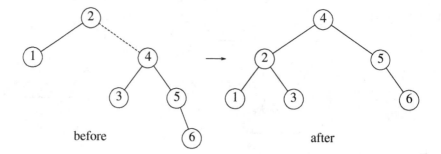

The rotation is performed by making 2 a child of 4 and 4's original left subtree the new right subtree of 2. Every item in this subtree must lie between 2 and 4, so this transformation makes sense. The next item we insert is 7, which causes another rotation:

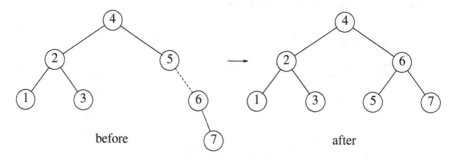

4.4.2 Double Rotation

The algorithm described above has one problem: As Figure 4.34 shows, it does not work for cases 2 or 3. The problem is that subtree Y is too deep, and a single rotation does not make it any less deep. The *double rotation* that solves the problem is shown in Figure 4.35.

The fact that subtree Y in Figure 4.34 has had an item inserted into it guarantees that it is nonempty. Thus, we may assume that it has a root and two subtrees. Consequently, the tree may be viewed as four subtrees connected by three nodes. As the diagram suggests,

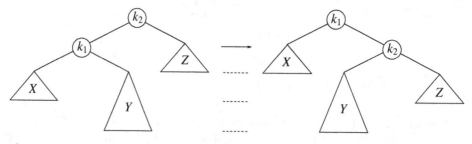

Figure 4.34   Single rotation fails to fix case 2

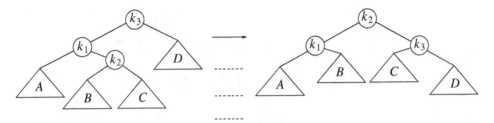

Figure 4.35 Left–right double rotation to fix case 2

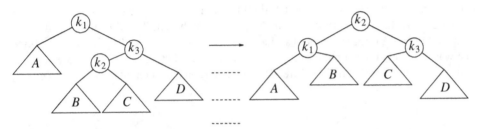

Figure 4.36 Right–left double rotation to fix case 3

exactly one of tree B or C is two levels deeper than D (unless all are empty), but we cannot be sure which one. It turns out not to matter; in Figure 4.35, both B and C are drawn at $1\frac{1}{2}$ levels below D.

To rebalance, we see that we cannot leave k_3 as the root, and a rotation between k_3 and k_1 was shown in Figure 4.34 to not work, so the only alternative is to place k_2 as the new root. This forces k_1 to be k_2's left child and k_3 to be its right child, and it also completely determines the resulting locations of the four subtrees. It is easy to see that the resulting tree satisfies the AVL tree property, and as was the case with the single rotation, it restores the height to what it was before the insertion, thus guaranteeing that all rebalancing and height updating is complete. Figure 4.36 shows that the symmetric case 3 can also be fixed by a double rotation. In both cases the effect is the same as rotating between α's child and grandchild, and then between α and its new child.

We will continue our previous example by inserting 10 through 16 in reverse order, followed by 8 and then 9. Inserting 16 is easy, since it does not destroy the balance property, but inserting 15 causes a height imbalance at node 7. This is case 3, which is solved by a right–left double rotation. In our example, the right–left double rotation will involve 7, 16, and 15. In this case, k_1 is the node with item 7, k_3 is the node with item 16, and k_2 is the node with item 15. Subtrees A, B, C, and D are empty.

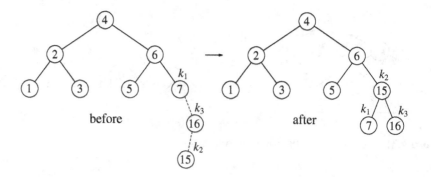

Next we insert 14, which also requires a double rotation. Here the double rotation that will restore the tree is again a right–left double rotation that will involve 6, 15, and 7. In this case, k_1 is the node with item 6, k_2 is the node with item 7, and k_3 is the node with item 15. Subtree A is the tree rooted at the node with item 5; subtree B is the empty subtree that was originally the left child of the node with item 7, subtree C is the tree rooted at the node with item 14, and finally, subtree D is the tree rooted at the node with item 16.

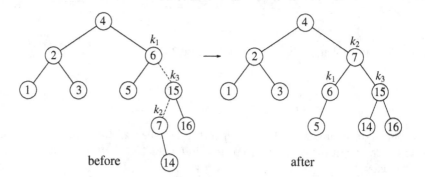

If 13 is now inserted, there is an imbalance at the root. Since 13 is not between 4 and 7, we know that the single rotation will work.

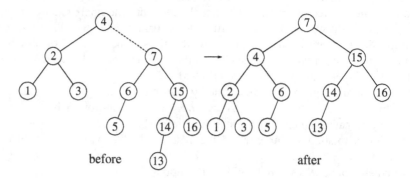

Insertion of 12 will also require a single rotation:

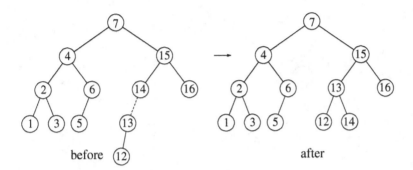

before   after

To insert 11, a single rotation needs to be performed, and the same is true for the subsequent insertion of 10. We insert 8 without a rotation creating an almost perfectly balanced tree:

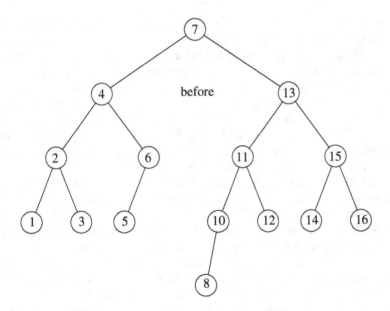

before

Finally, we will insert 9 to show the symmetric case of the double rotation. Notice that 9 causes the node containing 10 to become unbalanced. Since 9 is between 10 and 8 (which is 10's child on the path to 9), a double rotation needs to be performed, yielding the following tree:

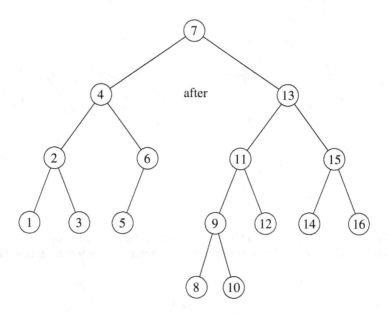

Let us summarize what happens. The programming details are fairly straightforward except that there are several cases. To insert a new node with item X into an AVL tree T, we recursively insert X into the appropriate subtree of T (let us call this T_{LR}). If the height of T_{LR} does not change, then we are done. Otherwise, if a height imbalance appears in T, we do the appropriate single or double rotation depending on X and the items in T and T_{LR}, update the heights (making the connection from the rest of the tree above), and are done. Since one rotation always suffices, a carefully coded nonrecursive version generally turns out to be faster than the recursive version, but on modern compilers the difference is not as significant as in the past. However, nonrecursive versions are quite difficult to code correctly, whereas a casual recursive implementation is easily readable.

Another efficiency issue concerns storage of the height information. Since all that is really required is the difference in height, which is guaranteed to be small, we could get by with two bits (to represent $+1$, 0, -1) if we really try. Doing so will avoid repetitive calculation of balance factors but results in some loss of clarity. The resulting code is some-what more complicated than if the height were stored at each node. If a recursive routine is written, then speed is probably not the main consideration. In this case, the slight speed advantage obtained by storing balance factors hardly seems worth the loss of clarity and relative simplicity. Furthermore, since most machines will align this to at least an 8-bit boundary anyway, there is not likely to be any difference in the amount of space used. An eight-bit **byte** will allow us to store absolute heights of up to 127. Since the tree is balanced, it is inconceivable that this would be insufficient (see the exercises).

With all this, we are ready to write the AVL routines. We show some of the code here; the rest is online. First, we need the `AvlNode` class. This is given in Figure 4.37. We also

```
1        private static class AvlNode<AnyType>
2        {
3                // Constructors
4            AvlNode( AnyType theElement )
5              { this( theElement, null, null ); }
6
7            AvlNode( AnyType theElement, AvlNode<AnyType> lt, AvlNode<AnyType> rt )
8              { element = theElement; left = lt; right = rt; height = 0; }
9
10           AnyType            element;       // The data in the node
11           AvlNode<AnyType>   left;          // Left child
12           AvlNode<AnyType>   right;         // Right child
13           int                height;        // Height
14       }
```

Figure 4.37 Node declaration for AVL trees

```
1        /**
2         * Return the height of node t, or -1, if null.
3         */
4        private int height( AvlNode<AnyType> t )
5        {
6            return t == null ? -1 : t.height;
7        }
```

Figure 4.38 Method to compute height of an AVL node

need a quick method to return the height of a node. This method is necessary to handle the annoying case of a null reference. This is shown in Figure 4.38. The basic insertion routine is easy to write (see Figure 4.39): It adds only a single line at the end that invokes a balancing method. The balancing method applies a single or double rotation if needed, updates the height, and returns the resulting tree.

For the trees in Figure 4.40, rotateWithLeftChild converts the tree on the left to the tree on the right, returning a reference to the new root. rotateWithRightChild is symmetric. The code is shown in Figure 4.41.

Similarly, the double rotation pictured in Figure 4.42 can be implemented by the code shown in Figure 4.43.

Since deletion in a binary search tree is somewhat more complicated than insertion, one can assume that deletion in an AVL tree is also more complicated. In a perfect world, one would hope that the deletion routine in Figure 4.25 could easily be modified by changing the last line to return after calling the **balance** method, as was done for insertion. This would yield the code in Figure 4.44. This change works! A deletion could cause one side of the tree to become two levels shallower than the other side. The case-by-case analysis is similar to the imbalances that are caused by insertion, but not exactly the same. For

```
1      /**
2       * Internal method to insert into a subtree.
3       * @param x the item to insert.
4       * @param t the node that roots the subtree.
5       * @return the new root of the subtree.
6       */
7      private AvlNode<AnyType> insert( AnyType x, AvlNode<AnyType> t )
8      {
9          if( t == null )
10             return new AvlNode<>( x, null, null );
11
12         int compareResult = x.compareTo( t.element );
13
14         if( compareResult < 0 )
15             t.left = insert( x, t.left );
16         else if( compareResult > 0 )
17             t.right = insert( x, t.right );
18         else
19             ;  // Duplicate; do nothing
20         return balance( t );
21     }
22
23     private static final int ALLOWED_IMBALANCE = 1;
24
25     // Assume t is either balanced or within one of being balanced
26     private AvlNode<AnyType> balance( AvlNode<AnyType> t )
27     {
28         if( t == null )
29             return t;
30
31         if( height( t.left ) - height( t.right ) > ALLOWED_IMBALANCE )
32             if( height( t.left.left ) >= height( t.left.right ) )
33                 t = rotateWithLeftChild( t );
34             else
35                 t = doubleWithLeftChild( t );
36         else
37         if( height( t.right ) - height( t.left ) > ALLOWED_IMBALANCE )
38             if( height( t.right.right ) >= height( t.right.left ) )
39                 t = rotateWithRightChild( t );
40             else
41                 t = doubleWithRightChild( t );
42
43         t.height = Math.max( height( t.left ), height( t.right ) ) + 1;
44         return t;
45     }
```

Figure 4.39 Insertion into an AVL tree

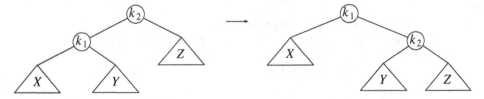

Figure 4.40 Single rotation

```
1      /**
2       * Rotate binary tree node with left child.
3       * For AVL trees, this is a single rotation for case 1.
4       * Update heights, then return new root.
5       */
6      private AvlNode<AnyType> rotateWithLeftChild( AvlNode<AnyType> k2 )
7      {
8          AvlNode<AnyType> k1 = k2.left;
9          k2.left = k1.right;
10         k1.right = k2;
11         k2.height = Math.max( height( k2.left ), height( k2.right ) ) + 1;
12         k1.height = Math.max( height( k1.left ), k2.height ) + 1;
13         return k1;
14     }
```

Figure 4.41 Routine to perform single rotation

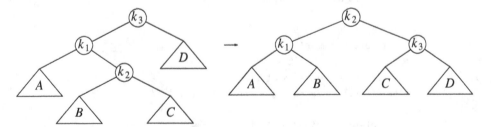

Figure 4.42 Double rotation

instance, case 1 in Figure 4.31, which would now reflect a deletion from tree Z (rather than an insertion into X), must be augmented with the possiblity that tree Y could be as deep as tree X. Even so, it is easy to see that the rotation rebalances this case and the symmetric case 4 in Figure 4.33. Thus the code for `balance` in Figure 4.39 lines 32 and 38 uses >= instead of > specifically to ensure that single rotations are done in these cases, rather than double rotations. We leave verification of the remaining cases as an exercise.

```
1        /**
2         * Double rotate binary tree node: first left child
3         * with its right child; then node k3 with new left child.
4         * For AVL trees, this is a double rotation for case 2.
5         * Update heights, then return new root.
6         */
7        private AvlNode<AnyType> doubleWithLeftChild( AvlNode<AnyType> k3 )
8        {
9            k3.left = rotateWithRightChild( k3.left );
10           return rotateWithLeftChild( k3 );
11       }
```

Figure 4.43   Routine to perform double rotation

```
1        /**
2         * Internal method to remove from a subtree.
3         * @param x the item to remove.
4         * @param t the node that roots the subtree.
5         * @return the new root of the subtree.
6         */
7        private AvlNode<AnyType> remove( AnyType x, AvlNode<AnyType> t )
8        {
9            if( t == null )
10               return t;    // Item not found; do nothing
11
12           int compareResult = x.compareTo( t.element );
13
14           if( compareResult < 0 )
15               t.left = remove( x, t.left );
16           else if( compareResult > 0 )
17               t.right = remove( x, t.right );
18           else if( t.left != null && t.right != null ) // Two children
19           {
20               t.element = findMin( t.right ).element;
21               t.right = remove( t.element, t.right );
22           }
23           else
24               t = ( t.left != null ) ? t.left : t.right;
25           return balance( t );
26       }
```

Figure 4.44   Deletion in an AVL tree

4.5 Splay Trees

We now describe a relatively simple data structure, known as a **splay tree**, that guarantees that any M consecutive tree operations starting from an empty tree take at most $O(M \log N)$ time. Although this guarantee does not preclude the possibility that any *single* operation might take $\theta(N)$ time, and thus the bound is not as strong as an $O(\log N)$ worst-case bound per operation, the net effect is the same: There are no bad input sequences. Generally, when a sequence of M operations has total worst-case running time of $O(Mf(N))$, we say that the **amortized** running time is $O(f(N))$. Thus, a splay tree has an $O(\log N)$ amortized cost per operation. Over a long sequence of operations, some may take more, some less.

Splay trees are based on the fact that the $O(N)$ worst-case time per operation for binary search trees is not bad, as long as it occurs relatively infrequently. Any one access, even if it takes $\theta(N)$, is still likely to be extremely fast. The problem with binary search trees is that it is possible, and not uncommon, for a whole sequence of bad accesses to take place. The cumulative running time then becomes noticeable. A search tree data structure with $O(N)$ worst-case time, but a *guarantee* of at most $O(M \log N)$ for any M consecutive operations, is certainly satisfactory, because there are no bad sequences.

If any particular operation is allowed to have an $O(N)$ worst-case time bound, and we still want an $O(\log N)$ amortized time bound, then it is clear that whenever a node is accessed, it must be moved. Otherwise, once we find a deep node, we could keep performing accesses on it. If the node does not change location, and each access costs $\theta(N)$, then a sequence of M accesses will cost $\theta(M \cdot N)$.

The basic idea of the splay tree is that after a node is accessed, it is pushed to the root by a series of AVL tree rotations. Notice that if a node is deep, there are many nodes on the path that are also relatively deep, and by restructuring we can make future accesses cheaper on all these nodes. Thus, if the node is unduly deep, then we want this restructuring to have the side effect of balancing the tree (to some extent). Besides giving a good time bound in theory, this method is likely to have practical utility, because in many applications, when a node is accessed, it is likely to be accessed again in the near future. Studies have shown that this happens much more often than one would expect. Splay trees also do not require the maintenance of height or balance information, thus saving space and simplifying the code to some extent (especially when careful implementations are written).

4.5.1 A Simple Idea (That Does Not Work)

One way of performing the restructuring described above is to perform single rotations, bottom up. This means that we rotate every node on the access path with its parent. As an example, consider what happens after an access (a find) on k_1 in the following tree.

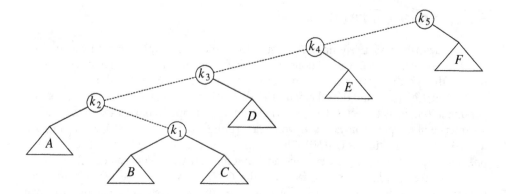

The access path is dashed. First, we would perform a single rotation between k_1 and its parent, obtaining the following tree.

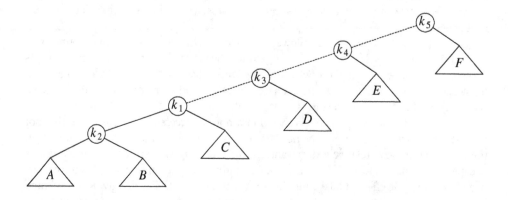

Then, we rotate between k_1 and k_3, obtaining the next tree.

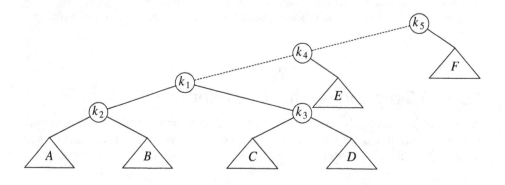

Then two more rotations are performed until we reach the root.

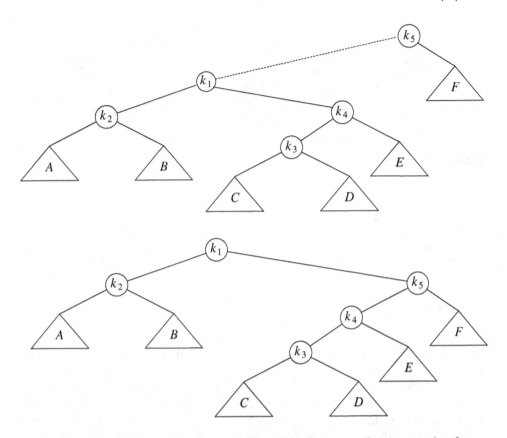

These rotations have the effect of pushing k_1 all the way to the root, so that future accesses on k_1 are easy (for a while). Unfortunately, it has pushed another node (k_3) almost as deep as k_1 used to be. An access on that node will then push another node deep, and so on. Although this strategy makes future accesses of k_1 cheaper, it has not significantly improved the situation for the other nodes on the (original) access path. It turns out that it is possible to prove that using this strategy, there is a sequence of M operations requiring $\Omega(M \cdot N)$ time, so this idea is not quite good enough. The simplest way to show this is to consider the tree formed by inserting keys $1, 2, 3, \ldots, N$ into an initially empty tree (work this example out). This gives a tree consisting of only left children. This is not necessarily bad, though, since the time to build this tree is $O(N)$ total. The bad part is that accessing the node with key 1 takes $N - 1$ units of time. After the rotations are complete, an access of the node with key 2 takes $N - 2$ units of time. The total for accessing all the keys in order is $\sum_{i=1}^{N-1} i - \Omega(N^2)$. After they are accessed, the tree reverts to its original state, and we can repeat the sequence.

4.5.2 Splaying

The splaying strategy is similar to the rotation idea above, except that we are a little more selective about how rotations are performed. We will still rotate bottom up along the access

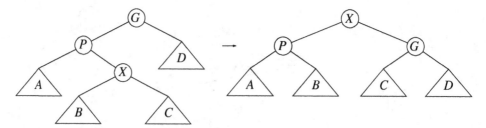

Figure 4.45 Zig-zag

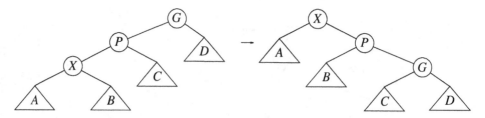

Figure 4.46 Zig-zig

path. Let X be a (nonroot) node on the access path at which we are rotating. If the parent of X is the root of the tree, we merely rotate X and the root. This is the last rotation along the access path. Otherwise, X has both a parent (P) and a grandparent (G), and there are two cases, plus symmetries, to consider. The first case is the **zig-zag** case (see Figure 4.45). Here, X is a right child and P is a left child (or vice versa). If this is the case, we perform a double rotation, exactly like an AVL double rotation. Otherwise, we have a **zig-zig** case: X and P are both left children (or, in the symmetric case, both right children). In that case, we transform the tree on the left of Figure 4.46 to the tree on the right.

As an example, consider the tree from the last example, with a contains on k_1:

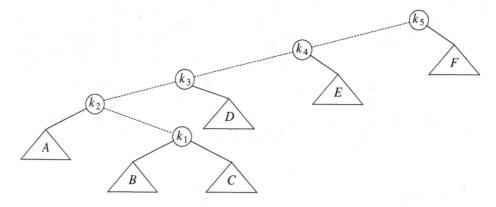

The first splay step is at k_1 and is clearly a *zig-zag*, so we perform a standard AVL double rotation using k_1, k_2, and k_3. The resulting tree follows.

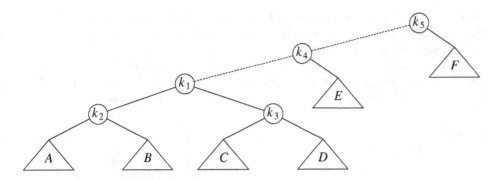

The next splay step at k_1 is a *zig-zig*, so we do the *zig-zig* rotation with k_1, k_4, and k_5, obtaining the final tree.

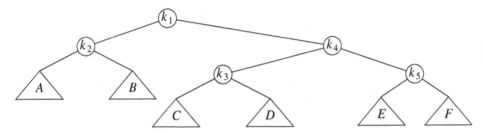

Although it is hard to see from small examples, splaying not only moves the accessed node to the root but also has the effect of roughly halving the depth of most nodes on the access path (some shallow nodes are pushed down at most two levels).

To see the difference that splaying makes over simple rotation, consider again the effect of inserting items $1, 2, 3, \ldots, N$ into an initially empty tree. This takes a total of $O(N)$, as before, and yields the same tree as simple rotations. Figure 4.47 shows the result of splaying

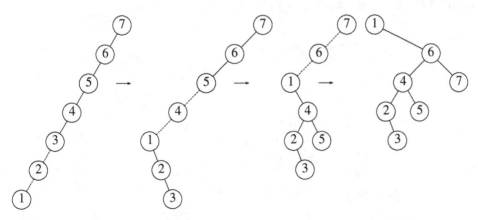

Figure 4.47 Result of splaying at node 1

at the node with item 1. The difference is that after an access of the node with item 1, which takes $N-1$ units, the access on the node with item 2 will only take about $N/2$ units instead of $N-2$ units; there are no nodes quite as deep as before.

An access on the node with item 2 will bring nodes to within $N/4$ of the root, and this is repeated until the depth becomes roughly $\log N$ (an example with $N=7$ is too small to see the effect well). Figures 4.48 to 4.56 show the result of accessing items 1 through 9 in

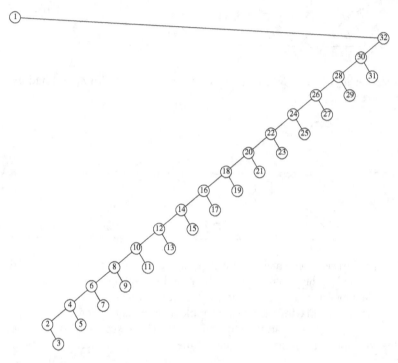

Figure 4.48   Result of splaying at node 1 a tree of all left children

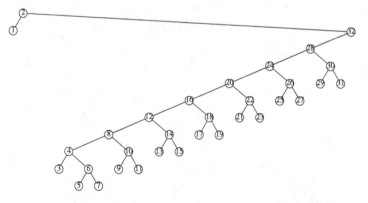

Figure 4.49   Result of splaying the previous tree at node 2

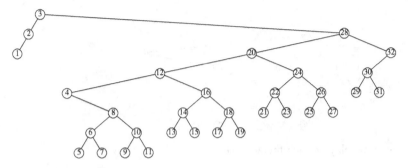

Figure 4.50   Result of splaying the previous tree at node 3

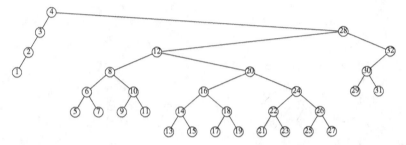

Figure 4.51   Result of splaying the previous tree at node 4

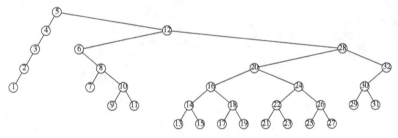

Figure 4.52   Result of splaying the previous tree at node 5

a 32-node tree that originally contains only left children. Thus we do not get the same bad behavior from splay trees that is prevalent in the simple rotation strategy. (Actually, this turns out to be a very good case. A rather complicated proof shows that for this example, the N accesses take a total of $O(N)$ time.)

These figures highlight the fundamental and crucial property of splay trees. When access paths are long, thus leading to a longer-than-normal search time, the rotations tend to be good for future operations. When accesses are cheap, the rotations are not as good and can be bad. The extreme case is the initial tree formed by the insertions. All the insertions were constant-time operations leading to a bad initial tree. At that point in time, we had a very bad tree, but we were running ahead of schedule and had the compensation of less total running time. Then a couple of really horrible accesses left a nearly balanced tree,

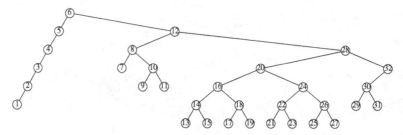

Figure 4.53   Result of splaying the previous tree at node 6

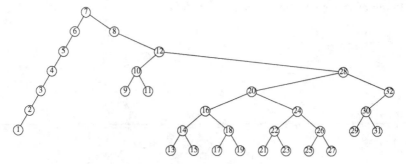

Figure 4.54   Result of splaying the previous tree at node 7

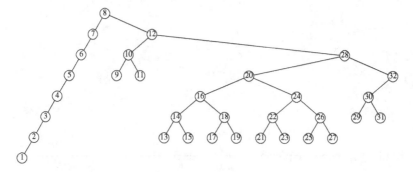

Figure 4.55   Result of splaying the previous tree at node 8

but the cost was that we had to give back some of the time that had been saved. The main theorem, which we will prove in Chapter 11, is that we never fall behind a pace of $O(\log N)$ per operation: We are always on schedule, even though there are occasionally bad operations.

We can perform deletion by accessing the node to be deleted. This puts the node at the root. If it is deleted, we get two subtrees T_L and T_R (left and right). If we find the largest element in T_L (which is easy), then this element is rotated to the root of T_L, and T_L will now have a root with no right child. We can finish the deletion by making T_R the right child.

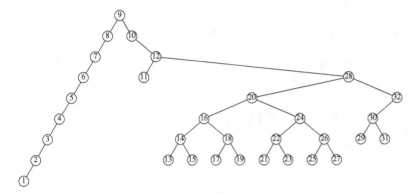

Figure 4.56   Result of splaying the previous tree at node 9

The analysis of splay trees is difficult, because it must take into account the ever-changing structure of the tree. On the other hand, splay trees are much simpler to program than AVL trees, since there are fewer cases to consider and no balance information to maintain. Some empirical evidence suggests that this translates into faster code in practice, although the case for this is far from complete. Finally, we point out that there are several variations of splay trees that can perform even better in practice. One variation is completely coded in Chapter 12.

4.6 Tree Traversals (Revisited)

Because of the ordering information in a binary search tree, it is simple to list all the items in sorted order. The recursive method in Figure 4.57 does the real work.

Convince yourself that this method works. As we have seen before, this kind of routine when applied to trees is known as an **inorder traversal** (which makes sense, since it lists the items in order). The general strategy of an inorder traversal is to process the left subtree first, then perform processing at the current node, and finally process the right subtree. The interesting part about this algorithm, aside from its simplicity, is that the total running time is $O(N)$. This is because there is constant work being performed at every node in the tree. Each node is visited once, and the work performed at each node is testing against null, setting up two method calls, and doing a println. Since there is constant work per node and N nodes, the running time is $O(N)$.

Sometimes we need to process both subtrees first before we can process a node. For instance, to compute the height of a node, we need to know the height of the subtrees first. The code in Figure 4.58 computes this. Since it is always a good idea to check the special cases—and crucial when recursion is involved—notice that the routine will declare the height of a leaf to be zero, which is correct. This general order of traversal, which we have also seen before, is known as a **postorder traversal**. Again, the total running time is $O(N)$, because constant work is performed at each node.

```
1      /**
2       * Print the tree contents in sorted order.
3       */
4      public void printTree( )
5      {
6          if( isEmpty( ) )
7              System.out.println( "Empty tree" );
8          else
9              printTree( root );
10     }
11
12     /**
13      * Internal method to print a subtree in sorted order.
14      * @param t the node that roots the subtree.
15      */
16     private void printTree( BinaryNode<AnyType> t )
17     {
18         if( t != null )
19         {
20             printTree( t.left );
21             System.out.println( t.element );
22             printTree( t.right );
23         }
24     }
```

Figure 4.57   Routine to print a binary search tree in order

```
1      /**
2       * Internal method to compute height of a subtree.
3       * @param t the node that roots the subtree.
4       */
5      private int height( BinaryNode<AnyType> t )
6      {
7          if( t == null )
8              return -1;
9          else
10             return 1 + Math.max( height( t.left ), height( t.right ) );
11     }
```

Figure 4.58   Routine to compute the height of a tree using a postorder traversal

The third popular traversal scheme that we have seen is **preorder traversal**. Here, the node is processed before the children. This could be useful, for example, if you wanted to label each node with its depth.

The common idea in all these routines is that you handle the `null` case first, and then the rest. Notice the lack of extraneous variables. These routines pass only the reference to the node that roots the subtree and do not declare or pass any extra variables. The more compact the code, the less likely that a silly bug will turn up. A fourth, less often used, traversal (which we have not seen yet) is **level-order traversal**. In a level-order traversal, all nodes at depth d are processed before any node at depth $d + 1$. Level-order traversal differs from the other traversals in that it is not done recursively; a queue is used, instead of the implied stack of recursion.

4.7 B-Trees

Thus far, we have assumed that we can store an entire data structure in the main memory of a computer. Suppose, however, that we have more data than can fit in main memory, meaning that we must have the data structure reside on disk. When this happens, the rules of the game change because the Big-Oh model is no longer meaningful.

The problem is that a Big-Oh analysis assumes that all operations are equal. However, this is not true, especially when disk I/O is involved. Modern computers execute billions of instructions per second. That is pretty fast, mainly because the speed depends largely on electrical properties. On the other hand, a disk is mechanical. Its speed depends largely on the time it takes to spin the disk and to move a disk head. Many disks spin at 7,200 RPM. Thus in 1 min, it makes 7,200 revolutions; hence, one revolution occurs in 1/120 of a second, or 8.3 ms. On average, we might expect that we have to spin a disk halfway to find what we are looking for, but this is compensated by the time to move the disk head, so we get an access time of 8.3 ms. (This is a very charitable estimate; 9–11 ms access times are more common.) Consequently, we can do approximately 120 disk accesses per second. This sounds pretty good, until we compare it with the processor speed. What we have is billions of instructions equal to 120 disk accesses. Of course, everything here is a rough calculation, but the relative speeds are pretty clear: Disk accesses are incredibly expensive. Furthermore, processor speeds are increasing at a much faster rate than disk speeds (it is disk *sizes* that are increasing quite quickly). So we are willing to do lots of calculations just to save a disk access. In almost all cases, it is the number of disk accesses that will dominate the running time. Thus, if we halve the number of disk accesses, the running time will also halve.

Here is how the typical search tree performs on disk. Suppose we want to access the driving records for citizens in the State of Florida. We assume that we have 10 million items, that each key is 32 bytes (representing a name), and that a record is 256 bytes. We assume this does not fit in main memory and that we are 1 of 20 users on a system (so we have 1/20 of the resources). Thus, in 1 sec, we can execute billions of instructions or perform six disk accesses.

The unbalanced binary search tree is a disaster. In the worst case, it has linear depth and thus could require 10 million disk accesses. On average, a successful search would require $1.38 \log N$ disk accesses, and since $\log 10000000 \approx 24$, an average search would require 32 disk accesses, or 5 sec. In a typical randomly constructed tree, we would expect

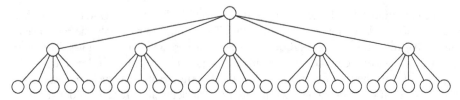

Figure 4.59   5-ary tree of 31 nodes has only three levels

that a few nodes are three times deeper; these would require about 100 disk accesses, or 16 sec. An AVL tree is somewhat better. The worst case of $1.44 \log N$ is unlikely to occur, and the typical case is very close to $\log N$. Thus an AVL tree would use about 25 disk accesses on average, requiring 4 sec.

We want to reduce the number of disk accesses to a very small constant, such as three or four; and we are willing to write complicated code to do this because machine instructions are essentially free, as long as we are not ridiculously unreasonable. It should probably be clear that a binary search tree will not work, since the typical AVL tree is close to optimal height. We cannot go below $\log N$ using a binary search tree. The solution is intuitively simple: If we have more branching, we have less height. Thus, while a perfect binary tree of 31 nodes has five levels, a 5-ary tree of 31 nodes has only three levels, as shown in Figure 4.59. An **M-ary search tree** allows M-way branching. As branching increases, the depth decreases. Whereas a complete binary tree has height that is roughly $\log_2 N$, a complete M-ary tree has height that is roughly $\log_M N$.

We can create an M-ary search tree in much the same way as a binary search tree. In a binary search tree, we need one key to decide which of two branches to take. In an M-ary search tree, we need $M - 1$ keys to decide which branch to take. To make this scheme efficient in the worst case, we need to ensure that the M-ary search tree is balanced in some way. Otherwise, like a binary search tree, it could degenerate into a linked list. Actually, we want an even more restrictive balancing condition. That is, we do not want an M-ary search tree to degenerate to even a binary search tree, because then we would be stuck with $\log N$ accesses.

One way to implement this is to use a **B-tree**. The basic B-tree[4] is described here. Many variations and improvements are possible, and an implementation is somewhat complex because there are quite a few cases. However, it is easy to see that, in principle, a B-tree guarantees only a few disk accesses.

A B-tree of order M is an M-ary tree with the following properties:[5]

1. The data items are stored at leaves.

2. The nonleaf nodes store up to $M - 1$ keys to guide the searching; key i represents the smallest key in subtree $i + 1$.

3. The root is either a leaf or has between two and M children.

---

[4] What is described is popularly known as a B^+ tree.

[5] Rules 3 and 5 must be relaxed for the first L insertions.

4. All nonleaf nodes (except the root) have between $\lceil M/2 \rceil$ and M children.

5. All leaves are at the same depth and have between $\lceil L/2 \rceil$ and L data items, for some L (the determination of L is described shortly).

An example of a B-tree of order 5 is shown in Figure 4.60. Notice that all nonleaf nodes have between three and five children (and thus between two and four keys); the root could possibly have only two children. Here, we have $L = 5$. (It happens that L and M are the same in this example, but this is not necessary.) Since L is 5, each leaf has between three and five data items. Requiring nodes to be half full guarantees that the B-tree does not degenerate into a simple binary tree. Although there are various definitions of B-trees that change this structure, mostly in minor ways, this definition is one of the popular forms.

Each node represents a disk block, so we choose M and L on the basis of the size of the items that are being stored. As an example, suppose one block holds 8,192 bytes. In our Florida example, each key uses 32 bytes. In a B-tree of order M, we would have $M - 1$ keys, for a total of $32M - 32$ bytes, plus M branches. Since each branch is essentially a number of another disk block, we can assume that a branch is 4 bytes. Thus the branches use $4M$ bytes. The total memory requirement for a nonleaf node is thus $36M - 32$. The largest value of M for which this is no more than 8,192 is 228. Thus we would choose $M = 228$. Since each data record is 256 bytes, we would be able to fit 32 records in a block. Thus we would choose $L = 32$. We are guaranteed that each leaf has between 16 and 32 data records and that each internal node (except the root) branches in at least 114 ways. Since there are 10 million records, there are at most 625,000 leaves. Consequently, in the worst case, leaves would be on level 4. In more concrete terms, the worst-case number of accesses is given by approximately $\log_{M/2} N$, give or take 1 (for example, the root and the next level could be cached in main memory, so that, over the long run, disk accesses would be needed only for level 3 and deeper).

The remaining issue is how to add and remove items from the B-tree; the ideas involved are sketched next. Note that many of the themes seen before recur.

We begin by examining insertion. Suppose we want to insert 57 into the B-tree in Figure 4.60. A search down the tree reveals that it is not already in the tree. We can then add it to the leaf as a fifth item. Note that we may have to reorganize all the data in the leaf

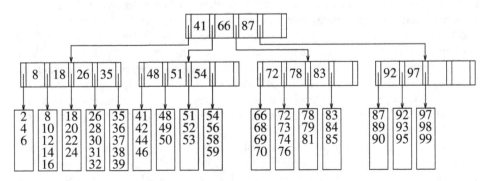

Figure 4.60 B-tree of order 5

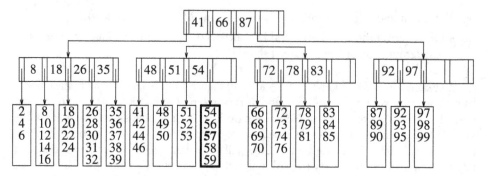

Figure 4.61   B-tree after insertion of 57 into the tree in Figure 4.60

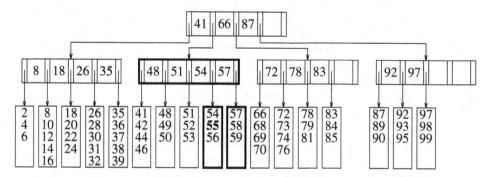

Figure 4.62   Insertion of 55 into the B-tree in Figure 4.61 causes a split into two leaves

to do this. However, the cost of doing this is negligible when compared to that of the disk access, which in this case also includes a disk write.

Of course, that was relatively painless because the leaf was not already full. Suppose we now want to insert 55. Figure 4.61 shows a problem: The leaf where 55 wants to go is already full. The solution is simple, however: Since we now have $L+1$ items, we split them into two leaves, both guaranteed to have the minimum number of data records needed. We form two leaves with three items each. Two disk accesses are required to write these leaves, and a third disk access is required to update the parent. Note that in the parent, both keys and branches change, but they do so in a controlled way that is easily calculated. The resulting B-tree is shown in Figure 4.62. Although splitting nodes is time-consuming because it requires at least two additional disk writes, it is a relatively rare occurrence. If L is 32, for example, then when a node is split, two leaves with 16 and 17 items, respectively, are created. For the leaf with 17 items, we can perform 15 more insertions without another split. Put another way, for every split, there are roughly $L/2$ nonsplits.

The node splitting in the previous example worked because the parent did not have its full complement of children. But what would happen if it did? Suppose, for example, that we were to insert 40 into the B-tree in Figure 4.62. We would then have to split the leaf containing the keys 35 through 39, and now 40, into two leaves. But doing this would give

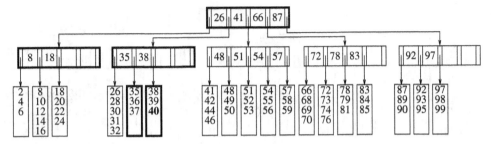

Figure 4.63   Insertion of 40 into the B-tree in Figure 4.62 causes a split into two leaves and then a split of the parent node

the parent six children, and it is allowed only five. Hence, the solution is to split the parent. The result of this is shown in Figure 4.63. When the parent is split, we must update the values of the keys and also the parent's parent, thus incurring an additional two disk writes (so this insertion costs five disk writes). However, once again, the keys change in a very controlled manner, although the code is certainly not simple because of a host of cases.

When a nonleaf node is split, as is the case here, its parent gains a child. What if the parent already has reached its limit of children? In that case, we continue splitting nodes up the tree until either we find a parent that does not need to be split or we reach the root. If we split the root, then we have two roots. Obviously, this is unacceptable, but we can create a new root that has the split roots as its two children. This is why the root is granted the special two-child minimum exemption. It also is the only way that a B-tree gains height. Needless to say, splitting all the way up to the root is an exceptionally rare event, because a tree with four levels indicates that the root has been split three times throughout the entire sequence of insertions (assuming no deletions have occurred). In fact, splitting of any nonleaf node is also quite rare.

There are other ways to handle the overflowing of children. One technique is to put a child up for adoption should a neighbor have room. To insert 29 into the B-tree in Figure 4.63, for example, we could make room by moving 32 to the next leaf. This technique requires a modification of the parent because the keys are affected. However, it tends to keep nodes fuller and thus saves space in the long run.

We can perform deletion by finding the item that needs to be removed and then removing it. The problem is that if the leaf it was in had the minimum number of data items, then it is now below the minimum. We can rectify this situation by adopting a neighboring item, if the neighbor is not itself at its minimum. If it is, then we can combine with the neighbor to form a full leaf. Unfortunately, this means that the parent has lost a child. If this loss causes the parent to fall below its minimum, then it follows the same strategy. This process could percolate all the way up to the root. The root cannot have just one child (and even if this were allowed, it would be silly). If a root is left with one child as a result of the adoption process, then we remove the root and make its child the new root of the tree. This is the only way for a B-tree to lose height. For example, suppose we want to remove 99 from the B-tree in Figure 4.63. Since the leaf has only two items, and its neighbor is already at its minimum of three, we combine the items into a new leaf of five items. As a

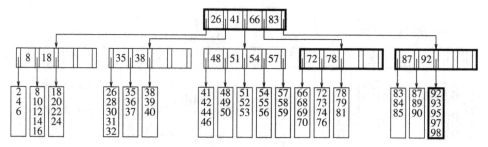

Figure 4.64   B-tree after the deletion of 99 from the B-tree in Figure 4.63

result, the parent has only two children. However, it can adopt from a neighbor because the neighbor has four children. As a result, both have three children. The result is shown in Figure 4.64.

4.8 Sets and Maps in the Standard Library

The List containers discussed in Chapter 3, namely ArrayList and LinkedList, are inefficient for searching. Consequently, the Collections API provides two additional containers, Set and Map, that provide efficient implementations for basic operations such as insertion, deletion, and searching.

4.8.1 Sets

The Set interface represents a Collection that does not allow duplicates. A special kind of Set, given by the SortedSet interface, guarantees that the items are maintained in sorted order. Because a Set IS-A Collection, the idioms used to access items in a List, which are inherited from Collection, also work for a Set. The print method described in Figure 3.6 will work if passed a Set.

The unique operations required by the Set are the abilities to insert, remove, and perform a basic search (efficiently). For a Set, the add method returns true if the add succeeds and false if it fails because the item being added is already present. The implementation of Set that maintains items in sorted order is a TreeSet. Basic operations in a TreeSet take logarithmic worst-case time.

By default, ordering assumes that the items in the TreeSet implement the Comparable interface. An alternative ordering can be specified by instantiating the TreeSet with a Comparator. For instance, we can create a TreeSet that stores String objects, ignoring case distinctions by using the CaseInsensitiveCompare function object coded in Figure 1.18. In the following code, the Set s has size 1.

```
Set<String> s = new TreeSet<>( new CaseInsensitiveCompare( ) );
s.add( "Hello" ); s.add( "HeLLo" );
System.out.println( "The size is: " + s.size( ) );
```

4.8.2 Maps

A Map is an interface that represents a collection of entries that consists of keys and their values. Keys must be unique, but several keys can map to the same values. Thus values need not be unique. In a SortedMap, the keys in the map are maintained in logically sorted order. An implementation of SortedMap is the TreeMap. The basic operations for a Map include methods such as isEmpty, clear, size, and most importantly, the following:

```
boolean containsKey( KeyType key )
ValueType get( KeyType key )
ValueType put( KeyType key, ValueType value )
```

get returns the value associated with key in the Map, or null if key is not present. If there are no null values in the Map, the value returned by get can be used to determine if key is in the Map. However, if there are null values, you have to use containsKey. Method put places a key/value pair into the Map, returning either null or the old value associated with key.

Iterating through a Map is trickier than a Collection because the Map does not provide an iterator. Instead, three methods are provided that return the view of a Map as a Collection. Since the views are themselves Collections, the views can be iterated. The three methods are:

```
Set<KeyType> keySet( )
Collection<ValueType> values( )
Set<Map.Entry<KeyType,ValueType>> entrySet( )
```

Methods keySet and values return simple collections (the keys contain no duplicates, thus the keys are returned in a Set). The entrySet is returned as a Set of entries (there are no duplicate entries, since the keys are unique). Each entry is represented by the nested interface Map.Entry. For an object of type Map.Entry, the available methods include accessing the key, the value, and changing the value:

```
KeyType getKey( )
ValueType getValue( )
ValueType setValue( ValueType newValue )
```

4.8.3 Implementation of TreeSet and TreeMap

Java requires that TreeSet and TreeMap support the basic add, remove, and contains operations in logarithmic worst-case time. Consequently, the underlying implementation is a balanced binary search tree. Typically, an AVL tree is not used; instead, top-down red-black trees, which are discussed in Section 12.2, are often used.

An important issue in implementing TreeSet and TreeMap is providing support for the iterator classes. Of course, internally, the iterator maintains a link to the "current" node in the iteration. The hard part is efficiently advancing to the next node. There are several possible solutions, some of which are listed here:

1. When the iterator is constructed, have each iterator store as its data an array containing the TreeSet items. This is lame, because we might as well use toArray and have no need for an iterator.

2. Have the iterator maintain a stack storing nodes on the path to the current node. With this information, one can deduce the next node in the iteration, which is either the node in the current node's right subtree that contains the minimum item, or the nearest ancestor that contains the current node in its left subtree. This makes the iterator somewhat large, and makes the iterator code clumsy.

3. Have each node in the search tree store its parent in addition to the children. The iterator is not as large, but there is now extra memory required in each node, and the code to iterate is still clumsy.

4. Have each node maintain extra links: one to the next smaller, and one to the next larger node. This takes space, but the iteration is very simple to do, and it is easy to maintain these links.

5. Maintain the extra links only for nodes that have null left or right links, by using extra Boolean variables to allow the routines to tell if a left link is being used as a standard binary search tree left link or a link to the next smaller node, and similarly for the right link (Exercise 4.50). This idea is called a *threaded tree,* and is used in many balanced binary search tree implementations.

4.8.4 An Example That Uses Several Maps

Many words are similar to other words. For instance, by changing the first letter, the word wine can become dine, fine, line, mine, nine, pine, or vine. By changing the third letter, wine can become wide, wife, wipe, or wire, among others. By changing the fourth letter, wine can become wind, wing, wink, or wins, among others. This gives 15 different words that can be obtained by changing only one letter in wine. In fact, there are over 20 different words, some more obscure. We would like to write a program to find all words that can be changed into at least 15 other words by a single one-character substitution. We assume that we have a dictionary consisting of approximately 89,000 different words of varying lengths. Most words are between 6 and 11 characters. The distribution includes 8,205 six-letter words, 11,989 seven-letter words, 13,672 eight-letter words, 13,014 nine-letter words, 11,297 ten-letter words, and 8,617 eleven-letter words. (In reality, the most changeable words are three-, four- and five-letter words, but the longer words are the time-consuming ones to check.)

The most straightforward strategy is to use a Map in which the keys are words and the values are lists containing the words that can be changed from the key with a one-character substitution. The routine in Figure 4.65 shows how the Map that is eventually produced (we have yet to write code for that part) can be used to print the required answers. The code obtains the entry set and uses the enhanced for loop to step through the entry set and view entries that are pairs consisting of a word and a list of words.

```
1      public static void printHighChangeables( Map<String,List<String>> adjWords,
2                                                int minWords )
3      {
4          for( Map.Entry<String,List<String>> entry : adjWords.entrySet( ) )
5          {
6              List<String> words = entry.getValue( );
7
8              if( words.size( ) >= minWords )
9              {
10                 System.out.print( entry.getKey( ) + " (" );
11                 System.out.print( words.size( ) + "):" );
12                 for( String w : words )
13                     System.out.print( " " + w );
14                 System.out.println( );
15             }
16         }
17     }
```

Figure 4.65 Given a map containing words as keys and a list of words that differ in only one character as values, output words that have `minWords` or more words obtainable by a one-character substitution

```
1      // Returns true if word1 and word2 are the same length
2      // and differ in only one character.
3      private static boolean oneCharOff( String word1, String word2 )
4      {
5          if( word1.length( ) != word2.length( ) )
6              return false;
7
8          int diffs = 0;
9
10         for( int i = 0; i < word1.length( ); i++ )
11             if( word1.charAt( i ) != word2.charAt( i ) )
12                 if( ++diffs > 1 )
13                     return false;
14
15         return diffs == 1;
16     }
```

Figure 4.66 Routine to check if two words differ in only one character

The main issue is how to construct the `Map` from an array that contains the 89,000 words. The routine in Figure 4.66 is a straightforward function to test if two words are identical except for a one-character substitution. We can use the routine to provide the simplest algorithm for the `Map` construction, which is a brute-force test of all pairs of words. This algorithm is shown in Figure 4.67.

To step through the collection of words, we could use an iterator, but because we are stepping through it with a nested loop (i.e., several times), we dump the collection into an array using `toArray` (lines 9 and 11). Among other things, this avoids repeated calls to cast from `Object` to `String`, which occur behind the scenes if generics are used. Instead, we are simply indexing a `String[]`.

```
 1      // Computes a map in which the keys are words and values are Lists of words
 2      // that differ in only one character from the corresponding key.
 3      // Uses a quadratic algorithm (with appropriate Map).
 4      public static Map<String,List<String>>
 5      computeAdjacentWords( List<String> theWords )
 6      {
 7          Map<String,List<String>> adjWords = new TreeMap<>( );
 8
 9          String [ ] words = new String[ theWords.size( ) ];
10
11          theWords.toArray( words );
12          for( int i = 0; i < words.length; i++ )
13              for( int j = i + 1; j < words.length; j++ )
14                  if( oneCharOff( words[ i ], words[ j ] ) )
15                  {
16                      update( adjWords, words[ i ], words[ j ] );
17                      update( adjWords, words[ j ], words[ i ] );
18                  }
19
20          return adjWords;
21      }
22
23      private static <KeyType> void update( Map<KeyType,List<String>> m,
24                                            KeyType key, String value )
25      {
26          List<String> lst = m.get( key );
27          if( lst == null )
28          {
29              lst = new ArrayList<>( );
30              m.put( key, lst );
31          }
32
33          lst.add( value );
34      }
```

Figure 4.67 Function to compute a map containing words as keys and a list of words that differ in only one character as values. This version runs in 75 seconds on an 89,000-word dictionary

If we find a pair of words that differ in only one character, we can update the Map at lines 16 and 17. In the private update method, at line 26 we see if there is already a list of words associated with the key. If we have previously seen key, because 1st is not null, then it is in the Map, and we need only add the new word to the List in the Map, and we do this by calling add at line 33. If we have never seen key before, then lines 29 and 30 place it in the Map, with a List of size 0, so the add updates the List to be size 1. All in all, this is a standard idiom for maintaining a Map, in which the value is a collection.

The problem with this algorithm is that it is slow, and takes 75 seconds on our computer. An obvious improvement is to avoid comparing words of different lengths. We can do this by grouping words by their length, and then running the previous algorithm on each of the separate groups.

To do this, we can use a second map! Here the key is an integer representing a word length, and the value is a collection of all the words of that length. We can use a List to store each collection, and the same idiom applies. The code is shown in Figure 4.68. Line 9 shows the declaration for the second Map, lines 12 and 13 populate the Map, and then an extra loop is used to iterate over each group of words. Compared to the first algorithm, the second algorithm is only marginally more difficult to code and runs in 16 seconds, or about five times as fast.

Our third algorithm is more complex, and uses additional maps! As before, we group the words by word length, and then work on each group separately. To see how this algorithm works, suppose we are working on words of length 4. Then first we want to find word pairs such as wine and nine that are identical except for the first letter. One way to do this, for each word of length 4, is to remove the first character, leaving a three-character word representative. Form a Map in which the key is the representative, and the value is a List of all words that have that representative. For instance, in considering the first character of the four-letter word group, representative "ine" corresponds to "dine", "fine", "wine", "nine", "mine", "vine", "pine", "line". Representative "oot" corresponds to "boot", "foot", "hoot", "loot", "soot", "zoot". Each individual List that is a value in this latest Map forms a clique of words in which any word can be changed to any other word by a one-character substitution, so after this latest Map is constructed, it is easy to traverse it and add entries to the original Map that is being computed. We would then proceed to the second character of the four-letter word group, with a new Map. And then the third character, and finally the fourth character.

The general outline is:

```
for each group g, containing words of length len
    for each position p (ranging from 0 to len-1)
    {
        Make an empty Map<String,List<String> > repsToWords
        for each word w
        {
            Obtain w's representative by removing position p
            Update repsToWords
        }
        Use cliques in repsToWords to update adjWords map
    }
```

Figure 4.69 contains an implementation of this algorithm. The running time improves to one second. It is interesting to note that although the use of the additional Maps makes the algorithm faster, and the syntax is relatively clean, the code makes no use of the fact that the keys of the Map are maintained in sorted order.

As such, it is possible that a data structure that supports the Map operations but does not guarantee sorted order can perform better, since it is being asked to do less. Chapter 5 explores this possibility and discusses the ideas behind the alternative Map implementation, known as a HashMap. A HashMap reduces the running time of the implementation from one second to roughly 0.8 seconds.

```
1     // Computes a map in which the keys are words and values are Lists of words
2     // that differ in only one character from the corresponding key.
3     // Uses a quadratic algorithm (with appropriate Map), but speeds things by
4     // maintaining an additional map that groups words by their length.
5     public static Map<String,List<String>>
6     computeAdjacentWords( List<String> theWords )
7     {
8         Map<String,List<String>> adjWords = new TreeMap<>( );
9         Map<Integer,List<String>> wordsByLength = new TreeMap<>( );
10
11         // Group the words by their length
12         for( String w : theWords )
13             update( wordsByLength, w.length( ), w );
14
15         // Work on each group separately
16         for( List<String> groupsWords : wordsByLength.values( ) )
17         {
18             String [ ] words = new String[ groupsWords.size( ) ];
19
20             groupsWords.toArray( words );
21             for( int i = 0; i < words.length; i++ )
22                 for( int j = i + 1; j < words.length; j++ )
23                     if( oneCharOff( words[ i ], words[ j ] ) )
24                     {
25                         update( adjWords, words[ i ], words[ j ] );
26                         update( adjWords, words[ j ], words[ i ] );
27                     }
28         }
29
30         return adjWords;
31     }
```

Figure 4.68 Function to compute a map containing words as keys and a list of words that differ in only one character as values. Splits words into groups by word length. This version runs in 16 seconds on an 89,000-word dictionary

```
1    // Computes a map in which the keys are words and values are Lists of words
2    // that differ in only one character from the corresponding key.
3    // Uses an efficient algorithm that is O(N log N) with a TreeMap.
4    public static Map<String,List<String>>
5    computeAdjacentWords( List<String> words )
6    {
7        Map<String,List<String>> adjWords = new TreeMap<>( );
8        Map<Integer,List<String>> wordsByLength = new TreeMap<>( );
9
10          // Group the words by their length
11       for( String w : words )
12           update( wordsByLength, w.length( ), w );
13
14          // Work on each group separately
15       for( Map.Entry<Integer,List<String>> entry : wordsByLength.entrySet( ) )
16       {
17           List<String> groupsWords = entry.getValue( );
18           int groupNum = entry.getKey( );
19
20              // Work on each position in each group
21           for( int i = 0; i < groupNum; i++ )
22           {
23               // Remove one character in specified position, computing
24               // representative.  Words with same representative are
25               // adjacent, so first populate a map ...
26               Map<String,List<String>> repToWord = new TreeMap<>( );
27
28               for( String str : groupsWords )
29               {
30                   String rep = str.substring( 0, i ) + str.substring( i + 1 );
31                   update( repToWord, rep, str );
32               }
33
34               // and then look for map values with more than one string
35               for( List<String> wordClique : repToWord.values( ) )
36                   if( wordClique.size( ) >= 2 )
37                       for( String s1 : wordClique )
38                           for( String s2 : wordClique )
39                               if( s1 != s2 )
40                                   update( adjWords, s1, s2 );
41           }
42       }
43
44       return adjWords;
45   }
```

Figure 4.69 Function to compute a map containing words as keys and a list of words that differ in only one character as values. Runs in 1 second on an 89,000-word dictionary

Summary

We have seen uses of trees in operating systems, compiler design, and searching. Expression trees are a small example of a more general structure known as a **parse tree**, which is a central data structure in compiler design. Parse trees are not binary but are relatively simple extensions of expression trees (although the algorithms to build them are not quite so simple).

Search trees are of great importance in algorithm design. They support almost all the useful operations, and the logarithmic average cost is very small. Nonrecursive implementations of search trees are somewhat faster, but the recursive versions are sleeker, more elegant, and easier to understand and debug. The problem with search trees is that their performance depends heavily on the input being random. If this is not the case, the running time increases significantly, to the point where search trees become expensive linked lists.

We saw several ways to deal with this problem. AVL trees work by insisting that all nodes' left and right subtrees differ in heights by at most one. This ensures that the tree cannot get too deep. The operations that do not change the tree, as insertion does, can all use the standard binary search tree code. Operations that change the tree must restore the tree. This can be somewhat complicated, especially in the case of deletion. We showed how to restore the tree after insertions in $O(\log N)$ time.

We also examined the splay tree. Nodes in splay trees can get arbitrarily deep, but after every access the tree is adjusted in a somewhat mysterious manner. The net effect is that any sequence of M operations takes $O(M \log N)$ time, which is the same as a balanced tree would take.

B-trees are balanced M-way (as opposed to 2-way or binary) trees, which are well suited for disks; a special case is the 2–3 tree ($M = 3$), which is another way to implement balanced search trees.

In practice, the running time of all the balanced tree schemes, while slightly faster for searching, is worse (by a constant factor) for insertions and deletions than the simple binary search tree, but this is generally acceptable in view of the protection being given against easily obtained worst-case input. Chapter 12 discusses some additional search tree data structures and provides detailed implementations.

A final note: By inserting elements into a search tree and then performing an inorder traversal, we obtain the elements in sorted order. This gives an $O(N \log N)$ algorithm to sort, which is a worst-case bound if any sophisticated search tree is used. We shall see better ways in Chapter 7, but none that have a lower time bound.

Exercises

Questions 4.1 to 4.3 refer to the tree in Figure 4.70.

4.1   For the tree in Figure 4.70:
    a. Which node is the root?
    b. Which nodes are leaves?

4.2   For each node in the tree of Figure 4.70:
   a. Name the parent node.
   b. List the children.
   c. List the siblings.
   d. Compute the depth.
   e. Compute the height.

4.3   What is the depth of the tree in Figure 4.70?

4.4   Show that in a binary tree of N nodes, there are $N + 1$ `null` links representing children.

4.5   Show that the maximum number of nodes in a binary tree of height h is $2^{h+1} - 1$.

4.6   A *full node* is a node with two children. Prove that the number of full nodes plus one is equal to the number of leaves in a nonempty binary tree.

4.7   Suppose a binary tree has leaves $l_1, l_2, \ldots, l_M$ at depths $d_1, d_2, \ldots, d_M$, respectively. Prove that $\sum_{i=1}^{M} 2^{-d_i} \leq 1$ and determine when the equality is true.

4.8   Give the prefix, infix, and postfix expressions corresponding to the tree in Figure 4.71.

4.9   a. Show the result of inserting 3, 1, 4, 6, 9, 2, 5, 7 into an initially empty binary search tree.
   b. Show the result of deleting the root.

4.10   Write a program that lists all files in a directory and their sizes. Mimic the routine in the online code.

4.11   Write an implementation of the `TreeSet` class, with associated iterators using a binary search tree. Add to each node a link to the parent node.

4.12   Write an implementation of the `TreeMap` class by storing a data member of type `TreeSet<Map.Entry<KeyType,ValueType>>`.

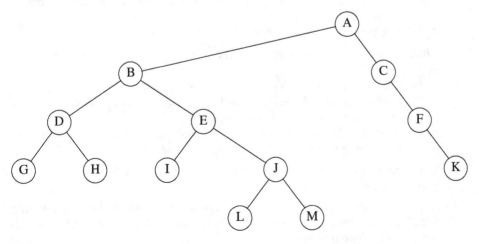

Figure 4.70   Tree for Exercises 4.1 to 4.3

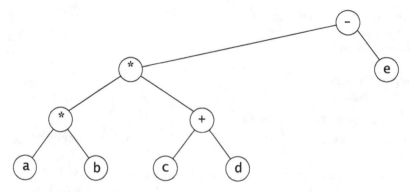

Figure 4.71   Tree for Exercise 4.8

4.13 Write an implementation of the `TreeSet` class, with associated iterators, using a binary search tree. Add to each node a link to the next smallest and next largest node. To make your code simpler, add a header and tail node which are not part of the binary search tree, but help make the linked list part of the code simpler.

4.14 Suppose you want to perform an experiment to verify the problems that can be caused by random `insert/remove` pairs. Here is a strategy that is not perfectly random, but close enough. You build a tree with N elements by inserting N elements chosen at random from the range 1 to $M = \alpha N$. You then perform N^2 pairs of insertions followed by deletions. Assume the existence of a routine, `randomInteger(a, b)`, which returns a uniform random integer between a and b inclusive.

    a. Explain how to generate a random integer between 1 and M that is not already in the tree (so a random insertion can be performed). In terms of N and α, what is the running time of this operation?

    b. Explain how to generate a random integer between 1 and M that is already in the tree (so a random deletion can be performed). What is the running time of this operation?

    c. What is a good choice of α? Why?

4.15 Write a program to evaluate empirically the following strategies for removing nodes with two children:

    a. Replace with the largest node, X, in T_L and recursively remove X.

    b. Alternately replace with the largest node in T_L and the smallest node in T_R, and recursively remove the appropriate node.

    c. Replace with either the largest node in T_L or the smallest node in T_R (recursively removing the appropriate node), making the choice randomly.

Which strategy seems to give the most balance? Which takes the least CPU time to process the entire sequence?

4.16 Redo the binary search tree class to implement lazy deletion. Note carefully that this affects all of the routines. Especially challenging are `findMin` and `findMax`, which must now be done recursively.

**4.17 Prove that the depth of a random binary search tree (depth of the deepest node) is $O(\log N)$, on average.

4.18 *a. Give a precise expression for the minimum number of nodes in an AVL tree of height h.

   b. What is the minimum number of nodes in an AVL tree of height 15?

4.19 Show the result of inserting 2, 1, 4, 5, 9, 3, 6, 7 into an initially empty AVL tree.

*4.20 Keys $1, 2, \ldots, 2^k - 1$ are inserted in order into an initially empty AVL tree. Prove that the resulting tree is perfectly balanced.

4.21 Write the remaining procedures to implement AVL single and double rotations.

4.22 Design a linear-time algorithm that verifies that the height information in an AVL tree is correctly maintained and that the balance property is in order.

4.23 Write a nonrecursive method to insert into an AVL tree.

4.24 Show that the deletion algorithm in Figure 4.44 is correct, and explain what happens if > is used instead of >= at lines 32 and 38 in Figure 4.39.

4.25 a. How many bits are required per node to store the height of a node in an N-node AVL tree?

   b. What is the smallest AVL tree that overflows an 8-bit height counter?

4.26 Write the methods to perform the double rotation without the inefficiency of doing two single rotations.

4.27 Show the result of accessing the keys 3, 9, 1, 5 in order in the splay tree in Figure 4.72.

4.28 Show the result of deleting the element with key 6 in the resulting splay tree for the previous exercise.

4.29 a. Show that if all nodes in a splay tree are accessed in sequential order, the resulting tree consists of a chain of left children.

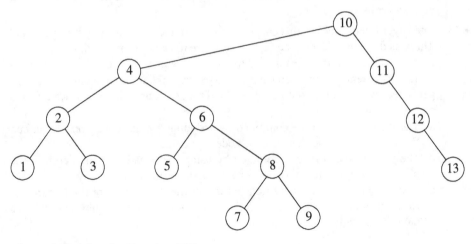

Figure 4.72 Tree for Exercise 4.27

**b. Show that if all nodes in a splay tree are accessed in sequential order, then the total access time is $O(N)$, regardless of the initial tree.

4.30 Write a program to perform random operations on splay trees. Count the total number of rotations performed over the sequence. How does the running time compare to AVL trees and unbalanced binary search trees?

4.31 Write efficient methods that take only a reference to the root of a binary tree, T, and compute:
a. The number of nodes in T.
b. The number of leaves in T.
c. The number of full nodes in T.
What is the running time of your routines?

4.32 Design a recursive linear-time algorithm that tests whether a binary tree satisfies the search tree order property at every node.

4.33 Write a recursive method that takes a reference to the root node of a tree T and returns a reference to the root node of the tree that results from removing all leaves from T.

4.34 Write a method to generate an N-node random binary search tree with distinct keys 1 through N. What is the running time of your routine?

4.35 Write a method to generate the AVL tree of height h with fewest nodes. What is the running time of your method?

4.36 Write a method to generate a perfectly balanced binary search tree of height h with keys 1 through $2^{h+1} - 1$. What is the running time of your method?

4.37 Write a method that takes as input a binary search tree, T, and two keys k_1 and k_2, which are ordered so that $k_1 \leq k_2$, and prints all elements X in the tree such that $k_1 \leq Key(X) \leq k_2$. Do not assume any information about the type of keys except that they can be ordered (consistently). Your program should run in $O(K + \log N)$ average time, where K is the number of keys printed. Bound the running time of your algorithm.

4.38 The larger binary trees in this chapter were generated automatically by a program. This was done by assigning an (x, y) coordinate to each tree node, drawing a circle around each coordinate (this is hard to see in some pictures), and connecting each node to its parent. Assume you have a binary search tree stored in memory (perhaps generated by one of the routines above) and that each node has two extra fields to store the coordinates.
a. The x coordinate can be computed by assigning the inorder traversal number. Write a routine to do this for each node in the tree.
b. The y coordinate can be computed by using the negative of the depth of the node. Write a routine to do this for each node in the tree.
c. In terms of some imaginary unit, what will the dimensions of the picture be? How can you adjust the units so that the tree is always roughly two-thirds as high as it is wide?

d. Prove that using this system no lines cross, and that for any node, X, all elements in X's left subtree appear to the left of X and all elements in X's right subtree appear to the right of X.

4.39 Write a general-purpose tree-drawing program that will convert a tree into the following graph-assembler instructions:
a. $Circle(X, Y)$
b. $DrawLine(i, j)$
The first instruction draws a circle at (X, Y), and the second instruction connects the ith circle to the jth circle (circles are numbered in the order drawn). You should either make this a program and define some sort of input language or make this a method that can be called from any program. What is the running time of your routine?

4.40 (This exercise assumes familiarity with the Java Swing Library.) Write a program that reads graph-assembler instructions and generates Java code that draws into a canvas. (Note that you have to scale the stored coordinates into pixels.)

4.41 Write a routine to list out the nodes of a binary tree in *level-order*. List the root, then nodes at depth 1, followed by nodes at depth 2, and so on. You must do this in linear time. Prove your time bound.

4.42 *a. Write a routine to perform insertion into a B-tree.
*b. Write a routine to perform deletion from a B-tree. When an item is deleted, is it necessary to update information in the internal nodes?
*c. Modify your insertion routine so that if an attempt is made to add into a node that already has M entries, a search is performed for a sibling with less than M children before the node is split.

4.43 A B^*-*tree* of order M is a B-tree in which each interior node has between $2M/3$ and M children. Describe a method to perform insertion into a B^*-tree.

4.44 Show how the tree in Figure 4.73 is represented using a child/sibling link implementation.

4.45 Write a procedure to traverse a tree stored with child/sibling links.

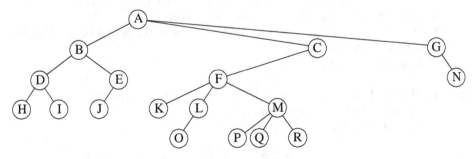

Figure 4.73 Tree for Exercise 4.44

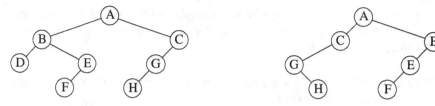

Figure 4.74 Two isomorphic trees

4.46 Two binary trees are *similar* if they are both empty or both nonempty and have similar left and right subtrees. Write a method to decide whether two binary trees are similar. What is the running time of your method?

4.47 Two trees, T_1 and T_2, are *isomorphic* if T_1 can be transformed into T_2 by swapping left and right children of (some of the) nodes in T_1. For instance, the two trees in Figure 4.74 are isomorphic because they are the same if the children of A, B, and G, but not the other nodes, are swapped.

    a. Give a polynomial time algorithm to decide if two trees are isomorphic.

    *b. What is the running time of your program (there is a linear solution)?

4.48 *a. Show that via AVL single rotations, any binary search tree T_1 can be transformed into another search tree T_2 (with the same items).

    *b. Give an algorithm to perform this transformation using $O(N \log N)$ rotations on average.

    **c. Show that this transformation can be done with $O(N)$ rotations, worst case.

4.49 Suppose we want to add the operation findKth to our repertoire. The operation findKth(k) returns the kth smallest item in the tree. Assume all items are distinct. Explain how to modify the binary search tree to support this operation in $O(\log N)$ average time, without sacrificing the time bounds of any other operation.

4.50 Since a binary search tree with N nodes has $N + 1$ null references, half the space allocated in a binary search tree for link information is wasted. Suppose that if a node has a null left child, we make its left child link to its inorder predecessor, and if a node has a null right child, we make its right child link to its inorder successor. This is known as a *threaded tree*, and the extra links are called *threads*.

    a. How can we distinguish threads from real children links?

    b. Write routines to perform insertion and deletion into a tree threaded in the manner described above.

    c. What is the advantage of using threaded trees?

4.51 Let $f(N)$ be the average number of full nodes in a binary search tree.

    a. Determine the values of $f(0)$ and $f(1)$.

    b. Show that for $N > 1$

$$f(N) = \frac{N-2}{N} + \frac{1}{N} \sum_{i=0}^{N-1} (f(i) + f(N - i - 1))$$

```
IX: {Series |(}         {2}
IX: {Series!geometric|(} {4}
IX: {Euler's constant}   {4}
IX: {Series!geometric|)} {4}
IX: {Series!arithmetic|(} {4}
IX: {Series!arithmetic|)} {5}
IX: {Series!harmonic|(}  {5}
IX: {Euler's constant}   {5}
IX: {Series!harmonic|)}  {5}
IX: {Series|)}           {5}
```

Figure 4.75 Sample input for Exercise 4.53

```
Euler's constant: 4, 5
Series: 2-5
    arithmetic: 4-5
    geometric: 4
    harmonic: 5
```

Figure 4.76 Sample output for Exercise 4.53

    c. Show (by induction) that $f(N) = (N − 2)/3$ is a solution to the equation in part (b), with the initial conditions in part (a).

    d. Use the results of Exercise 4.6 to determine the average number of leaves in an N node binary search tree.

4.52 Write a program that reads a Java source code file and outputs a list of all identifiers (that is, variable names but not keywords, not found in comments or string constants) in alphabetical order. Each identifier should be output with a list of line numbers on which it occurs.

4.53 Generate an index for a book. The input file consists of a set of index entries. Each line consists of the string IX:, followed by an index entry name enclosed in braces, followed by a page number that is enclosed in braces. Each ! in an index entry name represents a sub-level. A (|represents the start of a range, and a |) represents the end of the range. Occasionally, this range will be the same page. In that case, output only a single page number. Otherwise, do not collapse or expand ranges on your own. As an example, Figure 4.75 shows sample input, and Figure 4.76 shows the corresponding output.

References

More information on binary search trees, and in particular the mathematical properties of trees, can be found in the two books by Knuth, [22] and [23].

    Several papers deal with the lack of balance caused by biased deletion algorithms in binary search trees. Hibbard's paper [19] proposed the original deletion algorithm and established that one deletion preserves the randomness of the trees. A complete analysis has been performed only for trees with three nodes [20] and four nodes [5]. Eppinger's paper [14] provided early empirical evidence of nonrandomness, and the papers by Culberson and Munro [10], [11] provided some analytical evidence (but not a complete proof for the general case of intermixed insertions and deletions).

AVL trees were proposed by Adelson-Velskii and Landis [1]. Simulation results for AVL trees, and variants in which the height imbalance is allowed to be at most k for various values of k, are presented in [21]. Analysis of the average search cost in AVL trees is incomplete, but some results are contained in [24].

[3] and [8] considered self-adjusting trees like the type in Section 4.5.1. Splay trees are described in [28].

B-trees first appeared in [6]. The implementation described in the original paper allows data to be stored in internal nodes as well as leaves. The data structure we have described is sometimes known as a B^+-tree. A survey of the different types of B-trees is presented in [9]. Empirical results of the various schemes are reported in [17]. Analysis of 2–3 trees and B-trees can be found in [4], [13], and [32].

Exercise 4.17 is deceptively difficult. A solution can be found in [15]. Exercise 4.29 is from [31]. Information on B*-trees, described in Exercise 4.47, can be found in [12]. Exercise 4.47 is from [2]. A solution to Exercise 4.48 using $2N - 6$ rotations is given in [29]. Using threads, à la Exercise 4.50, was first proposed in [27]. k-d trees, which handle multidimensional data, were first proposed in [7] and are discussed in Chapter 12.

Other popular balanced search trees are red-black trees [18] and weight-balanced trees [26]. More balanced-tree schemes can be found in the books [16], [25], and [30].

1. G. M. Adelson-Velskii and E. M. Landis, "An Algorithm for the Organization of Information," *Soviet. Mat. Doklady,* 3 (1962), 1259–1263.

2. A. V. Aho, J. E. Hopcroft, and J. D. Ullman, *The Design and Analysis of Computer Algorithms,* Addison-Wesley, Reading, Mass., 1974.

3. B. Allen and J. I. Munro, "Self Organizing Search Trees," *Journal of the ACM,* 25 (1978), 526–535.

4. R. A. Baeza-Yates, "Expected Behaviour of B^+-trees under Random Insertions," *Acta Informatica,* 26 (1989), 439–471.

5. R. A. Baeza-Yates, "A Trivial Algorithm Whose Analysis Isn't: A Continuation," *BIT,* 29 (1989), 88–113.

6. R. Bayer and E. M. McCreight, "Organization and Maintenance of Large Ordered Indices," *Acta Informatica,* 1 (1972), 173–189.

7. J. L. Bentley, "Multidimensional Binary Search Trees Used for Associative Searching," *Communications of the ACM,* 18 (1975), 509–517.

8. J. R. Bitner, "Heuristics that Dynamically Organize Data Structures," *SIAM Journal on Computing,* 8 (1979), 82–110.

9. D. Comer, "The Ubiquitous B-tree," *Computing Surveys,* 11 (1979), 121–137.

10. J. Culberson and J. I. Munro, "Explaining the Behavior of Binary Search Trees under Prolonged Updates: A Model and Simulations," *Computer Journal,* 32 (1989), 68–75.

11. J. Culberson and J. I. Munro, "Analysis of the Standard Deletion Algorithms in Exact Fit Domain Binary Search Trees," *Algorithmica,* 5 (1990), 295–311.

12. K. Culik, T. Ottman, and D. Wood, "Dense Multiway Trees," *ACM Transactions on Database Systems,* 6 (1981), 486–512.

13. B. Eisenbath, N. Ziviana, G. H. Gonnet, K. Melhorn, and D. Wood, "The Theory of Fringe Analysis and Its Application to 2–3 Trees and B-trees," *Information and Control,* 55 (1982), 125–174.

14. J. L. Eppinger, "An Empirical Study of Insertion and Deletion in Binary Search Trees," *Communications of the ACM,* 26 (1983), 663–669.

15. P. Flajolet and A. Odlyzko, "The Average Height of Binary Trees and Other Simple Trees," *Journal of Computer and System Sciences,* 25 (1982), 171–213.

16. G. H. Gonnet and R. Baeza-Yates, *Handbook of Algorithms and Data Structures,* 2d ed., Addison-Wesley, Reading, Mass., 1991.

17. E. Gudes and S. Tsur, "Experiments with B-tree Reorganization," *Proceedings of ACM SIGMOD Symposium on Management of Data* (1980), 200–206.

18. L. J. Guibas and R. Sedgewick, "A Dichromatic Framework for Balanced Trees," *Proceedings of the Nineteenth Annual IEEE Symposium on Foundations of Computer Science* (1978), 8–21.

19. T. H. Hibbard, "Some Combinatorial Properties of Certain Trees with Applications to Searching and Sorting," *Journal of the ACM,* 9 (1962), 13–28.

20. A. T. Jonassen and D. E. Knuth, "A Trivial Algorithm Whose Analysis Isn't," *Journal of Computer and System Sciences,* 16 (1978), 301–322.

21. P. L. Karlton, S. H. Fuller, R. E. Scroggs, and E. B. Kaehler, "Performance of Height Balanced Trees," *Communications of the ACM,* 19 (1976), 23–28.

22. D. E. Knuth, *The Art of Computer Programming: Vol. 1: Fundamental Algorithms,* 3d ed., Addison-Wesley, Reading, Mass., 1997.

23. D. E. Knuth, *The Art of Computer Programming: Vol. 3: Sorting and Searching,* 2d ed., Addison-Wesley, Reading, Mass., 1998.

24. K. Melhorn, "A Partial Analysis of Height-Balanced Trees under Random Insertions and Deletions," *SIAM Journal of Computing,* 11 (1982), 748–760.

25. K. Melhorn, *Data Structures and Algorithms 1: Sorting and Searching,* Springer-Verlag, Berlin, 1984.

26. J. Nievergelt and E. M. Reingold, "Binary Search Trees of Bounded Balance," *SIAM Journal on Computing,* 2 (1973), 33–43.

27. A. J. Perlis and C. Thornton, "Symbol Manipulation in Threaded Lists," *Communications of the ACM,* 3 (1960), 195–204.

28. D. D. Sleator and R. E. Tarjan, "Self-adjusting Binary Search Trees," *Journal of ACM,* 32 (1985), 652–686.

29. D. D. Sleator, R. E. Tarjan, and W. P. Thurston, "Rotation Distance, Triangulations, and Hyperbolic Geometry," *Journal of AMS* (1988), 647–682.

30. H. F. Smith, *Data Structures—Form and Function,* Harcourt Brace Jovanovich, Orlando, Fla., 1987.

31. R. E. Tarjan, "Sequential Access in Splay Trees Takes Linear Time," *Combinatorica,* 5 (1985), 367–378.

32. A. C. Yao, "On Random 2–3 Trees," *Acta Informatica,* 9 (1978), 159–170.

CHAPTER 5

Hashing

In Chapter 4, we discussed the search tree ADT, which allowed various operations on a set of elements. In this chapter, we discuss the *hash table* ADT, which supports only a subset of the operations allowed by binary search trees.

The implementation of hash tables is frequently called **hashing**. Hashing is a technique used for performing insertions, deletions, and searches in constant average time. Tree operations that require any ordering information among the elements are not supported efficiently. Thus, operations such as findMin, findMax, and the printing of the entire table in sorted order in linear time are not supported.

The central data structure in this chapter is the **hash table.** We will

- See several methods of implementing the hash table.
- Compare these methods analytically.
- Show numerous applications of hashing.
- Compare hash tables with binary search trees.

5.1 General Idea

The ideal hash table data structure is merely an array of some fixed size, containing the items. As discussed in Chapter 4, generally a search is performed on some part (that is, data field) of the item. This is called the **key.** For instance, an item could consist of a string (that serves as the key) and additional data fields (for instance, a name that is part of a large employee structure). We will refer to the table size as *TableSize,* with the understanding that this is part of a hash data structure and not merely some variable floating around globally. The common convention is to have the table run from 0 to *TableSize* − 1; we will see why shortly.

Each key is mapped into some number in the range 0 to *TableSize* − 1 and placed in the appropriate cell. The mapping is called a **hash function,** which ideally should be simple to compute and should ensure that any two distinct keys get different cells. Since there are a finite number of cells and a virtually inexhaustible supply of keys, this is clearly impossible, and thus we seek a hash function that distributes the keys evenly among the cells. Figure 5.1 is typical of a perfect situation. In this example, *john* hashes to 3, *phil* hashes to 4, *dave* hashes to 6, and *mary* hashes to 7.

0	
1	
2	
3	john 25000
4	phil 31250
5	
6	dave 27500
7	mary 28200
8	
9	

Figure 5.1 An ideal hash table

This is the basic idea of hashing. The only remaining problems deal with choosing a function, deciding what to do when two keys hash to the same value (this is known as a **collision**), and deciding on the table size.

5.2 Hash Function

If the input keys are integers, then simply returning *Key* mod *TableSize* is generally a reasonable strategy, unless *Key* happens to have some undesirable properties. In this case, the choice of hash function needs to be carefully considered. For instance, if the table size is 10 and the keys all end in zero, then the standard hash function is a bad choice. For reasons we shall see later, and to avoid situations like the one above, it is often a good idea to ensure that the table size is prime. When the input keys are random integers, then this function is not only very simple to compute but also distributes the keys evenly.

Usually, the keys are strings; in this case, the hash function needs to be chosen carefully.

One option is to add up the ASCII (or Unicode) values of the characters in the string. The routine in Figure 5.2 implements this strategy.

The hash function depicted in Figure 5.2 is simple to implement and computes an answer quickly. However, if the table size is large, the function does not distribute the keys well. For instance, suppose that *TableSize* = 10,007 (10,007 is a prime number). Suppose all the keys are eight or fewer characters long. Since an ASCII character has an integer value that is always at most 127, the hash function typically can only assume values between 0 and 1,016, which is 127 ∗ 8. This is clearly not an equitable distribution!

Another hash function is shown in Figure 5.3. This hash function assumes that *Key* has at least three characters. The value 27 represents the number of letters in the English alphabet, plus the blank, and 729 is 27^2. This function examines only the first three characters,

```
1              public static int hash( String key, int tableSize )
2              {
3                  int hashVal = 0;
4
5                  for( int i = 0; i < key.length( ); i++ )
6                      hashVal += key.charAt( i );
7
8                  return hashVal % tableSize;
9              }
```

Figure 5.2 A simple hash function

```
1              public static int hash( String key, int tableSize )
2              {
3                  return ( key.charAt( 0 ) + 27 * key.charAt( 1 ) +
4                          729 * key.charAt( 2 ) ) % tableSize;
5              }
```

Figure 5.3 Another possible hash function—not too good

but if these are random and the table size is 10,007, as before, then we would expect a reasonably equitable distribution. Unfortunately, English is not random. Although there are $26^3 = 17,576$ possible combinations of three characters (ignoring blanks), a check of a reasonably large online dictionary reveals that the number of different combinations is actually only 2,851. Even if none of *these* combinations collide, only 28 percent of the table can actually be hashed to. Thus this function, although easily computable, is also not appropriate if the hash table is reasonably large.

Figure 5.4 shows a third attempt at a hash function. This hash function involves all characters in the key and can generally be expected to distribute well (it computes $\sum_{i=0}^{KeySize-1} Key[KeySize - i - 1] \cdot 37^i$ and brings the result into proper range). The code computes a polynomial function (of 37) by use of Horner's rule. For instance, another way of computing $h_k = k_0 + 37k_1 + 37^2k_2$ is by the formula $h_k = ((k_2) * 37 + k_1) * 37 + k_0$. Horner's rule extends this to an nth degree polynomial.

The hash function takes advantage of the fact that overflow is allowed. This may introduce a negative number; thus the extra test at the end.

The hash function described in Figure 5.4 is not necessarily the best with respect to table distribution but does have the merit of extreme simplicity and is reasonably fast. If the keys are very long, the hash function will take too long to compute. A common practice in this case is not to use all the characters. The length and properties of the keys would then influence the choice. For instance, the keys could be a complete street address. The hash function might include a couple of characters from the street address and perhaps a couple of characters from the city name and zip code. Some programmers implement their hash function by using only the characters in the odd spaces, with the idea that the time saved computing the hash function will make up for a slightly less evenly distributed function.

```
1      /**
2       * A hash routine for String objects.
3       * @param key the String to hash.
4       * @param tableSize the size of the hash table.
5       * @return the hash value.
6       */
7      public static int hash( String key, int tableSize )
8      {
9          int hashVal = 0;
10
11         for( int i = 0; i < key.length( ); i++ )
12             hashVal = 37 * hashVal + key.charAt( i );
13
14         hashVal %= tableSize;
15         if( hashVal < 0 )
16             hashVal += tableSize;
17
18         return hashVal;
19     }
```

Figure 5.4 A good hash function

The main programming detail left is collision resolution. If, when an element is inserted, it hashes to the same value as an already inserted element, then we have a **collision** and need to resolve it. There are several methods for dealing with this. We will discuss two of the simplest: separate chaining and open addressing; then we will look at some more recently discovered alternatives.

5.3 Separate Chaining

The first strategy, commonly known as **separate chaining**, is to keep a list of all elements that hash to the same value. We can use the standard library list implementations. If space is tight, it might be preferable to avoid their use (since those lists are doubly linked and waste space). We assume, for this section, that the keys are the first 10 perfect squares and that the hashing function is simply $hash(x) = x \bmod 10$. (The table size is not prime but is used here for simplicity.) Figure 5.5 should make this clear.

To perform a search, we use the hash function to determine which list to traverse. We then search the appropriate list. To perform an insert, we check the appropriate list to see whether the element is already in place (if duplicates are expected, an extra field is usually kept, and this field would be incremented in the event of a match). If the element turns out to be new, it is inserted at the front of the list, since it is convenient and also because frequently it happens that recently inserted elements are the most likely to be accessed in the near future.

The class skeleton required to implement separate chaining is shown in Figure 5.6. The hash table stores an array of linked lists, which are allocated in the constructor.

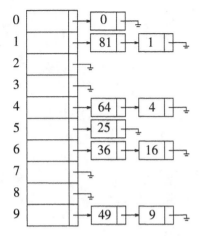

Figure 5.5   A separate chaining hash table

```
1    public class SeparateChainingHashTable<AnyType>
2    {
3        public SeparateChainingHashTable( )
4          { /* Figure 5.9 */ }
5        public SeparateChainingHashTable( int size )
6          { /* Figure 5.9 */ }
7
8        public void insert( AnyType x )
9          { /* Figure 5.10 */ }
10       public void remove( AnyType x )
11         { /* Figure 5.10 */ }
12       public boolean contains( AnyType x )
13         { /* Figure 5.10 */ }
14       public void makeEmpty( )
15         { /* Figure 5.9 */ }
16
17       private static final int DEFAULT_TABLE_SIZE = 101;
18
19       private List<AnyType> [ ] theLists;
20       private int currentSize;
21
22       private void rehash( )
23         { /* Figure 5.22 */ }
24       private int myhash( AnyType x )
25         { /* Figure 5.7 */ }
26
27       private static int nextPrime( int n )
28         { /* See online code */ }
29       private static boolean isPrime( int n )
30         { /* See online code */ }
31   }
```

Figure 5.6   Class skeleton for separate chaining hash table

```
1        private int myhash( AnyType x )
2        {
3            int hashVal = x.hashCode( );
4
5            hashVal %= theLists.length;
6            if( hashVal < 0 )
7                hashVal += theLists.length;
8
9            return hashVal;
10       }
```

Figure 5.7 myHash method for hash tables

```
1    public class Employee
2    {
3        public boolean equals( Object rhs )
4          { return rhs instanceof Employee && name.equals( ((Employee)rhs).name ); }
5
6        public int hashCode( )
7          { return name.hashCode( ); }
8
9        private String name;
10       private double salary;
11       private int seniority;
12
13           // Additional fields and methods
14   }
```

Figure 5.8 Example of Employee class that can be in a hash table

Just as the binary search tree works only for objects that are Comparable, the hash tables in this chapter work only for objects that follow a certain protocol. In Java such objects must provide an appropriate equals method and a hashCode method that returns an int. The hash table can then scale this int into a suitable array index via myHash, as shown in Figure 5.7. Figure 5.8 illustrates an Employee class that can be stored in a hash table. The Employee class provides an equals method and a hashCode method based on the Employee's name. The hashCode for the Employee class works by using the hashCode defined in the Standard String class. That hashCode is basically the code in Figure 5.4 with lines 14–16 removed.

Figure 5.9 shows the constructors and makeEmpty.

The code to implement contains, insert, and remove is shown in Figure 5.10.

```
1      /**
2       * Construct the hash table.
3       */
4      public SeparateChainingHashTable( )
5      {
6          this( DEFAULT_TABLE_SIZE );
7      }
8
9      /**
10      * Construct the hash table.
11      * @param size approximate table size.
12      */
13     public SeparateChainingHashTable( int size )
14     {
15         theLists = new LinkedList[ nextPrime( size ) ];
16         for( int i = 0; i < theLists.length; i++ )
17             theLists[ i ] = new LinkedList<>( );
18     }
19
20     /**
21      * Make the hash table logically empty.
22      */
23     public void makeEmpty( )
24     {
25         for( int i = 0; i < theLists.length; i++ )
26             theLists[ i ].clear( );
27         currentSize = 0;
28     }
```

Figure 5.9　Constructors and makeEmpty for separate chaining hash table

In the insertion routine, if the item to be inserted is already present, then we do nothing; otherwise, we place it in the list. The element can be placed anywhere in the list; using add is most convenient in our case.

Any scheme could be used besides linked lists to resolve the collisions; a binary search tree or even another hash table would work, but we expect that if the table is large and the hash function is good, all the lists should be short, so basic separate chaining makes no attempt to try anything complicated.

We define the load factor, λ, of a hash table to be the ratio of the number of elements in the hash table to the table size. In the example above, $\lambda = 1.0$. The average length of a list is λ. The effort required to perform a search is the constant time required to evaluate the hash function plus the time to traverse the list. In an unsuccessful search, the number of nodes to examine is λ on average. A successful search requires that about $1 + (\lambda/2)$ links be traversed. To see this, notice that the list that is being searched contains the one node that stores the match plus zero or more other nodes. The expected number of "other nodes"

```
1      /**
2       * Find an item in the hash table.
3       * @param x the item to search for.
4       * @return true if x is not found.
5       */
6      public boolean contains( AnyType x )
7      {
8          List<AnyType> whichList = theLists[ myhash( x ) ];
9          return whichList.contains( x );
10     }
11
12     /**
13      * Insert into the hash table. If the item is
14      * already present, then do nothing.
15      * @param x the item to insert.
16      */
17     public void insert( AnyType x )
18     {
19         List<AnyType> whichList = theLists[ myhash( x ) ];
20         if( !whichList.contains( x ) )
21         {
22             whichList.add( x );
23
24                 // Rehash; see Section 5.5
25             if( ++currentSize > theLists.length )
26                 rehash( );
27         }
28     }
29
30     /**
31      * Remove from the hash table.
32      * @param x the item to remove.
33      */
34     public void remove( AnyType x )
35     {
36         List<AnyType> whichList = theLists[ myhash( x ) ];
37         if( whichList.contains( x ) )
38         {
39             whichList.remove( x );
40             currentSize--;
41         }
42     }
```

Figure 5.10 contains, insert, and remove routines for separate chaining hash table

in a table of N elements and M lists is $(N - 1)/M = \lambda - 1/M$, which is essentially λ, since M is presumed to be large. On average, half the "other nodes" are searched, so combined with the matching node, we obtain an average search cost of $1 + \lambda/2$ nodes. This analysis shows that the table size is not really important, but the load factor is. The general rule for separate chaining hashing is to make the table size about as large as the number of elements expected (in other words, let $\lambda \approx 1$). In the code in Figure 5.10, if the load factor exceeds 1, we expand the table size by calling `rehash` at line 26. `rehash` is discussed in Section 5.5. It is also a good idea, as mentioned before, to keep the table size prime to ensure a good distribution.

5.4 Hash Tables Without Linked Lists

Separate chaining hashing has the disadvantage of using linked lists. This could slow the algorithm down a bit because of the time required to allocate new cells (especially in other languages), and also essentially requires the implementation of a second data structure. An alternative to resolving collisions with linked lists is to try alternative cells until an empty cell is found. More formally, cells $h_0(x), h_1(x), h_2(x), \ldots$ are tried in succession, where $h_i(x) = (hash(x) + f(i))$ mod *TableSize*, with $f(0) = 0$. The function, f, is the collision resolution strategy. Because all the data go inside the table, a bigger table is needed in such a scheme than for separate chaining hashing. Generally, the load factor should be below $\lambda = 0.5$ for a hash table that doesn't use separate chaining. We call such tables **probing hash tables**. We now look at three common collision resolution strategies.

5.4.1 Linear Probing

In linear probing, f is a linear function of i, typically $f(i) = i$. This amounts to trying cells sequentially (with wraparound) in search of an empty cell. Figure 5.11 shows the result of inserting keys {89, 18, 49, 58, 69} into a hash table using the same hash function as before and the collision resolution strategy, $f(i) = i$.

The first collision occurs when 49 is inserted; it is put in the next available spot, namely, spot 0, which is open. The key 58 collides with 18, 89, and then 49 before an empty cell is found three away. The collision for 69 is handled in a similar manner. As long as the table is big enough, a free cell can always be found, but the time to do so can get quite large. Worse, even if the table is relatively empty, blocks of occupied cells start forming. This effect, known as **primary clustering**, means that any key that hashes into the cluster will require several attempts to resolve the collision, and then it will add to the cluster.

Although we will not perform the calculations here, it can be shown that the expected number of probes using linear probing is roughly $\frac{1}{2}(1 + 1/(1 - \lambda)^2)$ for insertions and unsuccessful searches, and $\frac{1}{2}(1 + 1/(1 - \lambda))$ for successful searches. The calculations are somewhat involved. It is easy to see from the code that insertions and unsuccessful searches require the same number of probes. A moment's thought suggests that, on average, successful searches should take less time than unsuccessful searches.

The corresponding formulas, if clustering is not a problem, are fairly easy to derive. We will assume a very large table and that each probe is independent of the previous probes.

	Empty Table	After 89	After 18	After 49	After 58	After 69
0				49	49	49
1					58	58
2						69
3						
4						
5						
6						
7						
8			18	18	18	18
9		89	89	89	89	89

Figure 5.11   Hash table with linear probing, after each insertion

These assumptions are satisfied by a *random* collision resolution strategy and are reasonable unless λ is very close to 1. First, we derive the expected number of probes in an unsuccessful search. This is just the expected number of probes until we find an empty cell. Since the fraction of empty cells is $1 - \lambda$, the number of cells we expect to probe is $1/(1 - \lambda)$. The number of probes for a successful search is equal to the number of probes required when the particular element was inserted. When an element is inserted, it is done as a result of an unsuccessful search. Thus, we can use the cost of an unsuccessful search to compute the average cost of a successful search.

The caveat is that λ changes from 0 to its current value, so that earlier insertions are cheaper and should bring the average down. For instance, in the table in Figure 5.11, $\lambda = 0.5$, but the cost of accessing 18 is determined when 18 is inserted. At that point, $\lambda = 0.2$. Since 18 was inserted into a relatively empty table, accessing it should be easier than accessing a recently inserted element such as 69. We can estimate the average by using an integral to calculate the mean value of the insertion time, obtaining

$$I(\lambda) = \frac{1}{\lambda} \int_0^\lambda \frac{1}{1 - x} dx = \frac{1}{\lambda} \ln \frac{1}{1 - \lambda}$$

These formulas are clearly better than the corresponding formulas for linear probing. Clustering is not only a theoretical problem but actually occurs in real implementations. Figure 5.12 compares the performance of linear probing (dashed curves) with what would be expected from more random collision resolution. Successful searches are indicated by an S, and unsuccessful searches and insertions are marked with U and I, respectively.

If $\lambda = 0.75$, then the formula above indicates that 8.5 probes are expected for an insertion in linear probing. If $\lambda = 0.9$, then 50 probes are expected, which is unreasonable. This compares with 4 and 10 probes for the respective load factors if clustering were not a problem. We see from these formulas that linear probing can be a bad idea if the table is expected to be more than half full. If $\lambda = 0.5$, however, only 2.5 probes are required on average for insertion, and only 1.5 probes are required, on average, for a successful search.

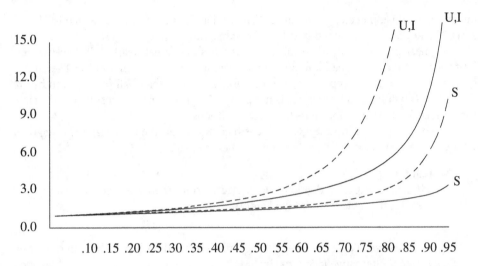

Figure 5.12 Number of probes plotted against load factor for linear probing (dashed) and random strategy (S is successful search, U is unsuccessful search, and I is insertion)

	Empty Table	After 89	After 18	After 49	After 58	After 69
0				49	49	49
1						
2					58	58
3						69
4						
5						
6						
7						
8			18	18	18	18
9		89	89	89	89	89

Figure 5.13 Hash table with quadratic probing, after each insertion

5.4.2 Quadratic Probing

Quadratic probing is a collision resolution method that eliminates the primary clustering problem of linear probing. Quadratic probing is what you would expect—the collision function is quadratic. The popular choice is $f(i) = i^2$. Figure 5.13 shows the resulting hash table with this collision function on the same input used in the linear probing example.

When 49 collides with 89, the next position attempted is one cell away. This cell is empty, so 49 is placed there. Next 58 collides at position 8. Then the cell one away is tried, but another collision occurs. A vacant cell is found at the next cell tried, which is $2^2 = 4$ away. The key 58 is thus placed in cell 2. The same thing happens for 69.

For linear probing it is a bad idea to let the hash table get nearly full, because performance degrades. For quadratic probing, the situation is even more drastic: There is no guarantee of finding an empty cell once the table gets more than half full, or even before the table gets half full if the table size is not prime. This is because at most half of the table can be used as alternative locations to resolve collisions.

Indeed, we prove now that if the table is half empty and the table size is prime, then we are always guaranteed to be able to insert a new element.

Theorem 5.1.

If quadratic probing is used, and the table size is prime, then a new element can always be inserted if the table is at least half empty.

Proof.

Let the table size, *TableSize*, be an (odd) prime greater than 3. We show that the first $\lceil TableSize/2 \rceil$ alternative locations (including the initial location $h_0(x)$) are all distinct. Two of these locations are $h(x)+i^2$ (mod *TableSize*) and $h(x)+j^2$ (mod *TableSize*), where $0 \le i,j \le \lfloor TableSize/2 \rfloor$. Suppose, for the sake of contradiction, that these locations are the same, but $i \ne j$. Then

$$h(x) + i^2 = h(x) + j^2 \qquad (\text{mod } TableSize)$$
$$i^2 = j^2 \qquad (\text{mod } TableSize)$$
$$i^2 - j^2 = 0 \qquad (\text{mod } TableSize)$$
$$(i - j)(i + j) = 0 \qquad (\text{mod } TableSize)$$

Since *TableSize* is prime, it follows that either $(i - j)$ or $(i + j)$ is equal to 0 (mod *TableSize*). Since i and j are distinct, the first option is not possible. Since $0 \le i,j \le \lfloor TableSize/2 \rfloor$, the second option is also impossible. Thus, the first $\lceil TableSize/2 \rceil$ alternative locations are distinct. If at most $\lfloor TableSize/2 \rfloor$ positions are taken, then an empty spot can always be found.

If the table is even one more than half full, the insertion could fail (although this is extremely unlikely). Therefore, it is important to keep this in mind. It is also crucial that the table size be prime.[1] If the table size is not prime, the number of alternative locations can be severely reduced. As an example, if the table size were 16, then the only alternative locations would be at distances 1, 4, or 9 away.

Standard deletion cannot be performed in a probing hash table, because the cell might have caused a collision to go past it. For instance, if we remove 89, then virtually all the remaining `contains` operations will fail. Thus, probing hash tables require lazy deletion, although in this case there really is no laziness implied.

---

[1] If the table size is a prime of the form $4k + 3$, and the quadratic collision resolution strategy $F(i) = \pm i^2$ is used, then the entire table can be probed. The cost is a slightly more complicated routine.

The class skeleton required to implement probing hash tables is shown in Figure 5.14. Instead of an array of lists, we have an array of hash table entry cells, which are also shown in Figure 5.14. Each entry in the array of HashEntry references is either

1. null.
2. Not null, and the entry is *active* (isActive is true).
3. Not null, and the entry is *marked deleted* (isActive is false).

Constructing the table (Figure 5.15) consists of allocating space and then setting each HashEntry reference to null.

contains(x), shown in Figure 5.16 (on page 186), invokes private methods isActive and findPos. The private method findPos performs the collision resolution. We ensure in the insert routine that the hash table is at least twice as large as the number of elements in the table, so quadratic resolution will always work. In the implementation in Figure 5.16, elements that are marked as deleted count as being in the table. This can cause problems, because the table can get too full prematurely. We shall discuss this item presently.

Lines 25 through 28 represent the fast way of doing quadratic resolution. From the definition of the quadratic resolution function, $f(i) = f(i - 1) + 2i - 1$, so the next cell to try is a distance from the previous cell tried and this distance increases by 2 on successive probes. If the new location is past the array, it can be put back in range by subtracting *TableSize*. This is faster than the obvious method, because it avoids the multiplication and division that seem to be required. An important warning: The order of testing at lines 22 and 23 is important. Don't switch it!

The final routine is insertion. As with separate chaining hashing, we do nothing if x is already present. It is a simple modification to do something else. Otherwise, we place it at the spot suggested by the findPos routine. The code is shown in Figure 5.17 (on page 187). If the load factor exceeds 0.5, the table is full and we enlarge the hash table. This is called rehashing, and is discussed in Section 5.5.

Although quadratic probing eliminates primary clustering, elements that hash to the same position will probe the same alternative cells. This is known as **secondary clustering.** Secondary clustering is a slight theoretical blemish. Simulation results suggest that it generally causes less than an extra half probe per search. The following technique eliminates this, but does so at the cost of computing an extra hash function.

5.4.3 Double Hashing

The last collision resolution method we will examine is **double hashing.** For double hashing, one popular choice is $f(i) = i \cdot hash_2(x)$. This formula says that we apply a second hash function to x and probe at a distance $hash_2(x), 2hash_2(x), \ldots$, and so on. A poor choice of $hash_2(x)$ would be disastrous. For instance, the obvious choice $hash_2(x) = x \bmod 9$ would not help if 99 were inserted into the input in the previous examples. Thus, the function must never evaluate to zero. It is also important to make sure all cells can be probed (this is not possible in the example below, because the table size is not prime). A function such as $hash_2(x) = R - (x \bmod R)$, with R a prime smaller than *TableSize*, will work well. If we choose $R = 7$, then Figure 5.18 shows the results of inserting the same keys as before.

```
1    public class QuadraticProbingHashTable<AnyType>
2    {
3        public QuadraticProbingHashTable( )
4          { /* Figure 5.15 */ }
5        public QuadraticProbingHashTable( int size )
6          { /* Figure 5.15 */ }
7        public void makeEmpty( )
8          { /* Figure 5.15 */ }
9
10       public boolean contains( AnyType x )
11         { /* Figure 5.16 */ }
12       public void insert( AnyType x )
13         { /* Figure 5.17 */ }
14       public void remove( AnyType x )
15         { /* Figure 5.17 */ }
16
17       private static class HashEntry<AnyType>
18       {
19           public AnyType  element;  // the element
20           public boolean isActive;  // false if marked deleted
21
22           public HashEntry( AnyType e )
23             { this( e, true ); }
24
25           public HashEntry( AnyType e, boolean i )
26             { element  = e; isActive = i; }
27       }
28
29       private static final int DEFAULT_TABLE_SIZE = 11;
30
31       private HashEntry<AnyType> [ ] array; // The array of elements
32       private int currentSize;              // The number of occupied cells
33
34       private void allocateArray( int arraySize )
35         { /* Figure 5.15 */ }
36       private boolean isActive( int currentPos )
37         { /* Figure 5.16 */ }
38       private int findPos( AnyType x )
39         { /* Figure 5.16 */ }
40       private void rehash( )
41         { /* Figure 5.22 */ }
42
43       private int myhash( AnyType x )
44         { /* See online code */ }
45       private static int nextPrime( int n )
46         { /* See online code */ }
47       private static boolean isPrime( int n )
48         { /* See online code */ }
49   }
```

Figure 5.14 Class skeleton for hash tables using probing strategies, including the nested HashEntry class

```
1       /**
2        * Construct the hash table.
3        */
4       public QuadraticProbingHashTable( )
5       {
6           this( DEFAULT_TABLE_SIZE );
7       }
8
9       /**
10       * Construct the hash table.
11       * @param size the approximate initial size.
12       */
13      public QuadraticProbingHashTable( int size )
14      {
15          allocateArray( size );
16          makeEmpty( );
17      }
18
19      /**
20       * Make the hash table logically empty.
21       */
22      public void makeEmpty( )
23      {
24          currentSize = 0;
25          for( int i = 0; i < array.length; i++ )
26              array[ i ] = null;
27      }
28
29      /**
30       * Internal method to allocate array.
31       * @param arraySize the size of the array.
32       */
33      private void allocateArray( int arraySize )
34      {
35          array = new HashEntry[ nextPrime( arraySize ) ];
36      }
```

Figure 5.15   Routines to initialize hash table

The first collision occurs when 49 is inserted. $hash_2(49) = 7 - 0 = 7$, so 49 is inserted in position 6. $hash_2(58) = 7 - 2 = 5$, so 58 is inserted at location 3. Finally, 69 collides and is inserted at a distance $hash_2(69) = 7 - 6 = 1$ away. If we tried to insert 60 in position 0, we would have a collision. Since $hash_2(60) = 7 - 4 = 3$, we would then try positions 3, 6, 9, and then 2 until an empty spot is found. It is generally possible to find some bad case, but there are not too many here.

```
1     /**
2      * Find an item in the hash table.
3      * @param x the item to search for.
4      * @return the matching item.
5      */
6     public boolean contains( AnyType x )
7     {
8         int currentPos = findPos( x );
9         return isActive( currentPos );
10    }
11
12    /**
13     * Method that performs quadratic probing resolution in half-empty table.
14     * @param x the item to search for.
15     * @return the position where the search terminates.
16     */
17    private int findPos( AnyType x )
18    {
19        int offset = 1;
20        int currentPos = myhash( x );
21
22        while( array[ currentPos ] != null &&
23               !array[ currentPos ].element.equals( x ) )
24        {
25            currentPos += offset;  // Compute ith probe
26            offset += 2;
27            if( currentPos >= array.length )
28                currentPos -= array.length;
29        }
30
31        return currentPos;
32    }
33
34    /**
35     * Return true if currentPos exists and is active.
36     * @param currentPos the result of a call to findPos.
37     * @return true if currentPos is active.
38     */
39    private boolean isActive( int currentPos )
40    {
41        return array[ currentPos ] != null && array[ currentPos ].isActive;
42    }
```

Figure 5.16 contains routine (and private helpers) for hashing with quadratic probing

```
1        /**
2         * Insert into the hash table. If the item is
3         * already present, do nothing.
4         * @param x the item to insert.
5         */
6        public void insert( AnyType x )
7        {
8                // Insert x as active
9            int currentPos = findPos( x );
10           if( isActive( currentPos ) )
11               return;
12
13           array[ currentPos ] = new HashEntry<>( x, true );
14
15               // Rehash; see Section 5.5
16           if( ++currentSize > array.length / 2 )
17               rehash( );
18       }
19
20       /**
21        * Remove from the hash table.
22        * @param x the item to remove.
23        */
24       public void remove( AnyType x )
25       {
26           int currentPos = findPos( x );
27           if( isActive( currentPos ) )
28               array[ currentPos ].isActive = false;
29       }
```

Figure 5.17 insert routine for hash tables with quadratic probing

As we have said before, the size of our sample hash table is not prime. We have done this for convenience in computing the hash function, but it is worth seeing why it is important to make sure the table size is prime when double hashing is used. If we attempt to insert 23 into the table, it would collide with 58. Since $hash_2(23) = 7 - 2 = 5$, and the table size is 10, we essentially have only one alternative location, and it is already taken. Thus, if the table size is not prime, it is possible to run out of alternative locations prematurely. However, if double hashing is correctly implemented, simulations imply that the expected number of probes is almost the same as for a random collision resolution strategy. This makes double hashing theoretically interesting. Quadratic probing, however, does not require the use of a second hash function and is thus likely to be simpler and faster in practice, especially for keys like strings whose hash functions are expensive to compute.

	Empty Table	After 89	After 18	After 49	After 58	After 69
0						69
1						
2						
3					58	58
4						
5						
6				49	49	49
7						
8			18	18	18	18
9		89	89	89	89	89

Figure 5.18   Hash table with double hashing, after each insertion

5.5 Rehashing

If the table gets too full, the running time for the operations will start taking too long and insertions might fail for open addressing hashing with quadratic resolution. This can happen if there are too many removals intermixed with insertions. A solution, then, is to build another table that is about twice as big (with an associated new hash function) and scan down the entire original hash table, computing the new hash value for each (nondeleted) element and inserting it in the new table.

As an example, suppose the elements 13, 15, 24, and 6 are inserted into a linear probing hash table of size 7. The hash function is $h(x) = x \bmod 7$. Suppose linear probing is used to resolve collisions. The resulting hash table appears in Figure 5.19.

If 23 is inserted into the table, the resulting table in Figure 5.20 will be over 70 percent full. Because the table is so full, a new table is created. The size of this table is 17, because this is the first prime that is twice as large as the old table size. The new hash function is then $h(x) = x \bmod 17$. The old table is scanned, and elements 6, 15, 23, 24, and 13 are inserted into the new table. The resulting table appears in Figure 5.21.

This entire operation is called **rehashing**. This is obviously a very expensive operation; the running time is $O(N)$, since there are N elements to rehash and the table size is roughly $2N$, but it is actually not all that bad, because it happens very infrequently. In particular, there must have been $N/2$ insertions prior to the last rehash, so it essentially adds a constant cost to each insertion.[2] If this data structure is part of the program, the effect is not noticeable. On the other hand, if the hashing is performed as part of an interactive system, then the unfortunate user whose insertion caused a rehash could see a slowdown.

Rehashing can be implemented in several ways with quadratic probing. One alternative is to rehash as soon as the table is half full. The other extreme is to rehash only when an

---

[2] This is why the new table is made twice as large as the old table.

0	6
1	15
2	
3	24
4	
5	
6	13

Figure 5.19 Hash table with linear probing with input 13, 15, 6, 24

0	6
1	15
2	23
3	24
4	
5	
6	13

Figure 5.20 Hash table with linear probing after 23 is inserted

insertion fails. A third, middle-of-the-road strategy is to rehash when the table reaches a certain load factor. Since performance does degrade as the load factor increases, the third strategy, implemented with a good cutoff, could be best.

Rehashing for separate chaining hash tables is similar. Figure 5.22 shows that rehashing is simple to implement, and provides an implementation for separate chaining rehashing also.

5.6 Hash Tables in the Standard Library

The Standard Library includes hash table implementations of Set and Map, namely HashSet and HashMap. The items in the HashSet (or the keys in the HashMap) must provide an equals and hashCode method, as described earlier in Section 5.3. The HashSet and HashMap are currently implemented using separate chaining hashing.

These classes can be used if it is not important for the entries to be viewable in sorted order. For instance, in the word-changing example in Section 4.8, there were three maps:

0	
1	
2	
3	
4	
5	
6	6
7	23
8	24
9	
10	
11	
12	
13	13
14	
15	15
16	

Figure 5.21   Linear probing hash table after rehashing

1. A map in which the key is a *word length*, and the value is a collection of all words of that word length.

2. A map in which the key is a *representative*, and the value is a collection of all words with that representative.

3. A map in which the key is a *word*, and the value is a collection of all words that differ in only one character from that word.

Because the order in which word lengths are processed does not matter, the first map can be a HashMap. Because the representatives are not even needed after the second map is built, the second map can be a HashMap. The third map can also be a HashMap, unless we want printHighChangeables to alphabetically list the subset of words that can be changed into a large number of other words.

The performance of a HashMap can often be superior to a TreeMap, but it is hard to know for sure without writing the code both ways. Thus, in cases where either a HashMap or TreeMap is acceptable, it is preferable to declare variables using the interface type Map and then change the instantiation from a TreeMap to a HashMap, and perform timing tests.

```
1     /**
2      * Rehashing for quadratic probing hash table.
3      */
4     private void rehash( )
5     {
6         HashEntry<AnyType> [ ] oldArray = array;
7
8             // Create a new double-sized, empty table
9         allocateArray( nextPrime( 2 * oldArray.length ) );
10        currentSize = 0;
11
12            // Copy table over
13        for( int i = 0; i < oldArray.length; i++ )
14            if( oldArray[ i ] != null && oldArray[ i ].isActive )
15                insert( oldArray[ i ].element );
16    }
17
18    /**
19     * Rehashing for separate chaining hash table.
20     */
21    private void rehash( )
22    {
23        List<AnyType> [ ]  oldLists = theLists;
24
25            // Create new double-sized, empty table
26        theLists = new List[ nextPrime( 2 * theLists.length ) ];
27        for( int j = 0; j < theLists.length; j++ )
28            theLists[ j ] = new LinkedList<>( );
29
30            // Copy table over
31        currentSize = 0;
32        for( int i = 0; i < oldLists.length; i++ )
33            for( AnyType item : oldLists[ i ] )
34                insert( item );
35    }
```

Figure 5.22 Rehashing for both separate chaining and probing hash tables

In Java, library types that can be reasonably inserted into a HashSet or as keys into a HashMap already have equals and hashCode defined. In particular the String class has a hashCode that is essentially the code in Figure 5.4 with lines 14–16 removed, and 37 replaced with 31. Because the expensive part of the hash table operations is computing the hashCode, the hashCode method in the String class contains an important optimization: Each String object stores internally the value of its hashCode. Initially it is 0, but if hashCode is invoked, the value is remembered. Thus if hashCode is computed on the same String

```
1      public final class String
2      {
3          public int hashCode( )
4          {
5              if( hash != 0 )
6                  return hash;
7
8              for( int i = 0; i < length( ); i++ )
9                  hash = hash * 31 + (int) charAt( i );
10             return hash;
11         }
12
13         private int hash = 0;
14     }
```

Figure 5.23   Excerpt of `String` class `hashCode`

object a second time, we can avoid the expensive recomputation. This technique is called **caching the hash code**, and represents another classic time-space tradeoff. Figure 5.23 shows an implementation of the `String` class that caches the hash code.

Caching the hash code works only because `String`s are immutable: If the `String` were allowed to change, it would invalidate the `hashCode`, and the `hashCode` would have to be reset back to 0. Although two `String` objects with the same state must have their hash codes computed independently, there are many situations in which the same `String` object keeps having its hash code queried. One situation where caching the hash code helps occurs during rehashing, because all the `String`s involved in the rehashing have already had their hash codes cached. On the other hand, caching the hash code does not help in the representative map for the word changing example. Each of the representatives is a different `String` computed by removing a character from a larger `String`, and thus each individual `String` has to have its hash code computed separately. However, in the third map, caching the hash code does help, because the keys are only `String`s that were stored in the original array of `String`s.

5.7   Hash Tables with Worst-Case $O(1)$ Access

The hash tables that we have examined so far all have the property that with reasonable load factors, and appropriate hash functions, we can expect $O(1)$ cost on average for insertions, removes, and searching. But what is the expected worst case for a search assuming a reasonably well-behaved hash function?

For separate chaining, assuming a load factor of 1, this is one version of the classic **balls and bins problem**: Given N balls placed randomly (uniformly) in N bins, what is the expected number of balls in the most occupied bin? The answer is well known to be $\Theta(\log N / \log \log N)$, meaning that on average, we expect some queries to take nearly logarithmic time. Similar types of bounds are observed (or provable) for the length of the longest expected probe sequence in a probing hash table.

We would like to obtain $O(1)$ worst-case cost. In some applications, such as hardware implementations of lookup tables for routers and memory caches, it is especially important that the search have a definite (i.e., constant) amount of completion time. Let us assume that N is known in advance, so no rehashing is needed. If we are allowed to rearrange items as they are inserted, then $O(1)$ worst-case cost is achievable for searches.

In the remainder of this section we describe the earliest solution to this problem, namely perfect hashing, and then two more recent approaches that appear to offer promising alternatives to the classic hashing schemes that have been prevalent for many years.

5.7.1   Perfect Hashing

Suppose, for purposes of simplification, that all N items are known in advance. If a separate chaining implementation could guarantee that each list had at most a constant number of items, we would be done. We know that as we make more lists, the lists will on average be shorter, so theoretically if we have enough lists, then with a reasonably high probability we might expect to have no collisions at all!

But there are two fundamental problems with this approach: First, the number of lists might be unreasonably large; second, even with lots of lists, we might still get unlucky.

The second problem is relatively easy to address in principle. Suppose we choose the number of lists to be M (i.e., *TableSize* is M), which is sufficiently large to guarantee that with probability at least $\frac{1}{2}$, there will be no collisions. Then if a collision is detected, we simply clear out the table and try again using a different hash function that is independent of the first. If we still get a collision, we try a third hash function, and so on. The expected number of trials will be at most 2 (since the success probability is $\frac{1}{2}$), and this is all folded into the insertion cost. Section 5.8 discusses the crucial issue of how to produce additional hash functions.

So we are left with determining how large M, the number of lists, needs to be. Unfortunately, M needs to be quite large; specifically $M = \Omega(N^2)$. However, if $M = N^2$, we can show that the table is collision free with probability at least $\frac{1}{2}$, and this result can be used to make a workable modification to our basic approach.

Theorem 5.2.
If N balls are placed into $M = N^2$ bins, the probability that no bin has more than one ball is less than $\frac{1}{2}$.

Proof.
If a pair (i, j) of balls are placed in the same bin, we call that a collision. Let $C_{i,j}$ be the expected number of collisions produced by any two balls (i, j). Clearly the probability that any two specified balls collide is $1/M$, and thus $C_{i,j}$ is $1/M$, since the number of collisions that involve the pair (i, j) is either 0 or 1. Thus the expected number of collisions in the entire table is $\sum_{(i,j),\ i<j} C_{i,j}$. Since there are $N(N-1)/2$ pairs, this sum is $N(N-1)/(2M) = N(N-1)/(2N^2) < \frac{1}{2}$. Since the expected number of collisions is below $\frac{1}{2}$, the probability that there is even one collision must also be below $\frac{1}{2}$.

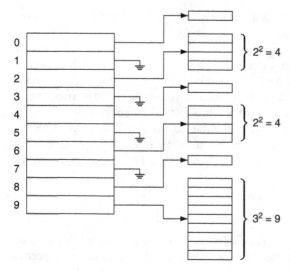

Figure 5.24   Perfect hashing table using secondary hash tables

Of course, using N^2 lists is impractical. However, the preceding analysis suggests the following alternative: Use only N bins, but resolve the collisions in each bin by using hash tables instead of linked lists. The idea is that because the bins are expected to have only a few items each, the hash table that is used for each bin can be quadratic in the bin size. Figure 5.24 shows the basic structure. Here, the primary hash table has ten bins. Bins 1, 3, 5, and 7 are all empty. Bins 0, 4, and 8 have one item, so they are resolved by a secondary hash table with one position. Bins 2 and 6 have two items, so they will be resolved into a secondary hash table with four (2^2) positions. And bin 9 has three items, so it is resolved into a secondary hash table with nine (3^2) positions.

As with the original idea, each secondary hash table will be constructed using a different hash function until it is collision free. The primary hash table can also be constructed several times if the number of collisions that are produced is higher than required. This scheme is known as **perfect hashing**. All that remains to be shown is that the total size of the secondary hash tables is indeed expected to be linear.

Theorem 5.3.
If N items are placed into a primary hash table containing N bins, then the total size of the secondary hash tables has expected value at most $2N$.

Proof.
Using the same logic as in the proof of Theorem 5.2, the expected number of pairwise collisions is at most $N(N - 1)/2N$, or $(N - 1)/2$. Let b_i be the number of items that hash to position i in the primary hash table; observe that b_i^2 space is used for this cell in the secondary hash table, and that this accounts for $b_i (b_i - 1)/2$ pairwise collisions, which we will call c_i. Thus the amount of space used for the ith secondary hash table is $2c_i + b_i$. The total space is then $2 \sum c_i + \sum b_i$. The total number of collisions is

$(N-1)/2$ (from the first sentence of this proof); the total number of items is of course N, so we obtain a total secondary space requirement of $2(N-1)/2 + N < 2N$.

Thus the probability that the total secondary space requirement is more than $4N$ is at most $\frac{1}{2}$ (since, otherwise the expected value would be higher than $2N$), so we can keep choosing hash functions for the primary table until we generate the appropriate secondary space requirement. Once that is done, each secondary hash table will itself require only an average of two trials to be collision free. After the tables are built, any lookup can be done in two probes.

Perfect hashing works if the items are all known in advance. There are dynamic schemes that allow insertions and deletions (**dynamic perfect hashing**), but instead we will investigate two newer alternatives that are relatively easy to code and appear to be competitive in practice with the classic hashing algorithms.

5.7.2 Cuckoo Hashing

From our previous discussion, we know that in the balls and bins problem, if N items are randomly tossed into N bins, the size of the largest bin is expected to be $\Theta(\log N / \log \log N)$. Since this bound has been known for a long time, and the problem has been well studied by mathematicians, it was surprising when in the mid 1990s, it was shown that if, at each toss, two bins were randomly chosen and the item was tossed into the more empty bin (at the time), then the size of the largest bin would only be $\Theta(\log \log N)$, a significantly lower number. Quickly, a host of potential algorithms and data structures arose out of this new concept of the "power of two choices."

One of the ideas is **cuckoo hashing**. In cuckoo hashing, suppose we have N items. We maintain two tables each more than half empty, and we have two independent has functions that can assign each item to a position in each table. Cuckoo hashing maintains the invariant that an item is always stored in one of these two locations.

As an example, Figure 5.25 shows a potential cuckoo hash table for six items, with two tables of size 5 (These tables are too small, but serve well as an example). Based on the randomly chosen hash functions, item A can be at either position 0 in Table 1, or position 2 in Table 2. Item F can be at either position 3 in Table 1, or position 4 in Table 2, and so on. Immediately, this implies that a search in a cuckoo hash table requires at most two

Table 1			Table 2				
0	B		0	D		A: 0, 2	
1	C		1			B: 0, 0	
2			2	A		C: 1, 4	
3	E		3			D: 1, 0	
4			4	F		E: 3, 2	
						F: 3, 4	

Figure 5.25 Potential cuckoo hash table. Hash functions are shown on the right. For these six items, there are only three valid positions in Table 1 and three valid positions in Table 2, so it is not clear that this arrangement can easily be found

table accesses, and a remove is trivial, once the item is located (lazy deletion is not needed now!).

But there is an important detail: How is the table built? For instance, in Figure 5.25, there are only three available locations in the first table for the six items, and there are only three available locations in the second table for the six items. So there are only six available locations for these six items, and thus we must find an ideal matching of slots for our six items. Clearly if there were a seventh item *G* with locations 1 for Table 1 and 2 for Table 2, it could not be inserted into the table by any algorithm (the seven items would be competing for six table locations). One could argue that this means that the table would simply be too loaded (*G* would yield a 0.70 load factor), but at the same time, if the table had thousands of items, and were lightly loaded, but we had *A, B, C, D, E, F, G* with these hash positions, it would still be impossible to insert all seven of those items. So it is not at all obvious that this scheme can be made to work. The answer in this situation would be to pick another hash function, and this can be fine as long as it is unlikely that this situation occurs.

The cuckoo hashing algorithm itself is simple: To insert a new item *x*, first make sure it is not already there. We can then use the first hash function and if the (first) table location is empty, the item can be placed. So Figure 5.26 shows the result of inserting *A* into an empty hash table.

Suppose now we want to insert *B*, which has hash locations 0 in Table 1 and 0 in Table 2. For the remainder of the algorithm we will use (h_1, h_2) to specify the two locations, so *B*'s locations are given by (0, 0). Table 1 is already occupied in position 0. At this point there are two options: One is to look in Table 2. The problem is that position 0 in Table 2 could also be occupied. It happens that in this case it is not, but the algorithm that the standard cuckoo hash table uses does not bother to look. Instead, it preemptively places the new item *B* in Table 1. In order to do so, it must displace *A*, so *A* moves to Table 2, using its Table 2 hash location, which is position 2. The result is shown in Figure 5.27. It is easy to insert *C*, and this is shown in Figure 5.28.

Next we want to insert *D*, with hash locations (1, 0). But the Table 1 location (position 1) is already taken. Note also, that the Table 2 location is not already taken, but we don't look there. Instead, we have *D* replace *C*, and then *C* goes into Table 2 at position 4, as suggested by its second hash function. The resulting tables are shown in Figure 5.29. After this is done, *E* can be easily inserted. So far, so good, but can we now insert *F*? Figures 5.30 to 5.33 show that this algorithm successfully inserts *F*, by displacing *E*, then *A*, and then *B*.

Table 1		Table 2	
0	A	0	
1		1	
2		2	
3		3	
4		4	

A: 0, 2

Figure 5.26   Cuckoo hash table after insertion of *A*

Table 1		Table 2	
0	B	0	
1		1	
2		2	A
3		3	
4		4	

A: 0, 2
B: 0, 0

Figure 5.27 Cuckoo hash table after insertion of B

Table 1		Table 2	
0	B	0	
1	C	1	
2		2	A
3		3	
4		4	

A: 0, 2
B: 0, 0
C: 1, 4

Figure 5.28 Cuckoo hash table after insertion of C

Table 1		Table 2	
0	B	0	
1	D	1	
2		2	A
3	E	3	
4		4	C

A: 0, 2
B: 0, 0
C: 1, 4
D: 1, 0
E: 3, 2

Figure 5.29 Cuckoo hash table after insertion of D

Table 1		Table 2	
0	B	0	
1	D	1	
2		2	A
3	F	3	
4		4	C

A: 0, 2
B: 0, 0
C: 1, 4
D: 1, 0
E: 3, 2
F: 3, 4

Figure 5.30 Cuckoo hash table starting the insertion of F into the table in Figure 5.29. First, F displaces E

Clearly as we mentioned before, we cannot successfully insert G, with hash locations (1, 2). If we were to try, we would displace D, then B, then A, E, F, and C, and then C would try to go back into Table 1, position 1, displacing G which was placed there at the start. This would get us to Figure 5.34. So now G would try its alternate in Table 2 (location 2) and then displace A, which would displace B, which would displace D, which

Table 1	
0	B
1	D
2	
3	F
4	

Table 2	
0	
1	
2	E
3	
4	C

A: 0, 2
B: 0, 0
C: 1, 4
D: 1, 0
E: 3, 2
F: 3, 4

Figure 5.31 Continuing the insertion of F into the table in Figure 5.29. Next, E displaces A

Table 1	
0	A
1	D
2	
3	F
4	

Table 2	
0	
1	
2	E
3	
4	C

A: 0, 2
B: 0, 0
C: 1, 4
D: 1, 0
E: 3, 2
F: 3, 4

Figure 5.32 Continuing the insertion of F into the table in Figure 5.29. Next, A displaces B

Table 1	
0	A
1	D
2	
3	F
4	

Table 2	
0	B
1	
2	E
3	
4	C

A: 0, 2
B: 0, 0
C: 1, 4
D: 1, 0
E: 3, 2
F: 3, 4

Figure 5.33 Completing the insertion of F into the table in Figure 5.29. Miraculously (?), B finds an empty position in Table 2

would displace C, which would displace F, which would displace E, which would now displace G from position 2. At this point, G would be in a cycle.

The central issue then concerns questions such as what is the probability of there being a cycle that prevents the insertion from completing, and what is the expected number of displacements required for a successful insertion? Fortunately, if the table's load factor is below 0.5, an analysis shows that the probability of a cycle is very low, that the expected number of displacements is a small constant, and that it is extremely unlikely that a successful insertion would require more than $O(\log N)$ displacements. As such, we can simply rebuild the tables with new hash functions after a certain number of displacements are detected. More precisely, the probability that a single insertion would require a new set of hash functions can be made to be $O(1/N^2)$; the new hash functions themselves generate N more insertions to rebuild the table, but even so, this means the rebuilding cost is minimal.

Table 1	
0	B
1	C
2	
3	E
4	

Table 2	
0	D
1	
2	A
3	
4	F

A: 0, 2
B: 0, 0
C: 1, 4
D: 1, 0
E: 3, 2
F: 3, 4
G: 1, 2

Figure 5.34 Inserting G into the table in Figure 5.33. G displaces D, which displaces B, which displaces A, which displaces E, which displaces F, which displaces C, which displaces G. It is not yet hopeless since when G is displaced, we would now try the other hash table, at position 2. However, while that could be successful in general, in this case there is a cycle and the insertion will not terminate

	1 item per cell	2 items per cell	4 items per cell
2 hash functions	0.49	0.86	0.93
3 hash functions	0.91	0.97	0.98
4 hash functions	0.97	0.99	0.999

Figure 5.35 Maximum load factors for cuckoo hashing variations

However, if the table's load factor is at 0.5 or higher, then the probability of a cycle becomes drastically higher, and this scheme is unlikely to work well at all.

After the publication of cuckoo hashing, numerous extensions were proposed. For instance, instead of two tables, we can use a higher number of tables, such as 3 or 4. While this increases the cost of a lookup, it also drastically increases the theoretical space utilization. In some applications the lookups through separate hash functions can be done in parallel and thus cost little to no additional time. Another extension is to allow each table to store multiple keys; again this can increase space utilization and also make it easier to do insertions and can be more cache-friendly. Various combinations are possible, as shown in Figure 5.35. And finally, often cuckoo hash tables are implemented as one giant table with two (or more) hash functions that probe the entire table, and some variations attempt to place an item in the second hash table immediately if there is an available spot, rather than starting a sequence of displacements.

Cuckoo Hash Table Implementation

Implementing cuckoo hashing requires a collection of hash functions; simply using `hashCode` to generate the collection of hash functions makes no sense, since any `hashCode` collisions will result in collisions in all the hash functions. Figure 5.36 shows a simple interface that can be used to send families of hash functions to the cuckoo hash table.

Figure 5.37 provides the class skeleton for cuckoo hashing. We will code a variant that will allow an arbitrary number of hash functions (specified by the `HashFamily` object that constructs the hash table) which uses a single array that is addressed by all the

```
1    public interface HashFamily<AnyType>
2    {
3        int hash( AnyType x, int which );
4        int getNumberOfFunctions( );
5        void generateNewFunctions( );
6    }
```

Figure 5.36   Generic HashFamily interface for cuckoo hashing

hash functions. Thus our implementation differs from the classic notion of two separately addressable hash tables. We can implement the classic version by making relatively minor changes to the code; however, this version provided in this section seems to perform better in tests using simple hash functions.

In Figure 5.37, we specify that the maximum load for the table is 0.4; if the load factor of the table is about to exceed this limit, an automatic table expansion is performed. We also define ALLOWED_REHASHES, which specifies how many rehashes we will perform if evictions take too long. In theory, ALLOWED_REHASHES can be infinite, since we expect only a small constant number of rehashes are needed; in practice, depending on several factors such as the number of hash functions, the quality of the hash functions, and the load factor, the rehashes could significantly slow things down, and it might be worthwhile to expand the table, even though this will cost space. The data representation for the cuckoo hash table is straightforward: We store a simple array, the current size, and the collections of hash functions, represented in a HashFamily instance. We also maintain the number of hash functions, even though that is always obtainable from the HashFamily instance.

Figure 5.38 shows the constructor and doClear methods, and these are straightforward. Figure 5.39 shows a pair of private methods. The first, myHash, is used to select the appropriate hash function and then scale it into a valid array index. The second, findPos, consults all the hash functions to return the index containing item x, or −1 if x is not found. findPos is then used by contains and remove in Figures 5.40 and 5.41, respectively, and we can see that those methods are easy to implement.

The difficult routine is insertion. In Figure 5.42, we can see that the basic plan is to check to see if the item is already present, returning if so. Otherwise, we check to see if the table is fully loaded, and if so, we expand it. Finally we call a helper routine to do all the dirty work.

The helper routine for insertion is shown in Figure 5.43. We declare a variable rehashes to keep track of how many attempts have been made to rehash in this insertion. Our insertion routine is mutually recursive: If needed, insert eventually calls rehash, which eventually calls back into insert. Thus rehashes is declared in an outer scope for code simplicity.

Our basic logic is different from the classic scheme. We have already tested that the item to insert is not already present. At lines 15–25, we check to see if any of the valid

```
1    public class CuckooHashTable<AnyType>
2    {
3        public CuckooHashTable( HashFamily<? super AnyType> hf )
4          { /* Figure 5.38 */ }
5        public CuckooHashTable( HashFamily<? super AnyType> hf, int size );
6          { /* Figure 5.38 */ }
7
8        public void makeEmpty( )
9          { doClear( ); }
10
11       public boolean contains( AnyType x )
12         { /* Figure 5.40 */ }
13
14       private int myhash( AnyType x, int which )
15         { /* Figure 5.39 */ }
16
17       private int findPos( AnyType x )
18         { /* Figure 5.39 */ }
19
20       public boolean remove( AnyType x )
21         { /* Figure 5.41 */ }
22
23       public boolean insert( AnyType x )
24         { /* Figure 5.42 */ }
25
26       private void expand( )
27         { /* Figure 5.44 */ }
28
29       private void rehash( )
30         { /* Figure 5.44 */ }
31
32       private void doClear( )
33         { /* Figure 5.38 */ }
34
35       private void allocateArray( int arraySize )
36         { array = (AnyType[]) new Object[ arraySize ]; }
37
38       private static final double MAX_LOAD = 0.4;
39       private static final int ALLOWED_REHASHES = 1;
40       private static final int DEFAULT_TABLE_SIZE = 101;
41
42       private final HashFamily<? super AnyType> hashFunctions;
43       private final int numHashFunctions;
44       private AnyType [ ] array;
45       private int currentSize;
46   }
```

Figure 5.37   Class skeleton for cuckoo hashing

```
1      /**
2       * Construct the hash table.
3       * @param hf the hash family
4       */
5      public CuckooHashTable( HashFamily<? super AnyType> hf )
6      {
7          this( hf, DEFAULT_TABLE_SIZE );
8      }
9
10     /**
11      * Construct the hash table.
12      * @param hf the hash family
13      * @param size the approximate initial size.
14      */
15     public CuckooHashTable( HashFamily<? super AnyType> hf, int size )
16     {
17         allocateArray( nextPrime( size ) );
18         doClear( );
19         hashFunctions = hf;
20         numHashFunctions = hf.getNumberOfFunctions( );
21     }
22
23     private void doClear( )
24     {
25         currentSize = 0;
26         for( int i = 0; i < array.length; i++ )
27             array[ i ] = null;
28     }
```

Figure 5.38 Routines to initialize the cuckoo hash table

positions are empty; if so, we place our item in the first available position and we are done. Otherwise, we evict one of the existing items. However, there are some tricky issues:

- Evicting the first item did not perform well in experiments.
- Evicting the last item did not perform well in experiments.
- Evicting the items in sequence (i.e., the first eviction uses hash function 0, the next uses hash function 1, etc.) did not perform well in experiments.
- Evicting the item purely randomly did not perform well in experiments: In particular, with only two hash functions, it tended to create cycles.

To alleviate the last problem, we maintain the last position that was evicted and if our random item was the last evicted item, we select a new random item. This will loop forever if used with two hash functions, and both hash functions happen to probe to the same

```
1      /**
2       * Compute the hash code for x using specified hash function
3       * @param x the item
4       * @param which the hash function
5       * @return the hash code
6       */
7      private int myhash( AnyType x, int which )
8      {
9          int hashVal = hashFunctions.hash( x, which );
10
11         hashVal %= array.length;
12         if( hashVal < 0 )
13             hashVal += array.length;
14
15         return hashVal;
16     }
17
18     /**
19      * Method that searches all hash function places.
20      * @param x the item to search for.
21      * @return the position where the search terminates, or -1 if not found.
22      */
23     private int findPos( AnyType x )
24     {
25         for( int i = 0; i < numHashFunctions; i++ )
26         {
27             int pos = myhash( x, i );
28             if( array[ pos ] != null && array[ pos ].equals( x ) )
29                 return pos;
30         }
31
32         return -1;
33     }
```

Figure 5.39 Routines to find the location of an item in the cuckoo hash table and to compute the hash code for a given table

location, and that location was a prior eviction, so we limit the loop to five iterations (deliberately using an odd number).

The code for expand and rehash is shown in Figure 5.44. expand creates a larger array but keeps the same hash functions. The zero-parameter rehash leaves the array size unchanged but creates a new array that is populated with newly chosen hash functions.

Finally, Figure 5.45 shows the StringHashFamily class that provides a set of simple hash functions for strings. These hash functions replace the constant 37 in Figure 5.4 with randomly chosen numbers (not necessarily prime).

```
1     /**
2      * Find an item in the hash table.
3      * @param x the item to search for.
4      * @return true if item is found.
5      */
6     public boolean contains( AnyType x )
7     {
8         return findPos( x ) != -1;
9     }
```

Figure 5.40   Routine to search a cuckoo hash table

```
1     /**
2      * Remove from the hash table.
3      * @param x the item to remove.
4      * @return true if item was found and removed
5      */
6     public boolean remove( AnyType x )
7     {
8         int pos = findPos( x );
9
10        if( pos != -1 )
11        {
12            array[ pos ] = null;
13            currentSize--;
14        }
15
16        return pos != -1;
17    }
```

Figure 5.41   Routine to remove from a cuckoo hash table

The benefits of cuckoo hashing include the worst-case constant lookup and deletion times, the avoidance of lazy deletion and extra data, and the potential for parallelism. However, cuckoo hashing is extremely sensitive to the choice of hash functions; the inventors of the cuckoo hash table reported that many of the standard hash functions that they attempted performed poorly in tests. Furthermore, although the insertion time is expected to be constant time as long as the load factor is below $\frac{1}{2}$, the bound that has been shown for the expected insertion cost for classic cuckoo hashing with two separate tables (both with load factor λ) is roughly $1/(1 - (4\lambda^2)^{1/3})$, which deteriorates rapidly as the load factor gets close to $\frac{1}{2}$ (the formula itself makes no sense when λ equals or exceeds $\frac{1}{2}$). Using lower load factors or more than two hash functions seems like a reasonable alternative.

```
1     /**
2      * Insert into the hash table. If the item is
3      * already present, return false.
4      * @param x the item to insert.
5      */
6     public boolean insert( AnyType x )
7     {
8         if( contains( x ) )
9             return false;
10
11        if( currentSize >= array.length * MAX_LOAD )
12            expand( );
13
14        return insertHelper1( x );
15    }
```

Figure 5.42 Public insert routine for cuckoo hashing

5.7.3 Hopscotch Hashing

Hopscotch hashing is a new algorithm that tries to improve on the classic linear probing algorithm. Recall that in linear probing, cells are tried in sequential order, starting from the hash location. Because of primary and secondary clustering, this sequence can be long on average as the table gets loaded, and thus many improvements such as quadratic probing, double hashing, and so forth, have been proposed to reduce the number of collisions. However, on some modern architectures, the locality produced by probing adjacent cells is a more significant factor than the extra probes, and linear probing can still be practical or even a best choice.

The idea of hopscotch hashing is to bound the maximal length of the probe sequence by a predetermined constant that is optimized to the underlying computer's architecture. Doing so would give constant-time lookups in the worst case, and like cuckoo hashing, the lookup could be parallelized to simultaneously check the bounded set of possible locations.

If an insertion would place a new item too far from its hash location, then we efficiently go backward toward the hash location, evicting potential items. If we are careful, the evictions can be done quickly and guarantee that those evicted are not placed too far from their hash locations. The algorithm is deterministic in that given a hash function, either the items can be evicted or they can't. The latter case implies that the table is likely too crowded, and a rehash is in order; but this would happen only at extremely high load factors, exceeding 0.9. For a table with a load factor of $\frac{1}{2}$, the failure probability is almost zero (Exercise 5.23).

Let MAX_DIST be the chosen bound on the maximum probe sequence. This means that item x must be found somewhere in the MAX_DIST positions listed in $hash(x)$, $hash(x) + 1, \ldots,$ $hash(x) + (MAX\_DIST - 1)$. In order to efficiently process evictions, we maintain information that tells for each position x, whether the item in the alternate position is occupied by an element that hashes to position x.

```
1        private int rehashes = 0;
2        private Random r = new Random( );
3
4        private boolean insertHelper1( AnyType x )
5        {
6            final int COUNT_LIMIT = 100;
7
8            while( true )
9            {
10               int lastPos = -1;
11               int pos;
12
13               for( int count = 0; count < COUNT_LIMIT; count++ )
14               {
15                   for( int i = 0; i < numHashFunctions; i++ )
16                   {
17                       pos = myhash( x, i );
18
19                       if( array[ pos ] == null )
20                       {
21                           array[ pos ] = x;
22                           currentSize++;
23                           return true;
24                       }
25                   }
26
27                   // none of the spots are available. Evict out a random one
28                   int i = 0;
29                   do
30                   {
31                       pos = myhash( x, r.nextInt( numHashFunctions ) ) ;
32                   } while( pos == lastPos && i++ < 5 );
33
34                   AnyType tmp = array[ lastPos = pos ];
35                   array[ pos ] = x;
36                   x = tmp;
37               }
38
39               if( ++rehashes > ALLOWED_REHASHES )
40               {
41                   expand( );        // Make the table bigger
42                   rehashes = 0;     // Reset the # of rehashes
43               }
44               else
45                   rehash( );        // Same table size, new hash functions
46           }
47       }
```

Figure 5.43 Insertion routine for cuckoo hashing uses a different algorithm that chooses the item to evict randomly, attempting not to re-evict the last item. The table will attempt to select new hash functions (rehash) if there are too many evictions and will expand if there are too many rehashes

```
1      private void expand( )
2      {
3          rehash( (int) ( array.length / MAX_LOAD ) );
4      }
5
6      private void rehash( )
7      {
8          hashFunctions.generateNewFunctions( );
9          rehash( array.length );
10     }
11
12     private void rehash( int newLength )
13     {
14         AnyType [ ] oldArray = array;
15         allocateArray( nextPrime( newLength ) );
16         currentSize = 0;
17
18             // Copy table over
19         for( AnyType str : oldArray )
20             if( str != null )
21                 insert( str );
22     }
```

Figure 5.44   Rehashing and expanding code for cuckoo hash tables

As an example, Figure 5.46 shows a fairly crowded hopscotch hash table, using *MAX_DIST* = 4. The bit array for position 6 shows that only position 6 has an item (*C*) with hash value 6: Only the first bit of Hop[6] is set. Hop[7] has the first two bits set, indicating that positions 7 and 8 (*A* and *D*) are occupied with items whose hash value is 7. And Hop[8] has only the third bit set, indicating that the item in position 10 (*E*) has hash value 8. If *MAX_DIST* is no more than 32, the Hop array is essentially an array of 32-bit integers, so the additional space requirement is not substantial. If Hop[*pos*] contains all 1's for some *pos*, then an attempt to insert an item whose hash value is *pos* will clearly fail, since there would now be *MAX_DIST* + 1 items trying to reside within *MAX_DIST* positions of *pos*—an impossibility.

Continuing the example, suppose we now insert item *H* with hash value 9. Our normal linear probing would try to place it in position 13, but that is too far from the hash value of 9. So instead, we look to evict an item and relocate it to position 13. The only candidates to go into position 13 would be items with hash value of 10, 11, 12, or 13. If we examine Hop[10], we see that there are no candidates with hash value 10. But Hop[11] produces a candidate, *G*, with value 11 that can be placed into position 13. Since position 11 is now close enough to the hash value of *H*, we can now insert *H*. These steps, along with the changes to the Hop information, are shown in Figure 5.47.

Finally, we will attempt to insert *I* whose hash value is 6. Linear probing suggests position 14, but of course that is too far away. Thus we look for in Hop[11], and it tells

```
1    public class StringHashFamily implements HashFamily<String>
2    {
3        private final int [ ] MULTIPLIERS;
4        private final java.util.Random r = new java.util.Random( );
5
6        public StringHashFamily( int d )
7        {
8            MULTIPLIERS = new int[ d ];
9            generateNewFunctions( );
10       }
11
12       public int getNumberOfFunctions( )
13       {
14           return MULTIPLIERS.length;
15       }
16
17       public void generateNewFunctions( )
18       {
19           for( int i = 0; i < MULTIPLIERS.length; i++ )
20               MULTIPLIERS[ i ] = r.nextInt( );
21       }
22
23       public int hash( String x, int which )
24       {
25           final int multiplier = MULTIPLIERS[ which ];
26           int hashVal = 0;
27
28           for( int i = 0; i < x.length( ); i++ )
29               hashVal = multiplier * hashVal + x.charAt( i );
30
31           return hashVal;
32       }
33   }
```

Figure 5.45 Casual string hashing for cuckoo hashing; these hash functions do not provably satisfy the requirements needed for cuckoo hashing but offer decent performance if the table is not highly loaded and the alternate insertion routine in Figure 5.43 is used

us that G can move down, freeing up position 13. Now that 13 is vacant, we can look in Hop[10] to find another element to evict. But Hop[10] has all zeros in the first three positions, so there are no items with hash value 10 that can be moved. So we examine Hop[11]. There we find all zeros in the first two positions.

So we try Hop[12], where we need the first position to be 1, which it is. Thus F can move down. These two steps are shown in Figure 5.48. Notice, that if this were not the case—for instance if hash(F) were 9 instead of 12—we would be stuck and have to rehash.

	Item	Hop
...		
6	C	1000
7	A	1100
8	D	0010
9	B	1000
10	E	0000
11	G	1000
12	F	1000
13		0000
14		0000
...		

A: 7
B: 9
C: 6
D: 7
E: 8
F: 12
G: 11

Figure 5.46 Hopscotch hashing table. The hops tell which of the positions in the block are occupied with cells containing this hash value. Thus Hop[8] = 0010 indicates that only position 10 currently contains items whose hash value is 8, while positions 8, 9, and 11 do not

	Item	Hop
...		
6	C	1000
7	A	1100
8	D	0010
9	B	1000
10	E	0000
11	G	1000
12	F	1000
13		0000
14		0000
...		

$\rightarrow$

	Item	Hop
...		
6	C	1000
7	A	1100
8	D	0010
9	B	1000
10	E	0000
11		0010
12	F	1000
13	G	0000
14		0000
...		

$\rightarrow$

	Item	Hop
...		
6	C	1000
7	A	1100
8	D	0010
9	B	**1010**
10	E	0000
11	**H**	0010
12	F	1000
13	G	0000
14		0000
...		

A: 7
B: 9
C: 6
D: 7
E: 8
F: 12
G: 11
H: 9

Figure 5.47 Hopscotch hashing table. Attempting to insert H. Linear probing suggests location 13, but that is too far, so we evict G from position 11 to find a closer position

However, that is not a problem with our algorithm; instead there would simply be no way to place all of C, I, A, D, E, B, H, and F (if F's hash value were 9); these items would all have hash values between 6 and 9, and would thus need to be placed in the seven spots between 6 and 12. But that would be eight items in seven spots—an impossibility. However, since this is not the case for our example, and we have evicted an item from position 12, we can now continue, and Figure 5.49 shows the remaining eviction from position 9, and subsequent placement of I.

	Item	Hop
...		
6	C	1000
7	A	1100
8	D	0010
9	B	1010
10	E	0000
11	H	0010
12	F	1000
13	G	0000
14		0000
...		

$\rightarrow$

	Item	Hop
...		
6	C	1000
7	A	1100
8	D	0010
9	B	1010
10	E	0000
11	H	**0001**
12	F	1000
13		0000
14	G	0000
...		

$\rightarrow$

	Item	Hop
...		
6	C	1000
7	A	1100
8	D	0010
9	B	1010
10	E	0000
11	H	0001
12		**0100**
13	F	0000
14	G	0000
...		

A: 7
B: 9
C: 6
D: 7
E: 8
F: 12
G: 11
H: 9
I: 6

Figure 5.48 Hopscotch hashing table. Attempting to insert I. Linear probing suggests location 14, but that is too far; consulting Hop[11], we see that G can move down, leaving position 13 open. Consulting Hop[10] gives no suggestions. Hop[11] does not help either (why?), so Hop[12] suggests moving F

	Item	Hop
...		
6	C	1000
7	A	1100
8	D	0010
9	B	1010
10	E	0000
11	H	0001
12		0100
13	F	0000
14	G	0000
...		

$\rightarrow$

	Item	Hop
...		
6	C	1000
7	A	1100
8	D	0010
9		**0011**
10	E	0000
11	H	0001
12	**B**	0100
13	F	0000
14	G	0000
...		

$\rightarrow$

	Item	Hop
...		
6	C	**1001**
7	A	1100
8	D	0010
9	I	0011
10	E	0000
11	H	0001
12	B	0100
13	F	0000
14	G	0000
...		

A: 7
B: 9
C: 6
D: 7
E: 8
F: 12
G: 11
H: 9
I: 6

Figure 5.49 Hopscotch hashing table. Insertion of I continues: Next B is evicted, and finally we have a spot that is close enough to the hash value and can insert I

Hopscotch hashing is a relatively new algorithm, but the initial experimental results are very promising, especially for applications that make use of multiple processors and require significant parallelism and concurrency. It remains to be seen if either cuckoo hashing or hopscotch hashing emerge as a practical alternative to the classic separate chaining and linear/quadratic probing schemes.

5.8 Universal Hashing

Although hash tables are very efficient and have average cost per operation, assuming appropriate load factors, their analysis and performance depend on the hash function having two fundamental properties:

1. The hash function must be computable in constant time (i.e., independent of the number of items in the hash table).
2. The hash function must distribute its items uniformly among the array slots.

In particular, if the hash function is poor, then all bets are off, and the cost per operation can be linear. In this section, we discuss **universal hash functions**, which allow us to choose the hash function randomly in such a way that condition 2 above is satisfied. As in Section 5.7, we use M to represent *Tablesize*. Although a strong motivation for the use of universal hash functions is to provide theoretical justification for the assumptions used in the classic hash table analyses, these functions can also be used in applications that require a high level of robustness, in which worst-case (or even substantially degraded) performance, perhaps based on inputs generated by a saboteur or hacker, simply cannot be tolerated.

As in Section 5.7, we use M to represent *TableSize*.

Definition 5.1.
A family H of hash functions is *universal*, if for any $x \neq y$, the number of hash functions h in H for which $h(x) = h(y)$ is at most $|H|/M$.

Notice that this definition holds for each pair of items, rather than being averaged over all pairs of items. The definition above means that if we choose a hash function randomly from a universal family H, then the probability of a collision between any two distinct items is at most $1/M$, and when adding into a table with N items, the probability of a collision at the initial point is at most N/M, or the load factor.

The use of a universal hash function for separate chaining or hopscotch hashing would be sufficient to meet the assumptions used in the analysis of those data structures. However, it is not sufficient for cuckoo hashing, which requires a stronger notion of independence. In cuckoo hashing, we first see if there is a vacant location; if there is not, and we do an eviction, a different item is now involved in looking for a vacant location. This repeats until we find the vacant location, or decide to rehash [generally within $O(\log N)$ steps]. In order for the analysis to work, each step must have a collision probability of N/M independently, with a different item x being subject to the hash function. We can formalize this independence requirement in the following definition.

Definition 5.2.
A family H of hash functions is *k-universal*, if for any $x_1 \neq y_1, x_2 \neq y_2, \ldots, x_k \neq y_k$, the number of hash functions h in H for which $h(x_1) = h(y_1)$, $h(x_2) = h(y_2)$, $\ldots$, and $h(x_k) = h(y_k)$ is at most $|H|/M^k$.

With this definition, we see that the analysis of cuckoo hashing requires an $O(\log N)$-universal hash function (after that many evictions, we give up and rehash). In this section we look only at universal hash functions.

To design a simple universal hash function, we will assume first that we are mapping very large integers into smaller integers ranging from 0 to $M - 1$. Let p be a prime larger than the largest input key.

Our universal family H will consist of the following set of functions, where a and b are chosen randomly:

$$H = \{H_{a,b}(x) = ((ax + b) \mod p) \mod M, \text{ where } 1 \le a \le p - 1, 0 \le b \le p - 1\}$$

For example, in this family, three of the possible random choices of (a, b) yield three different hash functions:

$$H_{3,7}(x) = ((3x + 7) \mod p) \mod M$$
$$H_{4,1}(x) = ((4x + 1) \mod p) \mod M$$
$$H_{8,0}(x) = ((8x) \mod p) \mod M$$

Observe that there are $p(p - 1)$ possible hash functions that can be chosen.

Theorem 5.4.
The hash family $H = \{H_{a,b}(x) = ((ax + b) \mod p) \mod M, \text{ where } 1 \le a \le p - 1, 0 \le b \le p - 1\}$ is universal.

Proof.
Let x and y be distinct values, with $x > y$, such that $H_{a,b}(x) = H_{a,b}(y)$.

Clearly if $(ax + b) \mod p$ is equal to $(ay + b) \mod p$, then we will have a collision. However, this cannot happen: Subtracting equations yields $a(x - y) \equiv 0 \pmod{p}$, which would mean that p divides a or p divides $x - y$, since p is prime. But neither can happen, since both a and $x - y$ are between 1 and $p - 1$.

So let $r = (ax + b) \mod p$ and let $s = (ay + b) \mod p$, and by the above argument, $r \neq s$. Thus there are p possible values for r, and for each r, there are $p - 1$ possible values for s, for a total of $p(p - 1)$ possible (r, s) pairs. Notice that the number of (a, b) pairs and the number of (r, s) pairs is identical; thus each (r, s) pair will correspond to exactly one (a, b) pair if we can solve for (a, b) in terms of r and s. But that is easy: As before, subtracting equations yields $a(x - y) \equiv (r - s) \pmod{p}$, which means that by multiplying both sides by the unique multiplicative inverse of $(x - y)$ (which must exist, since $x - y$ is not zero and p is prime), we obtain a, in terms of r and s. Then b follows.

Finally, this means that the probability that x and y collide is equal to the probability that $r \equiv s \pmod{M}$, and the above analysis allows us to assume that r and s are chosen randomly, rather than a and b. Immediate intuition would place this probability at $1/M$, but that would only be true if p were an exact multiple of M, and all possible

(r, s) pairs were equally likely. Since p is prime, and $r \neq s$, that is not exactly true, so a more careful analysis is needed.

For a given r, the number of values of s that can collide mod M is at most $\lceil p/M \rceil - 1$ (the -1 is because $r \neq s$). It is easy to see that this is at most $(p - 1)/M$. Thus the probability that r and s will generate a collision is at most $1/M$ (we divide by $p - 1$, because as mentioned earlier in the proof, there are only $p - 1$ choices for s given r). This implies that the hash family is universal.

Implementation of this hash function would seem to require two mod operations: one mod p and the second mod M. Figure 5.50 shows a simple implementation in Java, assuming that M is significantly less than the $2^{31} - 1$ limit of a Java integer. Because the computations must now be exactly as specified, and thus overflow is no longer acceptable, we promote to 64-bit long computations.

However, we are allowed to choose any prime p, as long as it is larger than M. Hence, it makes sense to choose a prime that is most favorable for computations. One such prime is $p = 2^{31} - 1$. Prime numbers of this form are known as Mersenne primes; other Mersenne primes include $2^5 - 1$, $2^{61} - 1$ and $2^{89} - 1$. Just as a multiplication by a Mersenne prime such as 31 can be implemented by a bit shift and a subtract, a mod operation involving a Mersenne prime can also be implemented by a bit shift and an addition:

Suppose $r \equiv y \pmod{p}$. If we divide y by $(p + 1)$, then $y = q'(p + 1) + r'$, where q' and r' are the quotient and remainder, respectively. Thus, $r \equiv q'(p + 1) + r' \pmod{p}$. And since $(p + 1) \equiv 1 \pmod{p}$, we obtain $r \equiv q' + r' \pmod{p}$.

Figure 5.51 implements this idea, which is known as the **Carter-Wegman trick**. On line 8, the bit shift computes the quotient and the bitwise and computes the remainder

```
1       public static int universalHash( int x, int A, int B, int P, int M )
2       {
3           return (int) ( ( ( (long) A * x ) + B ) % P ) % M;
4       }
```

Figure 5.50   Simple implementation of universal hashing

```
1       public static final int DIGS = 31;
2       public static final int mersennep = (1<<DIGS) - 1;
3
4       public static int universalHash( int x, int A, int B, int M )
5       {
6           long hashVal = (long) A * x + B;
7
8           hashVal = ( ( hashVal >> DIGS ) + ( hashVal & mersennep ) );
9           if( hashVal >= mersennep )
10              hashVal -= mersennep;
11
12          return (int) hashVal % M;
13      }
```

Figure 5.51   Simple implementation of universal hashing

when dividing by $(p + 1)$; these bitwise operations work because $(p + 1)$ is an exact power of two. Since the remainder could be almost as large as p, the resulting sum might be larger than p, so we scale it back down at lines 9 and 10.

Universal hash functions exist for strings also. First, choose any prime p, larger than M (and larger than the largest character code). Then use our standard string hashing function, choosing the multiplier randomly between 1 and $p - 1$ and returning an intermediate hash value between 0 and $p - 1$, inclusive. Finally, apply a universal hash function to generate the final hash value between 0 and $M - 1$.

5.9   Extendible Hashing

Our last topic in this chapter deals with the case where the amount of data is too large to fit in main memory. As we saw in Chapter 4, the main consideration then is the number of disk accesses required to retrieve data.

As before, we assume that at any point we have N records to store; the value of N changes over time. Furthermore, at most M records fit in one disk block. We will use $M = 4$ in this section.

If either probing hashing or separate chaining hashing is used, the major problem is that collisions could cause several blocks to be examined during a search, even for a well-distributed hash table. Furthermore, when the table gets too full, an extremely expensive rehashing step must be performed, which requires $O(N)$ disk accesses.

A clever alternative, known as **extendible hashing**, allows a search to be performed in two disk accesses. Insertions also require few disk accesses.

We recall from Chapter 4 that a B-tree has depth $O(\log_{M/2} N)$. As M increases, the depth of a B-tree decreases. We could in theory choose M to be so large that the depth of the B-tree would be 1. Then any search after the first would take one disk access, since, presumably, the root node could be stored in main memory. The problem with this strategy is that the branching factor is so high that it would take considerable processing to determine which leaf the data were in. If the time to perform this step could be reduced, then we would have a practical scheme. This is exactly the strategy used by extendible hashing.

Let us suppose, for the moment, that our data consist of several six-bit integers. Figure 5.52 shows an extendible hashing scheme for these data. The root of the "tree" contains four links determined by the leading two bits of the data. Each leaf has up to $M = 4$ elements. It happens that in each leaf the first two bits are identical; this is indicated by the number in parentheses. To be more formal, D will represent the number of bits used by the root, which is sometimes known as the **directory**. The number of entries in the directory is thus 2^D. d_L is the number of leading bits that all the elements of some leaf L have in common. d_L will depend on the particular leaf, and $d_L \leq D$.

Suppose that we want to insert the key 100100. This would go into the third leaf, but as the third leaf is already full, there is no room. We thus split this leaf into two leaves, which are now determined by the first *three* bits. This requires increasing the directory size to 3. These changes are reflected in Figure 5.53.

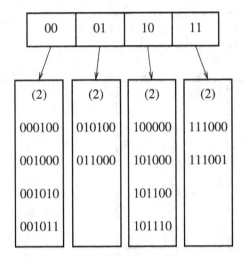

Figure 5.52   Extendible hashing: original data

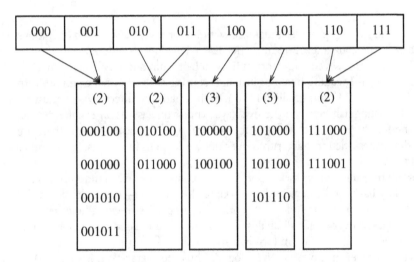

Figure 5.53   Extendible hashing: after insertion of 100100 and directory split

Notice that all the leaves not involved in the split are now pointed to by two adjacent directory entries. Thus, although an entire directory is rewritten, none of the other leaves is actually accessed.

If the key 000000 is now inserted, then the first leaf is split, generating two leaves with $d_L = 3$. Since $D = 3$, the only change required in the directory is the updating of the 000 and 001 links. See Figure 5.54.

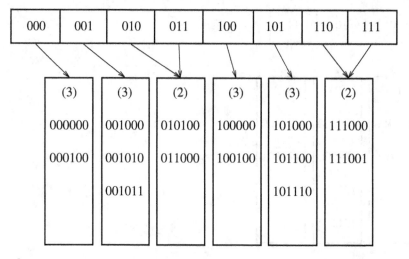

Figure 5.54   Extendible hashing: after insertion of 000000 and leaf split

This very simple strategy provides quick access times for insert and search operations on large databases. There are a few important details we have not considered.

First, it is possible that several directory splits will be required if the elements in a leaf agree in more than $D + 1$ leading bits. For instance, starting at the original example, with $D = 2$, if 111010, 111011, and finally 111100 are inserted, the directory size must be increased to 4 to distinguish between the five keys. This is an easy detail to take care of, but must not be forgotten. Second, there is the possibility of duplicate keys; if there are more than M duplicates, then this algorithm does not work at all. In this case, some other arrangements need to be made.

These possibilities suggest that it is important for the bits to be fairly random. This can be accomplished by hashing the keys into a reasonably long integer—hence the name.

We close by mentioning some of the performance properties of extendible hashing, which are derived after a very difficult analysis. These results are based on the reasonable assumption that the bit patterns are uniformly distributed.

The expected number of leaves is $(N/M) \log_2 e$. Thus the average leaf is $\ln 2 = 0.69$ full. This is the same as for B-trees, which is not entirely surprising, since for both data structures new nodes are created when the $(M + 1)$th entry is added.

The more surprising result is that the expected size of the directory (in other words, 2^D) is $O(N^{1+1/M}/M)$. If M is very small, then the directory can get unduly large. In this case, we can have the leaves contain links to the records instead of the actual records, thus increasing the value of M. This adds a second disk access to each search operation in order to maintain a smaller directory. If the directory is too large to fit in main memory, the second disk access would be needed anyway.

Summary

Hash tables can be used to implement the `insert` and search operations in constant average time. It is especially important to pay attention to details such as load factor when using hash tables, since otherwise the time bounds are not valid. It is also important to choose the hash function carefully when the key is not a short string or integer.

For separate chaining hashing, the load factor should be close to 1, although performance does not significantly degrade unless the load factor becomes very large. For probing hashing, the load factor should not exceed 0.5, unless this is completely unavoidable. If linear probing is used, performance degenerates rapidly as the load factor approaches 1. Rehashing can be implemented to allow the table to grow (and shrink), thus maintaining a reasonable load factor. This is important if space is tight and it is not possible just to declare a huge hash table.

Other alternatives such as cuckoo hashing and hopscotch hashing can also yield good results. Because all these algorithms are constant time, it is difficult to make strong statements about which hash table implementation is the "best"; recent simulation results provide conflicting guidance and suggest that the performance can depend strongly on the types of items being manipulated, the underlying computer hardware, and the programming language.

Binary search trees can also be used to implement `insert` and `contains` operations. Although the resulting average time bounds are $O(\log N)$, binary search trees also support routines that require order and are thus more powerful. Using a hash table, it is not possible to find the minimum element. It is not possible to search efficiently for a string unless the exact string is known. A binary search tree could quickly find all items in a certain range; this is not supported by hash tables. Furthermore, the $O(\log N)$ bound is not necessarily that much more than $O(1)$, especially since no multiplications or divisions are required by search trees.

On the other hand, the worst case for hashing generally results from an implementation error, whereas sorted input can make binary trees perform poorly. Balanced search trees are quite expensive to implement, so if no ordering information is required and there is any suspicion that the input might be sorted, then hashing is the data structure of choice.

Hashing applications are abundant. Compilers use hash tables to keep track of declared variables in source code. The data structure is known as a **symbol table.** Hash tables are the ideal application for this problem. Identifiers are typically short, so the hash function can be computed quickly, and alphabetizing the variables is often unnecessary.

A hash table is useful for any graph theory problem where the nodes have real names instead of numbers. Here, as the input is read, vertices are assigned integers from 1 onward by order of appearance. Again, the input is likely to have large groups of alphabetized entries. For example, the vertices could be computers. Then if one particular installation lists its computers as *ibm1, ibm2, ibm3,* . . . , there could be a dramatic effect on efficiency if a search tree is used.

A third common use of hash tables is in programs that play games. As the program searches through different lines of play, it keeps track of positions it has seen by computing a hash function based on the position (and storing its move for that position). If the same position reoccurs, usually by a simple transposition of moves, the program can avoid expensive recomputation. This general feature of all game-playing programs is known as the **transposition table.**

Yet another use of hashing is in online spelling checkers. If misspelling detection (as opposed to correction) is important, an entire dictionary can be prehashed and words can be checked in constant time. Hash tables are well suited for this, because it is not important to alphabetize words; printing out misspellings in the order they occurred in the document is certainly acceptable.

Hash tables are often used to implement caches, both in software (for instance, the cache in your Internet browser) and in hardware (for instance, the memory caches in modern computers). They are also used in hardware implementations of routers.

We close this chapter by returning to the word puzzle problem of Chapter 1. If the second algorithm described in Chapter 1 is used, and we assume that the maximum word size is some small constant, then the time to read in the dictionary containing W words and put it in a hash table is $O(W)$. This time is likely to be dominated by the disk I/O and not the hashing routines. The rest of the algorithm would test for the presence of a word for each ordered quadruple (*row, column, orientation, number of characters*). As each lookup would be $O(1)$, and there are only a constant number of orientations (8) and characters per word, the running time of this phase would be $O(R \cdot C)$. The total running time would be $O(R \cdot C + W)$, which is a distinct improvement over the original $O(R \cdot C \cdot W)$. We could make further optimizations, which would decrease the running time in practice; these are described in the exercises.

Exercises

5.1 Given input {4371, 1323, 6173, 4199, 4344, 9679, 1989} and a hash function $h(x) = x \bmod 10$, show the resulting:
a. Separate chaining hash table.
b. Hash table using linear probing.
c. Hash table using quadratic probing.
d. Hash table with second hash function $h_2(x) = 7 - (x \bmod 7)$.

5.2 Show the result of rehashing the hash tables in Exercise 5.1.

5.3 Write a program to compute the number of collisions required in a long random sequence of insertions using linear probing, quadratic probing, and double hashing.

5.4 A large number of deletions in a separate chaining hash table can cause the table to be fairly empty, which wastes space. In this case, we can rehash to a table half as large. Assume that we rehash to a larger table when there are twice as many elements as the table size. How empty should the table be before we rehash to a smaller table?

5.5 Reimplement separate chaining hash tables using singly linked lists instead of using `java.util.LinkedList`.

5.6 The `isEmpty` routine for quadratic probing has not been written. Can you implement it by returning the expression `currentSize==0`?

5.7   In the quadratic probing hash table, suppose that instead of inserting a new item into the location suggested by `findPos`, we insert it into the first inactive cell on the search path (thus, it is possible to reclaim a cell that is marked "deleted," potentially saving space).

  a. Rewrite the insertion algorithm to use this observation. Do this by having `find-Pos` maintain, with an additional variable, the location of the first inactive cell it encounters.

  b. Explain the circumstances under which the revised algorithm is faster than the original algorithm. Can it be slower?

5.8   Suppose instead of quadratic probing, we use "cubic probing"; here the ith probe is at $hash(x) + i^3$. Does cubic probing improve on quadratic probing?

5.9   The hash function in Figure 5.4 makes repeated calls to `key.length( )` in the for loop. Is it worth computing this once prior to entering the loop?

5.10 What are the advantages and disadvantages of the various collision resolution strategies?

5.11 Suppose that to mitigate the effects of secondary clustering we use as the collision resolution function $f(i) = i \cdot r(hash(x))$, where $hash(x)$ is the 32-bit hash value (not yet scaled to a suitable array index), and $r(y) = |48271y(\bmod (2^{31} - 1))|$ $\bmod$ *TableSize*. (Section 10.4.1 describes a method of performing this calculation without overflows, but it is unlikely that overflow matters in this case.) Explain why this strategy tends to avoid secondary clustering, and compare this strategy with both double hashing and quadratic probing.

5.12 Rehashing requires recomputing the hash function for all items in the hash table. Since computing the hash function is expensive, suppose objects provide a hash member function of their own, and each object stores the result in an additional data member the first time the hash function is computed for it. Show how such a scheme would apply for the `Employee` class in Figure 5.8, and explain under what circumstances the remembered hash value remains valid in each `Employee`.

5.13 Write a program to implement the following strategy for multiplying two sparse polynomials P_1, P_2 of size M and N, respectively. Each polynomial is represented as a linked list of objects consisting of a coefficient and an exponent (Exercise 3.12). We multiply each term in P_1 by a term in P_2 for a total of MN operations. One method is to sort these terms and combine like terms, but this requires sorting MN records, which could be expensive, especially in small-memory environments. Alternatively, we could merge terms as they are computed and then sort the result.

  a. Write a program to implement the alternative strategy.

  b. If the output polynomial has about $O(M + N)$ terms, what is the running time of both methods?

*5.14 Describe a procedure that avoids initializing a hash table (at the expense of memory).

5.15 Suppose we want to find the first occurrence of a string $P_1P_2 \cdots P_k$ in a long input string $A_1A_2 \cdots A_N$. We can solve this problem by hashing the pattern string, obtaining a hash value H_P, and comparing this value with the hash value formed from $A_1A_2 \cdots A_k, A_2A_3 \cdots A_{k+1}, A_3A_4 \cdots A_{k+2}$, and so on until $A_{N-k+1}A_{N-k+2} \cdots A_N$. If we have a match of hash values, we compare the strings character by character to verify the match. We return the position (in A) if the strings actually do match, and we continue in the unlikely event that the match is false.

  *a. Show that if the hash value of $A_iA_{i+1} \cdots A_{i+k-1}$ is known, then the hash value of $A_{i+1}A_{i+2} \cdots A_{i+k}$ can be computed in constant time.

  b. Show that the running time is $O(k + N)$ plus the time spent refuting false matches.

  *c. Show that the expected number of false matches is negligible.

  d. Write a program to implement this algorithm.

  **e. Describe an algorithm that runs in $O(k + N)$ worst-case time.

  **f. Describe an algorithm that runs in $O(N/k)$ average time.

5.16 Java 7 adds syntax that allows a switch statement to work with the String type (instead of the primitive integer types). Explain how hash tables can be used by the compiler to implement this language addition.

5.17 An (old style) BASIC program consists of a series of statements numbered in ascending order. Control is passed by use of a *goto* or *gosub* and a statement number. Write a program that reads in a legal BASIC program and renumbers the statements so that the first starts at number F and each statement has a number D higher than the previous statement. You may assume an upper limit of N statements, but the statement numbers in the input might be as large as a 32-bit integer. Your program must run in linear time.

5.18 a. Implement the word puzzle program using the algorithm described at the end of the chapter.

  b. We can get a big speed increase by storing, in addition to each word W, all of W's prefixes. (If one of W's prefixes is another word in the dictionary, it is stored as a real word.) Although this may seem to increase the size of the hash table drastically, it does not, because many words have the same prefixes. When a scan is performed in a particular direction, if the word that is looked up is not even in the hash table as a prefix, then the scan in that direction can be terminated early. Use this idea to write an improved program to solve the word puzzle.

  c. If we are willing to sacrifice the sanctity of the hash table ADT, we can speed up the program in part (b) by noting that if, for example, we have just computed the hash function for "excel," we do not need to compute the hash function for "excels" from scratch. Adjust your hash function so that it can take advantage of its previous calculation.

  d. In Chapter 2, we suggested using binary search. Incorporate the idea of using prefixes into your binary search algorithm. The modification should be simple. Which algorithm is faster?

```
1   class Map<KeyType,ValueType>
2   {
3       public Map( )
4
5       public void put( KeyType key, ValueType val )
6       public ValueType get( KeyType key )
7       public boolean isEmpty( )
8       public void makeEmpty( )
9
10      private QuadraticProbingHashTable<Entry<KeyType,ValueType>> items;
11
12      private static class Entry<KeyType,ValueType>
13      {
14          KeyType key;
15          ValueType value;
16          // Appropriate Constructors, etc.
17      }
18  }
```

Figure 5.55 Map skeleton for Exercise 5.20

5.19 Under certain assumptions, the expected cost of an insertion into a hash table with secondary clustering is given by $1/(1-\lambda)-\lambda-\ln(1-\lambda)$. Unfortunately, this formula is not accurate for quadratic probing. However, assuming that it is, determine the following:
  a. The expected cost of an unsuccessful search.
  b. The expected cost of a successful search.

5.20 Implement a generic Map that supports the put and get operations. The implementation will store a hash table of pairs (key, definition). Figure 5.55 provides the Map specification (minus some details).

5.21 Implement a spelling checker by using a hash table. Assume that the dictionary comes from two sources: an existing large dictionary and a second file containing a personal dictionary. Output all misspelled words and the line numbers in which they occur. Also, for each misspelled word, list any words in the dictionary that are obtainable by applying any of the following rules:
  a. Add one character.
  b. Remove one character.
  c. Exchange adjacent characters.

5.22 Prove **Markov's Inequality**: If X is any random variable and $a > 0$, then $\Pr(|X| \geq a) \leq E(|X|)/a$. Show how this inequality can be applied to Theorems 5.2 and 5.3.

5.23 If a hopscotch table with parameter MAX_DIST has load factor 0.5, what is the approximate probability that an insertion requires a rehash?

5.24 Implement a hopscotch hash table and compare its performance with linear probing, separate chaining, and cuckoo hashing.

5.25 Implement the classic cuckoo hash table in which two separate tables are maintained. The simplest way to do this is to use a single array and modify the hash function to access either the top half or the bottom half.

5.26 Extend the classic cuckoo hash table to use d hash functions.

5.27 Show the result of inserting the keys 10111101, 00000010, 10011011, 10111110, 01111111, 01010001, 10010110, 00001011, 11001111, 10011110, 11011011, 00101011, 01100001, 11110000, 01101111 into an initially empty extendible hashing data structure with $M = 4$.

5.28 Write a program to implement extendible hashing. If the table is small enough to fit in main memory, how does its performance compare with separate chaining and open addressing hashing?

References

Despite the apparent simplicity of hashing, much of the analysis is quite difficult, and there are still many unresolved questions. There are also many interesting theoretical issues.

Hashing dates to at least 1953, when H. P. Luhn wrote an internal IBM memorandum that used separate chaining hashing. Early papers on hashing are [11] and [32]. A wealth of information on the subject, including an analysis of hashing with linear probing under the assumption of totally random and independent hashing, can be found in [25]. More recent results have shown that linear probing requires only 5-independent hash functions [31]. An excellent survey on early classic hash tables methods is [28]; [29] contains suggestions, and pitfalls, for choosing hash functions. Precise analytic and simulation results for separate chaining, linear probing, quadratic probing, and double hashing can be found in [19]. However, due to changes (improvements) in computer architecture and compilers, simulation results tend to quickly become dated.

An analysis of double hashing can be found in [20] and [27]. Yet another collision resolution scheme is coalesced hashing, described in [33]. Yao [37] has shown the uniform hashing, in which no clustering exists, is optimal with respect to cost of a successful search, assuming that items cannot move once placed.

Universal hash functions were first described in [5] and [35]; the latter paper introduces the "Carter-Wegman trick" of using Mersenne prime numbers to avoid expensive mod operations. Perfect hashing is described in [16], and a dynamic version of perfect hashing was described in [8]. [12] is a survey of some classic dynamic hashing schemes.

The $\Theta(\log N/\log\log N)$ bound on the length of the longest list in separate chaining was shown (in precise form) in [18]. The "power of two choices," showing that when the shorter of two randomly selected lists in chosen, then the bound on the length of the longest list is lowered to only $\Theta(\log\log N)$, was first described in [2]. An early example of the power of two choices is [4]. The classic work on cuckoo hashing is [30]; since the initial paper, a host of new results have appeared that analyze the amount of independence needed in the hash functions, and describe alternative implementations [7], [34], [15], [10], [23], [24], [1], [6], [9] and [17]. Hopscotch hashing appeared in [21].

Extendible hashing appears in [13], with analysis in [14] and [36].

Exercise 5.15 (a–d) is from [22]. Part (e) is from [26], and part (f) is from [3].

1. Y. Arbitman, M. Naor, and G. Segev, "De-Amortized Cuckoo Hashing: Provable Worst-Case Performance and Experimental Results," *Proceedings of the 36th International Colloquium on Automata, Languages and Programming* (2009), 107–118.

2. Y. Azar, A. Broder, A. Karlin, and E. Upfal, "Balanced Allocations," *SIAM Journal of Computing*, 29 (1999), 180–200.

3. R. S. Boyer and J. S. Moore, "A Fast String Searching Algorithm," *Communications of the ACM*, 20 (1977), 762–772.

4. A. Broder and M. Mitzenmacher, "Using Multiple Hash Functions to Improve IP Lookups," *Proceedings of the Twentieth IEEE INFOCOM* (2001), 1454–1463.

5. J. L. Carter and M. N. Wegman, "Universal Classes of Hash Functions," *Journal of Computer and System Sciences*, 18 (1979), 143–154.

6. J. Cohen and D. Kane, "Bounds on the Independence Required for Cuckoo Hashing," preprint.

7. L. Devroye and P. Morin, "Cuckoo Hashing: Further Analysis," *Information Processing Letters*, 86 (2003), 215–219.

8. M. Dietzfelbinger, A. R. Karlin, K. Melhorn, F. Meyer auf der Heide, H. Rohnert, and R. E. Tarjan, "Dynamic Perfect Hashing: Upper and Lower Bounds," *SIAM Journal on Computing*, 23 (1994), 738–761.

9. M. Dietzfelbinger and U. Schellbach, "On Risks of Using Cuckoo Hashing with Simple Universal Hash Classes," *Proceedings of the Twentieth Annual ACM-SIAM Symposium on Discrete Algorithms* (2009), 795–804.

10. M. Dietzfelbinger and C. Weidling, "Balanced Allocation and Dictionaries with Tightly Packed Constant Size Bins," *Theoretical Computer Science*, 380 (2007), 47–68.

11. I. Dumey, "Indexing for Rapid Random-Access Memory," *Computers and Automation*, 5 (1956), 6–9.

12. R. J. Enbody and H. C. Du, "Dynamic Hashing Schemes," *Computing Surveys*, 20 (1988), 85–113.

13. R. Fagin, J. Nievergelt, N. Pippenger, and H. R. Strong, "Extendible Hashing—A Fast Access Method for Dynamic Files," *ACM Transactions on Database Systems*, 4 (1979), 315–344.

14. P. Flajolet, "On the Performance Evaluation of Extendible Hashing and Trie Searching," *Acta Informatica*, 20 (1983), 345–369.

15. D. Fotakis, R. Pagh, P. Sanders, and P. Spirakis, "Space Efficient Hash Tables with Worst Case Constant Access Time," *Theory of Computing Systems*, 38 (2005), 229–248.

16. M. L. Fredman, J. Komlos, and E. Szemeredi, "Storing a Sparse Table with $O(1)$ Worst Case Access Time," *Journal of the ACM*, 31 (1984), 538–544.

17. A. Frieze, P. Melsted, and M. Mitzenmacher, "An Analysis of Random-Walk Cuckoo Hashing," *Proceedings of the Twelfth International Workshop on Approximation Algorithms in Combinatorial Optimization (APPROX)* (2009), 350–364.

18. G. Gonnet, "Expected Length of the Longest Probe sequence in Hash Code Searching," *Journal of the Association for Computing Machinery*, 28 (1981), 289–304.

19. G. H. Gonnet and R. Baeza-Yates, *Handbook of Algorithms and Data Structures*, 2nd ed., Addison-Wesley, Reading, Mass., 1991.

20. L. J. Guibas and E. Szemeredi, "The Analysis of Double Hashing," *Journal of Computer and System Sciences*, 16 (1978), 226–274.

21. M. Herlihy, N. Shavit, and M. Tzafrir, "Hopscotch Hashing," *Proceedings of the Twenty-Second International Symposium on Distributed Computing* (2008), 350–364.

22. R. M. Karp and M. O. Rabin, "Efficient Randomized Pattern-Matching Algorithms," *Aiken Computer Laboratory Report TR-31-81*, Harvard University, Cambridge, Mass., 1981.

23. A. Kirsch and M. Mitzenmacher, "The Power of One Move: Hashing Schemes for Hardware," *Proceedings of the 27th IEEE International Conference on Computer Communications (INFOCOM)* (2008), 106–110.

24. A. Kirsch, M. Mitzenmacher, and U. Wieder, "More Robust Hashing: Cuckoo Hashing with a Stash," *Proceedings of the Sixteenth Annual European Symposium on Algorithms* (2008), 611–622.

25. D. E. Knuth, *The Art of Computer Programming, Vol. 3: Sorting and Searching*, 2nd ed., Addison-Wesley, Reading, Mass., 1998.

26. D. E. Knuth, J. H. Morris, and V. R. Pratt, "Fast Pattern Matching in Strings," *SIAM Journal on Computing*, 6 (1977), 323–350.

27. G. Lueker and M. Molodowitch, "More Analysis of Double Hashing," *Proceedings of the Twentieth ACM Symposium on Theory of Computing* (1988), 354–359.

28. W. D. Maurer and T. G. Lewis, "Hash Table Methods," *Computing Surveys*, 7 (1975), 5–20.

29. B. J. McKenzie, R. Harries, and T. Bell, "Selecting a Hashing Algorithm," *Software—Practice and Experience*, 20 (1990), 209–224.

30. R. Pagh and F. F. Rodler, "Cuckoo Hashing," *Journal of Algorithms*, 51 (2004), 122–144.

31. M. Pătrașcu and M. Thorup, "On the k-Independence Required by Linear Probing and Minwise Independence," *Proceedings of the 37th International Colloquium on Automata, Languages, and Programming* (2010), 715–726.

32. W. W. Peterson, "Addressing for Random Access Storage," *IBM Journal of Research and Development*, 1 (1957), 130–146.

33. J. S. Vitter, "Implementations for Coalesced Hashing," *Communications of the ACM*, 25 (1982), 911–926.

34. B. Vöcking, "How Asymmetry Helps Load Balancing," *Journal of the ACM*, 50 (2003), 568–589.

35. M. N. Wegman and J. Carter, "New Hash Functions and Their Use in Authentication and Set Equality," *Journal of Computer and System Sciences*, 22 (1981), 265–279.

36. A. C. Yao, "A Note on the Analysis of Extendible Hashing," *Information Processing Letters*, 11 (1980), 84–86.

37. A. C. Yao, "Uniform Hashing Is Optimal," *Journal of the ACM*, 32 (1985), 687–693.

CHAPTER 6

Priority Queues (Heaps)

Although jobs sent to a printer are generally placed on a queue, this might not always be the best thing to do. For instance, one job might be particularly important, so it might be desirable to allow that job to be run as soon as the printer is available. Conversely, if, when the printer becomes available, there are several 1-page jobs and one 100-page job, it might be reasonable to make the long job go last, even if it is not the last job submitted. (Unfortunately, most systems do not do this, which can be particularly annoying at times.)

Similarly, in a multiuser environment, the operating system scheduler must decide which of several processes to run. Generally a process is allowed to run only for a fixed period of time. One algorithm uses a queue. Jobs are initially placed at the end of the queue. The scheduler will repeatedly take the first job on the queue, run it until either it finishes or its time limit is up, and place it at the end of the queue if it does not finish. This strategy is generally not appropriate, because very short jobs will seem to take a long time because of the wait involved to run. Generally, it is important that short jobs finish as fast as possible, so these jobs should have precedence over jobs that have already been running. Furthermore, some jobs that are not short are still very important and should also have precedence.

This particular application seems to require a special kind of queue, known as a **priority queue**. In this chapter, we will discuss

- Efficient implementation of the priority queue ADT.
- Uses of priority queues.
- Advanced implementations of priority queues.

The data structures we will see are among the most elegant in computer science.

6.1 Model

A priority queue is a data structure that allows at least the following two operations: insert, which does the obvious thing; and deleteMin, which finds, returns, and removes the minimum element in the priority queue. The insert operation is the equivalent of enqueue, and deleteMin is the priority queue equivalent of the queue's dequeue operation.

As with most data structures, it is sometimes possible to add other operations, but these are extensions and not part of the basic model depicted in Figure 6.1.

Priority queues have many applications besides operating systems. In Chapter 7, we will see how priority queues are used for external sorting. Priority queues are also

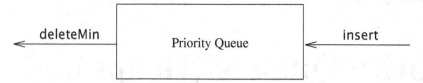

Figure 6.1   Basic model of a priority queue

important in the implementation of **greedy algorithms**, which operate by repeatedly finding a minimum; we will see specific examples in Chapters 9 and 10. In this chapter we will see a use of priority queues in discrete event simulation.

6.2 Simple Implementations

There are several obvious ways to implement a priority queue. We could use a simple linked list, performing insertions at the front in $O(1)$ and traversing the list, which requires $O(N)$ time, to delete the minimum. Alternatively, we could insist that the list be kept always sorted; this makes insertions expensive ($O(N)$) and deleteMins cheap ($O(1)$). The former is probably the better idea of the two, based on the fact that there are never more deleteMins than insertions.

Another way of implementing priority queues would be to use a binary search tree. This gives an $O(\log N)$ average running time for both operations. This is true in spite of the fact that although the insertions are random, the deletions are not. Recall that the only element we ever delete is the minimum. Repeatedly removing a node that is in the left subtree would seem to hurt the balance of the tree by making the right subtree heavy. However, the right subtree is random. In the worst case, where the deleteMins have depleted the left subtree, the right subtree would have at most twice as many elements as it should. This adds only a small constant to its expected depth. Notice that the bound can be made into a worst-case bound by using a balanced tree; this protects one against bad insertion sequences.

Using a search tree could be overkill because it supports a host of operations that are not required. The basic data structure we will use will not require links and will support both operations in $O(\log N)$ worst-case time. Insertion will actually take constant time on average, and our implementation will allow building a priority queue of N items in linear time, if no deletions intervene. We will then discuss how to implement priority queues to support efficient merging. This additional operation seems to complicate matters a bit and apparently requires the use of a linked structure.

6.3 Binary Heap

The implementation we will use is known as a **binary heap.** Its use is so common for priority queue implementations that, in the context of priority queues, when the word *heap* is used without a qualifier, it is generally assumed to be referring to this implementation of the data structure. In this section, we will refer to binary heaps merely as *heaps*. Like binary search trees, heaps have two properties, namely, a structure property and a heap-order property. As with AVL trees, an operation on a heap can destroy one of the properties, so a heap operation must not terminate until all heap properties are in order. This turns out to be simple to do.

6.3.1   Structure Property

A heap is a binary tree that is completely filled, with the possible exception of the bottom level, which is filled from left to right. Such a tree is known as a **complete binary tree**. Figure 6.2 shows an example.

It is easy to show that a complete binary tree of height h has between 2^h and $2^{h+1} - 1$ nodes. This implies that the height of a complete binary tree is $\lfloor \log N \rfloor$, which is clearly $O(\log N)$.

An important observation is that because a complete binary tree is so regular, it can be represented in an array and no links are necessary. The array in Figure 6.3 corresponds to the heap in Figure 6.2.

For any element in array position i, the left child is in position $2i$, the right child is in the cell after the left child $(2i + 1)$, and the parent is in position $\lfloor i/2 \rfloor$. Thus not only are links not required, but the operations required to traverse the tree are extremely simple

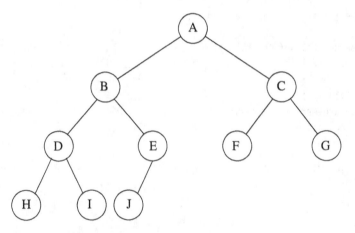

Figure 6.2   A complete binary tree

	A	B	C	D	E	F	G	H	I	J			
0	1	2	3	4	5	6	7	8	9	10	11	12	13

Figure 6.3   Array implementation of complete binary tree

```
 1    public class BinaryHeap<AnyType extends Comparable<? super AnyType>>
 2    {
 3        public BinaryHeap( )
 4            { /* See online code */ }
 5        public BinaryHeap( int capacity )
 6            { /* See online code */ }
 7        public BinaryHeap( AnyType [ ] items )
 8            { /* Figure 6.14 */ }
 9
10        public void insert( AnyType x )
11            { /* Figure 6.8 */ }
12        public AnyType findMin( )
13            { /* See online code */ }
14        public AnyType deleteMin( )
15            { /* Figure 6.12 */ }
16        public boolean isEmpty( )
17            { /* See online code */ }
18        public void makeEmpty( )
19            { /* See online code */ }
20
21        private static final int DEFAULT_CAPACITY = 10;
22
23        private int currentSize;      // Number of elements in heap
24        private AnyType [ ] array;    // The heap array
25
26        private void percolateDown( int hole )
27            { /* Figure 6.12 */ }
28        private void buildHeap( )
29            { /* Figure 6.14 */ }
30        private void enlargeArray( int newSize )
31            { /* See online code */ }
32    }
```

Figure 6.4 Class skeleton for priority queue

and likely to be very fast on most computers. The only problem with this implementation is that an estimate of the maximum heap size is required in advance, but typically this is not a problem (and we can resize if necessary). In Figure 6.3, the limit on the heap size is 13 elements. The array has a position 0; more on this later.

A heap data structure will, then, consist of an array (of Comparable objects) and an integer representing the current heap size. Figure 6.4 shows a priority queue skeleton.

Throughout this chapter, we shall draw the heaps as trees, with the implication that an actual implementation will use simple arrays.

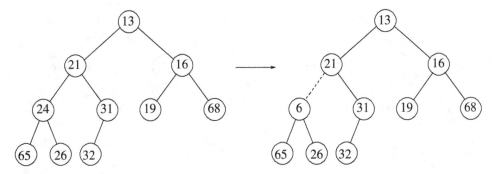

Figure 6.5   Two complete trees (only the left tree is a heap)

6.3.2 Heap-Order Property

The property that allows operations to be performed quickly is the **heap-order property**. Since we want to be able to find the minimum quickly, it makes sense that the smallest element should be at the root. If we consider that any subtree should also be a heap, then any node should be smaller than all of its descendants.

Applying this logic, we arrive at the heap-order property. In a heap, for every node X, the key in the parent of X is smaller than (or equal to) the key in X, with the exception of the root (which has no parent).[1] In Figure 6.5 the tree on the left is a heap, but the tree on the right is not (the dashed line shows the violation of heap order).

By the heap-order property, the minimum element can always be found at the root. Thus, we get the extra operation, findMin, in constant time.

6.3.3 Basic Heap Operations

It is easy (both conceptually and practically) to perform the two required operations. All the work involves ensuring that the heap-order property is maintained.

insert

To insert an element X into the heap, we create a hole in the next available location, since otherwise the tree will not be complete. If X can be placed in the hole without violating heap order, then we do so and are done. Otherwise we slide the element that is in the hole's parent node into the hole, thus bubbling the hole up toward the root. We continue this process until X can be placed in the hole. Figure 6.6 shows that to insert 14, we create a hole in the next available heap location. Inserting 14 in the hole would violate the heap-order property, so 31 is slid down into the hole. This strategy is continued in Figure 6.7 until the correct location for 14 is found.

---

[1] Analogously, we can declare a (*max*) heap, which enables us to efficiently find and remove the maximum element, by changing the heap-order property. Thus, a priority queue can be used to find *either* a minimum or a maximum, but this needs to be decided ahead of time.

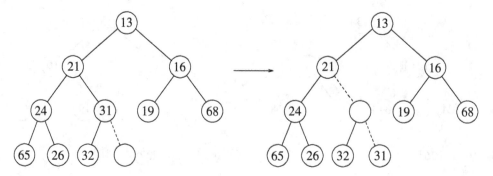

Figure 6.6 Attempt to insert 14: creating the hole and bubbling the hole up

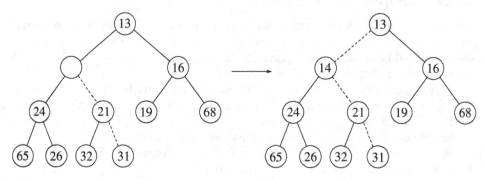

Figure 6.7 The remaining two steps to insert 14 in previous heap

This general strategy is known as a **percolate up**; the new element is percolated up the heap until the correct location is found. Insertion is easily implemented with the code shown in Figure 6.8.

We could have implemented the percolation in the insert routine by performing repeated swaps until the correct order was established, but a swap requires three assignment statements. If an element is percolated up d levels, the number of assignments performed by the swaps would be $3d$. Our method uses $d + 1$ assignments.

If the element to be inserted is the new minimum, it will be pushed all the way to the top. At some point, hole will be 1 and we will want to break out of the loop. We could do this with an explicit test, or we can put a reference to the inserted item in position 0 in order to make the loop terminate. We elect to place x into position 0 in our implementation.

The time to do the insertion could be as much as $O(\log N)$, if the element to be inserted is the new minimum and is percolated all the way to the root. On average, the percolation terminates early; it has been shown that 2.607 comparisons are required on average to perform an insert, so the average insert moves an element up 1.607 levels.

```
1       /**
2        * Insert into the priority queue, maintaining heap order.
3        * Duplicates are allowed.
4        * @param x the item to insert.
5        */
6       public void insert( AnyType x )
7       {
8           if( currentSize == array.length - 1 )
9               enlargeArray( array.length * 2 + 1 );
10
11          // Percolate up
12          int hole = ++currentSize;
13          for( array[ 0 ] = x; x.compareTo( array[ hole / 2 ] ) < 0; hole /= 2 )
14              array[ hole ] = array[ hole / 2 ];
15          array[ hole ] = x;
16      }
```

Figure 6.8   Procedure to insert into a binary heap

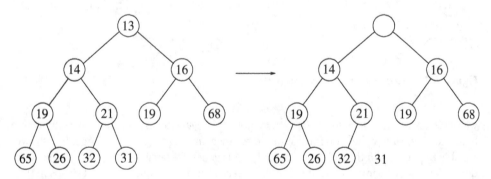

Figure 6.9   Creation of the hole at the root

deleteMin

deleteMins are handled in a similar manner as insertions. Finding the minimum is easy; the hard part is removing it. When the minimum is removed, a hole is created at the root. Since the heap now becomes one smaller, it follows that the last element X in the heap must move somewhere in the heap. If X can be placed in the hole, then we are done. This is unlikely, so we slide the smaller of the hole's children into the hole, thus pushing the hole down one level. We repeat this step until X can be placed in the hole. Thus, our action is to place X in its correct spot along a path from the root containing *minimum* children.

In Figure 6.9 the left figure shows a heap prior to the deleteMin. After 13 is removed, we must now try to place 31 in the heap. The value 31 cannot be placed in the hole, because this would violate heap order. Thus, we place the smaller child (14) in the hole, sliding the hole down one level (see Figure 6.10). We repeat this again, and since 31 is larger than 19,

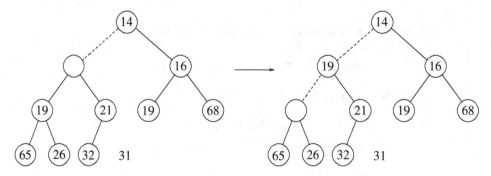

Figure 6.10 Next two steps in `deleteMin`

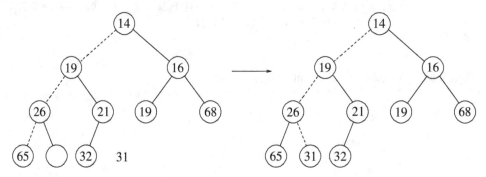

Figure 6.11 Last two steps in `deleteMin`

we place 19 into the hole and create a new hole one level deeper. We then place 26 in the hole and create a new hole on the bottom level since once again, 31 is too large. Finally, we are able to place 31 in the hole (Figure 6.11). This general strategy is known as a **percolate down**. We use the same technique as in the `insert` routine to avoid the use of swaps in this routine.

A frequent implementation error in heaps occurs when there are an even number of elements in the heap, and the one node that has only one child is encountered. You must make sure not to assume that there are always two children, so this usually involves an extra test. In the code depicted in Figure 6.12, we've done this test at line 29. One extremely tricky solution is always to ensure that your algorithm *thinks* every node has two children. Do this by placing a sentinel, of value higher than any in the heap, at the spot after the heap ends, at the start of each *percolate down* when the heap size is even. You should think very carefully before attempting this, and you must put in a prominent comment if you do use this technique. Although this eliminates the need to test for the presence of a right child, you cannot eliminate the requirement that you test when you reach the bottom, because this would require a sentinel for every leaf.

The worst-case running time for this operation is $O(\log N)$. On average, the element that is placed at the root is percolated almost to the bottom of the heap (which is the level it came from), so the average running time is $O(\log N)$.

```
1    /**
2     * Remove the smallest item from the priority queue.
3     * @return the smallest item, or throw UnderflowException, if empty.
4     */
5    public AnyType deleteMin( )
6    {
7        if( isEmpty( ) )
8            throw new UnderflowException( );
9
10       AnyType minItem = findMin( );
11       array[ 1 ] = array[ currentSize-- ];
12       percolateDown( 1 );
13
14       return minItem;
15   }
16
17   /**
18    * Internal method to percolate down in the heap.
19    * @param hole the index at which the percolate begins.
20    */
21   private void percolateDown( int hole )
22   {
23       int child;
24       AnyType tmp = array[ hole ];
25
26       for( ; hole * 2 <= currentSize; hole = child )
27       {
28           child = hole * 2;
29           if( child != currentSize &&
30                   array[ child + 1 ].compareTo( array[ child ] ) < 0 )
31               child++;
32           if( array[ child ].compareTo( tmp ) < 0 )
33               array[ hole ] = array[ child ];
34           else
35               break;
36       }
37       array[ hole ] = tmp;
38   }
```

Figure 6.12   Method to perform deleteMin in a binary heap

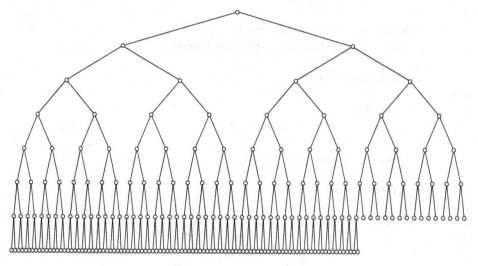

Figure 6.13   A very large complete binary tree

6.3.4 Other Heap Operations

Notice that although finding the minimum can be performed in constant time, a heap designed to find the minimum element (also known as a (*min*)heap) is of no help whatsoever in finding the maximum element. In fact, a heap has very little ordering information, so there is no way to find any particular element without a linear scan through the entire heap. To see this, consider the large heap structure (the elements are not shown) in Figure 6.13, where we see that the only information known about the maximum element is that it is at one of the leaves. Half the elements, though, are contained in leaves, so this is practically useless information. For this reason, if it is important to know where elements are, some other data structure, such as a hash table, must be used in addition to the heap. (Recall that the model does not allow looking inside the heap.)

If we assume that the position of every element is known by some other method, then several other operations become cheap. The first three operations below all run in logarithmic worst-case time.

decreaseKey

The decreaseKey(p, Δ) operation lowers the value of the item at position p by a positive amount Δ. Since this might violate the heap order, it must be fixed by a *percolate up*. This operation could be useful to system administrators: They can make their programs run with highest priority.

increaseKey

The increaseKey(p, Δ) operation increases the value of the item at position p by a positive amount Δ. This is done with a *percolate down*. Many schedulers automatically drop the priority of a process that is consuming excessive CPU time.

delete

The delete(p) operation removes the node at position p from the heap. This is done by first performing decreaseKey(p,∞) and then performing deleteMin(). When a process is terminated by a user (instead of finishing normally), it must be removed from the priority queue.

buildHeap

The binary heap is sometimes constructed from an initial collection of items. This constructor takes as input N items and places them into a heap. Obviously, this can be done with N successive inserts. Since each insert will take $O(1)$ average and $O(\log N)$ worst-case time, the total running time of this algorithm would be $O(N)$ average but $O(N \log N)$ worst-case. Since this is a special instruction and there are no other operations intervening, and we already know that the instruction can be performed in linear average time, it is reasonable to expect that with reasonable care a linear time bound can be guaranteed.

The general algorithm is to place the N items into the tree in any order, maintaining the structure property. Then, if percolateDown(i) percolates down from node i, the buildHeap routine in Figure 6.14 can be used by the constructor to create a heap-ordered tree.

The first tree in Figure 6.15 is the unordered tree. The seven remaining trees in Figures 6.15 through 6.18 show the result of each of the seven percolateDowns. Each dashed line corresponds to two comparisons: one to find the smaller child and one to compare the smaller child with the node. Notice that there are only 10 dashed lines in the entire algorithm (there could have been an 11th—where?) corresponding to 20 comparisons.

To bound the running time of buildHeap, we must bound the number of dashed lines. This can be done by computing the sum of the heights of all the nodes in the heap, which is the maximum number of dashed lines. What we would like to show is that this sum is $O(N)$.

Theorem 6.1.
For the perfect binary tree of height h containing $2^{h+1} - 1$ nodes, the sum of the heights of the nodes is $2^{h+1} - 1 - (h + 1)$.

Proof.
It is easy to see that this tree consists of 1 node at height h, 2 nodes at height $h - 1$, 2^2 nodes at height $h - 2$, and in general 2^i nodes at height $h - i$. The sum of the heights of all the nodes is then

$$S = \sum_{i=0}^{h} 2^i (h - i)$$

$$= h + 2(h - 1) + 4(h - 2) + 8(h - 3) + 16(h - 4) + \cdots + 2^{h-1}(1) \quad (6.1)$$

Multiplying by 2 gives the equation

$$2S = 2h + 4(h - 1) + 8(h - 2) + 16(h - 3) + \cdots + 2^h(1) \quad (6.2)$$

We subtract these two equations and obtain Equation (6.3). We find that certain terms almost cancel. For instance, we have $2h - 2(h - 1) = 2$, $4(h - 1) - 4(h - 2) = 4$, and so on. The last term in Equation (6.2), 2^h, does not appear in Equation (6.1);

```
1    /**
2     * Construct the binary heap given an array of items.
3     */
4    public BinaryHeap( AnyType [ ] items )
5    {
6        currentSize = items.length;
7        array = (AnyType[]) new Comparable[ ( currentSize + 2 ) * 11 / 10 ];
8
9        int i = 1;
10       for( AnyType item : items )
11           array[ i++ ] = item;
12       buildHeap( );
13   }
14
15   /**
16    * Establish heap order property from an arbitrary
17    * arrangement of items. Runs in linear time.
18    */
19   private void buildHeap( )
20   {
21       for( int i = currentSize / 2; i > 0; i-- )
22           percolateDown( i );
23   }
```

Figure 6.14 Sketch of `buildHeap`

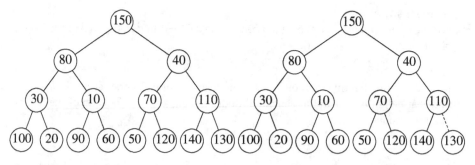

Figure 6.15 Left: initial heap; right: after `percolateDown(7)`

thus, it appears in Equation (6.3). The first term in Equation (6.1), h, does not appear in Equation (6.2); thus, $-h$ appears in Equation (6.3). We obtain

$$S = -h + 2 + 4 + 8 + \cdots + 2^{h-1} + 2^h = (2^{h+1} - 1) - (h + 1) \qquad (6.3)$$

which proves the theorem.

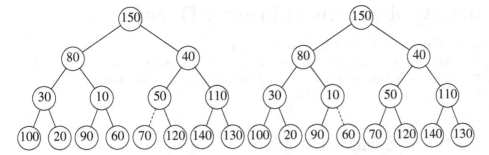

Figure 6.16 Left: after `percolateDown(6)`; right: after `percolateDown(5)`

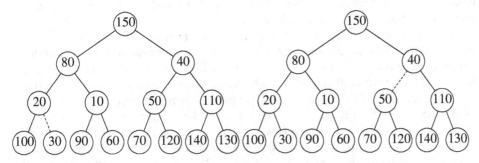

Figure 6.17 Left: after `percolateDown(4)`; right: after `percolateDown(3)`

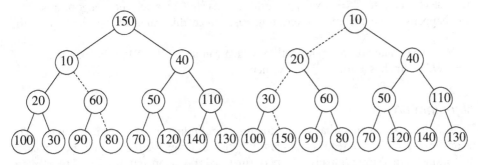

Figure 6.18 Left: after `percolateDown(2)`; right: after `percolateDown(1)`

A complete tree is not a perfect binary tree, but the result we have obtained is an upper bound on the sum of the heights of the nodes in a complete tree. Since a complete tree has between 2^h and 2^{h+1} nodes, this theorem implies that this sum is $O(N)$, where N is the number of nodes.

Although the result we have obtained is sufficient to show that `buildHeap` is linear, the bound on the sum of the heights is not as strong as possible. For a complete tree with $N = 2^h$ nodes, the bound we have obtained is roughly $2N$. The sum of the heights can be shown by induction to be $N - b(N)$, where $b(N)$ is the number of 1s in the binary representation of N.

6.4 Applications of Priority Queues

We have already mentioned how priority queues are used in operating systems design. In Chapter 9, we will see how priority queues are used to implement several graph algorithms efficiently. Here we will show how to use priority queues to obtain solutions to two problems.

6.4.1 The Selection Problem

The first problem we will examine is the *selection problem* from Chapter 1. Recall that the input is a list of N elements, which can be totally ordered, and an integer k. The selection problem is to find the kth largest element.

Two algorithms were given in Chapter 1, but neither is very efficient. The first algorithm, which we shall call algorithm 1A, is to read the elements into an array and sort them, returning the appropriate element. Assuming a simple sorting algorithm, the running time is $O(N^2)$. The alternative algorithm, 1B, is to read k elements into an array and sort them. The smallest of these is in the kth position. We process the remaining elements one by one. As an element arrives, it is compared with the kth element in the array. If it is larger, then the kth element is removed, and the new element is placed in the correct place among the remaining $k - 1$ elements. When the algorithm ends, the element in the kth position is the answer. The running time is $O(N \cdot k)$ (why?). If $k = \lceil N/2 \rceil$, then both algorithms are $O(N^2)$. Notice that for any k, we can solve the symmetric problem of finding the $(N - k + 1)$th smallest element, so $k = \lceil N/2 \rceil$ is really the hardest case for these algorithms. This also happens to be the most interesting case, since this value of k is known as the *median*.

We give two algorithms here, both of which run in $O(N \log N)$ in the extreme case of $k = \lceil N/2 \rceil$, which is a distinct improvement.

Algorithm 6A

For simplicity, we assume that we are interested in finding the kth *smallest* element. The algorithm is simple. We read the N elements into an array. We then apply the `buildHeap` algorithm to this array. Finally, we perform k `deleteMin` operations. The last element extracted from the heap is our answer. It should be clear that by changing the heap-order property, we could solve the original problem of finding the kth *largest* element.

The correctness of the algorithm should be clear. The worst-case timing is $O(N)$ to construct the heap, if `buildHeap` is used, and $O(\log N)$ for each `deleteMin`. Since there are k `deleteMins`, we obtain a total running time of $O(N + k \log N)$. If $k = O(N/\log N)$, then the running time is dominated by the `buildHeap` operation and is $O(N)$. For larger values of k, the running time is $O(k \log N)$. If $k = \lceil N/2 \rceil$, then the running time is $\Theta(N \log N)$.

Notice that if we run this program for $k = N$ and record the values as they leave the heap, we will have essentially sorted the input file in $O(N \log N)$ time. In Chapter 7, we will refine this idea to obtain a fast sorting algorithm known as *heapsort*.

Algorithm 6B

For the second algorithm, we return to the original problem and find the kth *largest* element. We use the idea from algorithm 1B. At any point in time we will maintain a set S of the k largest elements. After the first k elements are read, when a new element is read it is compared with the kth largest element, which we denote by S_k. Notice that S_k is the smallest element in S. If the new element is larger, then it replaces S_k in S. S will then have a new smallest element, which may or may not be the newly added element. At the end of the input, we find the smallest element in S and return it as the answer.

This is essentially the same algorithm described in Chapter 1. Here, however, we will use a heap to implement S. The first k elements are placed into the heap in total time $O(k)$ with a call to `buildHeap`. The time to process each of the remaining elements is $O(1)$, to test if the element goes into S, plus $O(\log k)$, to delete S_k and insert the new element if this is necessary. Thus, the total time is $O(k + (N - k) \log k) = O(N \log k)$. This algorithm also gives a bound of $\Theta(N \log N)$ for finding the median.

In Chapter 7, we will see how to solve this problem in $O(N)$ average time. In Chapter 10, we will see an elegant, albeit impractical, algorithm to solve this problem in $O(N)$ worst-case time.

6.4.2 Event Simulation

In Section 3.7.3, we described an important queuing problem. Recall that we have a system, such as a bank, where customers arrive and wait in a line until one of k tellers is available. Customer arrival is governed by a probability distribution function, as is the service time (the amount of time to be served once a teller is available). We are interested in statistics such as how long on average a customer has to wait or how long the line might be.

With certain probability distributions and values of k, these answers can be computed exactly. However, as k gets larger, the analysis becomes considerably more difficult, so it is appealing to use a computer to simulate the operation of the bank. In this way, the bank officers can determine how many tellers are needed to ensure reasonably smooth service.

A simulation consists of processing events. The two events here are (a) a customer arriving and (b) a customer departing, thus freeing up a teller.

We can use the probability functions to generate an input stream consisting of ordered pairs of arrival time and service time for each customer, sorted by arrival time. We do not need to use the exact time of day. Rather, we can use a quantum unit, which we will refer to as a **tick**.

One way to do this simulation is to start a simulation clock at zero ticks. We then advance the clock one tick at a time, checking to see if there is an event. If there is, then we process the event(s) and compile statistics. When there are no customers left in the input stream and all the tellers are free, then the simulation is over.

The problem with this simulation strategy is that its running time does not depend on the number of customers or events (there are two events per customer), but instead depends on the number of ticks, which is not really part of the input. To see why this is important, suppose we changed the clock units to milliticks and multiplied all the times in the input by 1,000. The result would be that the simulation would take 1,000 times longer!

The key to avoiding this problem is to advance the clock to the next event time at each stage. This is conceptually easy to do. At any point, the next event that can occur is either (a) the next customer in the input file arrives or (b) one of the customers at a teller leaves. Since all the times when the events will happen are available, we just need to find the event that happens nearest in the future and process that event.

If the event is a departure, processing includes gathering statistics for the departing customer and checking the line (queue) to see whether there is another customer waiting. If so, we add that customer, process whatever statistics are required, compute the time when that customer will leave, and add that departure to the set of events waiting to happen.

If the event is an arrival, we check for an available teller. If there is none, we place the arrival on the line (queue); otherwise we give the customer a teller, compute the customer's departure time, and add the departure to the set of events waiting to happen.

The waiting line for customers can be implemented as a queue. Since we need to find the event *nearest* in the future, it is appropriate that the set of departures waiting to happen be organized in a priority queue. The next event is thus the next arrival or next departure (whichever is sooner); both are easily available.

It is then straightforward, although possibly time-consuming, to write the simulation routines. If there are C customers (and thus $2C$ events) and k tellers, then the running time of the simulation would be $O(C \log(k + 1))$ because computing and processing each event takes $O(\log H)$, where $H = k + 1$ is the size of the heap.[2]

6.5 *d*-Heaps

Binary heaps are so simple that they are almost always used when priority queues are needed. A simple generalization is a **d-heap**, which is exactly like a binary heap except that all nodes have d children (thus, a binary heap is a 2-heap).

Figure 6.19 shows a 3-heap. Notice that a d-heap is much shallower than a binary heap, improving the running time of inserts to $O(\log_d N)$. However, for large d, the deleteMin operation is more expensive, because even though the tree is shallower, the minimum of d children must be found, which takes $d - 1$ comparisons using a standard algorithm. This raises the time for this operation to $O(d \log_d N)$. If d is a constant, both running times are, of course, $O(\log N)$. Although an array can still be used, the multiplications and divisions to find children and parents are now by d, which, unless d is a power of 2, seriously increases the running time, because we can no longer implement division by a bit shift. d-heaps are interesting in theory, because there are many algorithms where the number of insertions is much greater than the number of deleteMins (and thus a theoretical speedup is possible). They are also of interest when the priority queue is too large to fit entirely in main memory. In this case, a d-heap can be advantageous in much the same way as B-trees. Finally, there is evidence suggesting that 4-heaps may outperform binary heaps in practice.

The most glaring weakness of the heap implementation, aside from the inability to perform finds, is that combining two heaps into one is a hard operation. This extra operation

---

[2] We use $O(C \log(k + 1))$ instead of $O(C \log k)$ to avoid confusion for the $k = 1$ case.

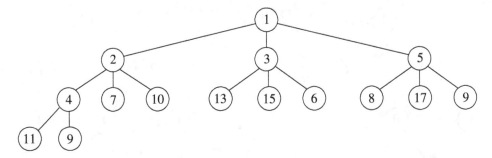

Figure 6.19 A *d*-heap

is known as a merge. There are quite a few ways of implementing heaps so that the running time of a merge is $O(\log N)$. We will now discuss three data structures, of various complexity, that support the merge operation efficiently. We will defer any complicated analysis until Chapter 11.

6.6 Leftist Heaps

It seems difficult to design a data structure that efficiently supports merging (that is, processes a merge in $o(N)$ time) and uses only an array, as in a binary heap. The reason for this is that merging would seem to require copying one array into another, which would take $\Theta(N)$ time for equal-sized heaps. For this reason, all the advanced data structures that support efficient merging require the use of a linked data structure. In practice, we can expect that this will make all the other operations slower.

Like a binary heap, a **leftist heap** has both a structural property and an ordering property. Indeed, a leftist heap, like virtually all heaps used, has the same heap-order property we have already seen. Furthermore, a leftist heap is also a binary tree. The only difference between a leftist heap and a binary heap is that leftist heaps are not perfectly balanced but actually attempt to be very unbalanced.

6.6.1 Leftist Heap Property

We define the **null path length**, $npl(X)$, of any node X to be the length of the shortest path from X to a node without two children. Thus, the *npl* of a node with zero or one child is 0, while $npl(\text{null}) = -1$. In the tree in Figure 6.20, the null path lengths are indicated inside the tree nodes.

Notice that the null path length of any node is 1 more than the minimum of the null path lengths of its children. This applies to nodes with less than two children because the null path length of null is -1.

The leftist heap property is that for every node X in the heap, the null path length of the left child is at least as large as that of the right child. This property is satisfied by only one of the trees in Figure 6.20, namely, the tree on the left. This property actually goes

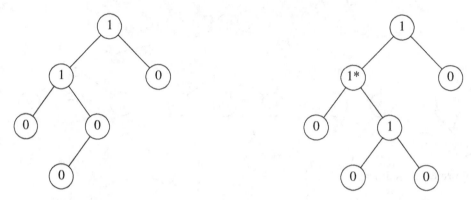

Figure 6.20   Null path lengths for two trees; only the left tree is leftist

out of its way to ensure that the tree is unbalanced, because it clearly biases the tree to get deep toward the left. Indeed, a tree consisting of a long path of left nodes is possible (and actually preferable to facilitate merging)—hence the name *leftist heap*.

Because leftist heaps tend to have deep left paths, it follows that the right path ought to be short. Indeed, the right path down a leftist heap is as short as any in the heap. Otherwise, there would be a path that goes through some node X and takes the left child. Then X would violate the leftist property.

Theorem 6.2.
A leftist tree with r nodes on the right path must have at least $2^r - 1$ nodes.

Proof.
The proof is by induction. If $r = 1$, there must be at least one tree node. Otherwise, suppose that the theorem is true for $1, 2, \ldots, r$. Consider a leftist tree with $r + 1$ nodes on the right path. Then the root has a right subtree with r nodes on the right path, and a left subtree with at least r nodes on the right path (otherwise it would not be leftist). Applying the inductive hypothesis to these subtrees yields a minimum of $2^r - 1$ nodes in each subtree. This plus the root gives at least $2^{r+1} - 1$ nodes in the tree, proving the theorem.

From this theorem, it follows immediately that a leftist tree of N nodes has a right path containing at most $\lfloor \log(N + 1) \rfloor$ nodes. The general idea for the leftist heap operations is to perform all the work on the right path, which is guaranteed to be short. The only tricky part is that performing inserts and merges on the right path could destroy the leftist heap property. It turns out to be extremely easy to restore the property.

6.6.2 Leftist Heap Operations

The fundamental operation on leftist heaps is merging. Notice that insertion is merely a special case of merging, since we may view an insertion as a merge of a one-node heap with a larger heap. We will first give a simple recursive solution and then show how this might

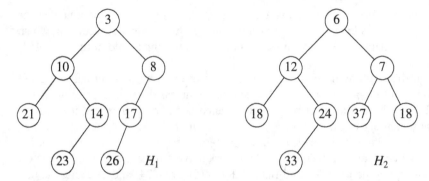

Figure 6.21 Two leftist heaps H_1 and H_2

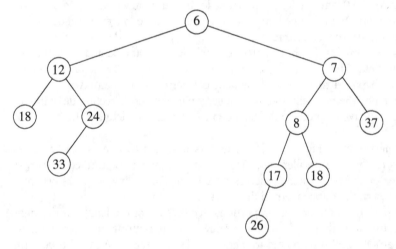

Figure 6.22 Result of merging H_2 with H_1's right subheap

be done nonrecursively. Our input is the two leftist heaps, H_1 and H_2, in Figure 6.21. You should check that these heaps really are leftist. Notice that the smallest elements are at the roots. In addition to space for the data and left and right references, each node will have an entry that indicates the null path length.

If either of the two heaps is empty, then we can return the other heap. Otherwise, to merge the two heaps, we compare their roots. First, we recursively merge the heap with the larger root with the right subheap of the heap with the smaller root. In our example, this means we recursively merge H_2 with the subheap of H_1 rooted at 8, obtaining the heap in Figure 6.22.

Since this tree is formed recursively, and we have not yet finished the description of the algorithm, we cannot at this point show how this heap was obtained. However, it is reasonable to assume that the resulting tree is a leftist heap, because it was obtained via a recursive step. This is much like the inductive hypothesis in a proof by induction. Since we can handle the base case (which occurs when one tree is empty), we can assume that the recursive step works as long as we can finish the merge; this is rule 3 of recursion, which we discussed in Chapter 1. We now make this new heap the right child of the root of H_1 (see Figure 6.23).

Although the resulting heap satisfies the heap-order property, it is not leftist because the left subtree of the root has a null path length of 1 whereas the right subtree has a null path length of 2. Thus, the leftist property is violated at the root. However, it is easy to see that the remainder of the tree must be leftist. The right subtree of the root is leftist, because of the recursive step. The left subtree of the root has not been changed, so it too must still be leftist. Thus, we need only to fix the root. We can make the entire tree leftist by merely swapping the root's left and right children (Figure 6.24) and updating the null path length—the new null path length is 1 plus the null path length of the new right child—completing the `merge`. Notice that if the null path length is not updated, then all null path lengths will be 0, and the heap will not be leftist but merely random. In this case, the algorithm will work, but the time bound we will claim will no longer be valid.

The description of the algorithm translates directly into code. The node class (Figure 6.25) is the same as the binary tree, except that it is augmented with the `npl` (null path length) field. The leftist heap stores a reference to the root as its data member. We have seen in Chapter 4 that when an element is inserted into an empty binary tree, the node referenced by the root will need to change. We use the usual technique of implementing `private` recursive methods to do the merging. The class skeleton is also shown in Figure 6.25.

The two `merge` routines (Figure 6.26) are drivers designed to remove special cases and ensure that H_1 has the smaller root. The actual merging is performed in `merge1` (Figure 6.27). The public `merge` method merges `rhs` into the controlling heap. `rhs` becomes empty. The alias test in the public method disallows `h.merge(h)`.

The time to perform the merge is proportional to the sum of the length of the right paths, because constant work is performed at each node visited during the recursive calls. Thus we obtain an $O(\log N)$ time bound to merge two leftist heaps. We can also perform this operation nonrecursively by essentially performing two passes. In the first pass, we create a new tree by merging the right paths of both heaps. To do this, we arrange the nodes on the right paths of H_1 and H_2 in sorted order, keeping their respective left children. In our example, the new right path is 3, 6, 7, 8, 18 and the resulting tree is shown in Figure 6.28. A second pass is made up the heap, and child swaps are performed at nodes that violate the leftist heap property. In Figure 6.28, there is a swap at nodes 7 and 3, and the same tree as before is obtained. The nonrecursive version is simpler to visualize but harder to code. We leave it to the reader to show that the recursive and nonrecursive procedures do the same thing.

As mentioned above, we can carry out insertions by making the item to be inserted a one-node heap and performing a `merge`. To perform a `deleteMin`, we merely destroy the root, creating two heaps, which can then be merged. Thus, the time to perform a `deleteMin` is $O(\log N)$. These two routines are coded in Figure 6.29 and Figure 6.30.

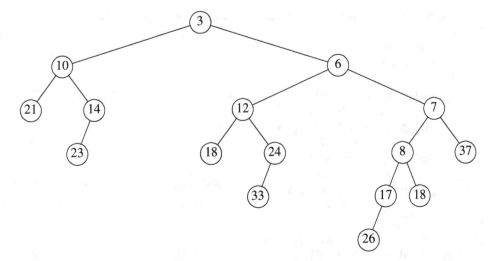

Figure 6.23   Result of attaching leftist heap of previous figure as H_1's right child

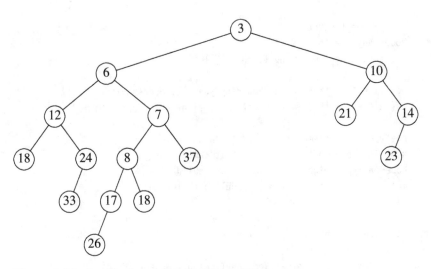

Figure 6.24   Result of swapping children of H_1's root

```
1   public class LeftistHeap<AnyType extends Comparable<? super AnyType>>
2   {
3       public LeftistHeap( )
4         { root = null; }
5
6       public void merge( LeftistHeap<AnyType> rhs )
7         { /* Figure 6.26 */ }
8       public void insert( AnyType x )
9         { /* Figure 6.29 */ }
10      public AnyType findMin( )
11        { /* See online code */ }
12      public AnyType deleteMin( )
13        { /* Figure 6.30 */ }
14
15      public boolean isEmpty( )
16        { return root == null; }
17      public void makeEmpty( )
18        { root = null; }
19
20      private static class Node<AnyType>
21        {
22              // Constructors
23          Node( AnyType theElement )
24            { this( theElement, null, null ); }
25
26          Node( AnyType theElement, Node<AnyType> lt, Node<AnyType> rt )
27            { element = theElement; left = lt; right = rt; npl = 0; }
28
29          AnyType      element;      // The data in the node
30          Node<AnyType> left;        // Left child
31          Node<AnyType> right;       // Right child
32          int           npl;         // null path length
33        }
34
35      private Node<AnyType> root;    // root
36
37      private Node<AnyType> merge( Node<AnyType> h1, Node<AnyType> h2 )
38        { /* Figure 6.26 */ }
39      private Node<AnyType> merge1( Node<AnyType> h1, Node<AnyType> h2 )
40        { /* Figure 6.27 */ }
41      private void swapChildren( Node<AnyType> t )
42        { /* See online code */ }
43  }
```

Figure 6.25 Leftist heap type declarations

```
1      /**
2       * Merge rhs into the priority queue.
3       * rhs becomes empty. rhs must be different from this.
4       * @param rhs the other leftist heap.
5       */
6      public void merge( LeftistHeap<AnyType> rhs )
7      {
8          if( this == rhs )    // Avoid aliasing problems
9              return;
10
11         root = merge( root, rhs.root );
12         rhs.root = null;
13     }
14
15     /**
16      * Internal method to merge two roots.
17      * Deals with deviant cases and calls recursive merge1.
18      */
19     private Node<AnyType> merge( Node<AnyType> h1, Node<AnyType> h2 )
20     {
21         if( h1 == null )
22             return h2;
23         if( h2 == null )
24             return h1;
25         if( h1.element.compareTo( h2.element ) < 0 )
26             return merge1( h1, h2 );
27         else
28             return merge1( h2, h1 );
29     }
```

Figure 6.26   Driving routines for merging leftist heaps

Finally, we can build a leftist heap in $O(N)$ time by building a binary heap (obviously using a linked implementation). Although a binary heap is clearly leftist, this is not necessarily the best solution, because the heap we obtain is the worst possible leftist heap. Furthermore, traversing the tree in reverse-level order is not as easy with links. The buildHeap effect can be obtained by recursively building the left and right subtrees and then percolating the root down. The exercises contain an alternative solution.

```
1      /**
2       * Internal method to merge two roots.
3       * Assumes trees are not empty, and h1's root contains smallest item.
4       */
5      private Node<AnyType> merge1( Node<AnyType> h1, Node<AnyType> h2 )
6      {
7          if( h1.left == null )   // Single node
8              h1.left = h2;       // Other fields in h1 already accurate
9          else
10         {
11             h1.right = merge( h1.right, h2 );
12             if( h1.left.npl < h1.right.npl )
13                 swapChildren( h1 );
14             h1.npl = h1.right.npl + 1;
15         }
16         return h1;
17     }
```

Figure 6.27   Actual routine to merge leftist heaps

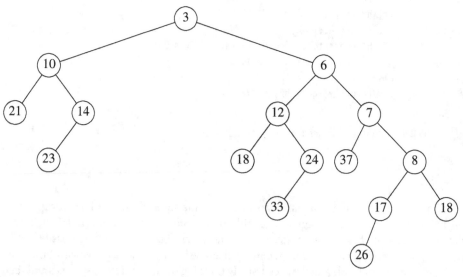

Figure 6.28   Result of merging right paths of H_1 and H_2

```
1      /**
2       * Insert into the priority queue, maintaining heap order.
3       * @param x the item to insert.
4       */
5      public void insert( AnyType x )
6      {
7          root = merge( new Node<>( x ), root );
8      }
```

Figure 6.29 Insertion routine for leftist heaps

```
1      /**
2       * Remove the smallest item from the priority queue.
3       * @return the smallest item, or throw UnderflowException if empty.
4       */
5      public AnyType deleteMin( )
6      {
7          if( isEmpty( ) )
8              throw new UnderflowException( );
9
10         AnyType minItem = root.element;
11         root = merge( root.left, root.right );
12
13         return minItem;
14     }
```

Figure 6.30 deleteMin routine for leftist heaps

6.7 Skew Heaps

A **skew heap** is a self-adjusting version of a leftist heap that is incredibly simple to implement. The relationship of skew heaps to leftist heaps is analogous to the relation between splay trees and AVL trees. Skew heaps are binary trees with heap order, but there is no structural constraint on these trees. Unlike leftist heaps, no information is maintained about the null path length of any node. The right path of a skew heap can be arbitrarily long at any time, so the worst-case running time of all operations is $O(N)$. However, as with splay trees, it can be shown (see Chapter 11) that for any M consecutive operations, the total worst-case running time is $O(M \log N)$. Thus, skew heaps have $O(\log N)$ amortized cost per operation.

As with leftist heaps, the fundamental operation on skew heaps is merging. The merge routine is once again recursive, and we perform the exact same operations as before, with one exception. The difference is that for leftist heaps, we check to see whether the left and right children satisfy the leftist heap structure property and swap them if they do not. For skew heaps, the swap is unconditional; we *always* do it, with the one exception that the largest of all the nodes on the right paths does not have its children swapped. This one exception is what happens in the natural recursive implementation, so it is not really a special case at all. Furthermore, it is not necessary to prove the bounds, but since this node is guaranteed not to have a right child, it would be silly to perform the swap and give it one. (In our example, there are no children of this node, so we do not worry about it.) Again, suppose our input is the same two heaps as before, Figure 6.31.

If we recursively merge H_2 with the subheap of H_1 rooted at 8, we will get the heap in Figure 6.32.

Again, this is done recursively, so by the third rule of recursion (Section 1.3) we need not worry about how it was obtained. This heap happens to be leftist, but there is no

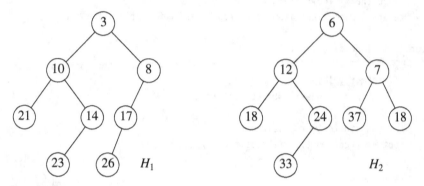

Figure 6.31   Two skew heaps H_1 and H_2

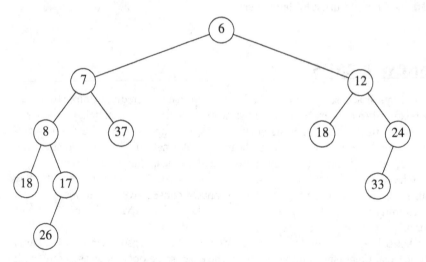

Figure 6.32   Result of merging H_2 with H_1's right subheap

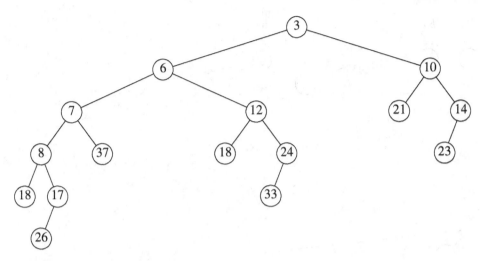

Figure 6.33 Result of merging skew heaps H_1 and H_2

guarantee that this is always the case. We make this heap the new left child of H_1, and the old left child of H_1 becomes the new right child (see Figure 6.33).

The entire tree is leftist, but it is easy to see that that is not always true: Inserting 15 into this new heap would destroy the leftist property.

We can perform all operations nonrecursively, as with leftist heaps, by merging the right paths and swapping left and right children for every node on the right path, with the exception of the last. After a few examples, it becomes clear that since all but the last node on the right path have their children swapped, the net effect is that this becomes the new left path (see the preceding example to convince yourself). This makes it very easy to merge two skew heaps visually.[3]

The implementation of skew heaps is left as a (trivial) exercise. Note that because a right path could be long, a recursive implementation could fail because of lack of stack space, even though performance would otherwise be acceptable. Skew heaps have the advantage that no extra space is required to maintain path lengths and no tests are required to determine when to swap children. It is an open problem to determine precisely the expected right path length of both leftist and skew heaps (the latter is undoubtedly more difficult). Such a comparison would make it easier to determine whether the slight loss of balance information is compensated by the lack of testing.

---

[3] This is not exactly the same as the recursive implementation (but yields the same time bounds). If we only swap children for nodes on the right path that are above the point where the merging of right paths terminated due to exhaustion of one heap's right path, we get the same result as the recursive version.

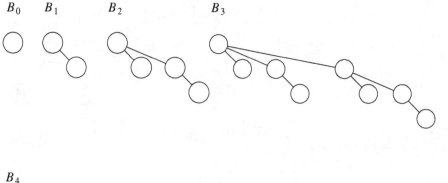

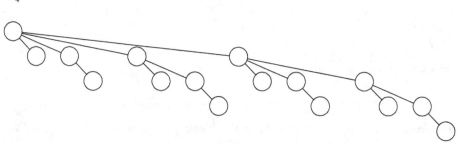

Figure 6.34   Binomial trees B_0, B_1, B_2, B_3, and B_4

6.8 Binomial Queues

Although both leftist and skew heaps support merging, insertion, and `deleteMin` all effec-
tively in $O(\log N)$ time per operation, there is room for improvement because we know
that binary heaps support insertion in *constant average* time per operation. Binomial queues
support all three operations in $O(\log N)$ worst-case time per operation, but insertions take
constant time on average.

6.8.1 Binomial Queue Structure

Binomial queues differ from all the priority queue implementations that we have seen in
that a binomial queue is not a heap-ordered tree but rather a *collection* of heap-ordered
trees, known as a **forest**. Each of the heap-ordered trees is of a constrained form known
as a **binomial tree** (the reason for the name will be obvious later). There is at most one
binomial tree of every height. A binomial tree of height 0 is a one-node tree; a binomial
tree, B_k, of height k is formed by attaching a binomial tree, B_{k-1}, to the root of another
binomial tree, B_{k-1}. Figure 6.34 shows binomial trees B_0, B_1, B_2, B_3, and B_4.

From the diagram we see that a binomial tree, B_k, consists of a root with children
$B_0, B_1, \ldots, B_{k-1}$. Binomial trees of height k have exactly 2^k nodes, and the number of
nodes at depth d is the binomial coefficient $\binom{k}{d}$. If we impose heap order on the binomial

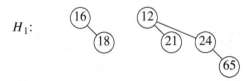

H_1:

Figure 6.35 Binomial queue H_1 with six elements

trees and allow at most one binomial tree of any height, we can represent a priority queue of any size by a collection of binomial trees. For instance, a priority queue of size 13 could be represented by the forest B_3, B_2, B_0. We might write this representation as 1101, which not only represents 13 in binary but also represents the fact that B_3, B_2, and B_0 are present in the representation and B_1 is not.

As an example, a priority queue of six elements could be represented as in Figure 6.35.

6.8.2 Binomial Queue Operations

The minimum element can then be found by scanning the roots of all the trees. Since there are at most $\log N$ different trees, the minimum can be found in $O(\log N)$ time. Alternatively, we can maintain knowledge of the minimum and perform the operation in $O(1)$ time, if we remember to update the minimum when it changes during other operations.

Merging two binomial queues is a conceptually easy operation, which we will describe by example. Consider the two binomial queues, H_1 and H_2, with six and seven elements, respectively, pictured in Figure 6.36.

The merge is performed by essentially adding the two queues together. Let H_3 be the new binomial queue. Since H_1 has no binomial tree of height 0 and H_2 does, we can just use the binomial tree of height 0 in H_2 as part of H_3. Next, we add binomial trees of height 1. Since both H_1 and H_2 have binomial trees of height 1, we merge them by making the larger root a subtree of the smaller, creating a binomial tree of height 2, shown in Figure 6.37. Thus, H_3 will not have a binomial tree of height 1. There are now three binomial trees of height 2, namely, the original trees of H_1 and H_2 plus the tree formed

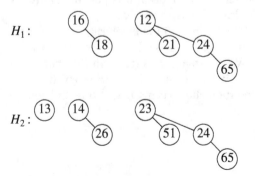

H_1:

H_2:

Figure 6.36 Two binomial queues H_1 and H_2

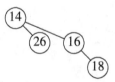

Figure 6.37   Merge of the two B_1 trees in H_1 and H_2

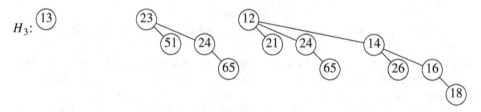

Figure 6.38   Binomial queue H_3: the result of merging H_1 and H_2

by the previous step. We keep one binomial tree of height 2 in H_3 and merge the other two, creating a binomial tree of height 3. Since H_1 and H_2 have no trees of height 3, this tree becomes part of H_3 and we are finished. The resulting binomial queue is shown in Figure 6.38.

Since merging two binomial trees takes constant time with almost any reasonable implementation, and there are $O(\log N)$ binomial trees, the merge takes $O(\log N)$ time in the worst case. To make this operation efficient, we need to keep the trees in the binomial queue sorted by height, which is certainly a simple thing to do.

Insertion is just a special case of merging, since we merely create a one-node tree and perform a merge. The worst-case time of this operation is likewise $O(\log N)$. More precisely, if the priority queue into which the element is being inserted has the property that the smallest nonexistent binomial tree is B_i, the running time is proportional to $i + 1$. For example, H_3 (Figure 6.38) is missing a binomial tree of height 1, so the insertion will terminate in two steps. Since each tree in a binomial queue is present with probability $\frac{1}{2}$, it follows that we expect an insertion to terminate in two steps, so the average time is constant. Furthermore, an analysis will show that performing N inserts on an initially empty binomial queue will take $O(N)$ worst-case time. Indeed, it is possible to do this operation using only $N - 1$ comparisons; we leave this as an exercise.

As an example, we show in Figures 6.39 through 6.45 the binomial queues that are formed by inserting 1 through 7 in order. Inserting 4 shows off a bad case. We merge 4 with B_0, obtaining a new tree of height 1. We then merge this tree with B_1, obtaining a tree of height 2, which is the new priority queue. We count this as three steps (two tree merges plus the stopping case). The next insertion after 7 is inserted is another bad case and would require three tree merges.

Figure 6.39   After 1 is inserted

Figure 6.40   After 2 is inserted

Figure 6.41   After 3 is inserted

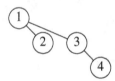

Figure 6.42   After 4 is inserted

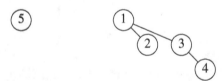

Figure 6.43   After 5 is inserted

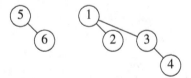

Figure 6.44   After 6 is inserted

A `deleteMin` can be performed by first finding the binomial tree with the smallest root. Let this tree be B_k, and let the original priority queue be H. We remove the binomial tree B_k from the forest of trees in H, forming the new binomial queue H'. We also remove the root of B_k, creating binomial trees $B_0, B_1, \ldots, B_{k-1}$, which collectively form priority queue H''. We finish the operation by merging H' and H''.

As an example, suppose we perform a `deleteMin` on H_3, which is shown again in Figure 6.46. The minimum root is 12, so we obtain the two priority queues H' and H'' in Figure 6.47 and Figure 6.48. The binomial queue that results from merging H' and H'' is the final answer and is shown in Figure 6.49.

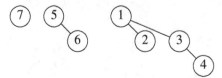

Figure 6.45   After 7 is inserted

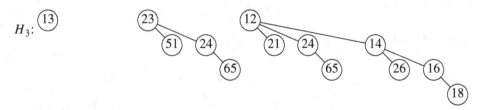

Figure 6.46   Binomial queue H_3

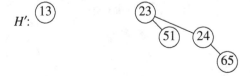

Figure 6.47   Binomial queue H', containing all the binomial trees in H_3 except B_3

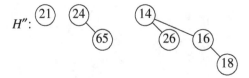

Figure 6.48   Binomial queue H'': B_3 with 12 removed

For the analysis, note first that the `deleteMin` operation breaks the original binomial queue into two. It takes $O(\log N)$ time to find the tree containing the minimum element and to create the queues H' and H''. Merging these two queues takes $O(\log N)$ time, so the entire `deleteMin` operation takes $O(\log N)$ time.

6.8.3 Implementation of Binomial Queues

The `deleteMin` operation requires the ability to find all the subtrees of the root quickly, so the standard representation of general trees is required: The children of each node are kept

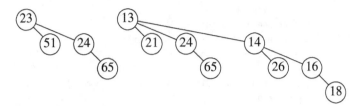

Figure 6.49   Result of applying `deleteMin` to H_3

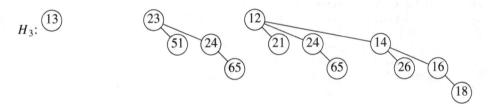

Figure 6.50   Binomial queue H_3 drawn as a forest

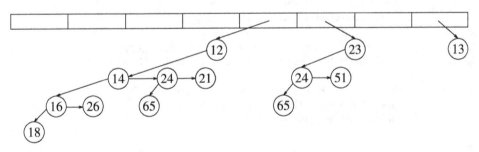

Figure 6.51   Representation of binomial queue H_3

in a linked list, and each node has a reference to its first child (if any). This operation also requires that the children be ordered by the size of their subtrees. We also need to make sure that it is easy to merge two trees. When two trees are merged, one of the trees is added as a child to the other. Since this new tree will be the largest subtree, it makes sense to maintain the subtrees in decreasing sizes. Only then will we be able to merge two binomial trees, and thus two binomial queues, efficiently. The binomial queue will be an array of binomial trees.

To summarize, then, each node in a binomial tree will contain the data, first child, and right sibling. The children in a binomial tree are arranged in decreasing rank.

Figure 6.51 shows how the binomial queue in Figure 6.50 is represented. Figure 6.52 shows the type declarations for a node in the binomial tree, and the binomial queue class skeleton.

```
1    public class BinomialQueue<AnyType extends Comparable<? super AnyType>>
2    {
3        public BinomialQueue( )
4          { /* See online code */ }
5        public BinomialQueue( AnyType item )
6          { /* See online code */ }
7
8        public void merge( BinomialQueue<AnyType> rhs )
9          { /* Figure 6.55 */ }
10       public void insert( AnyType x )
11         { merge( new BinomialQueue<>( x ) ); }
12       public AnyType findMin( )
13         { /* See online code */ }
14       public AnyType deleteMin( )
15         { /* Figure 6.56 */ }
16
17       public boolean isEmpty( )
18         { return currentSize == 0; }
19       public void makeEmpty( )
20         { /* See online code */ }
21
22       private static class Node<AnyType>
23       {
24             // Constructors
25           Node( AnyType theElement )
26             { this( theElement, null, null ); }
27
28           Node( AnyType theElement, Node<AnyType> lt, Node<AnyType> nt )
29             { element = theElement; leftChild = lt;  nextSibling = nt; }
30
31           AnyType         element;     // The data in the node
32           Node<AnyType> leftChild;   // Left child
33           Node<AnyType> nextSibling; // Right child
34       }
35
36       private static final int DEFAULT_TREES = 1;
37
38       private int currentSize;              // # items in priority queue
39       private Node<AnyType> [ ] theTrees;  // An array of tree roots
40
41       private void expandTheTrees( int newNumTrees )
42         { /* See online code */ }
43       private Node<AnyType> combineTrees( Node<AnyType> t1, Node<AnyType> t2 )
44         { /* Figure 6.54 */ }
45
46       private int capacity( )
47         { return ( 1 << theTrees.length ) - 1; }
48       private int findMinIndex( )
49         { /* See online code */ }
50   }
```

Figure 6.52 Binomial queue class skeleton and node definition

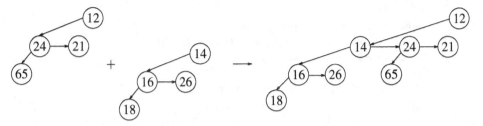

Figure 6.53 Merging two binomial trees

```
1      /**
2       * Return the result of merging equal-sized t1 and t2.
3       */
4      private Node<AnyType> combineTrees( Node<AnyType> t1, Node<AnyType> t2 )
5      {
6          if( t1.element.compareTo( t2.element ) > 0 )
7              return combineTrees( t2, t1 );
8          t2.nextSibling = t1.leftChild;
9          t1.leftChild = t2;
10         return t1;
11     }
```

Figure 6.54 Routine to merge two equal-sized binomial trees

In order to merge two binomial queues, we need a routine to merge two binomial trees of the same size. Figure 6.53 shows how the links change when two binomial trees are merged. The code to do this is simple and is shown in Figure 6.54.

We provide a simple implementation of the merge routine. H_1 is represented by the current object and H_2 is represented by rhs. The routine combines H_1 and H_2, placing the result in H_1 and making H_2 empty. At any point we are dealing with trees of rank i. t1 and t2 are the trees in H_1 and H_2, respectively, and carry is the tree carried from a previous step (it might be null). Depending on each of the eight possible cases, the tree that results for rank i and the carry tree of rank $i + 1$ is formed. This process proceeds from rank 0 to the last rank in the resulting binomial queue. The code is shown in Figure 6.55. Improvements to the code are suggested in Exercise 6.35.

The deleteMin routine for binomial queues is given in Figure 6.56.

We can extend binomial queues to support some of the nonstandard operations that binary heaps allow, such as decreaseKey and delete, when the position of the affected element is known. A decreaseKey is a percolateUp, which can be performed in $O(\log N)$ time if we add a field to each node that stores a parent link. An arbitrary delete can be performed by a combination of decreaseKey and deleteMin in $O(\log N)$ time.

```
1      /**
2       * Merge rhs into the priority queue.
3       * rhs becomes empty. rhs must be different from this.
4       * @param rhs the other binomial queue.
5       */
6      public void merge( BinomialQueue<AnyType> rhs )
7      {
8          if( this == rhs )    // Avoid aliasing problems
9              return;
10
11         currentSize += rhs.currentSize;
12
13         if( currentSize > capacity( ) )
14         {
15             int maxLength = Math.max( theTrees.length, rhs.theTrees.length );
16             expandTheTrees( maxLength + 1 );
17         }
18
19         Node<AnyType> carry = null;
20         for( int i = 0, j = 1; j <= currentSize; i++, j *= 2 )
21         {
22             Node<AnyType> t1 = theTrees[ i ];
23             Node<AnyType> t2 = i < rhs.theTrees.length ? rhs.theTrees[ i ] : null;
24
25             int whichCase = t1 == null ? 0 : 1;
26             whichCase += t2 == null ? 0 : 2;
27             whichCase += carry == null ? 0 : 4;
28
29             switch( whichCase )
30             {
31               case 0: /* No trees */
32               case 1: /* Only this */
33                 break;
34               case 2: /* Only rhs */
35                 theTrees[ i ] = t2;
36                 rhs.theTrees[ i ] = null;
37                 break;
38               case 4: /* Only carry */
39                 theTrees[ i ] = carry;
40                 carry = null;
41                 break;
42               case 3: /* this and rhs */
43                 carry = combineTrees( t1, t2 );
44                 theTrees[ i ] = rhs.theTrees[ i ] = null;
45                 break;
```

Figure 6.55   Routine to merge two priority queues

```
46              case 5: /* this and carry */
47                carry = combineTrees( t1, carry );
48                theTrees[ i ] = null;
49                break;
50              case 6: /* rhs and carry */
51                carry = combineTrees( t2, carry );
52                rhs.theTrees[ i ] = null;
53                break;
54              case 7: /* All three */
55                theTrees[ i ] = carry;
56                carry = combineTrees( t1, t2 );
57                rhs.theTrees[ i ] = null;
58                break;
59            }
60          }
61
62          for( int k = 0; k < rhs.theTrees.length; k++ )
63              rhs.theTrees[ k ] = null;
64          rhs.currentSize = 0;
65      }
```

Figure 6.55 *(continued)*

6.9 Priority Queues in the Standard Library

Prior to Java 1.5, there was no support in the Java library for priority queues. However, in Java 1.5, there is a generic PriorityQueue class. In this class, insert, findMin, and deleteMin are expressed via calls to add, element, and remove. The PriorityQueue can be constructed either with no parameters, a comparator, or another compatible collection.

Because there are many efficient implementations of priority queues, it is unfortunate that the library designers did not choose to make PriorityQueue an interface. Nonetheless, the PriorityQueue implementation in Java 1.5 is sufficient for most priority queue applications.

Summary

In this chapter we have seen various implementations and uses of the priority queue ADT. The standard binary heap implementation is elegant because of its simplicity and speed. It requires no links and only a constant amount of extra space, yet supports the priority queue operations efficiently.

```
1       /**
2        * Remove the smallest item from the priority queue.
3        * @return the smallest item, or throw UnderflowException if empty.
4        */
5       public AnyType deleteMin( )
6       {
7           if( isEmpty( ) )
8               throw new UnderflowException( );
9
10          int minIndex = findMinIndex( );
11          AnyType minItem = theTrees[ minIndex ].element;
12
13          Node<AnyType> deletedTree = theTrees[ minIndex ].leftChild;
14
15          // Construct H''
16          BinomialQueue<AnyType> deletedQueue = new BinomialQueue<>( );
17          deletedQueue.expandTheTrees( minIndex + 1 );
18
19          deletedQueue.currentSize = ( 1 << minIndex ) - 1;
20          for( int j = minIndex - 1; j >= 0; j-- )
21          {
22              deletedQueue.theTrees[ j ] = deletedTree;
23              deletedTree = deletedTree.nextSibling;
24              deletedQueue.theTrees[ j ].nextSibling = null;
25          }
26
27          // Construct H'
28          theTrees[ minIndex ] = null;
29          currentSize -= deletedQueue.currentSize + 1;
30
31          merge( deletedQueue );
32
33          return minItem;
34      }
```

Figure 6.56   deleteMin for binomial queues, with findMinIndex method

We considered the additional merge operation and developed three implementations, each of which is unique in its own way. The leftist heap is a wonderful example of the power of recursion. The skew heap represents a remarkable data structure because of the lack of balance criteria. Its analysis, which we will perform in Chapter 11, is interesting in its own right. The binomial queue shows how a simple idea can be used to achieve a good time bound.

We have also seen several uses of priority queues, ranging from operating systems scheduling to simulation. We will see their use again in Chapters 7, 9, and 10.

Exercises

6.1 Can both insert and findMin be implemented in constant time?

6.2 a. Show the result of inserting 10, 12, 1, 14, 6, 5, 8, 15, 3, 9, 7, 4, 11, 13, and 2, one at a time, into an initially empty binary heap.

b. Show the result of using the linear-time algorithm to build a binary heap using the same input.

6.3 Show the result of performing three deleteMin operations in the heap of the previous exercise.

6.4 A complete binary tree of N elements uses array positions 1 to N. Suppose we try to use an array representation of a binary tree that is not complete. Determine how large the array must be for the following:

a. a binary tree that has two extra levels (that is, it is very slightly unbalanced)

b. a binary tree that has a deepest node at depth $2 \log N$

c. a binary tree that has a deepest node at depth $4.1 \log N$

d. the worst-case binary tree

6.5 Rewrite the BinaryHeap insert method by placing a reference to the inserted item in position 0.

6.6 How many nodes are in the large heap in Figure 6.13?

6.7 a. Prove that for binary heaps, buildHeap does at most $2N - 2$ comparisons between elements.

b. Show that a heap of eight elements can be constructed in eight comparisons between heap elements.

**c. Give an algorithm to build a binary heap in $\frac{13}{8}N + O(\log N)$ element comparisons.

6.8 Show the following regarding the maximum item in the heap:

a. It must be at one of the leaves.

b. There are exactly $\lceil N/2 \rceil$ leaves.

c. Every leaf must be examined to find it.

****6.9** Show that the expected depth of the kth smallest element in a large complete heap (you may assume $N = 2^k - 1$) is bounded by $\log k$.

6.10 *a. Give an algorithm to find all nodes less than some value, X, in a binary heap. Your algorithm should run in $O(K)$, where K is the number of nodes output.

b. Does your algorithm extend to any of the other heap structures discussed in this chapter?

*c. Give an algorithm that finds an arbitrary item X in a binary heap using at most roughly $3N/4$ comparisons.

****6.11** Propose an algorithm to insert M nodes into a binary heap on N elements in $O(M + \log N \log \log N)$ time. Prove your time bound.

6.12 Write a program to take N elements and do the following:

a. Insert them into a heap one by one.

b. Build a heap in linear time.

Compare the running time of both algorithms for sorted, reverse-ordered, and random inputs.

6.13 Each deleteMin operation uses $2 \log N$ comparisons in the worst case.
  *a. Propose a scheme so that the deleteMin operation uses only $\log N + \log \log N + O(1)$ comparisons between elements. This need not imply less data movement.
  **b. Extend your scheme in part (a) so that only $\log N + \log \log \log N + O(1)$ comparisons are performed.
  **c. How far can you take this idea?
  d. Do the savings in comparisons compensate for the increased complexity of your algorithm?

6.14 If a d-heap is stored as an array, for an entry located in position i, where are the parents and children?

6.15 Suppose we need to perform M percolateUps and N deleteMins on a d-heap that initially has N elements.
  a. What is the total running time of all operations in terms of M, N, and d?
  b. If $d = 2$, what is the running time of all heap operations?
  c. If $d = \Theta(N)$, what is the total running time?
  *d. What choice of d minimizes the total running time?

6.16 Suppose that binary heaps are represented using explicit links. Give a simple algorithm to find the tree node that is at implicit position i.

6.17 Suppose that binary heaps are represented using explicit links. Consider the problem of merging binary heap lhs with rhs. Assume both heaps are perfect binary trees, containing $2^l - 1$ and $2^r - 1$ nodes, respectively.
  a. Give an $O(\log N)$ algorithm to merge the two heaps if $l = r$.
  b. Give an $O(\log N)$ algorithm to merge the two heaps if $|l - r| = 1$.
  c. Give an $O(\log^2 N)$ algorithm to merge the two heaps regardless of l and r.

6.18 A **min-max heap** is a data structure that supports both deleteMin and deleteMax in $O(\log N)$ per operation. The structure is identical to a binary heap, but the heap-order property is that for any node, X, at even depth, the element stored at X is smaller than the parent but larger than the grandparent (where this makes sense), and for any node X at odd depth, the element stored at X is larger than the parent but smaller than the grandparent. See Figure 6.57.
  a. How do we find the minimum and maximum elements?
  *b. Give an algorithm to insert a new node into the min-max heap.
  *c. Give an algorithm to perform deleteMin and deleteMax.
  *d. Can you build a min-max heap in linear time?
  **e. Suppose we would like to support deleteMin, deleteMax, and merge. Propose a data structure to support all operations in $O(\log N)$ time.

6.19 Merge the two leftist heaps in Figure 6.58.

6.20 Show the result of inserting keys 1 to 15 in order into an initially empty leftist heap.

6.21 Prove or disprove: A perfectly balanced tree forms if keys 1 to $2^k - 1$ are inserted in order into an initially empty leftist heap.

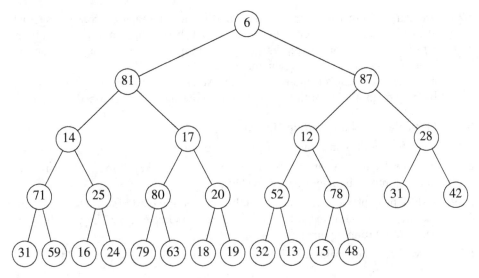

Figure 6.57 Min-max heap

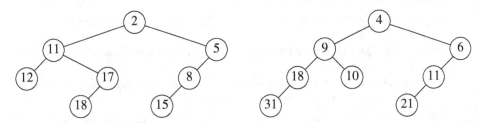

Figure 6.58 Input for Exercises 6.19 and 6.26

6.22 Give an example of input that generates the best leftist heap.

6.23 a. Can leftist heaps efficiently support decreaseKey?
b. What changes, if any (if possible), are required to do this?

6.24 One way to delete nodes from a known position in a leftist heap is to use a lazy strategy. To delete a node, merely mark it deleted. When a findMin or deleteMin is performed, there is a potential problem if the root is marked deleted, since then the node has to be actually deleted and the real minimum needs to be found, which may involve deleting other marked nodes. In this strategy, deletes cost one unit, but the cost of a deleteMin or findMin depends on the number of nodes that are marked deleted. Suppose that after a deleteMin or findMin there are k fewer marked nodes than before the operation.
  *a. Show how to perform the deleteMin in $O(k \log N)$ time.
  **b. Propose an implementation, with an analysis to show that the time to perform the deleteMin is $O(k \log(2N/k))$.

6.25 We can perform `buildHeap` in linear time for leftist heaps by considering each element as a one-node leftist heap, placing all these heaps on a queue, and performing the following step: Until only one heap is on the queue, dequeue two heaps, merge them, and enqueue the result.
a. Prove that this algorithm is $O(N)$ in the worst case.
b. Why might this algorithm be preferable to the algorithm described in the text?

6.26 Merge the two skew heaps in Figure 6.58.

6.27 Show the result of inserting keys 1 to 15 in order into a skew heap.

6.28 Prove or disprove: A perfectly balanced tree forms if the keys 1 to $2^k - 1$ are inserted in order into an initially empty skew heap.

6.29 A skew heap of N elements can be built using the standard binary heap algorithm. Can we use the same merging strategy described in Exercise 6.25 for skew heaps to get an $O(N)$ running time?

6.30 Prove that a binomial tree B_k has binomial trees $B_0, B_1, \ldots, B_{k-1}$ as children of the root.

6.31 Prove that a binomial tree of height k has $\binom{k}{d}$ nodes at depth d.

6.32 Merge the two binomial queues in Figure 6.59.

6.33 a. Show that N `inserts` into an initially empty binomial queue takes $O(N)$ time in the worst case.
b. Give an algorithm to build a binomial queue of N elements, using at most $N - 1$ comparisons between elements.
*c. Propose an algorithm to insert M nodes into a binomial queue of N elements in $O(M + \log N)$ worst-case time. Prove your bound.

6.34 Write an efficient routine to perform `insert` using binomial queues. Do not call `merge`.

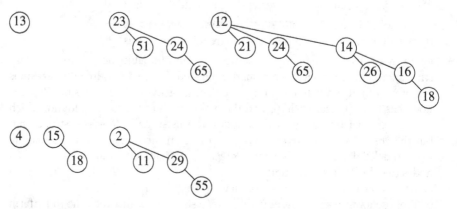

Figure 6.59 Input for Exercise 6.32

6.35   For the binomial queue:
        a. Modify the `merge` routine to terminate merging if there are no trees left in H_2 and the `carry` tree is `null`.
        b. Modify the `merge` so that the smaller tree is always merged into the larger.

**6.36   Suppose we extend binomial queues to allow at most two trees of the same height per structure. Can we obtain $O(1)$ worst-case time for insertion while retaining $O(\log N)$ for the other operations?

6.37   Suppose you have a number of boxes, each of which can hold total weight C and items $i_1, i_2, i_3, \ldots, i_N$, which weigh $w_1, w_2, w_3, \ldots, w_N$, respectively. The object is to pack all the items without placing more weight in any box than its capacity and using as few boxes as possible. For instance, if $C = 5$, and the items have weights 2, 2, 3, 3, then we can solve the problem with two boxes.

        In general, this problem is very hard, and no efficient solution is known. Write programs to implement efficiently the following approximation strategies:
        *a. Place the weight in the first box for which it fits (creating a new box if there is no box with enough room). (This strategy and all that follow would give three boxes, which is suboptimal.)
        b. Place the weight in the box with the most room for it.
        *c. Place the weight in the most filled box that can accept it without overflowing.
        **d. Are any of these strategies enhanced by presorting the items by weight?

6.38   Suppose we want to add the `decreaseAllKeys(Δ)` operation to the heap repertoire. The result of this operation is that all keys in the heap have their value decreased by an amount Δ. For the heap implementation of your choice, explain the necessary modifications so that all other operations retain their running times and `decreaseAllKeys` runs in $O(1)$.

6.39   Which of the two selection algorithms has the better time bound?

References

The binary heap was first described in [28]. The linear-time algorithm for its construction is from [14].

   The first description of d-heaps was in [19]. Recent results suggest that 4-heaps may improve binary heaps in some circumstances [22]. Leftist heaps were invented by Crane [11] and described in Knuth [21]. Skew heaps were developed by Sleator and Tarjan [24]. Binomial queues were invented by Vuillemin [27]; Brown provided a detailed analysis and empirical study showing that they perform well in practice [4], if carefully implemented.

   Exercise 6.7(b–c) is taken from [17]. Exercise 6.10(c) is from [6]. A method for constructing binary heaps that uses about $1.52N$ comparisons on average is described in [23]. Lazy deletion in leftist heaps (Exercise 6.24) is from [10]. A solution to Exercise 6.36 can be found in [8].

   Min-max heaps (Exercise 6.18) were originally described in [1]. A more efficient implementation of the operations is given in [18] and [25]. Alternative representations for double-ended priority queues are the *deap* and *diamond deque*. Details can be found in [5], [7], and [9]. Solutions to 6.18(e) are given in [12] and [20].

A theoretically interesting priority queue representation is the *Fibonacci heap* [16], which we will describe in Chapter 11. The Fibonacci heap allows all operations to be performed in $O(1)$ amortized time, except for deletions, which are $O(\log N)$. *Relaxed heaps* [13] achieve identical bounds in the worst case (with the exception of merge). The procedure of [3] achieves optimal worst-case bounds for all operations. Another interesting implementation is the *pairing heap* [15], which is described in Chapter 12. Finally, priority queues that work when the data consist of small integers are described in [2] and [26].

1. M. D. Atkinson, J. R. Sack, N. Santoro, and T. Strothotte, "Min-Max Heaps and Generalized Priority Queues," *Communications of the ACM,* 29 (1986), 996–1000.

2. J. D. Bright, "Range Restricted Mergeable Priority Queues," *Information Processing Letters,* 47 (1993), 159–164.

3. G. S. Brodal, "Worst-Case Efficient Priority Queues," *Proceedings of the Seventh Annual ACM-SIAM Symposium on Discrete Algorithms* (1996), 52–58.

4. M. R. Brown, "Implementation and Analysis of Binomial Queue Algorithms," *SIAM Journal on Computing,* 7 (1978), 298–319.

5. S. Carlsson, "The Deap—A Double-Ended Heap to Implement Double-Ended Priority Queues," *Information Processing Letters,* 26 (1987), 33–36.

6. S. Carlsson and J. Chen, "The Complexity of Heaps," *Proceedings of the Third Symposium on Discrete Algorithms* (1992), 393–402.

7. S. Carlsson, J. Chen, and T. Strothotte, "A Note on the Construction of the Data Structure 'Deap'," *Information Processing Letters,* 31 (1989), 315–317.

8. S. Carlsson, J. I. Munro, and P. V. Poblete, "An Implicit Binomial Queue with Constant Insertion Time," *Proceedings of First Scandinavian Workshop on Algorithm Theory* (1988), 1–13.

9. S. C. Chang and M. W. Due, "Diamond Deque: A Simple Data Structure for Priority Deques," *Information Processing Letters,* 46 (1993), 231–237.

10. D. Cheriton and R. E. Tarjan, "Finding Minimum Spanning Trees," *SIAM Journal on Computing,* 5 (1976), 724–742.

11. C. A. Crane, "Linear Lists and Priority Queues as Balanced Binary Trees," *Technical Report STAN-CS-72-259,* Computer Science Department, Stanford University, Stanford, Calif., 1972.

12. Y. Ding and M. A. Weiss, "The Relaxed Min-Max Heap: A Mergeable Double-Ended Priority Queue," *Acta Informatica,* 30 (1993), 215–231.

13. J. R. Driscoll, H. N. Gabow, R. Shrairman, and R. E. Tarjan, "Relaxed Heaps: An Alternative to Fibonacci Heaps with Applications to Parallel Computation," *Communications of the ACM,* 31 (1988), 1343–1354.

14. R. W. Floyd, "Algorithm 245: Treesort 3," *Communications of the ACM,* 7 (1964), 701.

15. M. L. Fredman, R. Sedgewick, D. D. Sleator, and R. E. Tarjan, "The Pairing Heap: A New Form of Self-adjusting Heap," *Algorithmica,* 1 (1986), 111–129.

16. M. L. Fredman and R. E. Tarjan, "Fibonacci Heaps and Their Uses in Improved Network Optimization Algorithms," *Journal of the ACM,* 34 (1987), 596–615.

17. G. H. Gonnet and J. I. Munro, "Heaps on Heaps," *SIAM Journal on Computing,* 15 (1986), 964–971.

18. A. Hasham and J. R. Sack, "Bounds for Min-max Heaps," *BIT,* 27 (1987), 315–323.

19. D. B. Johnson, "Priority Queues with Update and Finding Minimum Spanning Trees," *Information Processing Letters,* 4 (1975), 53–57.

20. C. M. Khoong and H. W. Leong, "Double-Ended Binomial Queues," *Proceedings of the Fourth Annual International Symposium on Algorithms and Computation* (1993), 128–137.

21. D. E. Knuth, *The Art of Computer Programming, Vol 3: Sorting and Searching,* 2d ed, Addison-Wesley, Reading, Mass., 1998.

22. A. LaMarca and R. E. Ladner, "The Influence of Caches on the Performance of Sorting," *Proceedings of the Eighth Annual ACM-SIAM Symposium on Discrete Algorithms* (1997), 370–379.

23. C. J. H. McDiarmid and B. A. Reed, "Building Heaps Fast," *Journal of Algorithms,* 10 (1989), 352–365.

24. D. D. Sleator and R. E. Tarjan, "Self-adjusting Heaps," *SIAM Journal on Computing,* 15 (1986), 52–69.

25. T. Strothotte, P. Eriksson, and S. Vallner, "A Note on Constructing Min-max Heaps," *BIT,* 29 (1989), 251–256.

26. P. van Emde Boas, R. Kaas, and E. Zijlstra, "Design and Implementation of an Efficient Priority Queue," *Mathematical Systems Theory,* 10 (1977), 99–127.

27. J. Vuillemin, "A Data Structure for Manipulating Priority Queues," *Communications of the ACM,* 21 (1978), 309–314.

28. J. W. J. Williams, "Algorithm 232: Heapsort," *Communications of the ACM,* 7 (1964), 347–348.

CHAPTER 7

Sorting

In this chapter we discuss the problem of sorting an array of elements. To simplify matters, we will assume in our examples that the array contains only integers, although our code will once again allow more general objects. For most of this chapter, we will also assume that the entire sort can be done in main memory, so that the number of elements is relatively small (less than a few million). Sorts that cannot be performed in main memory and must be done on disk or tape are also quite important. This type of sorting, known as external sorting, will be discussed at the end of the chapter.

Our investigation of internal sorting will show that

- There are several easy algorithms to sort in $O(N^2)$, such as insertion sort.
- There is an algorithm, Shellsort, that is very simple to code, runs in $o(N^2)$, and is efficient in practice.
- There are slightly more complicated $O(N \log N)$ sorting algorithms.
- Any general-purpose sorting algorithm requires $\Omega(N \log N)$ comparisons.

The rest of this chapter will describe and analyze the various sorting algorithms. These algorithms contain interesting and important ideas for code optimization as well as algorithm design. Sorting is also an example where the analysis can be precisely performed. Be forewarned that where appropriate, we will do as much analysis as possible.

7.1 Preliminaries

The algorithms we describe will all be interchangeable. Each will be passed an array containing the elements; we assume all array positions contain data to be sorted. We will assume that N is the number of elements passed to our sorting routines.

The objects being sorted are of type Comparable, as described in Section 1.4. We thus use the compareTo method to place a consistent ordering on the input. Besides (reference) assignments, this is the only operation allowed on the input data. Sorting under these conditions is known as **comparison-based sorting**. The sorting algorithms are easily rewritten to use Comparators, in the event that the default ordering is unavailable or unacceptable.

Original	34	8	64	51	32	21	Positions Moved
After $p = 1$	8	34	64	51	32	21	1
After $p = 2$	8	34	64	51	32	21	0
After $p = 3$	8	34	51	64	32	21	1
After $p = 4$	8	32	34	51	64	21	3
After $p = 5$	8	21	32	34	51	64	4

Figure 7.1 Insertion sort after each pass

7.2 Insertion Sort

7.2.1 The Algorithm

One of the simplest sorting algorithms is the **insertion sort**. Insertion sort consists of $N - 1$ **passes**. For pass $p = 1$ through $N - 1$, insertion sort ensures that the elements in positions 0 through p are in sorted order. Insertion sort makes use of the fact that elements in positions 0 through $p - 1$ are already known to be in sorted order. Figure 7.1 shows a sample array after each pass of insertion sort.

Figure 7.1 shows the general strategy. In pass p, we move the element in position p left until its correct place is found among the first $p + 1$ elements. The code in Figure 7.2 implements this strategy. Lines 12 through 15 implement that data movement without the explicit use of swaps. The element in position p is saved in tmp, and all larger elements (prior to position p) are moved one spot to the right. Then tmp is placed in the correct spot. This is the same technique that was used in the implementation of binary heaps.

7.2.2 Analysis of Insertion Sort

Because of the nested loops, each of which can take N iterations, insertion sort is $O(N^2)$. Furthermore, this bound is tight, because input in reverse order can achieve this bound. A precise calculation shows that the number of tests in the inner loop in Figure 7.2 is at most $p + 1$ times for each value of p. Summing over all p gives a total of

$$\sum_{i=2}^{N} i = 2 + 3 + 4 + \cdots + N = \Theta(N^2)$$

On the other hand, if the input is presorted, the running time is $O(N)$, because the test in the inner for loop always fails immediately. Indeed, if the input is almost sorted (this term will be more rigorously defined in the next section), insertion sort will run quickly. Because of this wide variation, it is worth analyzing the average-case behavior of this algorithm. It turns out that the average case is $\Theta(N^2)$ for insertion sort, as well as for a variety of other sorting algorithms, as the next section shows.

```
 1      /**
 2       * Simple insertion sort.
 3       * @param a an array of Comparable items.
 4       */
 5      public static <AnyType extends Comparable<? super AnyType>>
 6      void insertionSort( AnyType [ ] a )
 7      {
 8          int j;
 9
10          for( int p = 1; p < a.length; p++ )
11          {
12              AnyType tmp = a[ p ];
13              for( j = p; j > 0 && tmp.compareTo( a[ j - 1 ] ) < 0; j-- )
14                  a[ j ] = a[ j - 1 ];
15              a[ j ] = tmp;
16          }
17      }
```

Figure 7.2   Insertion sort routine

7.3   A Lower Bound for Simple Sorting Algorithms

An **inversion** in an array of numbers is any ordered pair (i, j) having the property that $i < j$ but $a[i] > a[j]$. In the example of the last section, the input list 34, 8, 64, 51, 32, 21 had nine inversions, namely $(34, 8)$, $(34, 32)$, $(34, 21)$, $(64, 51)$, $(64, 32)$, $(64, 21)$, $(51, 32)$, $(51, 21)$, and $(32, 21)$. Notice that this is exactly the number of swaps that needed to be (implicitly) performed by insertion sort. This is always the case, because swapping two adjacent elements that are out of place removes exactly one inversion, and a sorted array has no inversions. Since there is $O(N)$ other work involved in the algorithm, the running time of insertion sort is $O(I + N)$, where I is the number of inversions in the original array. Thus, insertion sort runs in linear time if the number of inversions is $O(N)$.

We can compute precise bounds on the average running time of insertion sort by computing the average number of inversions in a permutation. As usual, defining *average* is a difficult proposition. We will assume that there are no duplicate elements (if we allow duplicates, it is not even clear what the average number of duplicates is). Using this assumption, we can assume that the input is some permutation of the first N integers (since only relative ordering is important) and that all are equally likely. Under these assumptions, we have the following theorem:

Theorem 7.1.
The average number of inversions in an array of N distinct elements is $N(N - 1)/4$.

Proof.

For any list, L, of elements, consider L_r, the list in reverse order. The reverse list of the example is $21, 32, 51, 64, 8, 34$. Consider any pair of two elements in the list (x, y), with $y > x$. Clearly, in exactly one of L and L_r this ordered pair represents an inversion. The total number of these pairs in a list L and its reverse L_r is $N(N-1)/2$. Thus, an average list has half this amount, or $N(N-1)/4$ inversions.

This theorem implies that insertion sort is quadratic on average. It also provides a very strong lower bound about any algorithm that only exchanges adjacent elements.

Theorem 7.2.

Any algorithm that sorts by exchanging adjacent elements requires $\Omega(N^2)$ time on average.

Proof.

The average number of inversions is initially $N(N-1)/4 = \Omega(N^2)$. Each swap removes only one inversion, so $\Omega(N^2)$ swaps are required.

This is an example of a lower-bound proof. It is valid not only for insertion sort, which performs adjacent exchanges implicitly, but also for other simple algorithms such as bubble sort and selection sort, which we will not describe here. In fact, it is valid over an entire *class* of sorting algorithms, including those undiscovered, that perform only adjacent exchanges. Because of this, this proof cannot be confirmed empirically. Although this lower-bound proof is rather simple, in general proving lower bounds is much more complicated than proving upper bounds and in some cases resembles magic.

This lower bound shows us that in order for a sorting algorithm to run in subquadratic, or $o(N^2)$, time, it must do comparisons and, in particular, exchanges between elements that are far apart. A sorting algorithm makes progress by eliminating inversions, and to run efficiently, it must eliminate more than just one inversion per exchange.

7.4 Shellsort

Shellsort, named after its inventor, Donald Shell, was one of the first algorithms to break the quadratic time barrier, although it was not until several years after its initial discovery that a subquadratic time bound was proven. As suggested in the previous section, it works by comparing elements that are distant; the distance between comparisons decreases as the algorithm runs until the last phase, in which adjacent elements are compared. For this reason, Shellsort is sometimes referred to as **diminishing increment sort**.

Shellsort uses a sequence, $h_1, h_2, \ldots, h_t$, called the **increment sequence**. Any increment sequence will do as long as $h_1 = 1$, but some choices are better than others (we will discuss that issue later). After a *phase*, using some increment h_k, for every i, we have $a[i] \le a[i + h_k]$ (where this makes sense); all elements spaced h_k apart are sorted. The file is then said to be h_k**-sorted**. For example, Figure 7.3 shows an array after several phases of Shellsort. An important property of Shellsort (which we state without proof) is that an h_k-sorted file that is then h_{k-1}-sorted remains h_k-sorted. If this were not the case, the

Original	81	94	11	96	12	35	17	95	28	58	41	75	15
After 5-sort	35	17	11	28	12	41	75	15	96	58	81	94	95
After 3-sort	28	12	11	35	15	41	58	17	94	75	81	96	95
After 1-sort	11	12	15	17	28	35	41	58	75	81	94	95	96

Figure 7.3 Shellsort after each pass, using $\{1, 3, 5\}$ as the increment sequence

```
1      /**
2       * Shellsort, using Shell's (poor) increments.
3       * @param a an array of Comparable items.
4       */
5      public static <AnyType extends Comparable<? super AnyType>>
6      void shellsort( AnyType [ ] a )
7      {
8          int j;
9
10         for( int gap = a.length / 2; gap > 0; gap /= 2 )
11             for( int i = gap; i < a.length; i++ )
12             {
13                 AnyType tmp = a[ i ];
14                 for( j = i; j >= gap &&
15                             tmp.compareTo( a[ j - gap ] ) < 0; j -= gap )
16                     a[ j ] = a[ j - gap ];
17                 a[ j ] = tmp;
18             }
19     }
```

Figure 7.4 Shellsort routine using Shell's increments (better increments are possible)

algorithm would likely be of little value, since work done by early phases would be undone by later phases.

The general strategy to h_k-sort is for each position, i, in $h_k, h_k + 1, \ldots, N - 1$, place the element in the correct spot among $i, i - h_k, i - 2h_k$, and so on. Although this does not affect the implementation, a careful examination shows that the action of an h_k-sort is to perform an insertion sort on h_k independent subarrays. This observation will be important when we analyze the running time of Shellsort.

A popular (but poor) choice for increment sequence is to use the sequence suggested by Shell: $h_t = \lfloor N/2 \rfloor$, and $h_k = \lfloor h_{k+1}/2 \rfloor$ (This is not the sequence used in the example in Figure 7.3). Figure 7.4 contains a method that implements Shellsort using this sequence. We shall see later that there are increment sequences that give a significant improvement in the algorithm's running time; even a minor change can drastically affect performance (Exercise 7.10).

The program in Figure 7.4 avoids the explicit use of swaps in the same manner as our implementation of insertion sort.

7.4.1 Worst-Case Analysis of Shellsort

Although Shellsort is simple to code, the analysis of its running time is quite another story. The running time of Shellsort depends on the choice of increment sequence, and the proofs can be rather involved. The average-case analysis of Shellsort is a long-standing open problem, except for the most trivial increment sequences. We will prove tight worst-case bounds for two particular increment sequences.

Theorem 7.3.
The worst-case running time of Shellsort, using Shell's increments, is $\Theta(N^2)$.

Proof.
The proof requires showing not only an upper bound on the worst-case running time but also showing that there exists some input that actually takes $\Omega(N^2)$ time to run. We prove the lower bound first, by constructing a bad case. First, we choose N to be a power of 2. This makes all the increments even, except for the last increment, which is 1. Now, we will give as input an array with the $N/2$ largest numbers in the even positions and the $N/2$ smallest numbers in the odd positions (for this proof, the first position is position 1). As all the increments except the last are even, when we come to the last pass, the $N/2$ largest numbers are still all in even positions and the $N/2$ smallest numbers are still all in odd positions. The ith smallest number ($i \leq N/2$) is thus in position $2i - 1$ before the beginning of the last pass. Restoring the ith element to its correct place requires moving it $i-1$ spaces in the array. Thus, to merely place the $N/2$ smallest elements in the correct place requires at least $\sum_{i=1}^{N/2} i - 1 = \Omega(N^2)$ work. As an example, Figure 7.5 shows a bad (but not the worst) input when $N = 16$. The number of inversions remaining after the 2-sort is exactly $1 + 2 + 3 + 4 + 5 + 6 + 7 = 28$; thus, the last pass will take considerable time.

To finish the proof, we show the upper bound of $O(N^2)$. As we have observed before, a pass with increment h_k consists of h_k insertion sorts of about N/h_k elements. Since insertion sort is quadratic, the total cost of a pass is $O(h_k(N/h_k)^2) = O(N^2/h_k)$. Summing over all passes gives a total bound of $O(\sum_{i=1}^{t} N^2/h_i) = O(N^2 \sum_{i=1}^{t} 1/h_i)$. Because the increments form a geometric series with common ratio 2, and the largest term in the series is $h_1 = 1$, $\sum_{i=1}^{t} 1/h_i < 2$. Thus we obtain a total bound of $O(N^2)$.

Start	1	9	2	10	3	11	4	12	5	13	6	14	7	15	8	16
After 8-sort	1	9	2	10	3	11	4	12	5	13	6	14	7	15	8	16
After 4-sort	1	9	2	10	3	11	4	12	5	13	6	14	7	15	8	16
After 2-sort	1	9	2	10	3	11	4	12	5	13	6	14	7	15	8	16
After 1-sort	1	2	3	4	5	6	7	8	9	10	11	12	13	14	15	16

Figure 7.5 Bad case for Shellsort with Shell's increments (positions are numbered 1 to 16)

The problem with Shell's increments is that pairs of increments are not necessarily relatively prime, and thus the smaller increment can have little effect. Hibbard suggested a slightly different increment sequence, which gives better results in practice (and theoretically). His increments are of the form $1, 3, 7, \ldots, 2^k - 1$. Although these increments are almost identical, the key difference is that consecutive increments have no common factors. We now analyze the worst-case running time of Shellsort for this increment sequence. The proof is rather complicated.

Theorem 7.4.
The worst-case running time of Shellsort using Hibbard's increments is $\Theta(N^{3/2})$.

Proof.
We will prove only the upper bound and leave the proof of the lower bound as an exercise. The proof requires some well-known results from additive number theory. References to these results are provided at the end of the chapter.

For the upper bound, as before, we bound the running time of each pass and sum over all passes. For increments $h_k > N^{1/2}$, we will use the bound $O(N^2/h_k)$ from the previous theorem. Although this bound holds for the other increments, it is too large to be useful. Intuitively, we must take advantage of the fact that *this* increment sequence is *special*. What we need to show is that for any element $a[p]$ in position p, when it is time to perform an h_k-sort, there are only a few elements to the left of position p that are larger than $a[p]$.

When we come to h_k-sort the input array, we know that it has already been h_{k+1}- and h_{k+2}-sorted. Prior to the h_k-sort, consider elements in positions p and $p - i, i \le p$. If i is a multiple of h_{k+1} or h_{k+2}, then clearly $a[p - i] < a[p]$. We can say more, however. If i is expressible as a linear combination (in nonnegative integers) of h_{k+1} and h_{k+2}, then $a[p - i] < a[p]$. As an example, when we come to 3-sort, the file is already 7- and 15-sorted. 52 is expressible as a linear combination of 7 and 15, because $52 = 1 * 7 + 3 * 15$. Thus, $a[100]$ cannot be larger than $a[152]$ because $a[100] \le a[107] \le a[122] \le a[137] \le a[152]$.

Now, $h_{k+2} = 2h_{k+1} + 1$, so h_{k+1} and h_{k+2} cannot share a common factor. In this case, it is possible to show that all integers that are at least as large as $(h_{k+1} - 1)(h_{k+2} - 1) = 8h_k^2 + 4h_k$ can be expressed as a linear combination of h_{k+1} and h_{k+2} (see the reference at the end of the chapter).

This tells us that the body of the innermost for loop can be executed at most $8h_k + 4 = O(h_k)$ times for each of the $N - h_k$ positions. This gives a bound of $O(Nh_k)$ per pass.

Using the fact that about half the increments satisfy $h_k < \sqrt{N}$, and assuming that t is even, the total running time is then

$$O\left(\sum_{k=1}^{t/2} Nh_k + \sum_{k=t/2+1}^{t} N^2/h_k \right) = O\left(N \sum_{k=1}^{t/2} h_k + N^2 \sum_{k=t/2+1}^{t} 1/h_k \right)$$

Because both sums are geometric series, and since $h_{t/2} = \Theta(\sqrt{N})$, this simplifies to

$$= O\left(Nh_{t/2}\right) + O\left(\frac{N^2}{h_{t/2}}\right) = O(N^{3/2})$$

The average-case running time of Shellsort, using Hibbard's increments, is thought to be $O(N^{5/4})$, based on simulations, but nobody has been able to prove this. Pratt has shown that the $\Theta(N^{3/2})$ bound applies to a wide range of increment sequences.

Sedgewick has proposed several increment sequences that give an $O(N^{4/3})$ worst-case running time (also achievable). The average running time is conjectured to be $O(N^{7/6})$ for these increment sequences. Empirical studies show that these sequences perform significantly better in practice than Hibbard's. The best of these is the sequence $\{1, 5, 19, 41, 109, \ldots\}$, in which the terms are either of the form $9 \cdot 4^i - 9 \cdot 2^i + 1$ or $4^i - 3 \cdot 2^i + 1$. This is most easily implemented by placing these values in an array. This increment sequence is the best known in practice, although there is a lingering possibility that some increment sequence might exist that could give a significant improvement in the running time of Shellsort.

There are several other results on Shellsort that (generally) require difficult theorems from number theory and combinatorics and are mainly of theoretical interest. Shellsort is a fine example of a very simple algorithm with an extremely complex analysis.

The performance of Shellsort is quite acceptable in practice, even for N in the tens of thousands. The simplicity of the code makes it the algorithm of choice for sorting up to moderately large input.

7.5 Heapsort

As mentioned in Chapter 6, priority queues can be used to sort in $O(N \log N)$ time. The algorithm based on this idea is known as *heapsort* and gives the best Big-Oh running time we have seen so far.

Recall, from Chapter 6, that the basic strategy is to build a binary heap of N elements. This stage takes $O(N)$ time. We then perform N `deleteMin` operations. The elements leave the heap smallest first, in sorted order. By recording these elements in a second array and then copying the array back, we sort N elements. Since each `deleteMin` takes $O(\log N)$ time, the total running time is $O(N \log N)$.

The main problem with this algorithm is that it uses an extra array. Thus, the memory requirement is doubled. This could be a problem in some instances. Notice that the extra time spent copying the second array back to the first is only $O(N)$, so that this is not likely to affect the running *time* significantly. The problem is space.

A clever way to avoid using a second array makes use of the fact that after each `deleteMin`, the heap shrinks by 1. Thus the cell that was last in the heap can be used to store the element that was just deleted. As an example, suppose we have a heap with six elements. The first `deleteMin` produces a_1. Now the heap has only five elements, so we can place a_1 in position 6. The next `deleteMin` produces a_2. Since the heap will now only have four elements, we can place a_2 in position 5.

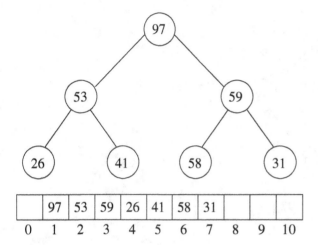

Figure 7.6 (*Max*) heap after `buildHeap` phase

Using this strategy, after the last `deleteMin` the array will contain the elements in *decreasing* sorted order. If we want the elements in the more typical *increasing* sorted order, we can change the ordering property so that the parent has a larger key than the child. Thus we have a (*max*)heap.

In our implementation, we will use a (*max*)heap but avoid the actual ADT for the purposes of speed. As usual, everything is done in an array. The first step builds the heap in linear time. We then perform $N - 1$ `deleteMax`es by swapping the last element in the heap with the first, decrementing the heap size, and percolating down. When the algorithm terminates, the array contains the elements in sorted order. For instance, consider the input sequence $31, 41, 59, 26, 53, 58, 97$. The resulting heap is shown in Figure 7.6.

Figure 7.7 shows the heap that results after the first `deleteMax`. As the figures imply, the last element in the heap is 31; 97 has been placed in a part of the heap array that is technically no longer part of the heap. After 5 more `deleteMax` operations, the heap will actually have only one element, but the elements left in the heap array will be in sorted order.

The code to perform heapsort is given in Figure 7.8. The slight complication is that, unlike the binary heap, where the data begin at array index 1, the array for heapsort contains data in position 0. Thus the code is a little different from the binary heap code. The changes are minor.

7.5.1 Analysis of Heapsort

As we saw in Chapter 6, the first phase, which constitutes the building of the heap, uses less than $2N$ comparisons. In the second phase, the ith `deleteMax` uses at most less than $2\lfloor \log(N - i + 1) \rfloor$ comparisons, for a total of at most $2N \log N - O(N)$ comparisons (assuming $N \geq 2$). Consequently, in the worst case, at most $2N \log N - O(N)$ comparisons are used by heapsort. Exercise 7.13 asks you to show that it is possible for all of the `deleteMax` operations to achieve their worst case simultaneously.

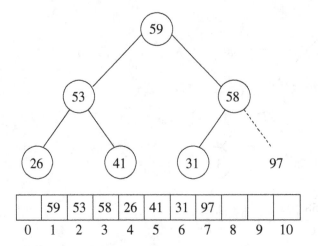

Figure 7.7 Heap after first `deleteMax`

Experiments have shown that the performance of heapsort is extremely consistent: On average it uses only slightly fewer comparisons than the worst-case bound suggests. For many years, nobody had been able to show nontrivial bounds on heapsort's average running time. The problem, it seems, is that successive `deleteMax` operations destroy the heap's randomness, making the probability arguments very complex. Eventually another approach proved successful.

Theorem 7.5.
The average number of comparisons used to heapsort a random permutation of N distinct items is $2N \log N - O(N \log \log N)$.

Proof.
The heap construction phase uses $\Theta(N)$ comparisons on average, and so we only need to prove the bound for the second phase. We assume a permutation of $\{1, 2, \ldots, N\}$.

Suppose the ith `deleteMax` pushes the root element down d_i levels. Then it uses $2d_i$ comparisons. For heapsort on any input, there is a cost sequence $D : d_1, d_2, \ldots, d_N$ that defines the cost of phase 2. That cost is given by $M_D = \sum_{i=1}^{N} d_i$; the number of comparisons used is thus $2M_D$.

Let $f(N)$ be the number of heaps of N items. One can show (Exercise 7.58) that $f(N) > (N/(4e))^N$ (where $e = 2.71828\ldots$). We will show that only an exponentially small fraction of these heaps (in particular $(N/16)^N$) have a cost smaller than $M = N(\log N - \log \log N - 4)$. When this is shown, it follows that the average value of M_D is at least M minus a term that is $o(1)$, and thus the average number of comparisons is at least $2M$. Consequently, our basic goal is to show that there are very few heaps that have small cost sequences.

```
1      /**
2       * Internal method for heapsort.
3       * @param i the index of an item in the heap.
4       * @return the index of the left child.
5       */
6      private static int leftChild( int i )
7      {
8          return 2 * i + 1;
9      }
10
11     /**
12      * Internal method for heapsort that is used in deleteMax and buildHeap.
13      * @param a an array of Comparable items.
14      * @int i the position from which to percolate down.
15      * @int n the logical size of the binary heap.
16      */
17     private static <AnyType extends Comparable<? super AnyType>>
18     void percDown( AnyType [ ] a, int i, int n )
19     {
20         int child;
21         AnyType tmp;
22
23         for( tmp = a[ i ]; leftChild( i ) < n; i = child )
24         {
25             child = leftChild( i );
26             if( child != n - 1 && a[ child ].compareTo( a[ child + 1 ] ) < 0 )
27                 child++;
28             if( tmp.compareTo( a[ child ] ) < 0 )
29                 a[ i ] = a[ child ];
30             else
31                 break;
32         }
33         a[ i ] = tmp;
34     }
35
36     /**
37      * Standard heapsort.
38      * @param a an array of Comparable items.
39      */
40     public static <AnyType extends Comparable<? super AnyType>>
41     void heapsort( AnyType [ ] a )
42     {
43         for( int i = a.length / 2 - 1; i >= 0; i-- )   /* buildHeap */
44             percDown( a, i, a.length );
45         for( int i = a.length - 1; i > 0; i-- )
46         {
47             swapReferences( a, 0, i );                  /* deleteMax */
48             percDown( a, 0, i );
49         }
50     }
```

Figure 7.8 Heapsort

Because level d_i has at most 2^{d_i} nodes, there are 2^{d_i} possible places that the root element can go for any d_i. Consequently, for any sequence D, the number of distinct corresponding deleteMax sequences is at most

$$S_D = 2^{d_1} 2^{d_2} \cdots 2^{d_N}$$

A simple algebraic manipulation shows that *for a given sequence D*

$$S_D = 2^{M_D}$$

Because each d_i can assume any value between 1 and $\lfloor \log N \rfloor$, there are at most $(\log N)^N$ possible sequences D. It follows that the number of distinct deleteMax sequences that require cost exactly equal to M is at most the number of cost sequences of total cost M times the number of deleteMax sequences for each of these cost sequences. A bound of $(\log N)^N 2^M$ follows immediately.

The total number of heaps with cost sequence less than M is at most

$$\sum_{i=1}^{M-1} (\log N)^N 2^i < (\log N)^N 2^M$$

If we choose $M = N(\log N - \log \log N - 4)$, then the number of heaps that have cost sequence less than M is at most $(N/16)^N$, and the theorem follows from our earlier comments.

Using a more complex argument, it can be shown that heapsort always uses at least $N \log N - O(N)$ comparisons, and that there are inputs that can achieve this bound. The average case analysis also can be improved to $2N \log N - O(N)$ comparisons (rather than the nonlinear second term in Theorem 7.5).

7.6 Mergesort

We now turn our attention to **mergesort**. Mergesort runs in $O(N \log N)$ worst-case running time, and the number of comparisons used is nearly optimal. It is a fine example of a recursive algorithm.

The fundamental operation in this algorithm is merging two sorted lists. Because the lists are sorted, this can be done in one pass through the input, if the output is put in a third list. The basic merging algorithm takes two input arrays A and B, an output array C, and three counters, *Actr*, *Bctr*, and *Cctr*, which are initially set to the beginning of their respective arrays. The smaller of $A[Actr]$ and $B[Bctr]$ is copied to the next entry in C, and the appropriate counters are advanced. When either input list is exhausted, the remainder of the other list is copied to C. An example of how the merge routine works is provided for the following input.

1	13	24	26

2	15	27	38

↑ *Actr* ↑ *Bctr* ↑ *Cctr*

If the array A contains 1, 13, 24, 26, and B contains 2, 15, 27, 38, then the algorithm proceeds as follows: First, a comparison is done between 1 and 2. 1 is added to C, and then 13 and 2 are compared.

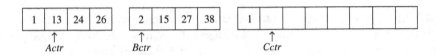

2 is added to C, and then 13 and 15 are compared.

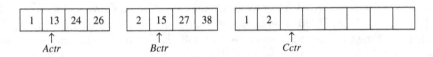

13 is added to C, and then 24 and 15 are compared. This proceeds until 26 and 27 are compared.

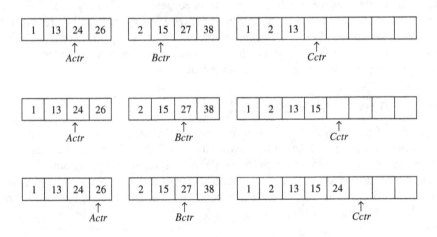

26 is added to C, and the A array is exhausted.

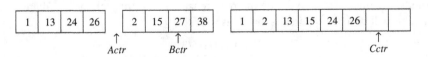

The remainder of the B array is then copied to C.

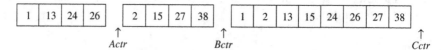

The time to merge two sorted lists is clearly linear, because at most $N - 1$ comparisons are made, where N is the total number of elements. To see this, note that every comparison adds an element to C, except the last comparison, which adds at least two.

The mergesort algorithm is therefore easy to describe. If $N = 1$, there is only one element to sort, and the answer is at hand. Otherwise, recursively mergesort the first half and the second half. This gives two sorted halves, which can then be merged together using the merging algorithm described above. For instance, to sort the eight-element array 24, 13, 26, 1, 2, 27, 38, 15, we recursively sort the first four and last four elements, obtaining 1, 13, 24, 26, 2, 15, 27, 38. Then we merge the two halves as above, obtaining the final list 1, 2, 13, 15, 24, 26, 27, 38. This algorithm is a classic divide-and-conquer strategy. The problem is *divided* into smaller problems and solved recursively. The *conquering* phase consists of patching together the answers. Divide-and-conquer is a very powerful use of recursion that we will see many times.

An implementation of mergesort is provided in Figure 7.9. The public `mergeSort` is just a driver for the private recursive method `mergeSort`.

The `merge` routine is subtle. If a temporary array is declared locally for each recursive call of `merge`, then there could be $\log N$ temporary arrays active at any point. A close examination shows that since `merge` is the last line of `mergeSort`, there only needs to be one temporary array active at any point, and that the temporary array can be created in the public `mergeSort` driver. Further, we can use any part of the temporary array; we will use the same portion as the input array `a`. This allows the improvement described at the end of this section. Figure 7.10 implements the `merge` routine.

7.6.1 Analysis of Mergesort

Mergesort is a classic example of the techniques used to analyze recursive routines: we have to write a recurrence relation for the running time. We will assume that N is a power of 2, so that we always split into even halves. For $N = 1$, the time to mergesort is constant, which we will denote by 1. Otherwise, the time to mergesort N numbers is equal to the time to do two recursive mergesorts of size $N/2$, plus the time to merge, which is linear. The following equations say this exactly:

$$T(1) = 1$$
$$T(N) = 2T(N/2) + N$$

This is a standard recurrence relation, which can be solved several ways. We will show two methods. The first idea is to divide the recurrence relation through by N. The reason for doing this will become apparent soon. This yields

$$\frac{T(N)}{N} = \frac{T(N/2)}{N/2} + 1$$

```
1     /**
2      * Internal method that makes recursive calls.
3      * @param a an array of Comparable items.
4      * @param tmpArray an array to place the merged result.
5      * @param left the left-most index of the subarray.
6      * @param right the right-most index of the subarray.
7      */
8     private static <AnyType extends Comparable<? super AnyType>>
9     void mergeSort( AnyType [ ] a, AnyType [ ] tmpArray, int left, int right )
10    {
11        if( left < right )
12        {
13            int center = ( left + right ) / 2;
14            mergeSort( a, tmpArray, left, center );
15            mergeSort( a, tmpArray, center + 1, right );
16            merge( a, tmpArray, left, center + 1, right );
17        }
18    }
19
20    /**
21     * Mergesort algorithm.
22     * @param a an array of Comparable items.
23     */
24    public static <AnyType extends Comparable<? super AnyType>>
25    void mergeSort( AnyType [ ] a )
26    {
27        AnyType [ ] tmpArray = (AnyType[]) new Comparable[ a.length ];
28
29        mergeSort( a, tmpArray, 0, a.length - 1 );
30    }
```

Figure 7.9 Mergesort routines

This equation is valid for any N that is a power of 2, so we may also write

$$\frac{T(N/2)}{N/2} = \frac{T(N/4)}{N/4} + 1$$

and

$$\frac{T(N/4)}{N/4} = \frac{T(N/8)}{N/8} + 1$$

$$\vdots$$

$$\frac{T(2)}{2} = \frac{T(1)}{1} + 1$$

```
1       /**
2        * Internal method that merges two sorted halves of a subarray.
3        * @param a an array of Comparable items.
4        * @param tmpArray an array to place the merged result.
5        * @param leftPos the left-most index of the subarray.
6        * @param rightPos the index of the start of the second half.
7        * @param rightEnd the right-most index of the subarray.
8        */
9       private static <AnyType extends Comparable<? super AnyType>>
10      void merge( AnyType [ ] a, AnyType [ ] tmpArray,
11                      int leftPos, int rightPos, int rightEnd )
12      {
13          int leftEnd = rightPos - 1;
14          int tmpPos = leftPos;
15          int numElements = rightEnd - leftPos + 1;
16
17          // Main loop
18          while( leftPos <= leftEnd && rightPos <= rightEnd )
19              if( a[ leftPos ].compareTo( a[ rightPos ] ) <= 0 )
20                  tmpArray[ tmpPos++ ] = a[ leftPos++ ];
21              else
22                  tmpArray[ tmpPos++ ] = a[ rightPos++ ];
23
24          while( leftPos <= leftEnd )     // Copy rest of first half
25              tmpArray[ tmpPos++ ] = a[ leftPos++ ];
26
27          while( rightPos <= rightEnd )  // Copy rest of right half
28              tmpArray[ tmpPos++ ] = a[ rightPos++ ];
29
30          // Copy tmpArray back
31          for( int i = 0; i < numElements; i++, rightEnd-- )
32              a[ rightEnd ] = tmpArray[ rightEnd ];
33      }
```

Figure 7.10 merge routine

Now add up all the equations. This means that we add all of the terms on the left-hand side and set the result equal to the sum of all of the terms on the right-hand side. Observe that the term $T(N/2)/(N/2)$ appears on both sides and thus cancels. In fact, virtually all the terms appear on both sides and cancel. This is called *telescoping* a sum. After everything is added, the final result is

$$\frac{T(N)}{N} = \frac{T(1)}{1} + \log N$$

because all of the other terms cancel and there are $\log N$ equations, and so all the 1's at the end of these equations add up to $\log N$. Multiplying through by N gives the final answer.

$$T(N) = N \log N + N = O(N \log N)$$

Notice that if we did not divide through by N at the start of the solutions, the sum would not telescope. This is why it was necessary to divide through by N.

An alternative method is to substitute the recurrence relation continually on the right-hand side. We have

$$T(N) = 2T(N/2) + N$$

Since we can substitute $N/2$ into the main equation,

$$2T(N/2) = 2(2(T(N/4)) + N/2) = 4T(N/4) + N$$

we have

$$T(N) = 4T(N/4) + 2N$$

Again, by substituting $N/4$ into the main equation, we see that

$$4T(N/4) = 4(2T(N/8) + N/4) = 8T(N/8) + N$$

So we have

$$T(N) = 8T(N/8) + 3N$$

Continuing in this manner, we obtain

$$T(N) = 2^k T(N/2^k) + k \cdot N$$

Using $k = \log N$, we obtain

$$T(N) = NT(1) + N \log N = N \log N + N$$

The choice of which method to use is a matter of taste. The first method tends to produce scrap work that fits better on a standard, $8^1/_2 \times 11$ sheet of paper, leading to fewer mathematical errors, but it requires a certain amount of experience to apply. The second method is more of a brute-force approach.

Recall that we have assumed $N = 2^k$. The analysis can be refined to handle cases when N is not a power of 2. The answer turns out to be almost identical (this is usually the case).

Although mergesort's running time is $O(N \log N)$, it has the significant problem that merging two sorted lists uses linear extra memory.[1] The additional work involved in copying to the temporary array and back, throughout the algorithm, slows the sort considerably. This copying can be avoided by judiciously switching the roles of a and tmpArray at alternate levels of the recursion. A variant of mergesort can also be implemented nonrecursively (Exercise 7.16).

---

[1] It is theoretically possible to use less extra memory, but the resulting algorithm is complex and impractical.

The running time of mergesort, when compared with other $O(N \log N)$ alternatives, depends heavily on the relative costs of comparing elements and moving elements in the array (and the temporary array). These costs are language dependent.

For instance, in Java, when performing a generic sort (using a `Comparator`), an element comparison can be expensive (because comparisons might not be easily inlined, and thus the overhead of dynamic dispatch could slow things down), but moving elements is cheap (because they are reference assignments, rather than copies of large objects). Mergesort uses the lowest number of comparisons of all the popular sorting algorithms, and thus is a good candidate for general-purpose sorting in Java. In fact, it is the algorithm used in the standard Java library for generic sorting.

On the other hand, in C++, in a generic sort, copying objects can be expensive if the objects are large, while comparing objects often is relatively cheap because of the ability of the compiler to aggressively perform inline optimization. In this scenario, it might be reasonable to have an algorithm use a few more comparisons, if we can also use significantly fewer data movements. *Quicksort,* which we discuss in the next section, achieves this tradeoff, and is the sorting routine commonly used in C++ libraries.

In Java, quicksort is also used as the standard library sort for primitive types. Here, the costs of comparisons and data moves are similar, so using significantly fewer data movements more than compensates for a few extra comparisons.

7.7 Quicksort

As its name implies, **quicksort** is a fast sorting algorithm in practice and is especially useful in C++, or for sorting primitive types in Java. Its average running time is $O(N \log N)$. It is very fast, mainly due to a very tight and highly optimized inner loop. It has $O(N^2)$ worst-case performance, but this can be made exponentially unlikely with a little effort. By combining quicksort with heapsort, we can achieve quicksort's fast running time on almost all inputs, with heapsort's $O(N \log N)$ worst-case running time. Exercise 7.27 describes this approach.

The quicksort algorithm is simple to understand and prove correct, although for many years it had the reputation of being an algorithm that could in theory be highly optimized but in practice was impossible to code correctly. Like mergesort, quicksort is a divide-and-conquer recursive algorithm.

Let us begin with the following simple sorting algorithm to sort a list. Arbitrarily choose any item, and then form three groups: those smaller than the chosen item, those equal to the chosen item, and those larger than the chosen item. Recursively sort the first and third groups, and then concatenate the three groups. The result is guaranteed by the basic principles of recursion to be a sorted arrangement of the original list. A direct implementation of this algorithm is shown in Figure 7.11, and its performance, is generally speaking, quite respectable on most inputs. In fact, if the list contains large numbers of duplicates with relatively few distinct items, as is sometimes the case, then the performance is extremely good.

The algorithm we have described forms the basis of the quicksort. However, by making the extra lists, and doing so recursively, it is hard to see how we have improved upon

```
1   public static void sort( List<Integer> items )
2   {
3       if( items.size( ) > 1 )
4       {
5           List<Integer> smaller = new ArrayList<>( );
6           List<Integer> same    = new ArrayList<>( );
7           List<Integer> larger  = new ArrayList<>( );
8
9           Integer chosenItem = items.get( items.size( ) / 2 );
10          for( Integer i : items )
11          {
12              if( i < chosenItem )
13                  smaller.add( i );
14              else if( i > chosenItem )
15                  larger.add( i );
16              else
17                  same.add( i );
18          }
19
20          sort( smaller );    // Recursive call!
21          sort( larger );     // Recursive call!
22
23          items.clear( );
24          items.addAll( smaller );
25          items.addAll( same );
26          items.addAll( larger );
27      }
28  }
```

Figure 7.11   Simple recursive sorting algorithm

mergesort. In fact, so far, we really haven't. In order to do better, we must avoid using significant extra memory and have inner loops that are clean. Thus quicksort is commonly written in a manner that avoids creating the second group (the equal items), and the algorithm has numerous subtle details that affect the performance; therein lies the complications.

We now describe the most common implementation of quicksort—"classic quicksort," in which the input is an array, and in which no extra arrays are created by the algorithm.

The classic quicksort algorithm to sort an array S consists of the following four easy steps:

1. If the number of elements in S is 0 or 1, then return.
2. Pick any element v in S. This is called the **pivot**.

3. **Partition** $S - \{v\}$ (the remaining elements in S) into two disjoint groups: $S_1 = \{x \in S - \{v\} | x \leq v\}$, and $S_2 = \{x \in S - \{v\} | x \geq v\}$.

4. Return $\{\text{quicksort}(S_1)$ followed by v followed by $\text{quicksort}(S_2)\}$.

Since the partition step ambiguously describes what to do with elements equal to the pivot, this becomes a design decision. Part of a good implementation is handling this case as efficiently as possible. Intuitively, we would hope that about half the keys that are equal to the pivot go into S_1 and the other half into S_2, much as we like binary search trees to be balanced.

Figure 7.12 shows the action of quicksort on a set of numbers. The pivot is chosen (by chance) to be 65. The remaining elements in the set are partitioned into two smaller sets. Recursively sorting the set of smaller numbers yields $0, 13, 26, 31, 43, 57$ (by rule 3 of recursion). The set of large numbers is similarly sorted. The sorted arrangement of the entire set is then trivially obtained.

It should be clear that this algorithm works, but it is not clear why it is any faster than mergesort. Like mergesort, it recursively solves two subproblems and requires linear additional work (step 3), but, unlike mergesort, the subproblems are not guaranteed to be of equal size, which is potentially bad. The reason that quicksort is faster is that the partitioning step can actually be performed in place and very efficiently. This efficiency can more than make up for the lack of equal-sized recursive calls.

The algorithm as described so far lacks quite a few details, which we now fill in. There are many ways to implement steps 2 and 3; the method presented here is the result of extensive analysis and empirical study and represents a very efficient way to implement quicksort. Even the slightest deviations from this method can cause surprisingly bad results.

7.7.1 Picking the Pivot

Although the algorithm as described works no matter which element is chosen as the pivot, some choices are obviously better than others.

A Wrong Way

The popular, uninformed choice is to use the first element as the pivot. This is acceptable if the input is random, but if the input is presorted or in reverse order, then the pivot provides a poor partition, because either all the elements go into S_1 or they go into S_2. Worse, this happens consistently throughout the recursive calls. The practical effect is that if the first element is used as the pivot and the input is presorted, then quicksort will take quadratic time to do essentially nothing at all, which is quite embarrassing. Moreover, presorted input (or input with a large presorted section) is quite frequent, so using the first element as the pivot *is an absolutely horrible idea* and should be discarded immediately. An alternative is choosing the larger of the first two distinct elements as the pivot, but this has the same bad properties as merely choosing the first element. Do not use that pivoting strategy either.

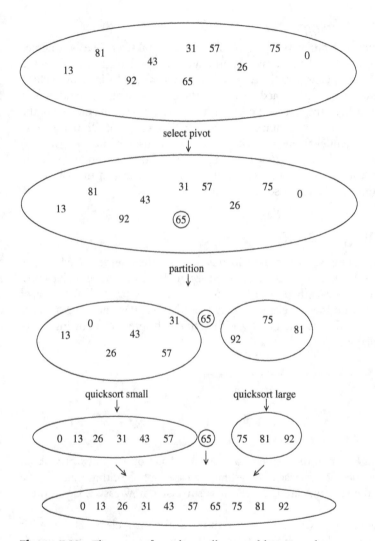

Figure 7.12 The steps of quicksort illustrated by example

A Safe Maneuver

A safe course is merely to choose the pivot randomly. This strategy is generally perfectly safe, unless the random number generator has a flaw (which is not as uncommon as you might think), since it is very unlikely that a random pivot would consistently provide a poor partition. On the other hand, random number generation is generally an expensive commodity and does not reduce the average running time of the rest of the algorithm at all.

Median-of-Three Partitioning

The median of a group of N numbers is the $\lceil N/2 \rceil$ th largest number. The best choice of pivot would be the median of the array. Unfortunately, this is hard to calculate and would slow down quicksort considerably. A good estimate can be obtained by picking three elements randomly and using the median of these three as the pivot. The randomness turns out not to help much, so the common course is to use as the pivot the median of the left, right, and center elements. For instance, with input 8, 1, 4, 9, 6, 3, 5, 2, 7, 0 as before, the left element is 8, the right element is 0, and the center (in position $\lfloor (left + right)/2 \rfloor$) element is 6. Thus, the pivot would be $v = 6$. Using median-of-three partitioning clearly eliminates the bad case for sorted input (the partitions become equal in this case) and actually reduces the number of comparisons by 14 percent.

7.7.2 Partitioning Strategy

There are several partitioning strategies used in practice, but the one described here is known to give good results. It is very easy, as we shall see, to do this wrong or inefficiently, but it is safe to use a known method. The first step is to get the pivot element out of the way by swapping it with the last element. i starts at the first element and j starts at the next-to-last element. If the original input was the same as before, the following figure shows the current situation.

```
8   1   4   9   0   3   5   2   7   6
↑                               ↑
i                               j
```

For now we will assume that all the elements are distinct. Later on we will worry about what to do in the presence of duplicates. As a limiting case, our algorithm must do the proper thing if *all* of the elements are identical. It is surprising how easy it is to do the *wrong* thing.

What our partitioning stage wants to do is to move all the small elements to the left part of the array and all the large elements to the right part. "Small" and "large" are, of course, relative to the pivot.

While i is to the left of j, we move i right, skipping over elements that are smaller than the pivot. We move j left, skipping over elements that are larger than the pivot. When i and j have stopped, i is pointing at a large element and j is pointing at a small element. If i is to the left of j, those elements are swapped. The effect is to push a large element to the right and a small element to the left. In the example above, i would not move and j would slide over one place. The situation is as follows.

```
8   1   4   9   0   3   5   2   7   6
↑                           ↑
i                           j
```

We then swap the elements pointed to by i and j and repeat the process until i and j cross.

		After First Swap							
2	1	4	9	0	3	5	8	7	6

$\uparrow$ (at position 0, labeled i) $\uparrow$ (labeled j)

		Before Second Swap							
2	1	4	9	0	3	5	8	7	6

$\uparrow$ (labeled i) $\uparrow$ (labeled j)

		After Second Swap							
2	1	4	5	0	3	9	8	7	6

$\uparrow$ (labeled i) $\uparrow$ (labeled j)

		Before Third Swap							
2	1	4	5	0	3	9	8	7	6

$\uparrow$ (labeled j) $\uparrow$ (labeled i)

At this stage, i and j have crossed, so no swap is performed. The final part of the partitioning is to swap the pivot element with the element pointed to by i.

		After Swap with Pivot							
2	1	4	5	0	3	6	8	7	9

$\uparrow$ (labeled i) $\uparrow$ (labeled pivot)

When the pivot is swapped with i in the last step, we know that every element in a position $p < i$ must be small. This is because either position p contained a small element to start with, or the large element originally in position p was replaced during a swap. A similar argument shows that elements in positions $p > i$ must be large.

One important detail we must consider is how to handle elements that are equal to the pivot. The questions are whether or not i should stop when it sees an element equal to the pivot and whether or not j should stop when it sees an element equal to the pivot. Intuitively, i and j ought to do the same thing, since otherwise the partitioning step is biased. For instance, if i stops and j does not, then all elements that are equal to the pivot will wind up in S_2.

To get an idea of what might be good, we consider the case where all the elements in the array are identical. If both i and j stop, there will be many swaps between identical elements. Although this seems useless, the positive effect is that i and j will cross in the middle, so when the pivot is replaced, the partition creates two nearly equal subarrays. The mergesort analysis tells us that the total running time would then be $O(N \log N)$.

If neither i nor j stops, and code is present to prevent them from running off the end of the array, no swaps will be performed. Although this seems good, a correct implementation would then swap the pivot into the last spot that i touched, which would be the next-to-last position (or last, depending on the exact implementation). This would create very uneven subarrays. If all the elements are identical, the running time is $O(N^2)$. The effect is the same as using the first element as a pivot for presorted input. It takes quadratic time to do nothing!

Thus, we find that it is better to do the unnecessary swaps and create even subarrays than to risk wildly uneven subarrays. Therefore, we will have both i and j stop if they encounter an element equal to the pivot. This turns out to be the only one of the four possibilities that does not take quadratic time for this input.

At first glance it may seem that worrying about an array of identical elements is silly. After all, why would anyone want to sort 50,000 identical elements? However, recall that quicksort is recursive. Suppose there are 1,000,000 elements, of which 50,000 are identical (or, more likely, complex elements whose sort keys are identical). Eventually, quicksort will make the recursive call on only these 50,000 elements. Then it really will be important to make sure that 50,000 identical elements can be sorted efficiently.

7.7.3 Small Arrays

For very small arrays ($N \leq 20$), quicksort does not perform as well as insertion sort. Furthermore, because quicksort is recursive, these cases will occur frequently. A common solution is not to use quicksort recursively for small arrays, but instead use a sorting algorithm that is efficient for small arrays, such as insertion sort. Using this strategy can actually save about 15 percent in the running time (over doing no cutoff at all). A good cutoff range is $N = 10$, although any cutoff between 5 and 20 is likely to produce similar results. This also saves nasty degenerate cases, such as taking the median of three elements when there are only one or two.

7.7.4 Actual Quicksort Routines

The driver for quicksort is shown in Figure 7.13.

The general form of the routines will be to pass the array and the range of the array (left and right) to be sorted. The first routine to deal with is pivot selection. The easiest way to do this is to sort a[left], a[right], and a[center] in place. This has the extra advantage that the smallest of the three winds up in a[left], which is where the partitioning step would put it anyway. The largest winds up in a[right], which is also the correct place, since it is larger than the pivot. Therefore, we can place the pivot in a[right-1] and initialize i and j to left+1 and right-2 in the partition phase. Yet another benefit is that because a[left] is smaller than the pivot, it will act as a sentinel for j. Thus, we do not need

```
1      /**
2       * Quicksort algorithm.
3       * @param a an array of Comparable items.
4       */
5      public static <AnyType extends Comparable<? super AnyType>>
6      void quicksort( AnyType [ ] a )
7      {
8          quicksort( a, 0, a.length - 1 );
9      }
```

Figure 7.13 Driver for quicksort

```
1      /**
2       * Return median of left, center, and right.
3       * Order these and hide the pivot.
4       */
5      private static <AnyType extends Comparable<? super AnyType>>
6      AnyType median3( AnyType [ ] a, int left, int right )
7      {
8          int center = ( left + right ) / 2;
9          if( a[ center ].compareTo( a[ left ] ) < 0 )
10             swapReferences( a, left, center );
11         if( a[ right ].compareTo( a[ left ] ) < 0 )
12             swapReferences( a, left, right );
13         if( a[ right ].compareTo( a[ center ] ) < 0 )
14             swapReferences( a, center, right );
15
16             // Place pivot at position right - 1
17         swapReferences( a, center, right - 1 );
18         return a[ right - 1 ];
19     }
```

Figure 7.14 Code to perform median-of-three partitioning

to worry about j running past the end. Since i will stop on elements equal to the pivot, storing the pivot in a[right-1] provides a sentinel for i. The code in Figure 7.14 does the median-of-three partitioning with all the side effects described. It may seem that it is only slightly inefficient to compute the pivot by a method that does not actually sort a[left], a[center], and a[right], but, surprisingly, this produces bad results (see Exercise 7.51).

The real heart of the quicksort routine is in Figure 7.15. It includes the partitioning and recursive calls. There are several things worth noting in this implementation. Line 16 initializes i and j to 1 past their correct values, so that there are no special cases to consider. This initialization depends on the fact that median-of-three partitioning has some side

```
1     /**
2      * Internal quicksort method that makes recursive calls.
3      * Uses median-of-three partitioning and a cutoff of 10.
4      * @param a an array of Comparable items.
5      * @param left the left-most index of the subarray.
6      * @param right the right-most index of the subarray.
7      */
8     private static <AnyType extends Comparable<? super AnyType>>
9     void quicksort( AnyType [ ] a, int left, int right )
10    {
11        if( left + CUTOFF <= right )
12        {
13            AnyType pivot = median3( a, left, right );
14
15                // Begin partitioning
16            int i = left, j = right - 1;
17            for( ; ; )
18            {
19                while( a[ ++i ].compareTo( pivot ) < 0 ) { }
20                while( a[ --j ].compareTo( pivot ) > 0 ) { }
21                if( i < j )
22                    swapReferences( a, i, j );
23                else
24                    break;
25            }
26
27            swapReferences( a, i, right - 1 );   // Restore pivot
28
29            quicksort( a, left, i - 1 );    // Sort small elements
30            quicksort( a, i + 1, right );   // Sort large elements
31        }
32        else  // Do an insertion sort on the subarray
33            insertionSort( a, left, right );
34    }
```

Figure 7.15 Main quicksort routine

effects; this program will not work if you try to use it without change with a simple pivoting strategy, because i and j start in the wrong place and there is no longer a sentinel for j.

The swapping action at line 22 is sometimes written explicitly, for speed purposes. For the algorithm to be fast, it may be necessary to force the compiler to compile this code inline. Many compilers will do this automatically if swapReferences is a final method, but for those that do not, the difference can be significant.

```
16              int i = left + 1, j = right ;
17              for( ; ; )
18              {
19                  while( a[ i ].compareTo( pivot ) < 0 ) i++;
20                  while( a[ j ].compareTo( pivot ) > 0 ) j--
21                  if( i < j )
22                      swapReferences( a, i, j );
23                  else
24                      break;
25              }
```

Figure 7.16   A small change to quicksort, which breaks the algorithm

Finally, lines 19 and 20 show why quicksort is so fast. The inner loop of the algorithm consists of an increment/decrement (by 1, which is fast), a test, and a jump. There is no extra juggling as there is in mergesort. This code is still surprisingly tricky. It is tempting to replace lines 16 through 25 with the statements in Figure 7.16. This does not work, because there would be an infinite loop if a[i] = a[j] = pivot.

7.7.5   Analysis of Quicksort

Like mergesort, quicksort is recursive, and hence, its analysis requires solving a recurrence formula. We will do the analysis for a quicksort, assuming a random pivot (no median-of-three partitioning) and no cutoff for small arrays. We will take $T(0) = T(1) = 1$, as in mergesort. The running time of quicksort is equal to the running time of the two recursive calls plus the linear time spent in the partition (the pivot selection takes only constant time). This gives the basic quicksort relation

$$T(N) = T(i) + T(N - i - 1) + cN \tag{7.1}$$

where $i = |S_1|$ is the number of elements in S_1. We will look at three cases.

Worst-Case Analysis

The pivot is the smallest element, all the time. Then $i = 0$ and if we ignore $T(0) = 1$, which is insignificant, the recurrence is

$$T(N) = T(N - 1) + cN, \quad N > 1 \tag{7.2}$$

We telescope, using Equation (7.2) repeatedly. Thus

$$T(N - 1) = T(N - 2) + c(N - 1) \tag{7.3}$$

$$T(N - 2) = T(N - 3) + c(N - 2) \tag{7.4}$$

$$\vdots$$

$$T(2) = T(1) + c(2) \tag{7.5}$$

Adding up all these equations yields

$$T(N) = T(1) + c\sum_{i=2}^{N} i = \Theta(N^2)$$

(7.6)

as claimed earlier. To see that this is the worst possible case, note that the total cost of all the partitions in recursive calls at depth d must be at most N. Since the recursion depth is at most N, this gives as $O(N^2)$ worst-case bound for quicksort.

Best-Case Analysis

In the best case, the pivot is in the middle. To simplify the math, we assume that the two subarrays are each exactly half the size of the original, and although this gives a slight overestimate, this is acceptable because we are only interested in a Big-Oh answer.

$$T(N) = 2T(N/2) + cN$$

(7.7)

Divide both sides of Equation (7.7) by N.

$$\frac{T(N)}{N} = \frac{T(N/2)}{N/2} + c$$

(7.8)

We will telescope using this equation.

$$\frac{T(N/2)}{N/2} = \frac{T(N/4)}{N/4} + c$$

(7.9)

$$\frac{T(N/4)}{N/4} = \frac{T(N/8)}{N/8} + c$$

(7.10)

$$\vdots$$

$$\frac{T(2)}{2} = \frac{T(1)}{1} + c$$

(7.11)

We add all the equations from (7.8) to (7.11) and note that there are $\log N$ of them:

$$\frac{T(N)}{N} = \frac{T(1)}{1} + c\log N$$

(7.12)

which yields

$$T(N) = cN\log N + N = \Theta(N\log N)$$

(7.13)

Notice that this is the exact same analysis as mergesort, hence we get the same answer. That this is the best case is implied by results in Section 7.8.

Average-Case Analysis

This is the most difficult part. For the average case, we assume that each of the sizes for S_1 is equally likely, and hence has probability $1/N$. This assumption is actually valid for our pivoting and partitioning strategy, but it is not valid for some others. Partitioning strategies that do not preserve the randomness of the subarrays cannot use this analysis. Interestingly, these strategies seem to result in programs that take longer to run in practice.

With this assumption, the average value of $T(i)$, and hence $T(N - i - 1)$, is $(1/N)\sum_{j=0}^{N-1} T(j)$. Equation (7.1) then becomes

$$T(N) = \frac{2}{N}\left[\sum_{j=0}^{N-1} T(j)\right] + cN \qquad (7.14)$$

If Equation (7.14) is multiplied by N, it becomes

$$NT(N) = 2\left[\sum_{j=0}^{N-1} T(j)\right] + cN^2 \qquad (7.15)$$

We need to remove the summation sign to simplify matters. We note that we can telescope with one more equation.

$$(N - 1)T(N - 1) = 2\left[\sum_{j=0}^{N-2} T(j)\right] + c(N - 1)^2 \qquad (7.16)$$

If we subtract Equation (7.16) from Equation (7.15), we obtain

$$NT(N) - (N - 1)T(N - 1) = 2T(N - 1) + 2cN - c \qquad (7.17)$$

We rearrange terms and drop the insignificant $-c$ on the right, obtaining

$$NT(N) = (N + 1)T(N - 1) + 2cN \qquad (7.18)$$

We now have a formula for $T(N)$ in terms of $T(N - 1)$ only. Again the idea is to telescope, but Equation (7.18) is in the wrong form. Divide Equation (7.18) by $N(N + 1)$:

$$\frac{T(N)}{N + 1} = \frac{T(N - 1)}{N} + \frac{2c}{N + 1} \qquad (7.19)$$

Now we can telescope.

$$\frac{T(N - 1)}{N} = \frac{T(N - 2)}{N - 1} + \frac{2c}{N} \qquad (7.20)$$

$$\frac{T(N - 2)}{N - 1} = \frac{T(N - 3)}{N - 2} + \frac{2c}{N - 1} \qquad (7.21)$$

$$\vdots$$

$$\frac{T(2)}{3} = \frac{T(1)}{2} + \frac{2c}{3} \qquad (7.22)$$

Adding Equations (7.19) through (7.22) yields

$$\frac{T(N)}{N+1} = \frac{T(1)}{2} + 2c \sum_{i=3}^{N+1} \frac{1}{i} \tag{7.23}$$

The sum is about $\log_e(N+1) + \gamma - \frac{3}{2}$, where $\gamma \approx 0.577$ is known as Euler's constant, so

$$\frac{T(N)}{N+1} = O(\log N) \tag{7.24}$$

And so

$$T(N) = O(N \log N) \tag{7.25}$$

Although this analysis seems complicated, it really is not—the steps are natural once you have seen some recurrence relations. The analysis can actually be taken further. The highly optimized version that was described above has also been analyzed, and this result gets extremely difficult, involving complicated recurrences and advanced mathematics. The effect of equal elements has also been analyzed in detail, and it turns out that the code presented does the right thing.

7.7.6 A Linear-Expected-Time Algorithm for Selection

Quicksort can be modified to solve the *selection problem,* which we have seen in Chapters 1 and 6. Recall that by using a priority queue, we can find the kth largest (or smallest) element in $O(N + k \log N)$. For the special case of finding the median, this gives an $O(N \log N)$ algorithm.

Since we can sort the array in $O(N \log N)$ time, one might expect to obtain a better time bound for selection. The algorithm we present to find the kth smallest element in a set S is almost identical to quicksort. In fact, the first three steps are the same. We will call this algorithm *quickselect.* Let $|S_i|$ denote the number of elements in S_i. The steps of quickselect are

1. If $|S| = 1$, then $k = 1$ and return the element in S as the answer. If a cutoff for small arrays is being used and $|S| \leq$ CUTOFF, then sort S and return the kth smallest element.
2. Pick a pivot element, $v \in S$.
3. Partition $S - \{v\}$ into S_1 and S_2, as was done with quicksort.
4. If $k \leq |S_1|$, then the kth smallest element must be in S_1. In this case, return quickselect (S_1, k). If $k = 1 + |S_1|$, then the pivot is the kth smallest element and we can return it as the answer. Otherwise, the kth smallest element lies in S_2, and it is the $(k - |S_1| - 1)$st smallest element in S_2. We make a recursive call and return quickselect $(S_2, k - |S_1| - 1)$.

In contrast to quicksort, quickselect makes only one recursive call instead of two. The worst case of quickselect is identical to that of quicksort and is $O(N^2)$. Intuitively, this is because quicksort's worst case is when one of S_1 and S_2 is empty; thus, quickselect is not really saving a recursive call. The average running time, however, is $O(N)$. The analysis is similar to quicksort's and is left as an exercise.

The implementation of quickselect is even simpler than the abstract description might imply. The code to do this is shown in Figure 7.17. When the algorithm terminates, the kth smallest element is in position $k - 1$ (because arrays start at index 0). This destroys the original ordering; if this is not desirable, then a copy must be made.

```
1      /**
2       * Internal selection method that makes recursive calls.
3       * Uses median-of-three partitioning and a cutoff of 10.
4       * Places the kth smallest item in a[k-1].
5       * @param a an array of Comparable items.
6       * @param left the left-most index of the subarray.
7       * @param right the right-most index of the subarray.
8       * @param k the desired index (1 is minimum) in the entire array.
9       */
10     private static <AnyType extends Comparable<? super AnyType>>
11     void quickSelect( AnyType [ ] a, int left, int right, int k )
12     {
13         if( left + CUTOFF <= right )
14         {
15             AnyType pivot = median3( a, left, right );
16
17                 // Begin partitioning
18             int i = left, j = right - 1;
19             for( ; ; )
20             {
21                 while( a[ ++i ].compareTo( pivot ) < 0 ) { }
22                 while( a[ --j ].compareTo( pivot ) > 0 ) { }
23                 if( i < j )
24                     swapReferences( a, i, j );
25                 else
26                     break;
27             }
28
29             swapReferences( a, i, right - 1 );   // Restore pivot
30
31             if( k <= i )
32                 quickSelect( a, left, i - 1, k );
33             else if( k > i + 1 )
34                 quickSelect( a, i + 1, right, k );
35         }
36         else  // Do an insertion sort on the subarray
37             insertionSort( a, left, right );
38     }
```

Figure 7.17 Main quickselect routine

Using a median-of-three pivoting strategy makes the chance of the worst case occurring almost negligible. By carefully choosing the pivot, however, we can eliminate the quadratic worst case and ensure an $O(N)$ algorithm. The overhead involved in doing this is considerable, so the resulting algorithm is mostly of theoretical interest. In Chapter 10, we will examine the linear-time worst-case algorithm for selection, and we shall also see an interesting technique of choosing the pivot that results in a somewhat faster selection algorithm in practice.

7.8 A General Lower Bound for Sorting

Although we have $O(N \log N)$ algorithms for sorting, it is not clear that this is as good as we can do. In this section, we prove that any algorithm for sorting that uses only comparisons requires $\Omega (N \log N)$ comparisons (and hence time) in the worst case, so that mergesort and heapsort are optimal to within a constant factor. The proof can be extended to show that $\Omega (N \log N)$ comparisons are required, even on average, for any sorting algorithm that uses only comparisons, which means that quicksort is optimal on average to within a constant factor.

Specifically, we will prove the following result: Any sorting algorithm that uses only comparisons requires $\lceil \log(N!) \rceil$ comparisons in the worst case and $\log(N!)$ comparisons on average. We will assume that all N elements are distinct, since any sorting algorithm must work for this case.

7.8.1 Decision Trees

A **decision tree** is an abstraction used to prove lower bounds. In our context, a decision tree is a binary tree. Each node represents a set of possible orderings, consistent with comparisons that have been made, among the elements. The results of the comparisons are the tree edges.

The decision tree in Figure 7.18 represents an algorithm that sorts the three elements a, b, and c. The initial state of the algorithm is at the root. (We will use the terms *state* and *node* interchangeably.) No comparisons have been done, so all orderings are legal. The first comparison that *this particular* algorithm performs compares a and b. The two results lead to two possible states. If $a < b$, then only three possibilities remain. If the algorithm reaches node 2, then it will compare a and c. Other algorithms might do different things; a different algorithm would have a different decision tree. If $a > c$, the algorithm enters state 5. Since there is only one ordering that is consistent, the algorithm can terminate and report that it has completed the sort. If $a < c$, the algorithm cannot do this, because there are two possible orderings and it cannot possibly be sure which is correct. In this case, the algorithm will require one more comparison.

Every algorithm that sorts by using only comparisons can be represented by a decision tree. Of course, it is only feasible to draw the tree for extremely small input sizes. The number of comparisons used by the sorting algorithm is equal to the depth of the deepest leaf. In our case, this algorithm uses three comparisons in the worst case. The average

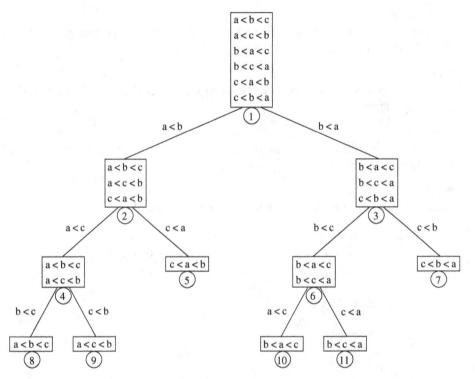

Figure 7.18   A decision tree for three-element sort

number of comparisons used is equal to the average depth of the leaves. Since a decision tree is large, it follows that there must be some long paths. To prove the lower bounds, all that needs to be shown are some basic tree properties.

Lemma 7.1.
Let T be a binary tree of depth d. Then T has at most 2^d leaves.

Proof.
The proof is by induction. If $d = 0$, then there is at most one leaf, so the basis is true. Otherwise, we have a root, which cannot be a leaf, and a left and right subtree, each of depth at most $d - 1$. By the induction hypothesis, they can each have at most 2^{d-1} leaves, giving a total of at most 2^d leaves. This proves the lemma.

Lemma 7.2.
A binary tree with L leaves must have depth at least $\lceil \log L \rceil$.

Proof.
Immediate from the preceding lemma.

Theorem 7.6.

Any sorting algorithm that uses only comparisons between elements requires at least $\lceil \log(N!) \rceil$ comparisons in the worst case.

Proof.

A decision tree to sort N elements must have $N!$ leaves. The theorem follows from the preceding lemma.

Theorem 7.7.

Any sorting algorithm that uses only comparisons between elements requires $\Omega(N \log N)$ comparisons.

Proof.

From the previous theorem, $\log(N!)$ comparisons are required.

$$
\begin{aligned}
\log(N!) &= \log(N(N-1)(N-2)\cdots(2)(1)) \\
&= \log N + \log(N-1) + \log(N-2) + \cdots + \log 2 + \log 1 \\
&\geq \log N + \log(N-1) + \log(N-2) + \cdots + \log(N/2) \\
&\geq \frac{N}{2} \log \frac{N}{2} \\
&\geq \frac{N}{2} \log N - \frac{N}{2} \\
&= \Omega(N \log N)
\end{aligned}
$$

This type of lower-bound argument, when used to prove a worst-case result, is sometimes known as an **information-theoretic** lower bound. The general theorem says that if there are P different possible cases to distinguish, and the questions are of the form YES/NO, then $\lceil \log P \rceil$ questions are always required in some case by any algorithm to solve the problem. It is possible to prove a similar result for the average-case running time of any comparison-based sorting algorithm. This result is implied by the following lemma, which is left as an exercise: Any binary tree with L leaves has an average depth of at least $\log L$.

Note that $\log(N!)$ is roughly $N \log N - O(N)$ (Exercise 7.34).

7.9 Decision-Tree Lower Bounds for Selection Problems

Section 7.8 employed a decision tree argument to show the fundamental lower bound that any comparison-based sorting algorithm must use roughly $N \log N$ comparisons. In this section we show additional lower bounds for selection in an N-element collection, specifically

1. $N - 1$ comparisons are necessary to find the smallest item
2. $N + \lceil \log N \rceil - 2$ comparisons are necessary to find the two smallest items
3. $\lceil 3N/2 \rceil - O(\log N)$ comparisons are necessary to find the median

The lower bounds for all these problems, with the exception of finding the median, are tight: Algorithms exist that use exactly the specified number of comparisons. In all our proofs, we assume all items are unique.

Lemma 7.3.
If all the leaves in a decision tree are at depth d or higher, the decision tree must have at least 2^d leaves.

Proof.
Note that all nonleaf nodes in a decision tree have two children. The proof is by induction and follows Lemma 7.1.

The first lower bound, for finding the smallest item, is the easiest and most trivial to show.

Theorem 7.8.
Any comparison-based algorithm to find the smallest element must use at least $N - 1$ comparisons.

Proof.
Every element, x, except the smallest element, must be involved in a comparison with some other element y, in which x is declared larger than y. Otherwise, if there were two different elements that had not been declared larger than any other elements, then either could be the smallest.

Lemma 7.4.
The decision tree for finding the smallest of N elements must have at least 2^{N-1} leaves.

Proof.
By Theorem 7.8, all leaves in this decision tree are at depth $N - 1$ or higher. Then this lemma follows from Lemma 7.3.

The bound for selection is a little more complicated and requires looking at the structure of the decision tree. It will allow us to prove lower bounds for problems 2 and 3 on our list.

Lemma 7.5.
The decision tree for finding the kth smallest of N elements must have at least $\binom{N}{k-1} 2^{N-k}$ leaves.

Proof.
Observe that any algorithm that correctly identifies the kth smallest element t must be able to prove that all other elements x are either larger than or smaller than t. Otherwise, it would be giving the same answer regardless of whether x was larger or smaller than t, and the answer cannot be the same in both circumstances. Thus each

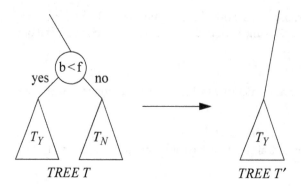

Figure 7.19 Smallest three elements are $S = \{\,a, b, c\,\}$; largest four elements are $R = \{\,d, e, f, g\,\}$; the comparison between b and f for this choice of R and S can be eliminated when forming tree T'

leaf in the tree, in addition to representing the kth smallest element, also represents the $k - 1$ smallest items that have been identified.

Let T be the decision tree, and consider two sets: $S = \{\,x_1, x_2, \ldots, x_{k-1}\,\}$, representing the $k - 1$ smallest items, and R which are the remaining items, including the kth smallest. Form a new decision tree T', by purging any comparisons in T between an element in S and an element in R. Since any element in S is smaller than an element in R, the comparison tree node and its right subtree may be removed from T, without any loss of information. Figure 7.19 shows how nodes can be pruned.

Any permutation of R that is fed into T' follows the same path of nodes and leads to the same leaf as a corresponding sequence consisting of a permutation of S followed by the same permutation of R. Since T identifies the overall kth smallest element, and the smallest element in R is that element, it follows that T' identifies the smallest element in R. Thus T' must have at least $2^{|R|-1} = 2^{N-k}$ leaves. These leaves in T' directly correspond to 2^{N-k} leaves representing S. Since there are $\binom{N}{k-1}$ choices for S, there must be at least $\binom{N}{k-1} 2^{N-k}$ leaves in T.

A direct application of Lemma 7.5 allows us to prove the lower bounds for finding the second smallest element, and also the median.

Theorem 7.9.

Any comparison-based algorithm to find the kth smallest element must use at least
$$N - k + \left\lceil \log\left(\binom{N}{k-1}\right)\right\rceil \text{ comparisons.}$$

Proof.

Immediate from Lemma 7.5 and Lemma 7.2.

Theorem 7.10.
Any comparison-based algorithm to find the second smallest element must use at least $N + \lceil \log N \rceil - 2$ comparisons.

Proof.
Applying Theorem 7.9, with $k = 2$ yields $N - 2 + \lceil \log N \rceil$.

Theorem 7.11.
Any comparison-based algorithm to find the median must use at least $\lceil 3N/2 \rceil - O(\log N)$ comparisons.

Proof.
Apply Theorem 7.9, with $k = \lceil N/2 \rceil$.

The lower bound for selection is not tight, nor is it the best known; see the references for details.

7.10 Adversary Lower Bounds

Although decision-tree arguments allowed us to show lower bounds for sorting and some selection problems, generally the bounds that result are not that tight, and sometimes are trivial.

For instance, consider the problem of finding the minimum item. Since there are N possible choices for the minimum, the information theory lower bound that is produced by a decision-tree argument is only $\log N$. In Theorem 7.8, we were able to show the $N - 1$ bound by using what is essentially an **adversary argument**. In this section, we expand on this argument and use it to prove the following lower bound:

4. $\lceil 3N/2 \rceil - 2$ comparisons are necessary to find both the smallest and largest item

Recall our proof that any algorithm to find the smallest item requires at least $N - 1$ comparisons:

Every element, x, except the smallest element, must be involved in a comparison with some other element y, in which x is declared larger than y. Otherwise, if there were two different elements that had not been declared larger than any other elements, then either could be the smallest.

This is the underlying idea of an adversary argument which has some basic steps:

1. Establish that some basic amount of information must be obtained by any algorithm that solves a problem

2. In each step of the algorithm, the adversary will maintain an input that is consistent with all the answers that have been provided by the algorithm thus far

3. Argue that with insufficient steps, there are multiple consistent inputs that would provide different answers to the algorithm; hence the algorithm has not done enough steps, because if the algorithm were to provide an answer at that point, the adversary would be able to show an input for which the answer is wrong.

To see how this works, we will reprove the lower bound for finding the smallest element using this proof template.

Theorem 7.8. *(restated)*

Any comparison-based algorithm to find the smallest element must use at least $N - 1$ comparisons.

New Proof.

Begin by marking each item as U (for unknown). When an item is declared larger than another item, we will change its marking to E (for eliminated). This change represents one unit of information. Initially each unknown item has a value of 0, but there have been no comparisons, so this ordering is consistent with prior answers.

A comparison between two items is either between two unknowns, or it involves at least one item eliminated from being the minimum. Figure 7.20 shows how our adversary will construct the input values, based on the questioning.

If the comparison is between two unknowns, the first is declared the smaller, and the second is automatically eliminated, providing one unit of information. We then assign it (irrevocably) a number larger than 0; the most convenient is the number of eliminated items. If a comparison is between an eliminated number and an unknown, the eliminated number (which is larger than 0 by the prior sentence) will be declared larger, and there will be no changes, no eliminations, and no information obtained. If two eliminated numbers are compared, then they will be different, and a consistent answer can be provided, again with no changes, and no information provided.

At the end, we need to obtain $N-1$ units of information, and each comparison provides only 1 unit at the most; hence, at least $N - 1$ comparisons are necessary.

Lower Bound for Finding the Minimum and Maximum

We can use this same technique to establish a lower bound for finding both the minimum and maximum item. Observe that all but one item must be eliminated from being the smallest, and all but one item must be eliminated from being the largest; thus the total information that any algorithm must acquire is $2N - 2$. However, a comparison $x < y$, eliminates both x from being the maximum and y from being the minimum; thus a comparison can provide two units of information. Consequently, this argument yields only the

x	y	Answer	Information	New x	New y
U	U	$x < y$	1	No change	Mark y as E Change y to #elim
All others		Consistently	0	No change	No change

Figure 7.20 Adversary constructs input for finding the minimum as algorithm runs

trivial $N - 1$ lower bound. Our adversary needs to do more work to ensure that it does not give out two units of information more than it needs to.

To achieve this, each item will initially be unmarked. If it "wins" a comparison (i.e., it is declared larger than some item), it obtains a W. If it "loses" a comparison (i.e., it is declared smaller than some item), it obtains an L. At the end, all but two items will be WL. Our adversary will ensure that it only hands out two units of information if it is comparing two unmarked items. That can happen only $\lfloor N/2 \rfloor$ times; then the remaining information has to be obtained one unit at a time, which will establish the bound.

Theorem 7.12.

Any comparison-based algorithm to find the minimum and maximum must use at least $\lceil 3N/2 \rceil - 2$ comparisons.

Proof.

The basic idea if that if two items are unmarked, the adversary must give out two pieces of information. Otherwise, one of the items has either a W or an L (perhaps both). In that case, with reasonable care, the adversary should be able to avoid giving out two units of information. For instance, if one item, x, has a W and the other item, y, is unmarked, the adversary lets x win again by saying $x > y$. This gives one unit of information for y, but no new information for x. It is easy to see that in principle, there is no reason that the adversary should have to give more than one unit of information out if there is at least one unmarked item involved in the comparison.

It remains to show that the adversary can maintain values that are consistent with its answers. If both items are unmarked, then obviously they can be safely assigned values consistent with the comparison answer; this case yields two units of information.

Otherwise, if one of the items involved in a comparison is unmarked, it can be assigned a value the first time, consistent with the other item in the comparison. This case yields one unit of information.

Otherwise both items involved in the comparison are marked. If both are WL, then we can answer consistently with the current assignment, yielding no information.[2]

Otherwise at least one of the items has only an L or only a W. We will allow that item to compare redundantly (if it is an L then it loses again, if it is a W then it wins again), and its value can be easily adjusted if needed, based on the other item in the comparison (an L can be lowered as needed; a W can be raised as needed). This yields at most one unit of information for the other item in the comparison, possibly zero. Figure 7.21 summarizes the action of the adversary, making y the primary element whose value changes in all cases.

At most $\lfloor N/2 \rfloor$ comparisons yield two units of information, meaning that the remaining information, namely $2N - 2 - 2\lfloor N/2 \rfloor$ units, must each be obtained one comparison at a time. Thus the total number of comparisons that are needed is at least $2N - 2 - \lfloor N/2 \rfloor = \lceil 3N/2 \rceil - 2$.

---

[2] It is possible that the current assignment for both items has the same number; in such a case we can increase all items whose current value is larger than y by 2, and then add 1 to y to break the tie.

x	y	Answer	Information	New x	New y
–	–	$x < y$	2	L 0	W 1
L	–	$x < y$	1	L unchanged	W $x + 1$
W or WL	–	$x > y$	1	W or WL unchanged	L $x - 1$
W or WL	W	$x < y$	1 or 0	WL unchanged	W $max(x + 1, y)$
L or W or WL	L	$x > y$	1 or 0 or 0	WL or W or WL unchanged	L $min(x - 1, y)$
WL	WL	consistent	0	unchanged	unchanged
–	W				
–	WL				
–	L		**SYMMETRIC TO AN ABOVE CASE**		
L	W				
L	WL				
W	WL				

Figure 7.21 Adversary constructs input for finding the maximum and minimum as algorithm runs

It is easy to see that this bound is achievable. Pair up the elements, and perform a comparison between each pair. Then find the maximum among the winners, and the minimum amoung the losers.

7.11 Linear-Time Sorts: Bucket Sort and Radix Sort

Although we proved in Section 7.8 that any general sorting algorithm that uses only comparisons requires $\Omega(N \log N)$ time in the worst case, recall that it is still possible to sort in linear time in some special cases.

A simple example is **bucket sort**. For bucket sort to work, extra information must be available. The input $A_1, A_2, \ldots, A_N$ must consist of only positive integers smaller than M. (Obviously extensions to this are possible.) If this is the case, then the algorithm is simple: Keep an array called count, of size M, which is initialized to all 0's. Thus, count has M cells, or buckets, which are initially empty. When A_i is read, increment count$[A_i]$ by 1. After all the input is read, scan the count array, printing out a representation of the sorted list. This algorithm takes $O(M + N)$; the proof is left as an exercise. If M is $O(N)$, then the total is $O(N)$.

Although this algorithm seems to violate the lower bound, it turns out that it does not because it uses a more powerful operation than simple comparisons. By incrementing the appropriate bucket, the algorithm essentially performs an M-way comparison in unit time. This is similar to the strategy used in extendible hashing (Section 5.9). This is clearly not in the model for which the lower bound was proven.

This algorithm does, however, question the validity of the model used in proving the lower bound. The model actually is a strong model, because a *general-purpose* sorting algorithm cannot make assumptions about the type of input it can expect to see, but must make decisions based on ordering information only. Naturally, if there is extra information available, we should expect to find a more efficient algorithm, since otherwise the extra information would be wasted.

Although bucket sort seems like much too trivial an algorithm to be useful, it turns out that there are many cases where the input is only small integers, so that using a method like quicksort is really overkill. One such example is **radix sort**.

Radix sort is sometimes known as card sort because it was used until the advent of modern computers to sort old-style punch cards. Suppose we have 10 numbers in the range 0 to 999 that we would like to sort. In general this is N numbers in the range 0 to $b^p - 1$ for some constant p. Obviously we cannot use bucket sort; there would be too many buckets. The trick is to use several passes of bucket sort. The natural algorithm would be to bucket sort by the most significant "digit" (digit is taken to base b), then next most significant, and so on. But a simpler idea is to perform bucket sorts in the reverse order, starting with the least significant "digit" first. Of course, more than one number could fall into the same bucket, and unlike the original bucket sort, these numbers could be different, so we keep them in a list. Each pass is stable: Items that agree in the current digit retain the ordering determined in prior passes. The trace in Figure 7.22 shows the result of sorting 64, 8, 216, 512, 27, 729, 0, 1, 343, 125, which is the first ten cubes arranged randomly (we use 0's to make clear the tens and hundreds digits). After the first pass, the items are sorted and in general, after the kth pass, the items are sorted on the k least significant digits. So at the end, the items are completely sorted. To see that the algorithm works, notice that the only possible failure would occur if two numbers came out of the same bucket in the wrong order. But the previous passes ensure that when several numbers enter a bucket, they enter in sorted order. The running time is $O(p(N + b))$ where p is the number of passes, N is the number of elements to sort, and b is the number of buckets.

One application of radix sort is sorting strings. If all the strings have the same length L, then by using buckets for each character, we can implement a radix sort in $O(NL)$

INITIAL ITEMS:	064, 008, 216, 512, 027, 729, 000, 001, 343, 125
SORTED BY 1's digit:	000, 001, 512, 343, 064, 125, 216, 027, 008, 729
SORTED BY 10's digit:	000, 001, 008, 512, 216, 125, 027, 729, 343, 064
SORTED BY 100's digit:	000, 001, 008, 027, 064, 125, 216, 343, 512, 729

Figure 7.22   Radix sort trace

```
1      /*
2       * Radix sort an array of Strings
3       * Assume all are all ASCII
4       * Assume all have same length
5       */
6      public static void radixSortA( String [ ] arr, int stringLen )
7      {
8          final int BUCKETS = 256;
9          ArrayList<String> [ ] buckets = new ArrayList<>[ BUCKETS ];
10
11         for( int i = 0; i < BUCKETS; i++ )
12             buckets[ i ] = new ArrayList<>( );
13
14         for( int pos = stringLen - 1; pos >= 0; pos-- )
15         {
16             for( String s : arr )
17                 buckets[ s.charAt( pos ) ].add( s );
18
19             int idx = 0;
20             for( ArrayList<String> thisBucket : buckets )
21             {
22                 for( String s : thisBucket )
23                     arr[ idx++ ] = s;
24
25                 thisBucket.clear( );
26             }
27         }
28     }
```

Figure 7.23 Simple implementation of radix sort for strings, using an `ArrayList` of buckets

time. The most straightforward way of doing this is shown in Figure 7.23. In our code, we assume that all characters are ASCII, residing in the first 256 positions of the Unicode character set. In each pass, we add an item to the appropriate bucket, and then after all the buckets are populated, we step through the buckets dumping everything back to the array. Notice that when a bucket is populated and emptied in the next pass, the order from the current pass is preserved.

Counting radix sort is an alternative implementation of radix sort that avoids using `ArrayList`s. Instead, we maintain a count of how many items would go in each bucket; this information can go into an array `count`, so that `count[k]` is the number of items that are in bucket k. Then we can use another array `offset`, so that `offset[k]` represents the number of items whose value is strictly smaller than k. Then when we see a value k for the first time in the final scan, `offset[k]` tells us a valid array spot where it can be written to (but we have to use a temporary array for the write), and after that is done, `offset[k]`

```
 1        /*
 2         * Counting radix sort an array of Strings
 3         * Assume all are all ASCII
 4         * Assume all have same length
 5         */
 6        public static void countingRadixSort( String [ ] arr, int stringLen )
 7        {
 8            final int BUCKETS = 256;
 9
10            int N = arr.length;
11            String [ ] buffer = new String [ N ];
12
13            String [ ] in = arr;
14            String [ ] out = buffer;
15
16            for( int pos = stringLen - 1; pos >= 0; pos-- )
17            {
18                int [ ] count = new int [ BUCKETS + 1 ];
19
20                for( int i = 0; i < N; i++ )
21                    count[ in[ i ].charAt( pos ) + 1 ]++;
22
23                for( int b = 1; b <= BUCKETS; b++ )
24                    count[ b ] += count[ b - 1 ];
25
26                for( int i = 0; i < N; i++ )
27                    out[ count[ in[ i ].charAt( pos ) ]++ ] = in[ i ];
28
29                  // swap in and out roles
30                String [ ] tmp = in;
31                in = out;
32                out = tmp;
33            }
34
35                // if odd number of passes, in is buffer, out is arr; so copy back
36            if( stringLen % 2 == 1 )
37                for( int i = 0; i < arr.length; i++ )
38                    out[ i ] = in[ i ];
39        }
```

Figure 7.24 Counting radix sort for fixed-length strings

```
1          /*
2           * Radix sort an array of Strings
3           * Assume all are all ASCII
4           * Assume all have length bounded by maxLen
5           */
6          public static void radixSort( String [ ] arr, int maxLen )
7          {
8              final int BUCKETS = 256;
9
10             ArrayList<String> [ ] wordsByLength = new ArrayList<>[ maxLen + 1 ];
11             ArrayList<String> [ ] buckets = new ArrayList<>[ BUCKETS ];
12
13             for( int i = 0; i < wordsByLength.length; i++ )
14                 wordsByLength[ i ] = new ArrayList<>( );
15
16             for( int i = 0; i < BUCKETS; i++ )
17                 buckets[ i ] = new ArrayList<>( );
18
19             for( String s : arr )
20                 wordsByLength[ s.length( ) ].add( s );
21
22             int idx = 0;
23             for( ArrayList<String> wordList : wordsByLength )
24                 for( String s : wordList )
25                     arr[ idx++ ] = s;
26
27             int startingIndex = arr.length;
28             for( int pos = maxLen - 1; pos >= 0; pos-- )
29             {
30                 startingIndex -= wordsByLength[ pos + 1 ].size( );
31
32                 for( int i = startingIndex; i < arr.length; i++ )
33                     buckets[ arr[ i ].charAt( pos ) ].add( arr[ i ] );
34
35                 idx = startingIndex;
36                 for( ArrayList<String> thisBucket : buckets )
37                 {
38                     for( String s : thisBucket )
39                         arr[ idx++ ] = s;
40
41                     thisBucket.clear( );
42                 }
43             }
44         }
```

Figure 7.25 Radix sort for variable length strings

can be incremented. Counting radix sort thus avoids the need to keep lists. As a further optimization, we can avoid using offset, by reusing the count array. The modification is that we initially have count[k+1] represent the number of items that are in bucket k. Then after that information is computed, we can scan the count array from the smallest to largest index, and increment count[k] by count[k-1]. It is easy to verify that after this scan, the count array stores exactly the same information that would have been stored in offset.

Figure 7.24 shows an implementation of counting radix sort. Lines 18 to 27 implement the logic above, assuming that the items are stored in array in, and the result of a single pass is placed in array out. Initially, in represents arr and out represents the temporary array, buffer. After each pass, we switch the roles of in and out. If there are an even number of passes, then at the end, out is referencing arr, so the sort is complete. Otherwise, we have to copy from the buffer back into arr.

Generally, counting radix sort is prefereable to using ArrayLists, but it can suffer from poor locality (out is filled in non-sequentially) and thus surprisingly, it is not always faster than using an array of ArrayLists.

We can extend either version of radix sort to work with variable-length strings. The basic algorithm is to first sort the strings by their length. Instead of looking at all the strings, we can then look only at strings that we know are long enough. Since the string lengths are small numbers, the initial sort by length can be done by—bucket sort! Figure 7.25 shows this implementation of radix sort, with ArrayLists. Here, the words are grouped into buckets by length at lines 19–20, and then placed back into the array at lines 22–25. Lines 32–33 look at only those strings that have a character at position pos, by making use of the variable startingIndex, which is maintained at lines 27 and 30. Except for that, lines 27–43 in Figure 7.25 are the same as lines 14–27 in Figure 7.23.

The running time of this version of radix sort is linear in the total number of characters in all the strings (each character appears exactly once at line 33, and the statement at line 39 executes precisiely as many times as the line 33). Radix sort for strings will perform especially well when the characters in the string are drawn from a reasonably small alphabet, and when the strings either are relatively short or are very similar. Because the $O(N \log N)$ comparison-based sorting algorithms will generally look only at a small number of characters in each string comparison, once the average string length starts getting large, radix sort's advantage is minimized or evaporates completely.

7.12 External Sorting

So far, all the algorithms we have examined require that the input fit into main memory. There are, however, applications where the input is much too large to fit into memory. This section will discuss **external sorting algorithms**, which are designed to handle very large inputs.

7.12.1 Why We Need New Algorithms

Most of the internal sorting algorithms take advantage of the fact that memory is directly addressable. Shellsort compares elements a[i] and a[i-h_k] in one time unit. Heapsort compares elements a[i] and a[i*2+1] in one time unit. Quicksort, with median-of-three partitioning, requires comparing a[left], a[center], and a[right] in a constant number of time units. If the input is on a tape, then all these operations lose their efficiency, since elements on a tape can only be accessed sequentially. Even if the data is on a disk, there is still a practical loss of efficiency because of the delay required to spin the disk and move the disk head.

To see how slow external accesses really are, create a random file that is large, but not too big to fit in main memory. Read the file in and sort it using an efficient algorithm. The time it takes to read the input is certain to be significant compared to the time to sort the input, even though sorting is an $O(N \log N)$ operation and reading the input is only $O(N)$.

7.12.2 Model for External Sorting

The wide variety of mass storage devices makes external sorting much more device dependent than internal sorting. The algorithms that we will consider work on tapes, which are probably the most restrictive storage medium. Since access to an element on tape is done by winding the tape to the correct location, tapes can be efficiently accessed only in sequential order (in either direction).

We will assume that we have at least three tape drives to perform the sorting. We need two drives to do an efficient sort; the third drive simplifies matters. If only one tape drive is present, then we are in trouble: Any algorithm will require $\Omega(N^2)$ tape accesses.

7.12.3 The Simple Algorithm

The basic external sorting algorithm uses the merging algorithm from mergesort. Suppose we have four tapes, T_{a1}, T_{a2}, T_{b1}, T_{b2}, which are two input and two output tapes. Depending on the point in the algorithm, the a and b tapes are either input tapes or output tapes. Suppose the data are initially on T_{a1}. Suppose further that the internal memory can hold (and sort) M records at a time. A natural first step is to read M records at a time from the input tape, sort the records internally, and then write the sorted records alternately to T_{b1} and T_{b2}. We will call each set of sorted records a **run**. When this is done, we rewind all the tapes. Suppose we have the same input as our example for Shellsort.

T_{a1}	81	94	11	96	12	35	17	99	28	58	41	75	15
T_{a2}													
T_{b1}													
T_{b2}													

If $M = 3$, then after the runs are constructed, the tapes will contain the data indicated in the following figure.

T_{a1}							
T_{a2}							
T_{b1}	11	81	94	17	28	99	15
T_{b2}	12	35	96	41	58	75	

Now T_{b1} and T_{b2} contain a group of runs. We take the first run from each tape and merge them, writing the result, which is a run twice as long, onto T_{a1}. Recall that merging two sorted lists is simple; we need almost no memory, since the merge is performed as T_{b1} and T_{b2} advance. Then we take the next run from each tape, merge these, and write the result to T_{a2}. We continue this process, alternating between T_{a1} and T_{a2}, until either T_{b1} or T_{b2} is empty. At this point either both are empty or there is one run left. In the latter case, we copy this run to the appropriate tape. We rewind all four tapes and repeat the same steps, this time using the a tapes as input and the b tapes as output. This will give runs of 4M. We continue the process until we get one run of length N.

This algorithm will require $\lceil \log(N/M) \rceil$ passes, plus the initial run-constructing pass. For instance, if we have 10 million records of 128 bytes each, and four megabytes of internal memory, then the first pass will create 320 runs. We would then need nine more passes to complete the sort. Our example requires $\lceil \log 13/3 \rceil = 3$ more passes, which are shown in the following figures.

T_{a1}	11	12	35	81	94	96	15				
T_{a2}	17	28	41	58	75	99					
T_{b1}											
T_{b2}											

T_{a1}											
T_{a2}											
T_{b1}	11	12	17	28	35	41	58	75	81	94	96	99
T_{b2}	15										

T_{a1}	11	12	15	17	28	35	41	58	75	81	94	96	99
T_{a2}													
T_{b1}													
T_{b2}													

7.12.4   Multiway Merge

If we have extra tapes, then we can expect to reduce the number of passes required to sort our input. We do this by extending the basic (two-way) merge to a k-way merge.

Merging two runs is done by winding each input tape to the beginning of each run. Then the smaller element is found, placed on an output tape, and the appropriate input tape is advanced. If there are k input tapes, this strategy works the same way, the only difference being that it is slightly more complicated to find the smallest of the k elements. We can find the smallest of these elements by using a priority queue. To obtain the next element to write on the output tape, we perform a deleteMin operation. The appropriate input tape is advanced, and if the run on the input tape is not yet completed, we insert the new element into the priority queue. Using the same example as before, we distribute the input onto the three tapes.

T_{a1}						
T_{a2}						
T_{a3}						
T_{b1}	11	81	94	41	58	75
T_{b2}	12	35	96	15		
T_{b3}	17	28	99			

We then need two more passes of three-way merging to complete the sort.

T_{a1}	11	12	17	28	35	81	94	96	99
T_{a2}	15	41	58	75					
T_{a3}									
T_{b1}									
T_{b2}									
T_{b3}									

T_{a1}													
T_{a2}													
T_{a3}													
T_{b1}	11	12	15	17	28	35	41	58	75	81	94	96	99
T_{b2}													
T_{b3}													

After the initial run construction phase, the number of passes required using k-way merging is $\lceil \log_k (N/M) \rceil$, because the runs get k times as large in each pass. For the example above, the formula is verified, since $\lceil \log_3 (13/3) \rceil = 2$. If we have 10 tapes, then $k = 5$, and our large example from the previous section would require $\lceil \log_5 320 \rceil = 4$ passes.

7.12.5 Polyphase Merge

The k-way merging strategy developed in the last section requires the use of $2k$ tapes. This could be prohibitive for some applications. It is possible to get by with only $k + 1$ tapes. As an example, we will show how to perform two-way merging using only three tapes.

Suppose we have three tapes, T_1, T_2, and T_3, and an input file on T_1 that will produce 34 runs. One option is to put 17 runs on each of T_2 and T_3. We could then merge this result onto T_1, obtaining one tape with 17 runs. The problem is that since all the runs are on one tape, we must now put some of these runs on T_2 to perform another merge. The logical way to do this is to copy the first eight runs from T_1 onto T_2 and then perform the merge. This has the effect of adding an extra half pass for every pass we do.

An alternative method is to split the original 34 runs unevenly. Suppose we put 21 runs on T_2 and 13 runs on T_3. We would then merge 13 runs onto T_1 before T_3 was empty. At this point, we could rewind T_1 and T_3 and merge T_1, with 13 runs, and T_2, which has 8 runs, onto T_3. We could then merge 8 runs until T_2 was empty, which would leave 5 runs left on T_1 and 8 runs on T_3. We could then merge T_1 and T_3, and so on. The following table shows the number of runs on each tape after each pass.

	Run Const.	After $T_3 + T_2$	After $T_1 + T_2$	After $T_1 + T_3$	After $T_2 + T_3$	After $T_1 + T_2$	After $T_1 + T_3$	After $T_2 + T_3$
T_1	0	13	5	0	3	1	0	1
T_2	21	8	0	5	2	0	1	0
T_3	13	0	8	3	0	2	1	0

The original distribution of runs makes a great deal of difference. For instance, if 22 runs are placed on T_2, with 12 on T_3, then after the first merge, we obtain 12 runs on T_3 and 10 runs on T_2. After another merge, there are 10 runs on T_1 and 2 runs on T_3. At this point the going gets slow, because we can only merge two sets of runs before T_3 is exhausted. Then T_1 has 8 runs and T_2 has 2 runs. Again, we can only merge two sets of runs, obtaining T_1 with 6 runs and T_3 with 2 runs. After three more passes, T_2 has two runs and the other tapes are empty. We must copy one run to another tape, and then we can finish the merge.

It turns out that the first distribution we gave is optimal. If the number of runs is a Fibonacci number F_N, then the best way to distribute them is to split them into two Fibonacci numbers F_{N-1} and F_{N-2}. Otherwise, it is necessary to pad the tape with dummy runs in order to get the number of runs up to a Fibonacci number. We leave the details of how to place the initial set of runs on the tapes as an exercise.

We can extend this to a k-way merge, in which case we need kth order Fibonacci numbers for the distribution, where the kth order Fibonacci number is defined as $F^{(k)}(N) = F^{(k)}(N-1) + F^{(k)}(N-2) + \cdots + F^{(k)}(N-k)$, with the appropriate initial conditions $F^{(k)}(N) = 0, 0 \le N \le k - 2, F^{(k)}(k-1) = 1$.

7.12.6 Replacement Selection

The last item we will consider is construction of the runs. The strategy we have used so far is the simplest possible: We read as many records as possible and sort them, writing the result to some tape. This seems like the best approach possible, until one realizes that as soon as the first record is written to an output tape, the memory it used becomes available for another record. If the next record on the input tape is larger than the record we have just output, then it can be included in the run.

Using this observation, we can give an algorithm for producing runs. This technique is commonly referred to as **replacement selection.** Initially, M records are read into memory and placed in a priority queue. We perform a `deleteMin`, writing the smallest (valued) record to the output tape. We read the next record from the input tape. If it is larger than the record we have just written, we can add it to the priority queue. Otherwise, it cannot go into the current run. Since the priority queue is smaller by one element, we can store this new element in the dead space of the priority queue until the run is completed and use the element for the next run. Storing an element in the dead space is similar to what is done in heapsort. We continue doing this until the size of the priority queue is zero, at which point the run is over. We start a new run by building a new priority queue, using all the elements in the dead space. Figure 7.26 shows the run construction for the small example we have been using, with $M = 3$. Dead elements are indicated by an asterisk.

In this example, replacement selection produces only three runs, compared with the five runs obtained by sorting. Because of this, a three-way merge finishes in one pass instead of two. If the input is randomly distributed, replacement selection can be shown to produce runs of average length $2M$. For our large example, we would expect 160 runs instead of 320 runs, so a five-way merge would require four passes. In this case, we have not saved a pass, although we might if we get lucky and have 125 runs or less. Since external sorts take so long, every pass saved can make a significant difference in the running time.

| | 3 Elements in Heap Array | | | Output | Next Element Read |
	h[1]	h[2]	h[3]		
Run 1	11	94	81	11	96
	81	94	96	81	12*
	94	96	12*	94	35*
	96	35*	12*	96	17*
	17*	35*	12*	End of Run	Rebuild Heap
Run 2	12	35	17	12	99
	17	35	99	17	28
	28	99	35	28	58
	35	99	58	35	41
	41	99	58	41	15*
	58	99	15*	58	End of Tape
	99		15*	99	
			15*	End of Run	Rebuild Heap
Run 3	15			15	

Figure 7.26   Example of run construction

As we have seen, it is possible for replacement selection to do no better than the standard algorithm. However, the input is frequently sorted or nearly sorted to start with, in which case replacement selection produces only a few very long runs. This kind of input is common for external sorts and makes replacement selection extremely valuable.

Summary

Sorting is one of the oldest and most well studied problems in computing. For most general internal sorting applications, an insertion sort, Shellsort, mergesort, or quicksort is the method of choice. The decision regarding which to use depends on the size of the input and on the underlying environment. Insertion sort is appropriate for very small amounts of input. Shellsort is a good choice for sorting moderate amounts of input. With a proper increment sequence, it gives excellent performance and uses only a few lines of code. Mergesort has $O(N \log N)$ worst-case performance but requires additional space. However, the number of comparisons that are used is nearly optimal, because any algorithm that sorts by using only element comparisons must use at least $\lceil \log (N!) \rceil$ comparisons for some input sequence. Quicksort does not by itself provide this worst-case guarantee and is tricky to code. However, it has almost certain $O(N \log N)$ performance and can be combined with heapsort to give an $O(N \log N)$ worst-case guarantee. Strings can be sorted in linear time using radix sort, and this may be a practical alternative to comparison-based sorts in some instances.

Exercises

7.1   Sort the sequence $3, 1, 4, 1, 5, 9, 2, 6, 5$ using insertion sort.

7.2   What is the running time of insertion sort if all elements are equal?

7.3   Suppose we exchange elements a[i] and a[i+k], which were originally out of order. Prove that at least 1 and at most $2k - 1$ inversions are removed.

7.4   Show the result of running Shellsort on the input $9, 8, 7, 6, 5, 4, 3, 2, 1$ using the increments $\{1, 3, 7\}$.

7.5   a. What is the running time of Shellsort using the two-increment sequence $\{1, 2\}$?
   b. Show that for any N, there exists a three-increment sequence such that Shellsort runs in $O(N^{5/3})$ time.
   c. Show that for any N, there exists a six-increment sequence such that Shellsort runs in $O(N^{3/2})$ time.

7.6   *a. Prove that the running time of Shellsort is $\Omega(N^2)$ using increments of the form $1, c, c^2, \ldots, c^i$ for any integer c.
   **b. Prove that for these increments, the average running time is $\Theta(N^{3/2})$.

7.7   Prove that if a k-sorted file is then h-sorted, it remains k-sorted.

**7.8   Prove that the running time of Shellsort, using the increment sequence suggested by Hibbard, is $\Omega(N^{3/2})$ in the worst case. *Hint:* You can prove the bound by considering the special case of what Shellsort does when all elements are either 0 or 1. Set $a[i] = 1$ if i is expressible as a linear combination of $h_t, h_{t-1}, \ldots, h_{\lfloor t/2 \rfloor + 1}$ and 0 otherwise.

7.9   Determine the running time of Shellsort for
  a. sorted input
  *b. reverse-ordered input

7.10   Do either of the following modifications to the Shellsort routine coded in Figure 7.4 affect the worst-case running time?
  a. Before line 11, subtract one from gap if it is even.
  b. Before line 11, add one to gap if it is even.

7.11   Show how heapsort processes the input $142, 543, 123, 65, 453, 879, 572, 434, 111, 242, 811, 102$.

7.12   What is the running time of heapsort for presorted input?

*7.13   Show that there are inputs that force every percolateDown in heapsort to go all the way to a leaf. (*Hint:* Work backward.)

7.14   Rewrite heapsort so that it sorts only items that are in the range low to high which are passed as additional parameters.

7.15   Sort $3, 1, 4, 1, 5, 9, 2, 6$ using mergesort.

7.16   How would you implement mergesort without using recursion?

7.17   Determine the running time of mergesort for
  a. sorted input
  b. reverse-ordered input
  c. random input

7.18   In the analysis of mergesort, constants have been disregarded. Prove that the number of comparisons used in the worst case by mergesort is $N\lceil \log N \rceil - 2^{\lceil \log N \rceil} + 1$.

7.19   Sort $3, 1, 4, 1, 5, 9, 2, 6, 5, 3, 5$ using quicksort with median-of-three partitioning and a cutoff of 3.

7.20   Using the quicksort implementation in this chapter, determine the running time of quicksort for
  a. sorted input
  b. reverse-ordered input
  c. random input

7.21   Repeat Exercise 7.20 when the pivot is chosen as
  a. the first element
  b. the larger of the first two distinct elements
  c. a random element
  *d. the average of all elements in the set

7.22 a. For the quicksort implementation in this chapter, what is the running time when all keys are equal?

   b. Suppose we change the partitioning strategy so that neither i nor j stops when an element with the same key as the pivot is found. What fixes need to be made in the code to guarantee that quicksort works, and what is the running time, when all keys are equal?

   c. Suppose we change the partitioning strategy so that i stops at an element with the same key as the pivot, but j does not stop in a similar case. What fixes need to be made in the code to guarantee that quicksort works, and when all keys are equal, what is the running time of quicksort?

7.23 Suppose we choose the element in the middle position of the array as the pivot. Does this make it unlikely that quicksort will require quadratic time?

7.24 Construct a permutation of 20 elements that is as bad as possible for quicksort using median-of-three partitioning and a cutoff of 3.

7.25 The quicksort in the text uses two recursive calls. Remove one of the calls as follows:

   a. Rewrite the code so that the second recursive call is unconditionally the last line in quicksort. Do this by reversing the if/else and returning after the call to insertionSort.

   b. Remove the tail recursion by writing a while loop and altering left.

7.26 Continuing from Exercise 7.25, after part (a),

   a. Perform a test so that the smaller subarray is processed by the first recursive call, while the larger subarray is processed by the second recursive call.

   b. Remove the tail recursion by writing a while loop and altering left or right, as necessary.

   c. Prove that the number of recursive calls is logarithmic in the worst case.

7.27 Suppose the recursive quicksort receives an int parameter, depth, from the driver that is initially approximately $2 \log N$.

   a. Modify the recursive quicksort to call heapsort on its current subarray if the level of recursion has reached depth. (*Hint:* Decrement depth as you make recursive calls; when it is 0, switch to heapsort.)

   b. Prove that the worst-case running time of this algorithm is $O(N \log N)$.

   c. Conduct experiments to determine how often heapsort gets called.

   d. Implement this technique in conjunction with tail-recursion removal in Exercise 7.25.

   e. Explain why the technique in Exercise 7.26 would no longer be needed.

7.28 When implementing quicksort, if the array contains lots of duplicates, it may be better to perform a three-way partition (into elements less than, equal to, and greater than the pivot), to make smaller recursive calls. Assume three-way comparisons, as provided by the compareTo method.

   a. Give an algorithm that performs a three-way in-place partition of an N-element subarray using only $N - 1$ three-way comparisons. If there are d items equal to the pivot, you may use d additional Comparable swaps, above and beyond the two-way partitioning algorithm. (*Hint:* As i and j move toward each other, maintain five groups of elements as shown below):

EQUAL   SMALL   UNKNOWN   LARGE   EQUAL
         i   j

b. Prove that using the algorithm above, sorting an N-element array that contains only d different values, takes $O(dN)$ time.

7.29 Write a program to implement the selection algorithm.

7.30 Solve the following recurrence:

$$T(N) = (1/N)\left[\sum_{i=0}^{N-1} T(i)\right] + cN, \qquad T(0) = 0.$$

7.31 A sorting algorithm is **stable** if elements with equal keys are left in the same order as they occur in the input. Which of the sorting algorithms in this chapter are stable and which are not? Why?

7.32 Suppose you are given a sorted list of N elements followed by $f(N)$ randomly ordered elements. How would you sort the entire list if
a. $f(N) = O(1)$?
b. $f(N) = O(\log N)$?
c. $f(N) = O(\sqrt{N})$?
*d. How large can $f(N)$ be for the entire list still to be sortable in $O(N)$ time?

7.33 Prove that any algorithm that finds an element X in a sorted list of N elements requires $\Omega(\log N)$ comparisons.

7.34 Using Stirling's formula, $N! \approx (N/e)^N \sqrt{2\pi N}$, give a precise estimate for $\log(N!)$.

7.35 *a. In how many ways can two sorted arrays of N elements be merged?
*b. Give a nontrivial lower bound on the number of comparisons required to merge two sorted lists of N elements, by taking the logarithm of your answer in part (a).

7.36 Prove that merging two sorted arrays of N items requires at least $2N - 1$ comparisons. You must show that if two elements in the merged list are consecutive and from different lists, then they must be compared.

7.37 Consider the following algorithm for sorting six numbers:

- Sort the first three numbers using Algorithm A.
- Sort the second three numbers using Algorithm B.
- Merge the two sorted groups using Algorithm C.

Show that this algorithm is suboptimal, regardless of the choices for Algorithms A, B, and C.

7.38 Write a program that reads N points in a plane and outputs any group of four or more colinear points (i.e., points on the same line). The obvious brute-force algorithm requires $O(N^4)$ time. However, there is a better algorithm that makes use of sorting and runs in $O(N^2 \log N)$ time.

7.39 Show that the two smallest elements among N can be found in $N + \lceil \log N \rceil - 2$ comparisons.

7.40 The following divide-and-conquer algorithm is proposed for finding the simultaneous maximum and minimum: If there is one item, it is the maximum and minimum, and if there are two items, then compare them and in one comparison you can find the maximum and minimum. Otherwise, split the input into two halves, divided as evenly as possibly (if N is odd, one of the two halves will have one more element than the other). Recursively find the maximum and minimum of each half, and then in two additional comparisons produce the maximum and minimum for the entire problem.

  a. Suppose N is a power of 2. What is the exact number of comparisons used by this algorithm?
  b. Suppose N is of the form $3 \cdot 2^k$. What is the exact number of comparisons used by this algorithm?
  c. Modify the algorithm as follows: When N is even, but not divisible by four, split the input into sizes of $N/2 - 1$ and $N/2 + 1$. What is the exact number of comparisons used by this algorithm?

7.41 Suppose we want to partition N items into G equal-sized groups of size N/G, such that the smallest N/G items are in group 1, the next smallest N/G items are in group 2, and so on. The groups themselves do not have to be sorted. For simplicity, you may assume that N and G are powers of two.

  a. Give an $O(N \log G)$ algorithm to solve this problem.
  b. Prove an $\Omega(N \log G)$ lower bound to solve this problem using comparison-based algorithms.

*7.42 Give a linear-time algorithm to sort N fractions, each of whose numerators and denominators are integers between 1 and N.

7.43 Suppose arrays A and B are both sorted and both contain N elements. Give an $O(\log N)$ algorithm to find the median of $A \cup B$.

7.44 Suppose you have an array of N elements containing only two distinct keys, true and false. Give an $O(N)$ algorithm to rearrange the list so that all false elements precede the true elements. You may use only constant extra space.

7.45 Suppose you have an array of N elements, containing three distinct keys, true, false, and maybe. Give an $O(N)$ algorithm to rearrange the list so that all false elements precede maybe elements, which in turn precede true elements. You may use only constant extra space.

7.46 a. Prove that any comparison-based algorithm to sort 4 elements requires 5 comparisons.
  b. Give an algorithm to sort 4 elements in 5 comparisons.

7.47 a. Prove that 7 comparisons are required to sort 5 elements using any comparison-based algorithm.
  *b. Give an algorithm to sort 5 elements with 7 comparisons.

7.48   Write an efficient version of Shellsort and compare performance when the following increment sequences are used:
a. Shell's original sequence
b. Hibbard's increments
c. Knuth's increments: $h_i = \frac{1}{2}(3^i + 1)$
d. Gonnet's increments: $h_t = \lfloor \frac{N}{2.2} \rfloor$, and $h_k = \lfloor \frac{h_{k+1}}{2.2} \rfloor$ (with $h_1 = 1$ if $h_2 = 2$)
e. Sedgewick's increments.

7.49   Implement an optimized version of quicksort and experiment with combinations of the following:
a. Pivot: first element, middle element, random element, median of three, median of five.
b. Cutoff values from 0 to 20.

7.50   Write a routine that reads in two alphabetized files and merges them together, forming a third, alphabetized, file.

7.51   Suppose we implement the median of three routine as follows: Find the median of a[left], a[center], a[right], and swap it with a[right]. Proceed with the normal partitioning step starting i at left and j at right-1 (instead of left+1 and right-2).
a. Suppose the input is $2, 3, 4, \ldots, N-1, N, 1$. For this input, what is the running time of this version of quicksort?
b. Suppose the input is in reverse order. For this input, what is the running time of this version of quicksort?

7.52   Prove that any comparison-based sorting algorithm requires $\Omega(N \log N)$ comparisons on average.

7.53   We are given an array that contains N numbers. We want to determine if there are two numbers whose sum equals a given number K. For instance, if the input is 8, 4, 1, 6, and K is 10, then the answer is yes (4 and 6). A number may be used twice. Do the following:
a. Give an $O(N^2)$ algorithm to solve this problem.
b. Give an $O(N \log N)$ algorithm to solve this problem. (*Hint:* Sort the items first. After that is done, you can solve the problem in linear time.)
c. Code both solutions and compare the running times of your algorithms.

7.54   Repeat Exercise 7.53 for four numbers. Try to design an $O(N^2 \log N)$ algorithm. (*Hint:* Compute all possible sums of two elements. Sort these possible sums. Then proceed as in Exercise 7.53.)

7.55   Repeat Exercise 7.53 for three numbers. Try to design an $O(N^2)$ algorithm.

7.56   Consider the following strategy for percolateDown. We have a hole at node X. The normal routine is to compare X's children and then move the child up to X if it is larger (in the case of a (*max*)heap) than the element we are trying to place, thereby pushing the hole down; we stop when it is safe to place the new element in the hole. The alternative strategy is to move elements up and the hole down as far as possible, without testing whether the new cell can be inserted. This would place the new cell in a leaf and probably violate the heap order; to fix the heap order, percolate the new cell up in the normal manner. Write a routine to include this idea, and compare the running time with a standard implementation of heapsort.

7.57 Propose an algorithm to sort a large file using only two tapes.

7.58 a. Show that a lower bound of $N!/2^{2N}$ on the number of heaps is implied by the fact that `buildHeap` uses at most $2N$ comparisons.
   b. Use Stirling's formula to expand this bound.

7.59 M is an N-by-N matrix in which the entries in each rows are in increasing order and the entries in each column are in increasing order (reading top to bottom). Consider the problem of determining if x is in M using three-way comparisons (i.e., one comparison of x with $M[i][j]$ tells you either that x is less than, equal to, or greater than $M[i][j]$).
   a. Give an algorithm that uses at most $2N - 1$ comparisons.
   b. Prove that any algorithm must use at least $2N - 1$ comparisons.

7.60 There is a prize hidden in a box; the value of the prize is a positive integer between 1 and N, and you are given N. To win the prize, you have to guess its value. Your goal is to do it in as few guesses as possible; however, among those guesses, you may only make at most g guesses that are too high. The value g will be specified at the start of the game, and if you make more than g guesses that are too high, you lose. So, for example, if $g = 0$, you then can win in N guesses by simply guessing the sequence $1, 2, 3, \ldots$.
   a. Suppose $g = \lceil \log N \rceil$. What strategy minimizes the number of guesses?
   b. Suppose $g = 1$. Show that you can always win in $O(N^{1/2})$ guesses.
   c. Suppose $g = 1$. Show that any algorithm that wins the prize must use $\Omega(N^{1/2})$ guesses.
   *d. Give an algorithm and matching lower bound for any constant g.

References

Knuth's book [16] is a comprehensive reference for sorting. Gonnet and Baeza-Yates [5] has some more results, as well as a huge bibliography.

The original paper detailing Shellsort is [29]. The paper by Hibbard [9] suggested the use of the increments $2^k - 1$ and tightened the code by avoiding swaps. Theorem 7.4 is from [19]. Pratt's lower bound, which uses a more complex method than that suggested in the text, can be found in [22]. Improved increment sequences and upper bounds appear in [13], [28], and [31]; matching lower bounds have been shown in [32]. It has been shown that no increment sequence gives an $O(N \log N)$ worst-case running time [20]. The average-case running time for Shellsort is still unresolved. Yao [34] has performed an extremely complex analysis for the three-increment case. The result has yet to be extended to more increments, but has been slightly improved [14]. The paper by Jiang, Li, and Vityani [15] has shown an $\Omega(p N^{1+1/p})$ lower bound on the average-case running time of p-pass Shellsort. Experiments with various increment sequences appear in [30].

Heapsort was invented by Williams [33]; Floyd [4] provided the linear-time algorithm for heap construction. Theorem 7.5 is from [23].

An exact average-case analysis of mergesort has been described in [7]. An algorithm to perform merging in linear time without extra space is described in [12].

Quicksort is from Hoare [10]. This paper analyzes the basic algorithm, describes most of the improvements, and includes the selection algorithm. A detailed analysis and empirical study was the subject of Sedgewick's dissertation [27]. Many of the important results appear in the three papers [24], [25], and [26]. [1] provides a detailed C implementation with some additional improvements, and points out that many implementations of the UNIX qsort library routine are easily driven to quadratic behavior. Exercise 7.27 is from [18].

Decision trees and sorting optimality are discussed in Ford and Johnson [5]. This paper also provides an algorithm that almost meets the lower bound in terms of number of comparisons (but not other operations). This algorithm was eventually shown to be slightly suboptimal by Manacher [17].

The selection lower bounds obtained in Theorem 7.9 are from [6]. The lower bound for finding the maximum and minimum simultaneously is from Pohl [21]. The current best lower bound for finding the median is slightly above $2N$ comparisons due to Dor and Zwick [3]; they also have the best upper bound, which is roughly $2.95N$ comparisons [2].

External sorting is covered in detail in [16]. Stable sorting, described in Exercise 7.31, has been addressed by Horvath [11].

1. J. L. Bentley and M. D. McElroy, "Engineering a Sort Function," *Software—Practice and Experience,* 23 (1993), 1249–1265.

2. D. Dor and U. Zwick, "Selecting the Median," *SIAM Journal on Computing,* 28 (1999), 1722–1758.

3. D. Dor and U. Zwick, "Median Selection Requires $(2 + \varepsilon)n$ Comparisons," *SIAM Journal on Discrete Math,* 14 (2001), 312–325.

4. R. W. Floyd, "Algorithm 245: Treesort 3," *Communications of the ACM,* 7 (1964), 701.

5. L. R. Ford and S. M. Johnson, "A Tournament Problem," *American Mathematics Monthly,* 66 (1959), 387–389.

6. F. Fussenegger and H. Gabow, "A Counting Approach to Lower Bounds for Selection Problems," *Journal of the ACM,* 26 (1979), 227–238.

7. M. Golin and R. Sedgewick, "Exact Analysis of Mergesort," *Fourth SIAM Conference on Discrete Mathematics,* 1988.

8. G. H. Gonnet and R. Baeza-Yates, *Handbook of Algorithms and Data Structures,* 2d ed., Addison-Wesley, Reading, Mass., 1991.

9. T. H. Hibbard, "An Empirical Study of Minimal Storage Sorting," *Communications of the ACM,* 6 (1963), 206–213.

10. C. A. R. Hoare, "Quicksort," *Computer Journal,* 5 (1962), 10–15.

11. E. C. Horvath, "Stable Sorting in Asymptotically Optimal Time and Extra Space," *Journal of the ACM,* 25 (1978), 177–199.

12. B. Huang and M. Langston, "Practical In-place Merging," *Communications of the ACM,* 31 (1988), 348–352.

13. J. Incerpi and R. Sedgewick, "Improved Upper Bounds on Shellsort," *Journal of Computer and System Sciences,* 31 (1985), 210–224.

14. S. Janson and D. E. Knuth, "Shellsort with Three Increments," *Random Structures and Algorithms*, 10 (1997), 125–142.

15. T. Jiang, M. Li, and P. Vitanyi, "A Lower Bound on the Average-Case Complexity of Shellsort," *Journal of the ACM*, 47 (2000), 905–911.

16. D. E. Knuth, *The Art of Computer Programming. Volume 3: Sorting and Searching*, 2d ed., Addison-Wesley, Reading, Mass., 1998.

17. G. K. Manacher, "The Ford-Johnson Sorting Algorithm Is Not Optimal," *Journal of the ACM*, 26 (1979), 441–456.

18. D. R. Musser, "Introspective Sorting and Selection Algorithms," *Software—Practice and Experience*, 27 (1997), 983–993.

19. A. A. Papernov and G. V. Stasevich, "A Method of Information Sorting in Computer Memories," *Problems of Information Transmission*, 1 (1965), 63–75.

20. C. G. Plaxton, B. Poonen, and T. Suel, "Improved Lower Bounds for Shellsort," Proceedings of the Thirty-third Annual Symposium on the Foundations of Computer Science (1992), 226–235.

21. I. Pohl, "A Sorting Problem and Its Complexity," *Communications of the ACM*, 15 (1972), 462–464.

22. V. R. Pratt, *Shellsort and Sorting Networks*, Garland Publishing, New York, 1979. (Originally presented as the author's Ph.D. thesis, Stanford University, 1971.)

23. R. Schaffer and R. Sedgewick, "The Analysis of Heapsort," *Journal of Algorithms*, 14 (1993), 76–100.

24. R. Sedgewick, "Quicksort with Equal Keys," *SIAM Journal on Computing*, 6 (1977), 240–267.

25. R. Sedgewick, "The Analysis of Quicksort Programs," *Acta Informatica*, 7 (1977), 327–355.

26. R. Sedgewick, "Implementing Quicksort Programs," *Communications of the ACM*, 21 (1978), 847–857.

27. R. Sedgewick, *Quicksort*, Garland Publishing, New York, 1978. (Originally presented as the author's Ph.D. thesis, Stanford University, 1975.)

28. R. Sedgewick, "A New Upper Bound for Shellsort," *Journal of Algorithms*, 7 (1986), 159–173.

29. D. L. Shell, "A High-Speed Sorting Procedure," *Communications of the ACM*, 2 (1959), 30–32.

30. M. A. Weiss, "Empirical Results on the Running Time of Shellsort," *Computer Journal*, 34 (1991), 88–91.

31. M. A. Weiss and R. Sedgewick, "More on Shellsort Increment Sequences," *Information Processing Letters*, 34 (1990), 267–270.

32. M. A. Weiss and R. Sedgewick, "Tight Lower Bounds for Shellsort," *Journal of Algorithms*, 11 (1990), 242–251.

33. J. W. J. Williams, "Algorithm 232: Heapsort," *Communications of the ACM*, 7 (1964), 347–348.

34. A. C. Yao, "An Analysis of $(h, k, 1)$ Shellsort," *Journal of Algorithms*, 1 (1980), 14–50.

CHAPTER 8

The Disjoint Set Class

In this chapter, we describe an efficient data structure to solve the equivalence problem. The data structure is simple to implement. Each routine requires only a few lines of code, and a simple array can be used. The implementation is also extremely fast, requiring constant average time per operation. This data structure is also very interesting from a theoretical point of view, because its analysis is extremely difficult; the functional form of the worst case is unlike any we have yet seen. For the disjoint set data structure, we will

- Show how it can be implemented with minimal coding effort.
- Greatly increase its speed, using just two simple observations.
- Analyze the running time of a fast implementation.
- See a simple application.

8.1 Equivalence Relations

A **relation** R is defined on a set S if for every pair of elements (a, b), $a, b \in S$, $a R b$ is either true or false. If $a R b$ is true, then we say that a is related to b.

An **equivalence relation** is a relation R that satisfies three properties:

1. (*Reflexive*) $a R a$, for all $a \in S$.
2. (*Symmetric*) $a R b$ if and only if $b R a$.
3. (*Transitive*) $a R b$ and $b R c$ implies that $a R c$.

We will consider several examples.

The $\leq$ relationship is not an equivalence relationship. Although it is reflexive, since $a \leq a$, and transitive, since $a \leq b$ and $b \leq c$ implies $a \leq c$, it is not symmetric, since $a \leq b$ does not imply $b \leq a$.

Electrical connectivity, where all connections are by metal wires, is an equivalence relation. The relation is clearly reflexive, as any component is connected to itself. If a is electrically connected to b, then b must be electrically connected to a, so the relation is symmetric. Finally, if a is connected to b and b is connected to c, then a is connected to c. Thus electrical connectivity is an equivalence relation.

Two cities are related if they are in the same country. It is easily verified that this is an equivalence relation. Suppose town a is related to b if it is possible to travel from a to b by taking roads. This relation is an equivalence relation if all the roads are two-way.

8.2 The Dynamic Equivalence Problem

Given an equivalence relation $\sim$, the natural problem is to decide, for any a and b, if $a \sim b$. If the relation is stored as a two-dimensional array of Boolean variables, then, of course, this can be done in constant time. The problem is that the relation is usually not explicitly, but rather implicitly, defined.

As an example, suppose the equivalence relation is defined over the five-element set $\{a_1, a_2, a_3, a_4, a_5\}$. Then there are 25 pairs of elements, each of which is either related or not. However, the information $a_1 \sim a_2$, $a_3 \sim a_4$, $a_5 \sim a_1$, $a_4 \sim a_2$ implies that all pairs are related. We would like to be able to infer this quickly.

The **equivalence class** of an element $a \in S$ is the subset of S that contains all the elements that are related to a. Notice that the equivalence classes form a partition of S: Every member of S appears in exactly one equivalence class. To decide if $a \sim b$, we need only to check whether a and b are in the same equivalence class. This provides our strategy to solve the equivalence problem.

The input is initially a collection of N sets, each with one element. This initial representation is that all relations (except reflexive relations) are false. Each set has a different element, so that $S_i \cap S_j = \emptyset$; this makes the sets **disjoint**.

There are two permissible operations. The first is find, which returns the name of the set (that is, the equivalence class) containing a given element. The second operation adds relations. If we want to add the relation $a \sim b$, then we first see if a and b are already related. This is done by performing finds on both a and b and checking whether they are in the same equivalence class. If they are not, then we apply union. This operation merges the two equivalence classes containing a and b into a new equivalence class. From a set point of view, the result of $\cup$ is to create a new set $S_k = S_i \cup S_j$, destroying the originals and preserving the disjointness of all the sets. The algorithm to do this is frequently known as the disjoint set **union/find algorithm** for this reason.

This algorithm is *dynamic* because, during the course of the algorithm, the sets can change via the union operation. The algorithm must also operate **online**: When a find is performed, it must give an answer before continuing. Another possibility would be an **off-line** algorithm. Such an algorithm would be allowed to see the entire sequence of unions and finds. The answer it provides for each find must still be consistent with all the unions that were performed up until the find, but the algorithm can give all its answers after it has seen *all* the questions. The difference is similar to taking a written exam (which is generally off-line—you only have to give the answers before time expires), and an oral exam (which is online, because you must answer the current question before proceeding to the next question).

Notice that we do not perform any operations comparing the relative values of elements but merely require knowledge of their location. For this reason, we can assume that all the elements have been numbered sequentially from 0 to $N - 1$ and that the numbering can be determined easily by some hashing scheme. Thus, initially we have $S_i = \{i\}$ for $i = 0$ through $N - 1$.[1]

---

[1] This reflects the fact that array indices start at 0.

Our second observation is that the name of the set returned by find is actually fairly arbitrary. All that really matters is that find(a)==find(b) is true if and only if a and b are in the same set.

These operations are important in many graph theory problems and also in compilers that process equivalence (or type) declarations. We will see an application later.

There are two strategies to solve this problem. One ensures that the find instruction can be executed in constant worst-case time, and the other ensures that the union instruction can be executed in constant worst-case time. It has been shown that both cannot be done simultaneously in constant worst-case time.

We will now briefly discuss the first approach. For the find operation to be fast, we could maintain, in an array, the name of the equivalence class for each element. Then find is just a simple $O(1)$ lookup. Suppose we want to perform union(a,b). Suppose that a is in equivalence class i and b is in equivalence class j. Then we scan down the array, changing all i's to j. Unfortunately, this scan takes $\Theta(N)$. Thus, a sequence of $N - 1$ unions (the maximum, since then everything is in one set) would take $\Theta(N^2)$ time. If there are $\Omega(N^2)$ find operations, this performance is fine, since the total running time would then amount to $O(1)$ for each union or find operation over the course of the algorithm. If there are fewer finds, this bound is not acceptable.

One idea is to keep all the elements that are in the same equivalence class in a linked list. This saves time when updating, because we do not have to search through the entire array. This by itself does not reduce the asymptotic running time, because it is still possible to perform $\Theta(N^2)$ equivalence class updates over the course of the algorithm.

If we also keep track of the size of each equivalence class, and when performing unions we change the name of the smaller equivalence class to the larger, then the total time spent for $N - 1$ merges is $O(N \log N)$. The reason for this is that each element can have its equivalence class changed at most $\log N$ times, since every time its class is changed, its new equivalence class is at least twice as large as its old. Using this strategy, any sequence of M finds and up to $N - 1$ unions takes at most $O(M + N \log N)$ time.

In the remainder of this chapter, we will examine a solution to the union/find problem that makes unions easy but finds hard. Even so, the running time for any sequence of at most M finds and up to $N - 1$ unions will be only a little more than $O(M + N)$.

8.3 Basic Data Structure

Recall that the problem does not require that a find operation return any specific name, just that finds on two elements return the same answer if and only if they are in the same set. One idea might be to use a tree to represent each set, since each element in a tree has the same root. Thus, the root can be used to name the set. We will represent each set by a tree. (Recall that a collection of trees is known as a **forest**.) Initially, each set contains one element. The trees we will use are not necessarily binary trees, but their representation is easy, because the only information we will need is a parent link. The name of a set is given by the node at the root. Since only the name of the parent is required, we can assume that this tree is stored implicitly in an array: Each entry s[i] in the array represents the parent

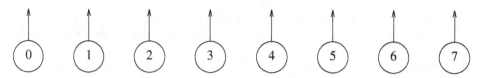

Figure 8.1   Eight elements, initially in different sets

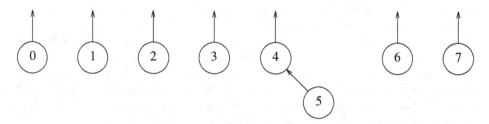

Figure 8.2   After union(4,5)

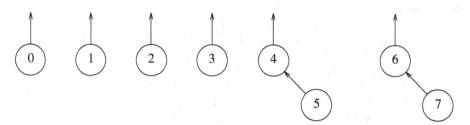

Figure 8.3   After union(6,7)

of element i. If i is a root, then $s[i] = -1$. In the forest in Figure 8.1, $s[i] = -1$ for $0 \le i < 8$. As with binary heaps, we will draw the trees explicitly, with the understanding that an array is being used. Figure 8.1 shows the explicit representation. We will draw the root's parent link vertically for convenience.

To perform a union of two sets, we merge the two trees by making the parent link of one tree's root link to the root node of the other tree. It should be clear that this operation takes constant time. Figures 8.2, 8.3, and 8.4 represent the forest after each of union(4,5), union(6,7), union(4,6), where we have adopted the convention that the new root after the union(x,y) is x. The implicit representation of the last forest is shown in Figure 8.5.

A find(x) on element x is performed by returning the root of the tree containing x. The time to perform this operation is proportional to the depth of the node representing x, assuming, of course, that we can find the node representing x in constant time. Using the strategy above, it is possible to create a tree of depth $N - 1$, so the worst-case running time of a find is $\Theta(N)$. Typically, the running time is computed for a *sequence* of M intermixed instructions. In this case, M consecutive operations could take $\Theta(MN)$ time in the worst case.

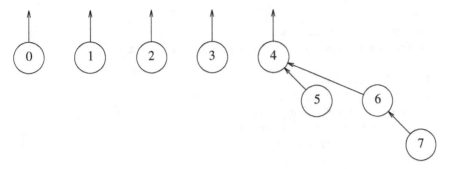

Figure 8.4   After union(4,6)

−1	−1	−1	−1	−1	4	4	6
0	1	2	3	4	5	6	7

Figure 8.5   Implicit representation of previous tree

```
1    public class DisjSets
2    {
3        public DisjSets( int numElements )
4          { /* Figure 8.7 */ }
5        public void union( int root1, int root2 )
6          { /* Figures 8.8 and 8.14 */ }
7        public int find( int x )
8          { /* Figures 8.9 and 8.16 */ }
9
10       private int [ ] s;
11   }
```

Figure 8.6   Disjoint set class skeleton

The code in Figures 8.6 through 8.9 represents an implementation of the basic algorithm, assuming that error checks have already been performed. In our routine, unions are performed on the roots of the trees. Sometimes the operation is performed by passing any two elements, and having the union perform two finds to determine the roots.

The average-case analysis is quite hard to do. The least of the problems is that the answer depends on how to define *average* (with respect to the union operation). For instance, in the forest in Figure 8.4, we could say that since there are five trees, there are $5 \cdot 4 = 20$ equally likely results of the next union (as any two different trees can be unioned). Of course, the implication of this model is that there is only a $\frac{2}{5}$ chance that the next union will involve the large tree. Another model might say that all unions between any two

```
1      /**
2       * Construct the disjoint sets object.
3       * @param numElements the initial number of disjoint sets.
4       */
5      public DisjSets( int numElements )
6      {
7          s = new int [ numElements ];
8          for( int i = 0; i < s.length; i++ )
9              s[ i ] = -1;
10     }
```

Figure 8.7 Disjoint set initialization routine

```
1      /**
2       * Union two disjoint sets.
3       * For simplicity, we assume root1 and root2 are distinct
4       * and represent set names.
5       * @param root1 the root of set 1.
6       * @param root2 the root of set 2.
7       */
8      public void union( int root1, int root2 )
9      {
10         s[ root2 ] = root1;
11     }
```

Figure 8.8 union (not the best way)

```
1      /**
2       * Perform a find.
3       * Error checks omitted again for simplicity.
4       * @param x the element being searched for.
5       * @return the set containing x.
6       */
7      public int find( int x )
8      {
9          if( s[ x ] < 0 )
10             return x;
11         else
12             return find( s[ x ] );
13     }
```

Figure 8.9 A simple disjoint set find algorithm

elements in different trees are equally likely, so a larger tree is more likely to be involved in the next union than a smaller tree. In the example above, there is an $\frac{8}{11}$ chance that the large tree is involved in the next union, since (ignoring symmetries) there are 6 ways in which to merge two elements in $\{0, 1, 2, 3\}$, and 16 ways to merge an element in $\{4, 5, 6, 7\}$ with an element in $\{0, 1, 2, 3\}$. There are still more models and no general agreement on which is the best. The average running time depends on the model; $\Theta(M)$, $\Theta(M \log N)$, and $\Theta(MN)$ bounds have actually been shown for three different models, although the latter bound is thought to be more realistic.

Quadratic running time for a sequence of operations is generally unacceptable. Fortunately, there are several ways of easily ensuring that this running time does not occur.

8.4 Smart Union Algorithms

The unions above were performed rather arbitrarily, by making the second tree a subtree of the first. A simple improvement is always to make the smaller tree a subtree of the larger, breaking ties by any method; we call this approach **union-by-size**. The three unions in the preceding example were all ties, and so we can consider that they were performed by size. If the next operation were union(3,4), then the forest in Figure 8.10 would form. Had the size heuristic not been used, a deeper tree would have been formed (Figure 8.11).

We can prove that if unions are done by size, the depth of any node is never more than $\log N$. To see this, note that a node is initially at depth 0. When its depth increases as a result of a union, it is placed in a tree that is at least twice as large as before. Thus, its depth can be increased at most $\log N$ times. (We used this argument in the quick-find algorithm at the end of Section 8.2.) This implies that the running time for a find operation is $O(\log N)$, and a sequence of M operations takes $O(M \log N)$. The tree in Figure 8.12 shows the worst tree possible after 16 unions and is obtained if all unions are between equal-sized trees (the worst-case trees are binomial trees, discussed in Chapter 6).

To implement this strategy, we need to keep track of the size of each tree. Since we are really just using an array, we can have the array entry of each root contain the *negative* of

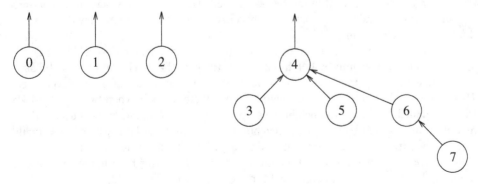

Figure 8.10 Result of union-by-size

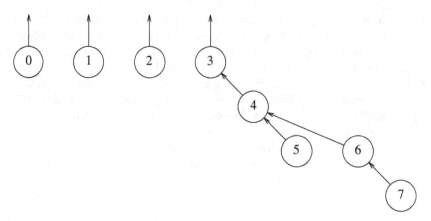

Figure 8.11   Result of an arbitrary `union`

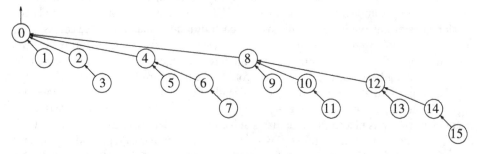

Figure 8.12   Worst-case tree for $N = 16$

the size of its tree. Thus, initially the array representation of the tree is all -1's. When a `union` is performed, check the sizes; the new size is the sum of the old. Thus, union-by-size is not at all difficult to implement and requires no extra space. It is also fast, on average. For virtually all reasonable models, it has been shown that a sequence of M operations requires $O(M)$ average time if union-by-size is used. This is because when random `unions` are performed, generally very small (usually one-element) sets are merged with large sets throughout the algorithm.

An alternative implementation, which also guarantees that all the trees will have depth at most $O(\log N)$, is **union-by-height**. We keep track of the height, instead of the size, of each tree and perform `unions` by making the shallow tree a subtree of the deeper tree. This is an easy algorithm, since the height of a tree increases only when two equally deep trees are joined (and then the height goes up by one). Thus, union-by-height is a trivial modification of union-by-size. Since heights of zero would not be negative, we actually store the negative of height, minus an additional 1. Initially, all entries are -1.

Figure 8.13 show a forest and its implicit representation for both union-by-size and union-by-height. The code in Figure 8.14 implements union-by-height.

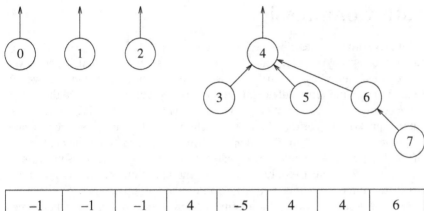

−1	−1	−1	4	−5	4	4	6
0	1	2	3	4	5	6	7

−1	−1	−1	4	−3	4	4	6
0	1	2	3	4	5	6	7

Figure 8.13   Forest with implicit representation for union-by-size and union-by-height

```
1      /**
2       * Union two disjoint sets using the height heuristic.
3       * For simplicity, we assume root1 and root2 are distinct
4       * and represent set names.
5       * @param root1 the root of set 1.
6       * @param root2 the root of set 2.
7       */
8      public void union( int root1, int root2 )
9      {
10         if( s[ root2 ] < s[ root1 ] )  // root2 is deeper
11             s[ root1 ] = root2;        // Make root2 new root
12         else
13         {
14             if( s[ root1 ] == s[ root2 ] )
15                 s[ root1 ]--;          // Update height if same
16             s[ root2 ] = root1;        // Make root1 new root
17         }
18     }
```

Figure 8.14   Code for union-by-height (rank)

8.5 Path Compression

The union/find algorithm, as described so far, is quite acceptable for most cases. It is very simple and linear on average for a sequence of M instructions (under all models). However, the worst case of $O(M \log N)$ can occur fairly easily and naturally. For instance, if we put all the sets on a queue and repeatedly dequeue the first two sets and enqueue the union, the worst case occurs. If there are many more finds than unions, this running time is worse than that of the quick-find algorithm. Moreover, it should be clear that there are probably no more improvements possible for the union algorithm. This is based on the observation that any method to perform the unions will yield the same worst-case trees, since it must break ties arbitrarily. Therefore, the only way to speed the algorithm up, without reworking the data structure entirely, is to do something clever on the find operation.

The clever operation is known as **path compression**. Path compression is performed during a find operation and is independent of the strategy used to perform unions. Suppose the operation is find(x). Then the effect of path compression is that *every* node on the path from x to the root has its parent changed to the root. Figure 8.15 shows the effect of path compression after find(14) on the generic worst tree of Figure 8.12.

The effect of path compression is that with an extra two link changes, nodes 12 and 13 are now one position closer to the root and nodes 14 and 15 are now two positions closer. Thus, the fast future accesses on these nodes will pay (we hope) for the extra work to do the path compression.

As the code in Figure 8.16 shows, path compression is a trivial change to the basic find algorithm. The only change to the find routine is that s[x] is made equal to the value returned by find; thus after the root of the set is found recursively, x's parent link references it. This occurs recursively to every node on the path to the root, so this implements path compression.

When unions are done arbitrarily, path compression is a good idea, because there is an abundance of deep nodes and these are brought near the root by path compression. It has been proven that when path compression is done in this case, a sequence of M

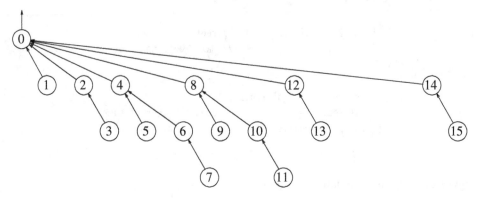

Figure 8.15 An example of path compression

```
1        /**
2         * Perform a find with path compression.
3         * Error checks omitted again for simplicity.
4         * @param x the element being searched for.
5         * @return the set containing x.
6         */
7        public int find( int x )
8        {
9            if( s[ x ] < 0 )
10               return x;
11           else
12               return s[ x ] = find( s[ x ] );
13       }
```

Figure 8.16 Code for the disjoint set find with path compression

operations requires at most $O(M \log N)$ time. It is still an open problem to determine what the average-case behavior is in this situation.

Path compression is perfectly compatible with union-by-size, and thus both routines can be implemented at the same time. Since doing union-by-size by itself is expected to execute a sequence of M operations in linear time, it is not clear that the extra pass involved in path compression is worthwhile on average. Indeed, this problem is still open. However, as we shall see later, the combination of path compression and a smart union rule guarantees a very efficient algorithm in all cases.

Path compression is not entirely compatible with union-by-height, because path compression can change the heights of the trees. It is not at all clear how to recompute them efficiently. The answer is do not!! Then the heights stored for each tree become estimated heights (sometimes known as **ranks**), but it turns out that **union-by-rank** (which is what this has now become) is just as efficient in theory as union-by-size. Furthermore, heights are updated less often than sizes. As with union-by-size, it is not clear whether path compression is worthwhile on average. What we will show in the next section is that with either union heuristic, path compression significantly reduces the worst-case running time.

8.6 Worst Case for Union-by-Rank and Path Compression

When both heuristics are used, the algorithm is almost linear in the worst case. Specifically, the time required in the worst case is $\Theta(M\alpha(M, N))$ (provided $M \geq N$), where $\alpha(M, N)$ is an incredibly slowly growing function that for all intents and purposes is at most 5 for any problem instance. However, $\alpha(M, N)$ is not a constant, so the running time is not linear.

In the remainder of this section, we first look at some very slow-growing functions, and then in Sections 8.6.2 to 8.6.4, we establish a bound on the worst-case for a sequence of at most $N - 1$ unions, and M find operations in an N-element universe in which union is by rank and finds use path compression. The same bound holds if union-by-rank is replaced with union-by-size.

8.6.1 Slowly Growing Functions

Consider the recurrence:

$$T(N) = \begin{cases} 0 & N \le 1 \\ T(\lfloor f(N) \rfloor) + 1 & N > 1 \end{cases} \tag{8.1}$$

In this equation, $T(N)$ represents the number of times, starting at N, that we must iteratively apply $f(N)$ until we reach 1 (or less). We assume that $f(N)$ is a nicely defined function that reduces N. Call the solution to the equation $f^*(N)$.

We have already encountered this recurrence when analyzing binary search. There, $f(N) = N/2$; each step halves N. We know that this can happen at most $\log N$ times until N reaches 1; hence we have $f^*(N) = \log N$ (we ignore low-order terms, etc.). Observe that in this case, $f^*(N)$ is much less than $f(N)$.

Figure 8.17 shows the solution for $T(N)$, for various $f(N)$. In our case, we are most interested in $f(N) = \log N$. The solution $T(N) = \log^* N$ is known as the **iterated logarithm**. The iterated logarithm, which represents the number of times the logarithm needs to be iteratively applied until we reach one, is a very slowly growing function. Observe that $\log^* 2 = 1$, $\log^* 4 = 2$, $\log^* 16 = 3$, $\log^* 65536 = 4$, and $\log^* 2^{65536} = 5$. But keep in mind that 2^{65536} is a 20,000-digit number. So while $\log^* N$ is a growing function, for all intents and purposes, it is at most 5. But we can still produce even more slowly growing functions. For instance, if $f(N) = \log^* N$, then $T(N) = \log^{**} N$. In fact, we can add stars at will to produce functions that grow slower and slower.

$f(N)$	$f^*(N)$
$N-1$	$N-1$
$N-2$	$N/2$
$N-c$	N/c
$N/2$	$\log N$
N/c	$\log_c N$
$\sqrt{N}$	$\log \log N$
$\log N$	$\log^* N$
$\log^* N$	$\log^{**} N$
$\log^{**} N$	$\log^{***} N$

Figure 8.17   Different values of the iterated function

8.6.2   An Analysis By Recursive Decomposition

We now establish a tight bound on the running time of a sequence of $M = \Omega(N)$ union/find operations. The unions and finds may occur in any order, but unions are done by rank and finds are done with path compression.

We begin by establishing two lemmas concerning the properties of the ranks. Figure 8.18 gives a visual picture of both lemmas.

Lemma 8.1.
When executing a sequence of union instructions, a node of rank $r > 0$ must have at least one child of rank $0, 1, \ldots, r - 1$.

Proof.
By induction. The basis $r = 1$ is clearly true. When a node grows from rank $r - 1$ to rank r, it obtains a child of rank $r - 1$. By the inductive hypothesis, it already has children of ranks $0, 1, \ldots, r - 2$, thus establishing the lemma.

The next lemma seems somewhat obvious but is used implicitly in the analysis.

Lemma 8.2.
At any point in the union/find algorithm, the ranks of the nodes on a path from the leaf to a root increase monotonically.

Proof.
The lemma is obvious if there is no path compression. If, after path compression, some node v is a descendant of w, then clearly v must have been a descendant of w when only unions were considered. Hence the rank of v is less than the rank of w.

Suppose we have two algorithms A and B. Algorithm A works and computes all the answers correctly, but algorithm B does not compute correctly, or even produce useful answers. Suppose, however, that every step in algorithm A can be mapped to an equivalent step in algorithm B. Then it is easy to see that the running time for algorithm B describes the running time for algorithm A, exactly.

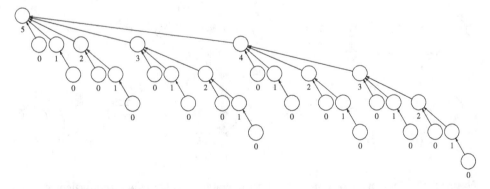

Figure 8.18   A large disjoint set tree (numbers below nodes are ranks)

We can use this idea to analyze the running time of the disjoint sets data structure. We will describe an algorithm *B* whose running time is exactly the same as the disjoint sets structure, and then algorithm *C*, whose running time is exactly the same as algorithm *B*. Thus any bound for algorithm *C* will be a bound for the disjoint sets data structure.

Partial Path Compression

Algorithm *A* is our standard sequence of union-by-rank and find with path compression operations. We design an algorithm *B* that will perform the exact same sequence of path compression operations as algorithm *A*. In algorithm *B*, we perform all the unions *prior to* any find. Then each find operation in algorithm *A* is replaced by a **partial find** operation in algorithm *B*. A partial find operation specifies the search item and the node up to which the path compression is performed. The node that will be used is the node that would have been the root at the time the matching find was performed in algorithm *A*.

Figure 8.19 shows that algorithm *A* and algorithm *B* will get equivalent trees (forests) at the end, and it is easy to see that the exact same amount of parent changes are performed by algorithm *A*'s finds, compared to algorithm *B*'s partial finds. But algorithm *B* should be simpler to analyze, since we have removed the mixing of unions and finds from the equation. The basic quantity to analyze is the number of parent changes that can occur in any sequence of partial finds, since all but the top two nodes in any find with path compression will obtain new parents.

A Recursive Decomposition

What we would like to do next is to divide each tree into two halves: a top half and a bottom half. We would then like to ensure that the number of partial find operations in the top half plus the number of partial find operations in the bottom half is exactly the same as the total number of partial find operations. We would then like to write a formula for the total path compression cost in the tree in terms of the path compression cost in the top half plus the path compression cost in the bottom half. Without specifying how we

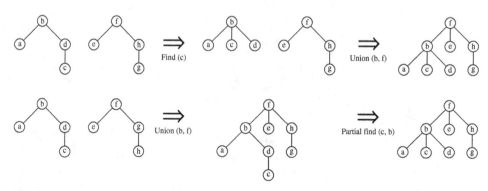

Figure 8.19 Sequences of union and find operations replaced with equivalent cost of union and partial find operations

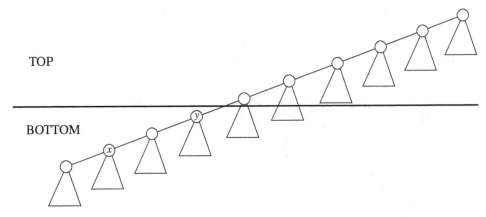

Figure 8.20   Recursive decomposition, Case 1: Partial find is entirely in bottom

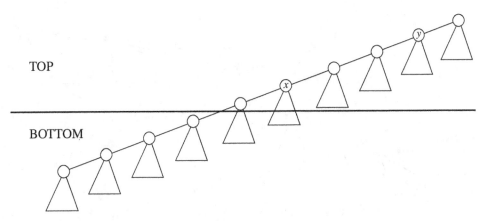

Figure 8.21   Recursive decomposition, Case 2: Partial find is entirely in top

decide which nodes are in the top half, and which nodes are in the bottom half, we can look at Figures 8.20, 8.21, and 8.22, to see how most of what we want to do can work immediately.

In Figure 8.20, the partial find resides entirely in the bottom half. Thus one partial find in the bottom half corresponds to one original partial find, and the charges can be recursively assigned to the bottom half.

In Figure 8.21, the partial find resides entirely in the top half. Thus one partial find in the top half corresponds to one original partial find, and the charges can be recursively assigned to the top half.

However, we run into lots of trouble when we reach Figure 8.22. Here x is in the bottom half, and y is in the top half. The path compression would require that all nodes from x to y's child acquire y as its parent. For nodes in the top half, that is no problem, but for nodes in the bottom half this is a deal breaker: Any recursive charges to the bottom

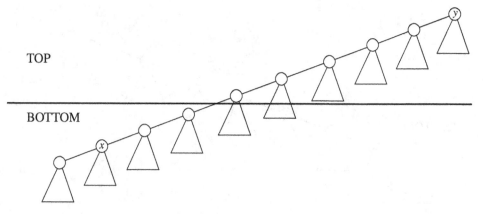

Figure 8.22   Recursive decomposition, Case 3: Partial find goes from bottom to top

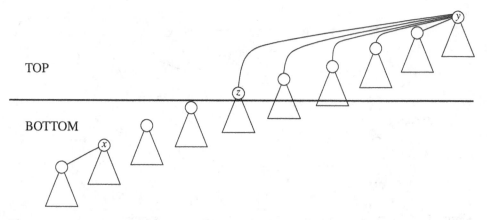

Figure 8.23   Recursive decomposition, Case 3: Path compression can be performed on the top nodes, but the bottom nodes must get new parents; the parents cannot be top parents, and they cannot be other bottom nodes

have to keep everything in the bottom. So as Figure 8.23 shows, we can perform the path compression on the top, but while some nodes in the bottom will need new parents, it is not clear what to do, because the new parents for those bottom nodes cannot be top nodes, and the new parents cannot be other bottom nodes.

The only option is to make a loop where these nodes' parents are themselves and make sure these parent changes are correctly charged in our accounting. Although this is a new algorithm because it can no longer be used to generate an identical tree, we don't need identical trees; we only need to be sure that each original partial find can be mapped into a new partial find operation, and that the charges are identical. Figure 8.24 shows what the new tree will look like, and so the big remaining issue is the accounting.

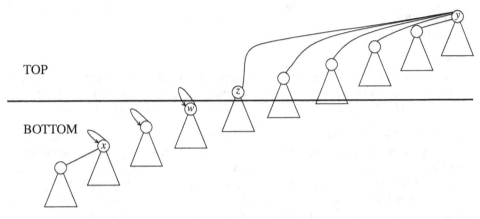

TOP

BOTTOM

Figure 8.24   Recursive decomposition, Case 3: The bottom node new parents are the nodes themselves

Looking at Figure 8.24, we see that the path compression charges from x to y can be split into three parts. First there is the path compression from z (the first top node on the upward path) to y. Clearly those charges are already accounted for recursively. Then there is the charge from the topmost-bottom node w to z. But that is only one unit, and there can be at most one of those per partial find operation. In fact, we can do a little better: There can be at most one of those per partial find operation on the top half. But how do we account for the parent changes on the path from x to w? One idea would be to argue that those changes would be exactly the same cost as if there were a partial find from x to w. But there is a big problem with that argument: It converts an original partial find into a partial find on the top *plus* a partial find on the bottom, which means the number of operations, M, would no longer be the same. Fortunately, there is a simpler argument: Since each node on the bottom can have its parent set to itself only once, the number of charges are limited by the number of nodes on the bottom, whose parents are also in the bottom (i.e. w is excluded).

There is one important detail that we must verify. Can we get in trouble on a subsequent partial find given that our reformulation detaches the nodes between x and w from the path to y? The answer is no. In the original partial find, suppose any of the nodes between x and w are involved in a subsequent original partial find. In that case, it will be with one of y's ancestors, and when that happens, any of those nodes will be the topmost "bottom node" in our reformulation. Thus on the subsequent partial find, the original partial find's parent change will have a corresponding one unit charge in our reformulation.

We can now proceed with the analysis. Let M be the total number of original partial find operations. Let M_t be the total number of partial find operations performed exclusively on the top half, and let M_b be the total number of partial find operations performed exclusively on the bottom half. Let N be the total number of nodes. Let N_t be the total number of top-half nodes, and let N_b be the total number of bottom half nodes, and let N_{nrb} be the total number of non-root bottom nodes (i.e the number of bottom nodes whose parents are also bottom nodes prior to any partial finds).

Lemma 8.3.

$M = M_t + M_b$.

Proof.

In cases 1 and 3, each original partial find operation is replaced by a partial find on the top half, and in case 2, it is replaced by a partial find on the bottom half. Thus each partial find is replaced by exactly one partial find operation on one of the halves.

Our basic idea is that we are going to partition the nodes so that all nodes with rank s or lower are in the bottom, and the remaining nodes are in the top. The choice of s will be made later in the proof. The next lemma shows that we can provide a recursive formula for the number of parent changes, by splitting the charges into the top and bottom groups. One of the key ideas is that a recursive formula is written not only in terms of M and N, which would be obvious, but also in terms of the maximum rank in the group.

Lemma 8.4.

Let $C(M, N, r)$ be the number of parent changes for a sequence of M finds with path compression on N items, whose maximum rank is r. Suppose we partition so that all nodes with rank at s or lower are in the bottom, and the remaining nodes are in the top. Assuming appropriate initial conditions,

$$C(M, N, r) < C(M_t, N_t, r) + C(M_b, N_b, s) + M_t + N_{nrb}$$

Proof.

The path compression that is performed in each of the three cases is covered by $C(M_t, N_t, r) + C(M_b, N_b, s)$. Node w in case 3 is accounted for by M_t. Finally, all the other bottom nodes on the path are non-root nodes that can have their parent set to themselves at most once in the entire sequence of compressions. They are accounted for by N_{nrb}.

If union-by-rank is used, then by Lemma 8.1, every top node has children of ranks $0, 1, \ldots, s$ prior to the commencement of the partial find operations. Each of those children are definitely root nodes in the bottom (their parent is a top node). So for each top node, $s + 2$ nodes (the $s + 1$ children plus the top node itself) are definitely not included in N_{nrb}. Thus, we can reformulate Lemma 8.4 as follows:

Lemma 8.5.

Let $C(M, N, r)$ be the number of parent changes for a sequence of M finds with path compression on N items, whose maximum rank is r. Suppose we partition so that all nodes with rank at s or lower are in the bottom, and the remaining nodes are in the top. Assuming appropriate initial conditions,

$$C(M, N, r) < C(M_t, N_t, r) + C(M_b, N_b, s) + M_t + N - (s + 2)N_t$$

Proof.

Substitute $N_{nrb} < N - (s + 2)N_t$ into Lemma 8.4.

If we look at Lemma 8.5, we see that $C(M, N, r)$ is recursively defined in terms of two smaller instances. Our basic goal at this point is to remove one of these instances, by providing a bound for it. What we would like to do is to remove $C(M_t, N_t, r)$. Why? Because, if we do so, what is left is $C(M_b, N_b, s)$. In that case, we have a recursive formula in which r is reduced to s. If s is small enough, we can make use of a variation of Equation 8.1, namely that the solution to

$$T(N) = \begin{cases} 0 & N \leq 1 \\ T(\lfloor f(N) \rfloor) + M & N > 1 \end{cases} \tag{8.2}$$

is $O(M f^*(N))$. So, let's start with a simple bound for $C(M, N, r)$:

Theorem 8.1.
$C(M, N, r) < M + N \log r$.

Proof.
We start with Lemmas 8.5:

$$C(M, N, r) < C(M_t, N_t, r) + C(M_b, N_b, s) + M_t + N - (s + 2)N_t \tag{8.3}$$

Observe that in the top half, there are only nodes of rank $s+1, s+2, \ldots, r$, and thus no node can have its parent change more than $(r-s-2)$ times. This yields a trivial bound of $N_t(r-s-2)$ for $C(M_t, N_t, r)$. Thus,

$$C(M, N, r) < N_t(r - s - 2) + C(M_b, N_b, s) + M_t + N - (s + 2)N_t \tag{8.4}$$

Combining terms,

$$C(M, N, r) < N_t(r - 2s - 4) + C(M_b, N_b, s) + M_t + N \tag{8.5}$$

Select $s = \lfloor r/2 \rfloor$. Then $r - 2s - 4 < 0$, so

$$C(M, N, r) < C(M_b, N_b, \lfloor r/2 \rfloor) + M_t + N \tag{8.6}$$

Equivalently, since according to Lemma 8.3, $M = M_b + M_t$ (the proof falls apart without this),

$$C(M, N, r) - M < C(M_b, N_b, \lfloor r/2 \rfloor) - M_b + N \tag{8.7}$$

Let $D(M, N, r) = C(M, N, r) - M$; then

$$D(M, N, r) < D(M_b, N_b, \lfloor r/2 \rfloor) + N \tag{8.8}$$

which implies $D(M, N, r) < N \log r$. This yields $C(M, N, r) < M + N \log r$.

Theorem 8.2.
Any sequence of $N - 1$ unions and M finds with path compression makes at most $M + N \log \log N$ parent changes during the finds.

Proof.
The bound is immediate from Theorem 8.1 since $r \leq \log N$.

8.6.3 An $O(M \log^* N)$ Bound

The bound in Theorem 8.2 is pretty good, but with a little work, we can do even better. Recall, that a central idea of the recursive decomposition is choosing s to be as small as possible. But to do this, the other terms must also be small, and as s gets smaller, we would expect $C(M_t, N_t, r)$ to get larger. But the bound for $C(M_t, N_t, r)$ used a primitive estimate, and Theorem 8.1 itself can now be used to give a better estimate for this term. Since the $C(M_t, N_t, r)$ estimate will now be lower, we will be able to use a lower s.

Theorem 8.3.
$C(M, N, r) < 2M + N \log^* r.$

Proof.
From Lemma 8.5 we have,

$$C(M, N, r) < C(M_t, N_t, r) + C(M_b, N_b, s) + M_t + N - (s + 2)N_t \tag{8.9}$$

and by Theorem 8.1, $C(M_t, N_t, r) < M_t + N_t \log r$. Thus,

$$C(M, N, r) < M_t + N_t \log r + C(M_b, N_b, s) + M_t + N - (s + 2)N_t \tag{8.10}$$

Rearranging and combining terms yields

$$C(M, N, r) < C(M_b, N_b, s) + 2M_t + N - (s - \log r + 2)N_t \tag{8.11}$$

So choose $s = \lfloor \log r \rfloor$. Clearly, this choice implies that $(s - \log r + 2) > 0$, and thus we obtain

$$C(M, N, r) < C(M_b, N_b, \lfloor \log r \rfloor) + 2M_t + N \tag{8.12}$$

Rearranging as in Theorem 8.1, we obtain,

$$C(M, N, r) - 2M < C(M_b, N_b, \lfloor \log r \rfloor) - 2M_b + N \tag{8.13}$$

This time, let $D(M, N, r) = C(M, N, r) - 2M$; then

$$D(M, N, r) < D(M_b, N_b, \lfloor \log r \rfloor) + N \tag{8.14}$$

which implies $D(M, N, r) < N \log^* r$. This yields $C(M, N, r) < 2M + N \log^* r$.

8.6.4 An $O(M \alpha(M, N))$ Bound

Not surprisingly, we can now use Theorem 8.3 to improve Theorem 8.3:

Theorem 8.4.
$C(M, N, r) < 3M + N \log^{**} r.$

Proof.
Following the steps in the proof of Theorem 8.3, we have

$$C(M, N, r) < C(M_t, N_t, r) + C(M_b, N_b, s) + M_t + N - (s+2)N_t \qquad (8.15)$$

and by Theorem 8.3, $C(M_t, N_t, r) < 2M_t + N_t \log^* r$. Thus,

$$C(M, N, r) < 2M_t + N_t \log^* r + C(M_b, N_b, s) + M_t + N - (s+2)N_t \qquad (8.16)$$

Rearranging and combining terms yields

$$C(M, N, r) < C(M_b, N_b, s) + 3M_t + N - (s - \log^* r + 2)N_t \qquad (8.17)$$

So choose $s = \log^* r$ to obtain

$$C(M, N, r) < C(M_b, N_b, \log^* r) + 3M_t + N \qquad (8.18)$$

Rearranging as in Theorems 8.1 and 8.3, we obtain,

$$C(M, N, r) - 3M < C(M_b, N_b, \log^* r) - 3M_b + N \qquad (8.19)$$

This time, let $D(M, N, r) = C(M, N, r) - 3M$; then

$$D(M, N, r) < D(M_b, N_b, \log^* r) + N \qquad (8.20)$$

which implies $D(M, N, r) < N \log^{**} r$. This yields $C(M, N, r) < 3M + N \log^{**} r$.

Needless to say, we could continue this ad-infinitim. Thus with a bit of math, we get a progression of bounds:

$$C(M, N, r) < 2M + N \log^* r$$
$$C(M, N, r) < 3M + N \log^{**} r$$
$$C(M, N, r) < 4M + N \log^{***} r$$
$$C(M, N, r) < 5M + N \log^{****} r$$
$$C(M, N, r) < 6M + N \log^{*****} r$$

Each of these bounds would seem to be better than the previous since, after all, the more *s the slower $\log^{**\cdots**} r$ grows. However, this ignores the fact that while $\log^{*****} r$ is smaller than $\log^{****} r$, the $6M$ term is NOT smaller than the $5M$ term.

Thus what we would like to do is to optimize the number of *s that are used.

Define $\alpha(M, N)$ to represent the optimal number of *s that will be used. Specifically,

$$\alpha(M, N) = min \left\{ i \geq 1 \,\Big|\, \log^{\overbrace{****}^{i \text{ times}}} (\log N) \leq (M/N) \right\}$$

Then, the running time of the union/find algorithm can be bounded by $O(M\alpha(M, N))$.

Theorem 8.5.

Any sequence of $N - 1$ unions and M finds with path compression makes at most

$$(i + 1)M + N \log^{\overbrace{****}^{i \text{ times}}} (\log N)$$

parent changes during the finds.

Proof.
This follows from the above discussion, and the fact that $r \leq \log N$.

Theorem 8.6.
Any sequence of $N - 1$ unions and M finds with path compression makes at most $M\alpha(M, N) + 2M$ parent changes during the finds.

Proof.
In Theorem 8.5, choose i to be $\alpha(M, N)$; thus we obtain a bound of $(i+1)M+N(M/N)$, or $M\alpha(M, N) + 2M$.

8.7 An Application

An example of the use of the union/find data structure is the generation of mazes, such as the one shown in Figure 8.25. In Figure 8.25, the starting point is the top-left corner, and the ending point is the bottom-right corner. We can view the maze as a 50-by-88 rectangle of cells in which the top-left cell is connected to the bottom-right cell, and cells are separated from their neighboring cells via walls.

A simple algorithm to generate the maze is to start with walls everywhere (except for the entrance and exit). We then continually choose a wall randomly, and knock it down if the cells that the wall separates are not already connected to each other. If we repeat this process until the starting and ending cells are connected, then we have a maze. It is actually

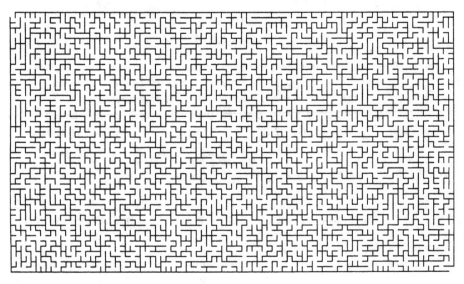

Figure 8.25   A 50-by-88 maze

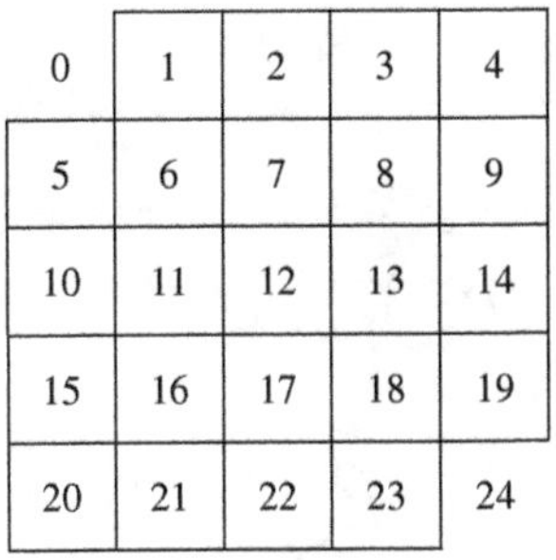

0	1	2	3	4
5	6	7	8	9
10	11	12	13	14
15	16	17	18	19
20	21	22	23	24

{0} {1} {2} {3} {4} {5} {6} {7} {8} {9} {10} {11} {12} {13} {14} {15} {16} {17} {18} {19} {20} {21} {22} {23} {24}

Figure 8.26 Initial state: all walls up, all cells in their own set

better to continue knocking down walls until every cell is reachable from every other cell (this generates more false leads in the maze).

We illustrate the algorithm with a 5-by-5 maze. Figure 8.26 shows the initial configuration. We use the union/find data structure to represent sets of cells that are connected to each other. Initially, walls are everywhere, and each cell is in its own equivalence class.

Figure 8.27 shows a later stage of the algorithm, after a few walls have been knocked down. Suppose, at this stage, the wall that connects cells 8 and 13 is randomly targeted. Because 8 and 13 are already connected (they are in the same set), we would not remove the wall, as it would simply trivialize the maze. Suppose that cells 18 and 13 are randomly targeted next. By performing two find operations, we see that these are in different sets;

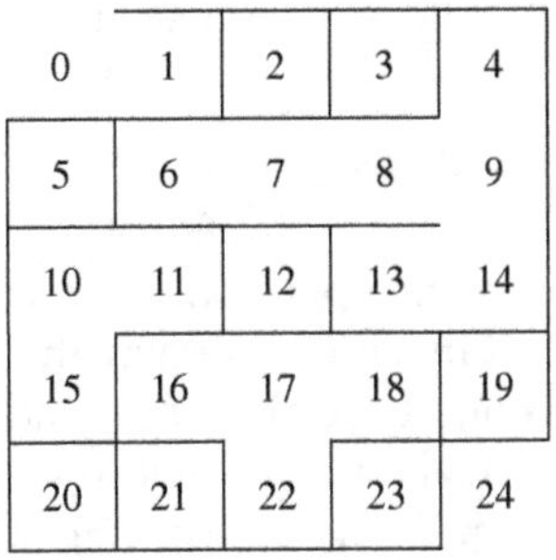

0	1	2	3	4
5	6	7	8	9
10	11	12	13	14
15	16	17	18	19
20	21	22	23	24

{0, 1} {2} {3} {4, 6, 7, 8, 9, 13, 14} {5} {10, 11, 15} {12} {16, 17, 18, 22} {19} {20} {21} {23} {24}

Figure 8.27 At some point in the algorithm: Several walls down, sets have merged; if at this point the wall between 8 and 13 is randomly selected, this wall is not knocked down, because 8 and 13 are already connected

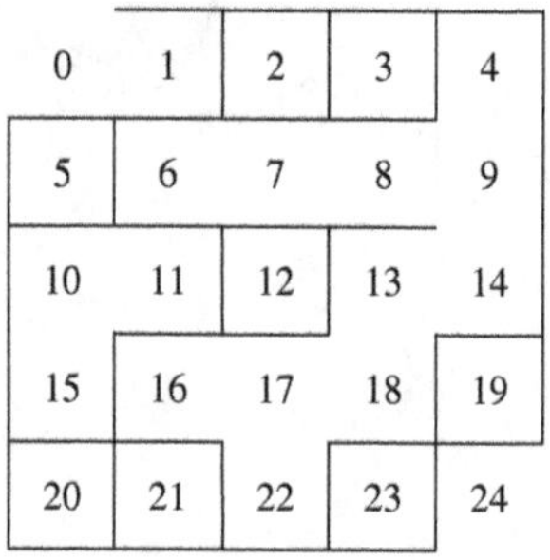

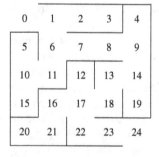

{0, 1} {2} {3} {4, 6, 7, 8, 9, 13, 14, 16, 17, 18, 22} {5} {10, 11, 15} {12} {19} {20} {21} {23} {24}

Figure 8.28 Wall between squares 18 and 13 is randomly selected in Figure 8.22; this wall is knocked down, because 18 and 13 are not already connected; their sets are merged

{0, 1, 2, 3, 4, 5, 6, 7, 8, 9, 10, 11, 12, 13, 14, 15, 16, 17, 18, 19, 20, 21, 22, 23, 24}

Figure 8.29 Eventually, 24 walls are knocked down; all elements are in the same set

thus 18 and 13 are not already connected. Therefore, we knock down the wall that separates them, as shown in Figure 8.28. Notice that as a result of this operation, the sets containing 18 and 13 are combined via a union operation. This is because everything that was connected to 18 is now connected to everything that was connected to 13. At the end of the algorithm, depicted in Figure 8.29, everything is connected, and we are done.

The running time of the algorithm is dominated by the union/find costs. The size of the union/find universe is equal to the number of cells. The number of find operations is proportional to the number of cells, since the number of removed walls is one less than the number of cells, while with care, we see that there are only about twice the number of walls as cells in the first place. Thus, if N is the number of cells, since there are two finds per randomly targeted wall, this gives an estimate of between (roughly) $2N$ and $4N$ find operations throughout the algorithm. Therefore, the algorithm's running time can be taken as $O(N \log^* N)$, and this algorithm quickly generates a maze.

Summary

We have seen a very simple data structure to maintain disjoint sets. When the union operation is performed, it does not matter, as far as correctness is concerned, which set retains its name. A valuable lesson that should be learned here is that it can be very important to consider the alternatives when a particular step is not totally specified. The union step is flexible; by taking advantage of this, we are able to get a much more efficient algorithm.

Path compression is one of the earliest forms of *self-adjustment,* which we have seen elsewhere (splay trees, skew heaps). Its use is extremely interesting, especially from a theoretical point of view, because it was one of the first examples of a simple algorithm with a not-so-simple worst-case analysis.

Exercises

8.1 Show the result of the following sequence of instructions: union(1,2), union(3,4), union(3,5), union(1,7), union(3,6), union(8,9), union(1,8), union(3,10), union (3,11), union(3,12), union(3,13), union(14,15), union(16,0), union(14,16), union (1,3), union(1, 14) when the unions are:
  a. Performed arbitrarily.
  b. Performed by height.
  c. Performed by size.

8.2 For each of the trees in the previous exercise, perform a find with path compression on the deepest node.

8.3 Write a program to determine the effects of path compression and the various unioning strategies. Your program should process a long sequence of equivalence operations using all six of the possible strategies.

8.4 Show that if unions are performed by height, then the depth of any tree is $O(\log N)$.

8.5 Suppose $f(N)$ is a nicely defined function that reduces N to a smaller integer. What is the solution to the recurrence $T(N) = \frac{N}{f(N)} T(f(N)) + N$ with appropriate initial conditions?

8.6 a. Show that if $M = N^2$, then the running time of M union/find operations is $O(M)$.
  b. Show that if $M = N \log N$, then the running time of M union/find operations is $O(M)$.
  *c. Suppose $M = \Theta(N \log \log N)$. What is the running time of M union/find operations?
  *d. Suppose $M = \Theta(N \log^* N)$. What is the running time of M union/find operations?

8.7   Tarjan's original bound for the union/find algorithm defined
$\alpha(M,N) = min\{i \geq 1 \big| (A\ (i, \lfloor M/N \rfloor) > \log N)\}$, where

$$A(1,j) = 2^j \qquad\qquad j \geq 1$$
$$A(i,1) = A(i-1,2) \qquad\qquad i \geq 2$$
$$A(i,j) = A(i-1,A(i,j-1)) \qquad i,j \geq 2$$

Here, $A(m,n)$ is one version of the **Ackermann function**. Are the two definitions of α asymptotically equivalent?

8.8   Prove that for the mazes generated by the algorithm in Section 8.7, the path from the starting to ending points is unique.

8.9   Design an algorithm that generates a maze that contains no path from start to finish but has the property that the removal of a **prespecified** wall creates a unique path.

*8.10   Suppose we want to add an extra operation, deunion, which undoes the last union operation that has not been already undone.
   a. Show that if we do union-by-height and finds without path compression, then deunion is easy and a sequence of M union, find, and deunion operations takes $O(M \log N)$ time.
   b. Why does path compression make deunion hard?
   **c. Show how to implement all three operations so that the sequence of M operations takes $O(M \log N/\log \log N)$ time.

*8.11   Suppose we want to add an extra operation, remove(x), which removes x from its current set and places it in its own. Show how to modify the union/find algorithm so that the running time of a sequence of M union, find, and remove operations is $O(M\alpha(M,N))$.

*8.12   Show that if all of the unions precede the finds, then the disjoint set algorithm with path compression requires linear time, even if the unions are done arbitrarily.

**8.13   Prove that if unions are done arbitrarily, but path compression is performed on the finds, then the worst-case running time is $\Theta(M \log N)$.

*8.14   Prove that if unions are done by size and path compression is performed, the worst-case running time is $O(M\alpha(M,N))$.

8.15   The disjoint sets analysis in Section 8.6 can be refined to provide tight bounds for small N.
   a. Show that $C(M,N,0)$ and $C(M,N,1)$ are both 0.
   b. Show that $C(M,N,2)$ is at most M.
   c. Let $r \leq 8$. Choose $s = 2$ and show that $C(M,N,r)$ is at most $M + N$.

8.16   Suppose we implement partial path compression on find(i) by making every other node on the path from i to the root link to its grandparent (where this makes sense). This is known as *path halving*.
   a. Write a procedure to do this.
   b. Prove that if path halving is performed on the finds and either union-by-height or union-by-size is used, the worst-case running time is $O(M\alpha(M,N))$.

8.17   Write a program that generates mazes of arbitrary size. Use Swing to generate a maze similar to that in Figure 8.25.

References

Various solutions to the union/find problem can be found in [6], [9], and [11]. Hopcroft and Ullman showed an $O(M \log^* N)$ bound using a nonrecursive decomposition. Tarjan [16] obtained the bound $O(M\alpha(M,N))$, where $\alpha(M,N)$ is as defined in Exercise 8.7. A more precise (but asymptotically identical) bound for $M < N$ appears in [2] and [19]. The analysis in Section 8.6 is due to Seidel and Sharir [15]. Various other strategies for path compression and unions also achieve the same bound; see [19] for details.

A lower bound showing that under certain restrictions $\Omega(M\alpha(M,N))$ time is required to process M union/find operations was given by Tarjan [17]. Identical bounds under less restrictive conditions have been shown in [7] and [14].

Applications of the union/find data structure appear in [1] and [10]. Certain special cases of the union/find problem can be solved in $O(M)$ time [8]. This reduces the running time of several algorithms, such as [1], graph dominance, and reducibility (see references in Chapter 9) by a factor of $\alpha(M,N)$. Others, such as [10] and the graph connectivity problem in this chapter, are unaffected. The paper lists 10 examples. Tarjan has used path compression to obtain efficient algorithms for several graph problems [18].

Average case results for the union/find problem appear in [5], [12], [22], and [3]. Results bounding the running time of any single operation (as opposed to the entire sequence) appear in [4] and [13].

Exercise 8.10 is solved in [21]. A general union/find structure, supporting more operations, is given in [20].

1. A. V. Aho, J. E. Hopcroft, and J. D. Ullman, "On Finding Lowest Common Ancestors in Trees," *SIAM Journal on Computing,* 5 (1976), 115–132.

2. L. Banachowski, "A Complement to Tarjan's Result about the Lower Bound on the Complexity of the Set Union Problem," *Information Processing Letters,* 11 (1980), 59–65.

3. B. Bollobás and I. Simon, "Probabilistic Analysis of Disjoint Set Union Algorithms," *SIAM Journal on Computing,* 22 (1993), 1053–1086.

4. N. Blum, "On the Single-Operation Worst-Case Time Complexity of the Disjoint Set Union Problem," *SIAM Journal on Computing,* 15 (1986), 1021–1024.

5. J. Doyle and R. L. Rivest, "Linear Expected Time of a Simple Union Find Algorithm," *Information Processing Letters,* 5 (1976), 146–148.

6. M. J. Fischer, "Efficiency of Equivalence Algorithms," in *Complexity of Computer Computation* (eds. R. E. Miller and J. W. Thatcher), Plenum Press, New York, 1972, 153–168.

7. M. L. Fredman and M. E. Saks, "The Cell Probe Complexity of Dynamic Data Structures," *Proceedings of the Twenty-first Annual Symposium on Theory of Computing* (1989), 345–354.

8. H. N. Gabow and R. E. Tarjan, "A Linear-Time Algorithm for a Special Case of Disjoint Set Union," *Journal of Computer and System Sciences,* 30 (1985), 209–221.

9. B. A. Galler and M. J. Fischer, "An Improved Equivalence Algorithm," *Communications of the ACM,* 7 (1964), 301–303.

10. J. E. Hopcroft and R. M. Karp, "An Algorithm for Testing the Equivalence of Finite Automata," *Technical Report TR-71-114,* Department of Computer Science, Cornell University, Ithaca, N.Y., 1971.

11. J. E. Hopcroft and J. D. Ullman, "Set Merging Algorithms," *SIAM Journal on Computing,* 2 (1973), 294–303.

12. D. E. Knuth and A. Schonhage, "The Expected Linearity of a Simple Equivalence Algorithm," *Theoretical Computer Science,* 6 (1978), 281–315.

13. J. A. LaPoutre, "New Techniques for the Union-Find Problem," *Proceedings of the First Annual ACM–SIAM Symposium on Discrete Algorithms* (1990), 54–63.

14. J. A. LaPoutre, "Lower Bounds for the Union-Find and the Split-Find Problem on Pointer Machines," *Proceedings of the Twenty-Second Annual ACM Symposium on Theory of Computing* (1990), 34–44.

15. R. Seidel and M. Sharir, "Top-Down Analysis of Path Compression," *SIAM Journal on Computing,* 34 (2005), 515–525.

16. R. E. Tarjan, "Efficiency of a Good but Not Linear Set Union Algorithm," *Journal of the ACM,* 22 (1975), 215–225.

17. R. E. Tarjan, "A Class of Algorithms Which Require Nonlinear Time to Maintain Disjoint Sets," *Journal of Computer and System Sciences,* 18 (1979), 110–127.

18. R. E. Tarjan, "Applications of Path Compression on Balanced Trees," *Journal of the ACM,* 26 (1979), 690–715.

19. R. E. Tarjan and J. van Leeuwen, "Worst Case Analysis of Set Union Algorithms," *Journal of the ACM,* 31 (1984), 245–281.

20. M. J. van Kreveld and M. H. Overmars, "Union-Copy Structures and Dynamic Segment Trees," *Journal of the ACM,* 40 (1993), 635–652.

21. J. Westbrook and R. E. Tarjan, "Amortized Analysis of Algorithms for Set Union with Back-tracking," *SIAM Journal on Computing,* 18 (1989), 1–11.

22. A. C. Yao, "On the Average Behavior of Set Merging Algorithms," *Proceedings of Eighth Annual ACM Symposium on the Theory of Computation* (1976), 192–195.

CHAPTER 9

Graph Algorithms

In this chapter we discuss several common problems in graph theory. Not only are these algorithms useful in practice, they are also interesting because in many real-life applications they are too slow unless careful attention is paid to the choice of data structures. We will

- Show several real-life problems, which can be converted to problems on graphs.
- Give algorithms to solve several common graph problems.
- Show how the proper choice of data structures can drastically reduce the running time of these algorithms.
- See an important technique, known as depth-first search, and show how it can be used to solve several seemingly nontrivial problems in linear time.

9.1 Definitions

A **graph** $G = (V, E)$ consists of a set of **vertices**, V, and a set of **edges**, E. Each edge is a pair (v, w), where $v, w \in V$. Edges are sometimes referred to as **arcs**. If the pair is ordered, then the graph is **directed**. Directed graphs are sometimes referred to as **digraphs**. Vertex w is **adjacent** to v if and only if $(v, w) \in E$. In an undirected graph with edge (v, w), and hence (w, v), w is adjacent to v and v is adjacent to w. Sometimes an edge has a third component, known as either a **weight** or a **cost**.

A **path** in a graph is a sequence of vertices $w_1, w_2, w_3, \ldots, w_N$ such that $(w_i, w_{i+1}) \in E$ for $1 \le i < N$. The **length** of such a path is the number of edges on the path, which is equal to $N - 1$. We allow a path from a vertex to itself; if this path contains no edges, then the path length is 0. This is a convenient way to define an otherwise special case. If the graph contains an edge (v, v) from a vertex to itself, then the path v, v is sometimes referred to as a **loop**. The graphs we will consider will generally be loopless. A **simple path** is a path such that all vertices are distinct, except that the first and last could be the same.

A **cycle** in a directed graph is a path of length at least 1 such that $w_1 = w_N$; this cycle is simple if the path is simple. For undirected graphs, we require that the edges be distinct. The logic of these requirements is that the path u, v, u in an undirected graph should not be considered a cycle, because (u, v) and (v, u) are the same edge. In a directed graph, these are different edges, so it makes sense to call this a cycle. A directed graph is **acyclic** if it has no cycles. A directed acyclic graph is sometimes referred to by its abbreviation, **DAG**.

An undirected graph is **connected** if there is a path from every vertex to every other vertex. A directed graph with this property is called **strongly connected**. If a directed graph is not strongly connected, but the underlying graph (without direction to the arcs) is connected, then the graph is said to be **weakly connected**. A **complete graph** is a graph in which there is an edge between every pair of vertices.

An example of a real-life situation that can be modeled by a graph is the airport system. Each airport is a vertex, and two vertices are connected by an edge if there is a nonstop flight from the airports that are represented by the vertices. The edge could have a weight, representing the time, distance, or cost of the flight. It is reasonable to assume that such a graph is directed, since it might take longer or cost more (depending on local taxes, for example) to fly in different directions. We would probably like to make sure that the airport system is strongly connected, so that it is always possible to fly from any airport to any other airport. We might also like to quickly determine the best flight between any two airports. "Best" could mean the path with the fewest number of edges or could be taken with respect to one, or all, of the weight measures.

Traffic flow can be modeled by a graph. Each street intersection represents a vertex, and each street is an edge. The edge costs could represent, among other things, a speed limit or a capacity (number of lanes). We could then ask for the shortest route or use this information to find the most likely location for bottlenecks.

In the remainder of this chapter, we will see several more applications of graphs. Many of these graphs can be quite large, so it is important that the algorithms we use be efficient.

9.1.1 Representation of Graphs

We will consider directed graphs (undirected graphs are similarly represented).

Suppose, for now, that we can number the vertices, starting at 1. The graph shown in Figure 9.1 represents 7 vertices and 12 edges.

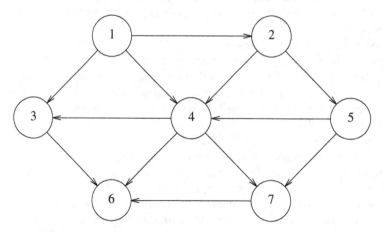

Figure 9.1 A directed graph

One simple way to represent a graph is to use a two-dimensional array. This is known as an **adjacency matrix** representation. For each edge (u, v), we set $A[u][v]$ to `true`; otherwise the entry in the array is `false`. If the edge has a weight associated with it, then we can set $A[u][v]$ equal to the weight and use either a very large or a very small weight as a sentinel to indicate nonexistent edges. For instance, if we were looking for the cheapest airplane route, we could represent nonexistent flights with a cost of ∞. If we were looking, for some strange reason, for the most expensive airplane route, we could use $-\infty$ (or perhaps 0) to represent nonexistent edges.

Although this has the merit of extreme simplicity, the space requirement is $\Theta(|V|^2)$, which can be prohibitive if the graph does not have very many edges. An adjacency matrix is an appropriate representation if the graph is **dense**: $|E| = \Theta(|V|^2)$. In most of the applications that we shall see, this is not true. For instance, suppose the graph represents a street map. Assume a Manhattan-like orientation, where almost all the streets run either north–south or east–west. Therefore, any intersection is attached to roughly four streets, so if the graph is directed and all streets are two-way, then $|E| \approx 4|V|$. If there are 3,000 intersections, then we have a 3,000-vertex graph with 12,000 edge entries, which would require an array of size 9,000,000. Most of these entries would contain zero. This is intuitively bad, because we want our data structures to represent the data that are actually there and not the data that are not present.

If the graph is not dense, in other words, if the graph is **sparse**, a better solution is an **adjacency list** representation. For each vertex, we keep a list of all adjacent vertices. The space requirement is then $O(|E| + |V|)$, which is linear in the size of the graph.[1] The abstract representation should be clear from Figure 9.2. If the edges have weights, then this additional information is also stored in the adjacency lists.

Adjacency lists are the standard way to represent graphs. Undirected graphs can be similarly represented; each edge (u, v) appears in two lists, so the space usage essentially doubles. A common requirement in graph algorithms is to find all vertices adjacent to some given vertex v, and this can be done, in time proportional to the number of such vertices found, by a simple scan down the appropriate adjacency list.

There are several alternatives for maintaining the adjacency lists. First, observe that the lists themselves can be maintained in any kind of `List`, namely `ArrayLists` or `LinkedLists`. However, for very sparse graphs, when using `ArrayLists`, the programmer may need to start the `ArrayLists` with a smaller capacity than the default; otherwise there could be significant wasted space.

Because it is important to be able to quickly obtain the list of adjacent vertices for any vertex, the two basic options are to use a map in which the keys are vertices and the values are adjacency lists, or to maintain each adjacency list as a data member of a `Vertex` class. The first option is arguably simpler, but the second option can be faster, because it avoids repeated lookups in the map.

In the second scenario, if the vertex is a `String` (for instance, an airport name, or the name of a street intersection), then a map can be used in which the key is the vertex name and the value is a `Vertex` and each `Vertex` object keeps a list of adjacent vertices, and perhaps also the original `String` name.

---

[1] When we speak of linear-time graph algorithms, $O(|E| + |V|)$ is the running time we require.

1	2, 4, 3
2	4, 5
3	6
4	6, 7, 3
5	4, 7
6	(empty)
7	6

Figure 9.2 An adjacency list representation of a graph

In most of the chapter, we present the graph algorithms using pseudocode. We will do this to save space and, of course, to make the presentation of the algorithms much clearer. At the end of Section 9.3, we provide a working Java implementation of a routine that makes underlying use of a shortest-path algorithm to obtain its answers.

9.2 Topological Sort

A **topological sort** is an ordering of vertices in a directed acyclic graph, such that if there is a path from v_i to v_j, then v_j appears *after* v_i in the ordering. The graph in Figure 9.3 represents the course prerequisite structure at a state university in Miami. A directed edge (v, w) indicates that course v must be completed before course w may be attempted. A topological ordering of these courses is any course sequence that does not violate the prerequisite requirement.

It is clear that a topological ordering is not possible if the graph has a cycle, since for two vertices v and w on the cycle, v precedes w and w precedes v. Furthermore, the ordering is not necessarily unique; any legal ordering will do. In the graph in Figure 9.4, $v_1, v_2, v_5, v_4, v_3, v_7, v_6$ and $v_1, v_2, v_5, v_4, v_7, v_3, v_6$ are both topological orderings.

A simple algorithm to find a topological ordering is first to find any vertex with no incoming edges. We can then print this vertex, and remove it, along with its edges, from the graph. Then we apply this same strategy to the rest of the graph.

To formalize this, we define the **indegree** of a vertex v as the number of edges (u, v). We compute the indegrees of all vertices in the graph. Assuming that the indegree for each

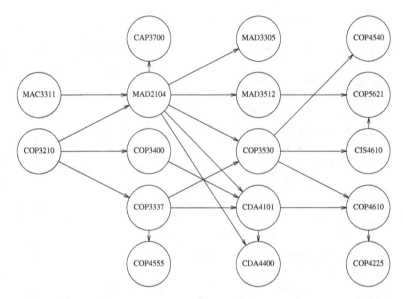

Figure 9.3 An acyclic graph representing course prerequisite structure

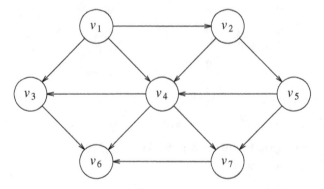

Figure 9.4 An acyclic graph

vertex is stored and that the graph is read into an adjacency list, we can then apply the algorithm in Figure 9.5 to generate a topological ordering.

The method `findNewVertexOfIndegreeZero` scans the array looking for a vertex with indegree 0 that has not already been assigned a topological number. It returns `null` if no such vertex exists; this indicates that the graph has a cycle.

Because `findNewVertexOfIndegreeZero` is a simple sequential scan of the array of vertices, each call to it takes $O(|V|)$ time. Since there are $|V|$ such calls, the running time of the algorithm is $O(|V|^2)$.

By paying more careful attention to the data structures, it is possible to do better. The cause of the poor running time is the sequential scan through the array of vertices. If the

```
void topsort( ) throws CycleFoundException
{
    for( int counter = 0; counter < NUM_VERTICES; counter++ )
    {
        Vertex v = findNewVertexOfIndegreeZero( );
        if( v == null )
            throw new CycleFoundException( );
        v.topNum = counter;
        for each Vertex w adjacent to v
            w.indegree--;
    }
}
```

Figure 9.5 Simple topological sort pseudocode

	Indegree Before Dequeue #						
Vertex	1	2	3	4	5	6	7
v_1	0	0	0	0	0	0	0
v_2	1	0	0	0	0	0	0
v_3	2	1	1	1	0	0	0
v_4	3	2	1	0	0	0	0
v_5	1	1	0	0	0	0	0
v_6	3	3	3	3	2	1	0
v_7	2	2	2	1	0	0	0
Enqueue	v_1	v_2	v_5	v_4	v_3, v_7		v_6
Dequeue	v_1	v_2	v_5	v_4	v_3	v_7	v_6

Figure 9.6 Result of applying topological sort to the graph in Figure 9.4

graph is sparse, we would expect that only a few vertices have their indegrees updated during each iteration. However, in the search for a vertex of indegree 0, we look at (potentially) all the vertices, even though only a few have changed.

We can remove this inefficiency by keeping all the (unassigned) vertices of indegree 0 in a special *box*. The findNewVertexOfIndegreeZero method then returns (and removes) any vertex in the box. When we decrement the indegrees of the adjacent vertices, we check each vertex and place it in the box if its indegree falls to 0.

To implement the box, we can use either a stack or a queue; we will use a queue. First, the indegree is computed for every vertex. Then all vertices of indegree 0 are placed on an initially empty queue. While the queue is not empty, a vertex v is removed, and all vertices adjacent to v have their indegrees decremented. A vertex is put on the queue as soon as its indegree falls to 0. The topological ordering then is the order in which the vertices dequeue. Figure 9.6 shows the status after each phase.

```
void topsort( ) throws CycleFoundException
{
    Queue<Vertex> q = new Queue<Vertex>( );
    int counter = 0;

    for each Vertex v
        if( v.indegree == 0 )
            q.enqueue( v );

    while( !q.isEmpty( ) )
    {
        Vertex v = q.dequeue( );
        v.topNum = ++counter;  // Assign next number

        for each Vertex w adjacent to v
            if( --w.indegree == 0 )
                q.enqueue( w );
    }
    if( counter != NUM_VERTICES )
        throw new CycleFoundException( );

}
```

Figure 9.7   Pseudocode to perform topological sort

A pseudocode implementation of this algorithm is given in Figure 9.7. As before, we will assume that the graph is already read into an adjacency list and that the indegrees are computed and stored with the vertices. We also assume each vertex has a field named topNum, in which to place its topological numbering.

The time to perform this algorithm is $O(|E| + |V|)$ if adjacency lists are used. This is apparent when one realizes that the body of the for loop is executed at most once per edge. Computing the indegrees can be done with the following code; this same logic shows that the cost of this computation is $O(|E| + |V|)$, even though there are nested loops.

```
for each Vertex v
    v.indegree = 0;

for each Vertex v
    for each Vertex w adjacent to v
        w.indegree++;
```

The queue operations are done at most once per vertex, and the other initialization steps, including the computation of indegrees, also take time proportional to the size of the graph.

9.3 Shortest-Path Algorithms

In this section we examine various shortest-path problems. The input is a weighted graph: associated with each edge (v_i, v_j) is a cost $c_{i,j}$ to traverse the edge. The cost of a path $v_1 v_2 \ldots v_N$ is $\sum_{i=1}^{N-1} c_{i,i+1}$. This is referred to as the **weighted path length**. The **unweighted path length** is merely the number of edges on the path, namely, $N - 1$.

Single-Source Shortest-Path Problem.
Given as input a weighted graph, $G = (V, E)$, and a distinguished vertex, s, find the shortest weighted path from s to every other vertex in G.

For example, in the graph in Figure 9.8, the shortest weighted path from v_1 to v_6 has a cost of 6 and goes from v_1 to v_4 to v_7 to v_6. The shortest unweighted path between these vertices is 2. Generally, when it is not specified whether we are referring to a weighted or an unweighted path, the path is weighted if the graph is. Notice also that in this graph there is no path from v_6 to v_1.

The graph in the preceding example has no edges of negative cost. The graph in Figure 9.9 shows the problems that negative edges can cause. The path from v_5 to v_4 has cost 1, but a shorter path exists by following the loop v_5, v_4, v_2, v_5, v_4, which has cost -5. This path is still not the shortest, because we could stay in the loop arbitrarily long. Thus, the shortest path between these two points is undefined. Similarly, the shortest path from v_1 to v_6 is undefined, because we can get into the same loop. This loop is known as a **negative-cost cycle**; when one is present in the graph, the shortest paths are not defined. Negative-cost edges are not necessarily bad, as the cycles are, but their presence seems to make the problem harder. For convenience, in the absence of a negative-cost cycle, the shortest path from s to s is zero.

There are many examples where we might want to solve the shortest-path problem. If the vertices represent computers; the edges represent a link between computers; and the costs represent communication costs (phone bill per megabyte of data), delay costs (number of seconds required to transmit a megabyte), or a combination of these and other

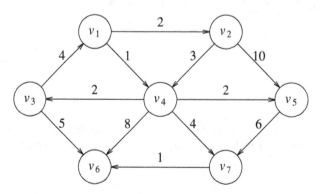

Figure 9.8 A directed graph G

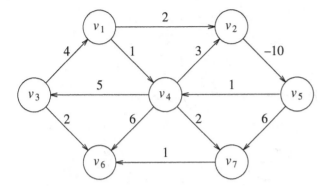

Figure 9.9 A graph with a negative-cost cycle

factors, then we can use the shortest-path algorithm to find the cheapest way to send electronic news from one computer to a set of other computers.

We can model airplane or other mass transit routes by graphs and use a shortest-path algorithm to compute the best route between two points. In this and many practical applications, we might want to find the shortest path from one vertex, s, to only one other vertex, t. Currently there are no algorithms in which finding the path from s to one vertex is any faster (by more than a constant factor) than finding the path from s to all vertices.

We will examine algorithms to solve four versions of this problem. First, we will consider the unweighted shortest-path problem and show how to solve it in $O(|E|+|V|)$. Next, we will show how to solve the weighted shortest-path problem if we assume that there are no negative edges. The running time for this algorithm is $O(|E| \log |V|)$ when implemented with reasonable data structures.

If the graph has negative edges, we will provide a simple solution, which unfortunately has a poor time bound of $O(|E| \cdot |V|)$. Finally, we will solve the weighted problem for the special case of acyclic graphs in linear time.

9.3.1 Unweighted Shortest Paths

Figure 9.10 shows an unweighted graph, G. Using some vertex, s, which is an input parameter, we would like to find the shortest path from s to all other vertices. We are only interested in the number of edges contained on the path, so there are no weights on the edges. This is clearly a special case of the weighted shortest-path problem, since we could assign all edges a weight of 1.

For now, suppose we are interested only in the length of the shortest paths, not in the actual paths themselves. Keeping track of the actual paths will turn out to be a matter of simple bookkeeping.

Suppose we choose s to be v_3. Immediately, we can tell that the shortest path from s to v_3 is then a path of length 0. We can mark this information, obtaining the graph in Figure 9.11.

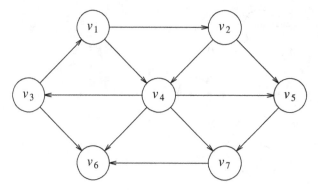

Figure 9.10   An unweighted directed graph G

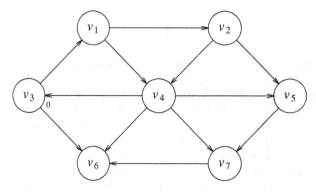

Figure 9.11   Graph after marking the start node as reachable in zero edges

Now we can start looking for all vertices that are a distance 1 away from s. These can be found by looking at the vertices that are adjacent to s. If we do this, we see that v_1 and v_6 are one edge from s. This is shown in Figure 9.12.

We can now find vertices whose shortest path from s is exactly 2, by finding all the vertices adjacent to v_1 and v_6 (the vertices at distance 1), whose shortest paths are not already known. This search tells us that the shortest path to v_2 and v_4 is 2. Figure 9.13 shows the progress that has been made so far.

Finally we can find, by examining vertices adjacent to the recently evaluated v_2 and v_4, that v_5 and v_7 have a shortest path of three edges. All vertices have now been calculated, and so Figure 9.14 shows the final result of the algorithm.

This strategy for searching a graph is known as **breadth-first search**. It operates by processing vertices in layers: The vertices closest to the start are evaluated first, and the most distant vertices are evaluated last. This is much the same as a level-order traversal for trees.

Given this strategy, we must translate it into code. Figure 9.15 shows the initial configuration of the table that our algorithm will use to keep track of its progress.

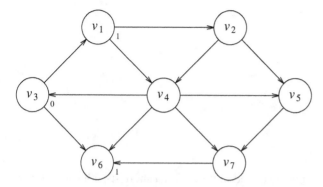

Figure 9.12   Graph after finding all vertices whose path length from *s* is 1

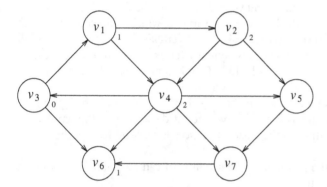

Figure 9.13   Graph after finding all vertices whose shortest path is 2

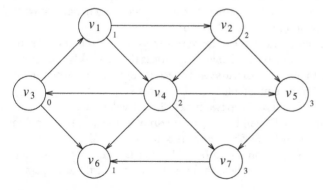

Figure 9.14   Final shortest paths

v	known	d_v	p_v
v_1	F	∞	0
v_2	F	∞	0
v_3	F	0	0
v_4	F	∞	0
v_5	F	∞	0
v_6	F	∞	0
v_7	F	∞	0

Figure 9.15 Initial configuration of table used in unweighted shortest-path computation

For each vertex, we will keep track of three pieces of information. First, we will keep its distance from s in the entry d_v. Initially all vertices are unreachable except for s, whose path length is 0. The entry in p_v is the bookkeeping variable, which will allow us to print the actual paths. The entry *known* is set to **true** after a vertex is processed. Initially, all entries are not *known*, including the start vertex. When a vertex is marked *known*, we have a guarantee that no cheaper path will ever be found, and so processing for that vertex is essentially complete.

The basic algorithm can be described in Figure 9.16. The algorithm in Figure 9.16 mimics the diagrams by declaring as *known* the vertices at distance $d = 0$, then $d = 1$, then $d = 2$, and so on, and setting all the adjacent vertices w that still have $d_w = \infty$ to a distance $d_w = d + 1$.

By tracing back through the p_v variable, the actual path can be printed. We will see how when we discuss the weighted case.

The running time of the algorithm is $O(|V|^2)$, because of the doubly nested **for** loops. An obvious inefficiency is that the outside loop continues until NUM_VERTICES $- 1$, even if all the vertices become *known* much earlier. Although an extra test could be made to avoid this, it does not affect the worst-case running time, as can be seen by generalizing what happens when the input is the graph in Figure 9.17 with start vertex v_9.

We can remove the inefficiency in much the same way as was done for topological sort. At any point in time, there are only two types of *unknown* vertices that have $d_v \neq \infty$. Some have $d_v =$ currDist and the rest have $d_v =$ currDist $+ 1$. Because of this extra structure, it is very wasteful to search through the entire table to find a proper vertex.

A very simple but abstract solution is to keep two boxes. Box #1 will have the unknown vertices with $d_v =$ currDist, and box #2 will have $d_v =$ currDist $+ 1$. The test to find an appropriate vertex v can be replaced by finding any vertex in box #1. After updating w (inside the innermost **if** block), we can add w to box #2. After the outermost **for** loop terminates, box #1 is empty, and box #2 can be transferred to box #1 for the next pass of the **for** loop.

We can refine this idea even further by using just one queue. At the start of the pass, the queue contains only vertices of distance currDist. When we add adjacent vertices of distance currDist $+ 1$, since they enqueue at the rear, we are guaranteed that they will not be processed until after all the vertices of distance currDist have been processed. After the

```
void unweighted( Vertex s )
{
    for each Vertex v
    {
        v.dist = INFINITY;
        v.known = false;
    }

    s.dist = 0;

    for( int currDist = 0; currDist < NUM_VERTICES; currDist++ )
        for each Vertex v
            if( !v.known && v.dist == currDist )
            {
                v.known = true;
                for each Vertex w adjacent to v
                    if( w.dist == INFINITY )
                    {
                        w.dist = currDist + 1;
                        w.path = v;
                    }
            }
}
```

Figure 9.16 Pseudocode for unweighted shortest-path algorithm

Figure 9.17 A bad case for unweighted shortest-path algorithm using Figure 9.16

last vertex at distance currDist dequeues and is processed, the queue only contains vertices of distance currDist + 1, so this process perpetuates. We merely need to begin the process by placing the start node on the queue by itself.

The refined algorithm is shown in Figure 9.18. In the pseudocode, we have assumed that the start vertex, s, is passed as a parameter. Also, it is possible that the queue might empty prematurely, if some vertices are unreachable from the start node. In this case, a distance of INFINITY will be reported for these nodes, which is perfectly reasonable. Finally, the known field is not used; once a vertex is processed it can never enter the queue again, so the fact that it need not be reprocessed is implicitly marked. Thus, the known field can be discarded. Figure 9.19 shows how the values on the graph we have been using are changed during the algorithm. (it includes changes that would occur to known if we had kept it).

Using the same analysis as was performed for topological sort, we see that the running time is $O(|E| + |V|)$, as long as adjacency lists are used.

```
void unweighted( Vertex s )
{
    Queue<Vertex> q = new Queue<Vertex>( );

    for each Vertex v
        v.dist = INFINITY;

    s.dist = 0;
    q.enqueue( s );

    while( !q.isEmpty( ) )
    {
        Vertex v = q.dequeue( );

        for each Vertex w adjacent to v
            if( w.dist == INFINITY )
            {
                w.dist = v.dist + 1;
                w.path = v;
                q.enqueue( w );
            }
    }
}
```

Figure 9.18　Pseudocode for unweighted shortest-path algorithm

9.3.2　Dijkstra's Algorithm

If the graph is weighted, the problem (apparently) becomes harder, but we can still use the ideas from the unweighted case.

We keep all of the same information as before. Thus, each vertex is marked as either *known* or *unknown*. A tentative distance d_v is kept for each vertex, as before. This distance turns out to be the shortest path length from s to v using only *known* vertices as intermediates. As before, we record p_v, which is the last vertex to cause a change to d_v.

The general method to solve the single-source shortest-path problem is known as **Dijkstra's algorithm**. This thirty-year-old solution is a prime example of a **greedy algorithm**. Greedy algorithms generally solve a problem in stages by doing what appears to be the best thing at each stage. For example, to make change in U.S. currency, most people count out the quarters first, then the dimes, nickels, and pennies. This greedy algorithm gives change using the minimum number of coins. The main problem with greedy algorithms is that they do not always work. The addition of a 12-cent piece breaks the coin-changing algorithm for returning 15 cents, because the answer it gives (one 12-cent piece and three pennies) is not optimal (one dime and one nickel).

Dijkstra's algorithm proceeds in stages, just like the unweighted shortest-path algorithm. At each stage, Dijkstra's algorithm selects a vertex v, which has the smallest d_v

	Initial State			v_3 Dequeued			v_1 Dequeued			v_6 Dequeued		
v	known	d_v	p_v	known	d_v	p_v	known	d_v	p_v	known	d_v	p_v
v_1	F	∞	0	F	1	v_3	T	1	v_3	T	1	v_3
v_2	F	∞	0	F	∞	0	F	2	v_1	F	2	v_1
v_3	F	0	0	T	0	0	T	0	0	T	0	0
v_4	F	∞	0	F	∞	0	F	2	v_1	F	2	v_1
v_5	F	∞	0	F	∞	0	F	∞	0	F	∞	0
v_6	F	∞	0	F	1	v_3	F	1	v_3	T	1	v_3
v_7	F	∞	0	F	∞	0	F	∞	0	F	∞	0
Q:	v_3			v_1, v_6			v_6, v_2, v_4			v_2, v_4		

	v_2 Dequeued			v_4 Dequeued			v_5 Dequeued			v_7 Dequeued		
v	known	d_v	p_v	known	d_v	p_v	known	d_v	p_v	known	d_v	p_v
v_1	T	1	v_3	T	1	v_3	T	1	v_3	T	1	v_3
v_2	T	2	v_1	T	2	v_1	T	2	v_1	T	2	v_1
v_3	T	0	0	T	0	0	T	0	0	T	0	0
v_4	F	2	v_1	T	2	v_1	T	2	v_1	T	2	v_1
v_5	F	3	v_2	F	3	v_2	T	3	v_2	T	3	v_2
v_6	T	1	v_3	T	1	v_3	T	1	v_3	T	1	v_3
v_7	F	∞	0	F	3	v_4	F	3	v_4	T	3	v_4
Q:	v_4, v_5			v_5, v_7			v_7			empty		

Figure 9.19 How the data change during the unweighted shortest-path algorithm

among all the *unknown* vertices, and declares that the shortest path from s to v is *known*. The remainder of a stage consists of updating the values of d_w.

In the unweighted case, we set $d_w = d_v + 1$ if $d_w = \infty$. Thus, we essentially lowered the value of d_w if vertex v offered a shorter path. If we apply the same logic to the weighted case, then we should set $d_w = d_v + c_{v,w}$ if this new value for d_w would be an improvement. Put simply, the algorithm decides whether or not it is a good idea to use v on the path to w. The original cost, d_w, is the cost without using v; the cost calculated above is the cheapest path using v (and only *known* vertices).

The graph in Figure 9.20 is our example. Figure 9.21 represents the initial configuration, assuming that the start node, s, is v_1. The first vertex selected is v_1, with path length 0. This vertex is marked *known*. Now that v_1 is *known*, some entries need to be adjusted. The vertices adjacent to v_1 are v_2 and v_4. Both these vertices get their entries adjusted, as indicated in Figure 9.22.

Next, v_4 is selected and marked *known*. Vertices v_3, v_5, v_6, and v_7 are adjacent, and it turns out that all require adjusting, as shown in Figure 9.23.

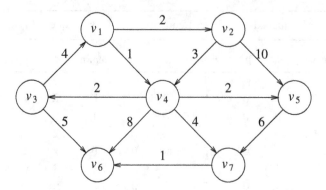

Figure 9.20   The directed graph G (again)

v	known	d_v	p_v
v_1	F	0	0
v_2	F	∞	0
v_3	F	∞	0
v_4	F	∞	0
v_5	F	∞	0
v_6	F	∞	0
v_7	F	∞	0

Figure 9.21   Initial configuration of table used in Dijkstra's algorithm

v	known	d_v	p_v
v_1	T	0	0
v_2	F	2	v_1
v_3	F	∞	0
v_4	F	1	v_1
v_5	F	∞	0
v_6	F	∞	0
v_7	F	∞	0

Figure 9.22   After v_1 is declared *known*

Next, v_2 is selected. v_4 is adjacent but already *known*, so no work is performed on it. v_5 is adjacent but not adjusted, because the cost of going through v_2 is $2 + 10 = 12$ and a path of length 3 is already known. Figure 9.24 shows the table after these vertices are selected.

v	known	d_v	p_v
v_1	T	0	0
v_2	F	2	v_1
v_3	F	3	v_4
v_4	T	1	v_1
v_5	F	3	v_4
v_6	F	9	v_4
v_7	F	5	v_4

Figure 9.23 After v_4 is declared *known*

v	known	d_v	p_v
v_1	T	0	0
v_2	T	2	v_1
v_3	F	3	v_4
v_4	T	1	v_1
v_5	F	3	v_4
v_6	F	9	v_4
v_7	F	5	v_4

Figure 9.24 After v_2 is declared *known*

v	known	d_v	p_v
v_1	T	0	0
v_2	T	2	v_1
v_3	T	3	v_4
v_4	T	1	v_1
v_5	T	3	v_4
v_6	F	8	v_3
v_7	F	5	v_4

Figure 9.25 After v_5 and then v_3 are declared *known*

The next vertex selected is v_5 at cost 3. v_7 is the only adjacent vertex, but it is not adjusted, because $3 + 6 > 5$. Then v_3 is selected, and the distance for v_6 is adjusted down to $3 + 5 = 8$. The resulting table is depicted in Figure 9.25.

Next v_7 is selected; v_6 gets updated down to $5 + 1 = 6$. The resulting table is Figure 9.26.

Finally, v_6 is selected. The final table is shown in Figure 9.27. Figure 9.28 graphically shows how edges are marked *known* and vertices updated during Dijkstra's algorithm.

v	known	d_v	p_v
v_1	T	0	0
v_2	T	2	v_1
v_3	T	3	v_4
v_4	T	1	v_1
v_5	T	3	v_4
v_6	F	6	v_7
v_7	T	5	v_4

Figure 9.26   After v_7 is declared *known*

v	known	d_v	p_v
v_1	T	0	0
v_2	T	2	v_1
v_3	T	3	v_4
v_4	T	1	v_1
v_5	T	3	v_4
v_6	T	6	v_7
v_7	T	5	v_4

Figure 9.27   After v_6 is declared *known* and algorithm terminates

To print out the actual path from a start vertex to some vertex v, we can write a recursive routine to follow the trail left in the p variables.

We now give pseudocode to implement Dijkstra's algorithm. Each **Vertex** stores various data fields that are used in the algorithm. This is shown in Figure 9.29.

The path can be printed out using the recursive routine in Figure 9.30. The routine recursively prints the path all the way up to the vertex before v on the path and then just prints v. This works because the path is simple.

Figure 9.31 shows the main algorithm, which is just a **for** loop to fill up the table using the greedy selection rule.

A proof by contradiction will show that this algorithm always works as long as no edge has a negative cost. If any edge has negative cost, the algorithm could produce the wrong answer (see Exercise 9.7(a)). The running time depends on how the vertices are manipulated, which we have yet to consider. If we use the obvious algorithm of sequentially scanning the vertices to find the minimum d_v, each phase will take $O(|V|)$ time to find the minimum, and thus $O(|V|^2)$ time will be spent finding the minimum over the course of the algorithm. The time for updating d_w is constant per update, and there is at most one update per edge for a total of $O(|E|)$. Thus, the total running time is $O(|E| + |V|^2) = O(|V|^2)$. If the graph is dense, with $|E| = \Theta(|V|^2)$, this algorithm is not only simple but also essentially optimal, since it runs in time linear in the number of edges.

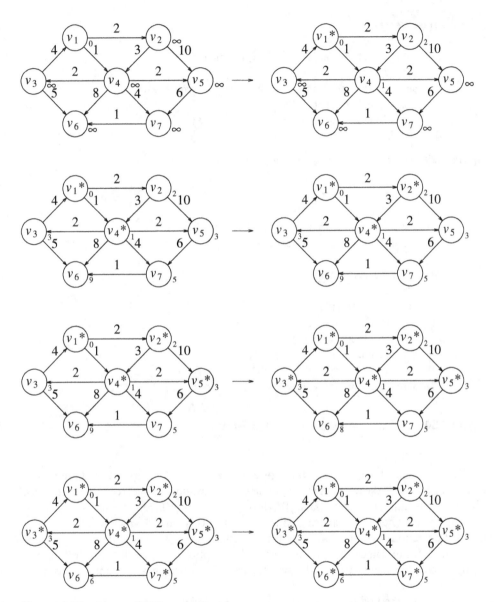

Figure 9.28 Stages of Dijkstra's algorithm

If the graph is sparse, with $|E| = \Theta(|V|)$, this algorithm is too slow. In this case, the distances would need to be kept in a priority queue. There are actually two ways to do this; both are similar.

Selection of the vertex v is a deleteMin operation, since once the unknown minimum vertex is found, it is no longer unknown and must be removed from future consideration. The update of w's distance can be implemented two ways.

```
class Vertex
{
    public List     adj;     // Adjacency list
    public boolean  known;
    public DistType dist;    // DistType is probably int
    public Vertex   path;
        // Other fields and methods as needed
}
```

Figure 9.29   Vertex class for Dijkstra's algorithm

```
/*
 * Print shortest path to v after dijkstra has run.
 * Assume that the path exists.
 */
void printPath( Vertex v )
{
    if( v.path != null )
    {
        printPath( v.path );
        System.out.print( " to " );
    }
    System.out.print( v );
}
```

Figure 9.30   Routine to print the actual shortest path

One way treats the update as a decreaseKey operation. The time to find the minimum is then $O(\log |V|)$, as is the time to perform updates, which amount to decreaseKey operations. This gives a running time of $O(|E| \log |V| + |V| \log |V|) = O(|E| \log |V|)$, an improvement over the previous bound for sparse graphs. Since priority queues do not efficiently support the find operation, the location in the priority queue of each value of d_i will need to be maintained and updated whenever d_i changes in the priority queue. If the priority queue is implemented by a binary heap, this will be messy. If a pairing heap (Chapter 12) is used, the code is not too bad.

An alternate method is to insert w and the new value d_w into the priority queue every time w's distance changes. Thus, there may be more than one representative for each vertex in the priority queue. When the deleteMin operation removes the smallest vertex from the priority queue, it must be checked to make sure that it is not already *known* and, if it is, it is simply ignored and another deleteMin is performed. Although this method is superior from a software point of view, and is certainly much easier to code, the size of the priority queue could get to be as large as $|E|$. This does not affect the asymptotic time bounds, since $|E| \le |V|^2$ implies that $\log |E| \le 2 \log |V|$. Thus, we still get an $O(|E| \log |V|)$ algorithm. However, the space requirement does increase, and this could be important in

```
void dijkstra( Vertex s )
{
    for each Vertex v
    {
        v.dist = INFINITY;
        v.known = false;
    }

    s.dist = 0;

    while( there is an unknown distance vertex )
    {
        Vertex v = smallest unknown distance vertex;

        v.known = true;

        for each Vertex w adjacent to v
            if( !w.known )
            {
                DistType cvw = cost of edge from v to w;

                if( v.dist + cvw < w.dist )
                {
                    // Update w
                    decrease( w.dist to v.dist + cvw );
                    w.path = v;
                }
            }
    }
}
```

Figure 9.31 Pseudocode for Dijkstra's algorithm

some applications. Moreover, because this method requires $|E|$ deleteMins instead of only $|V|$, it is likely to be slower in practice.

Notice that for the typical problems, such as computer mail and mass transit commutes, the graphs are typically very sparse because most vertices have only a couple of edges, so it is important in many applications to use a priority queue to solve this problem.

There are better time bounds possible using Dijkstra's algorithm if different data structures are used. In Chapter 11, we will see another priority queue data structure called the Fibonacci heap. When this is used, the running time is $O(|E| + |V| \log |V|)$. Fibonacci heaps have good theoretical time bounds but a fair amount of overhead, so it is not clear whether using Fibonacci heaps is actually better in practice than Dijkstra's algorithm with binary heaps. To date, there are no meaningful average-case results for this problem.

9.3.3   Graphs with Negative Edge Costs

If the graph has negative edge costs, then Dijkstra's algorithm does not work. The problem is that once a vertex u is declared *known,* it is possible that from some other, *unknown* vertex v there is a path back to u that is very negative. In such a case, taking a path from s to v back to u is better than going from s to u without using v. Exercise 9.7(a) asks you to construct an explicit example.

A tempting solution is to add a constant Δ to each edge cost, thus removing negative edges, calculate a shortest path on the new graph, and then use that result on the original. The naive implementation of this strategy does not work because paths with many edges become more weighty than paths with few edges.

A combination of the weighted and unweighted algorithms will solve the problem, but at the cost of a drastic increase in running time. We forget about the concept of *known* vertices, since our algorithm needs to be able to change its mind. We begin by placing s on a queue. Then, at each stage, we dequeue a vertex v. We find all vertices w adjacent to v such that $d_w > d_v + c_{v,w}$. We update d_w and p_w, and place w on a queue if it is not already there. A bit can be set for each vertex to indicate presence in the queue. We repeat the process until the queue is empty. Figure 9.32 (almost) implements this algorithm.

Although the algorithm works if there are no negative-cost cycles, it is no longer true that the code in the inner for loop is executed once per edge. Each vertex can dequeue at most $|V|$ times, so the running time is $O(|E| \cdot |V|)$ if adjacency lists are used (Exercise 9.7(b)). This is quite an increase from Dijkstra's algorithm, so it is fortunate that, in practice, edge costs are nonnegative. If negative-cost cycles are present, then the algorithm as written will loop indefinitely. By stopping the algorithm after any vertex has dequeued $|V| + 1$ times, we can guarantee termination.

9.3.4   Acyclic Graphs

If the graph is known to be acyclic, we can improve Dijkstra's algorithm by changing the order in which vertices are declared *known,* otherwise known as the vertex selection rule. The new rule is to select vertices in topological order. The algorithm can be done in one pass, since the selections and updates can take place as the topological sort is being performed.

This selection rule works because when a vertex v is selected, its distance, d_v, can no longer be lowered, since by the topological ordering rule it has no incoming edges emanating from *unknown* nodes.

There is no need for a priority queue with this selection rule; the running time is $O(|E| + |V|)$, since the selection takes constant time.

An acyclic graph could model some downhill skiing problem—we want to get from point a to b but can only go downhill, so clearly there are no cycles. Another possible application might be the modeling of (nonreversible) chemical reactions. We could have each vertex represent a particular state of an experiment. Edges would represent a transition from one state to another, and the edge weights might represent the energy released. If only transitions from a higher energy state to a lower are allowed, the graph is acyclic.

A more important use of acyclic graphs is **critical path analysis.** The graph in Figure 9.33 will serve as our example. Each node represents an activity that must be

```
void weightedNegative( Vertex s )
{
    Queue<Vertex> q = new Queue<Vertex>( );

    for each Vertex v
        v.dist = INFINITY;

    s.dist = 0;
    q.enqueue( s );

    while( !q.isEmpty( ) )
    {
        Vertex v = q.dequeue( );

        for each Vertex w adjacent to v
            if( v.dist + cvw < w.dist )
            {
                // Update w
                w.dist = v.dist + cvw;
                w.path = v;
                if( w is not already in q )
                    q.enqueue( w );
            }
    }
}
```

Figure 9.32   Pseudocode for weighted shortest-path algorithm with negative edge costs

performed, along with the time it takes to complete the activity. This graph is thus known as an *activity-node* graph. The edges represent precedence relationships: An edge (v, w) means that activity v must be completed before activity w may begin. Of course, this implies that the graph must be acyclic. We assume that any activities that do not depend (either directly or indirectly) on each other can be performed in parallel by different servers.

This type of a graph could be (and frequently is) used to model construction projects. In this case, there are several important questions which would be of interest to answer. First, what is the earliest completion time for the project? We can see from the graph that 10 time units are required along the path A, C, F, H. Another important question is to determine which activities can be delayed, and by how long, without affecting the minimum completion time. For instance, delaying any of $A, C, F,$ or H would push the completion time past 10 units. On the other hand, activity B is less critical and can be delayed up to two time units without affecting the final completion time.

To perform these calculations, we convert the activity-node graph to an **event-node graph**. Each event corresponds to the completion of an activity and all its dependent activities. Events reachable from a node v in the event-node graph may not commence until after the event v is completed. This graph can be constructed automatically or by hand. Dummy edges and nodes may need to be inserted in the case where an activity depends on

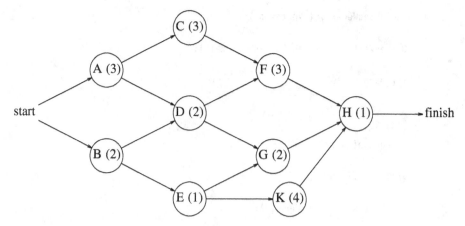

Figure 9.33   Activity-node graph

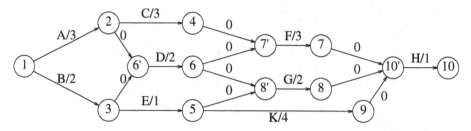

Figure 9.34   Event-node graph

several others. This is necessary in order to avoid introducing false dependencies (or false lack of dependencies). The event-node graph corresponding to the graph in Figure 9.33 is shown in Figure 9.34.

To find the earliest completion time of the project, we merely need to find the length of the *longest* path from the first event to the last event. For general graphs, the longest-path problem generally does not make sense, because of the possibility of **positive-cost cycles**. These are the equivalent of negative-cost cycles in shortest-path problems. If positive-cost cycles are present, we could ask for the longest *simple* path, but no satisfactory solution is known for this problem. Since the event-node graph is acyclic, we need not worry about cycles. In this case, it is easy to adapt the shortest-path algorithm to compute the earliest completion time for all nodes in the graph. If EC_i is the earliest completion time for node i, then the applicable rules are

$$EC_1 = 0$$
$$EC_w = \max_{(v,w)\in E} (EC_v + c_{v,w})$$

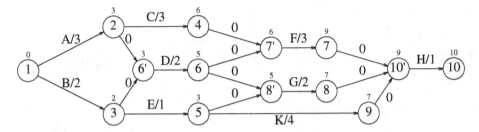

Figure 9.35   Earliest completion times

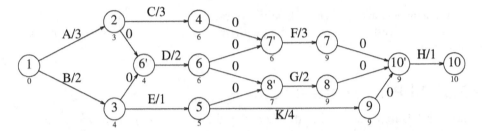

Figure 9.36   Latest completion times

Figure 9.35 shows the earliest completion time for each event in our example event-node graph.

We can also compute the latest time, LC_i, that each event can finish without affecting the final completion time. The formulas to do this are

$$LC_n = EC_n$$
$$LC_v = \min_{(v,w) \in E} (LC_w - c_{v,w})$$

These values can be computed in linear time by maintaining, for each vertex, a list of all adjacent and preceding vertices. The earliest completion times are computed for vertices by their topological order, and the latest completion times are computed by reverse topological order. The latest completion times are shown in Figure 9.36.

The **slack time** for each edge in the event-node graph represents the amount of time that the completion of the corresponding activity can be delayed without delaying the overall completion. It is easy to see that

$$Slack_{(v,w)} = LC_w - EC_v - c_{v,w}$$

Figure 9.37 shows the slack (as the third entry) for each activity in the event-node graph. For each node, the top number is the earliest completion time and the bottom entry is the latest completion time.

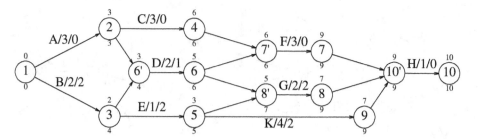

Figure 9.37 Earliest completion time, latest completion time, and slack

Some activities have zero slack. These are critical activities, which must finish on schedule. There is at least one path consisting entirely of zero-slack edges; such a path is a **critical path**.

9.3.5 All-Pairs Shortest Path

Sometimes it is important to find the shortest paths between all pairs of vertices in the graph. Although we could just run the appropriate single-source algorithm $|V|$ times, we might expect a somewhat faster solution, especially on a dense graph, if we compute all the information at once.

In Chapter 10, we will see an $O(|V|^3)$ algorithm to solve this problem for weighted graphs. Although, for dense graphs, this is the same bound as running a simple (non–priority queue) Dijkstra's algorithm $|V|$ times, the loops are so tight that the specialized all-pairs algorithm is likely to be faster in practice. On sparse graphs, of course, it is faster to run $|V|$ Dijkstra's algorithms coded with priority queues.

9.3.6 Shortest-Path Example

In this section we write some Java routines to compute word ladders. In a word ladder each word is formed by changing one character in the ladder's previous word. For instance, we can convert zero to five by a sequence of one-character substitutions as follows: zero hero here hire fire five.

This is an unweighted shortest problem in which each word is a vertex, and two vertices have edges (in both directions) between them if they can be converted to each other with a one-character substitution.

In Section 4.8, we described and wrote a Java routine that would create a Map in which the keys are words, and the values are Lists containing the words that can result from a one-character transformation. As such, this Map represents the graph, in adjacency list format, and we only need to write one routine to run the single-source unweighted shortest-path algorithm and a second routine to output the sequence of words, after the single-source shortest-path algorithm has completed. These two routines are both shown in Figure 9.38.

```
1    // Runs the shortest path calculation from the adjacency map, returns a List
2    // that contains the sequence of word changes to get from first to second.
3    // Returns null if no sequence can be found for any reason.
4    public static List<String>
5    findChain( Map<String,List<String>> adjacentWords, String first, String second )
6    {
7        Map<String,String> previousWord = new HashMap<String,String>( );
8        LinkedList<String> q = new LinkedList<String>( );
9
10       q.addLast( first );
11       while( !q.isEmpty( ) )
12       {
13           String current = q.removeFirst( );
14           List<String> adj = adjacentWords.get( current );
15
16           if( adj != null )
17               for( String adjWord : adj )
18                   if( previousWord.get( adjWord ) == null )
19                   {
20                       previousWord.put( adjWord, current );
21                       q.addLast( adjWord );
22                   }
23       }
24
25       previousWord.put( first, null );
26
27       return getChainFromPreviousMap( previousWord, first, second );
28   }
29
30   // After the shortest path calculation has run, computes the List that
31   // contains the sequence of word changes to get from first to second.
32   // Returns null if there is no path.
33   public static List<String> getChainFromPreviousMap( Map<String,String> prev,
34                                               String first, String second )
35   {
36       LinkedList<String> result = null;
37
38       if( prev.get( second ) != null )
39       {
40           result = new LinkedList<String>( );
41           for( String str = second; str != null; str = prev.get( str ) )
42               result.addFirst( str );
43       }
44
45       return result;
46   }
```

Figure 9.38 Java code to find word ladders

The first routine is findChain, which takes the Map representing the adjacency lists and the two words to be connected and returns a Map in which the keys are words, and the corresponding value is the word prior to the key on the shortest ladder starting at first. In other words, in the example above, if the starting word is zero, the value for key five is fire, the value for key fire is hire, the value for key hire is here, and so on. Clearly this provides enough information for the second routine, getChainFromPreviousMap, which can work its way backward.

findChain is a direct implementation of the pseudocode in Figure 9.18. It assumes that first is a valid word, which is an easily testable condition prior to the call. The basic loop incorrectly assigns a previous entry for first (when the initial word adjacent to first is processed), so at line 25 that entry is repaired.

getChainFromPrevMap uses the prev Map and second, which presumably is a key in the Map and returns the words used to form the word ladder, by working its way backward through prev. By using a LinkedList and inserting at the front, we obtain the word ladder in the correct order.

It is possible to generalize this problem to allow single-character substitutions that include the deletion of a character or the addition of a character. To compute the adjacency list requires only a little more effort: in the last algorithm in Section 4.8, every time a representative for word w in group g is computed, we check if the representative is a word in group $g - 1$. If it is, then the representative is adjacent to w (it is a single-character deletion), and w is adjacent to the representative (it is a single-character addition). It is also possible to assign a cost to a character deletion or insertion (that is higher than a simple substitution), and this yields a weighted shortest-path problem that can be solved with Dijkstra's algorithm.

9.4 Network Flow Problems

Suppose we are given a directed graph $G = (V, E)$ with edge capacities $c_{v,w}$. These capacities could represent the amount of water that could flow through a pipe or the amount of traffic that could flow on a street between two intersections. We have two vertices: s, which we call the **source**, and t, which is the **sink**. Through any edge, (v, w), at most $c_{v,w}$ units of "flow" may pass. At any vertex, v, that is not either s or t, the total flow coming in must equal the total flow going out. The maximum flow problem is to determine the maximum amount of flow that can pass from s to t. As an example, for the graph in Figure 9.39 on the left the maximum flow is 5, as indicated by the graph on the right. Although this example graph is acyclic, this is not a requirement; our (eventual) algorithm will work even if the graph has a cycle.

As required by the problem statement, no edge carries more flow than its capacity. Vertex a has three units of flow coming in, which it distributes to c and d. Vertex d takes three units of flow from a and b and combines this, sending the result to t. A vertex can combine and distribute flow in any manner that it likes, as long as edge capacities are not violated and as long as flow conservation is maintained (what goes in must come out).

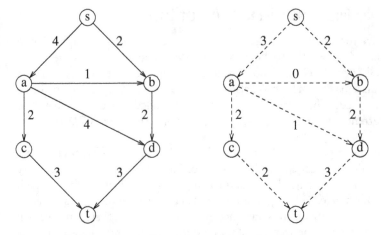

Figure 9.39   A graph (left) and its maximum flow

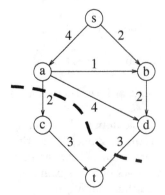

Figure 9.40   A cut in graph *G* partitions the vertices with *s* and *t* in different groups. The total edge cost across the cut is 5, proving that a flow of 5 is maximum

Looking at the graph, we see that *s* has edges of capacities 4 and 2 leaving it, and *t* has edges of capacities 3 and 3 entering it. So perhaps the maximum flow could be 6 instead of 5. However, Figure 9.40 shows how we can prove that the maximum flow is 5. We cut the graph into two parts; one part contains *s* and some other vertices; the other part contains *t*. Since flow must cross through the cut, the total capacity of all edges (u, v) where *u* is in *s*'s partition and *v* is in *t*'s partition is a bound on the maximum flow. These edges are (a, c) and (d, t), with total capacity 5, so the maximum flow cannot exceed 5. Any graph has a large number of cuts; the cut with minimum total capacity provides a bound on the maximum flow, and as it turns out (but it is not immediately obvious), the minimum cut capacity is exactly equal to the maximum flow.

9.4.1   A Simple Maximum-Flow Algorithm

A first attempt to solve the problem proceeds in stages. We start with our graph, G, and construct a flow graph G_f. G_f tells the flow that has been attained at any stage in the algorithm. Initially all edges in G_f have no flow, and we hope that when the algorithm terminates, G_f contains a maximum flow. We also construct a graph, G_r, called the **residual graph**. G_r tells, for each edge, how much more flow can be added. We can calculate this by subtracting the current flow from the capacity for each edge. An edge in G_r is known as a **residual edge**.

At each stage, we find a path in G_r from s to t. This path is known as an **augmenting path**. The minimum edge on this path is the amount of flow that can be added to every edge on this path. We do this by adjusting G_f and recomputing G_r. When we find no path from s to t in G_r, we terminate. This algorithm is nondeterministic, in that we are free to choose *any* path from s to t; obviously some choices are better than others, and we will address this issue later. We will run this algorithm on our example. The graphs below are G, G_f, G_r, respectively. Keep in mind that there is a slight flaw in this algorithm. The initial configuration is in Figure 9.41.

There are many paths from s to t in the residual graph. Suppose we select s, b, d, t. Then we can send two units of flow through every edge on this path. We will adopt the convention that once we have filled (**saturated**) an edge, it is removed from the residual graph. We then obtain Figure 9.42.

Next, we might select the path s, a, c, t, which also allows two units of flow. Making the required adjustments gives the graphs in Figure 9.43.

The only path left to select is s, a, d, t, which allows one unit of flow. The resulting graphs are shown in Figure 9.44.

The algorithm terminates at this point, because t is unreachable from s. The resulting flow of 5 happens to be the maximum. To see what the problem is, suppose that with our initial graph, we chose the path s, a, d, t. This path allows three units of flow and thus seems to be a good choice. The result of this choice, however, leaves only one path from s to t in the residual graph; it allows one more unit of flow, and thus, our algorithm has

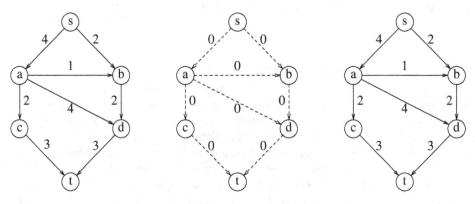

Figure 9.41   Initial stages of the graph, flow graph, and residual graph

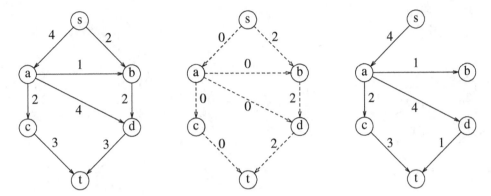

Figure 9.42 G, G_f, G_r after two units of flow added along s, b, d, t

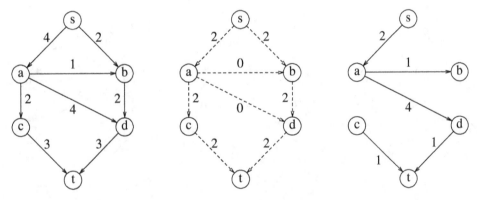

Figure 9.43 G, G_f, G_r after two units of flow added along s, a, c, t

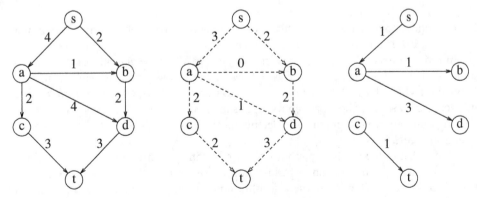

Figure 9.44 G, G_f, G_r after one unit of flow added along s, a, d, t—algorithm terminates

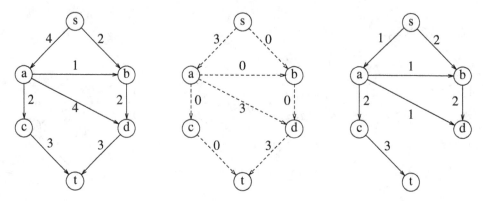

Figure 9.45 G, G_f, G_r if initial action is to add three units of flow along $s, a, d,$ t—algorithm terminates after one more step with suboptimal solution

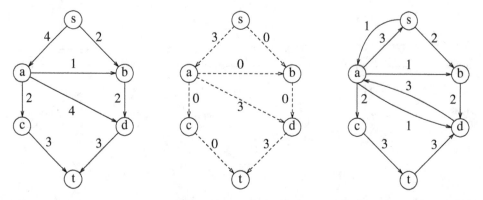

Figure 9.46 Graphs after three units of flow added along s, a, d, t using correct algorithm

failed to find an optimal solution. This is an example of a greedy algorithm that does not work. Figure 9.45 shows why the algorithm fails.

In order to make this algorithm work, we need to allow the algorithm to change its mind. To do this, for every edge (v, w) with flow $f_{v,w}$ in the flow graph, we will add an edge in the residual graph (w, v) of capacity $f_{v,w}$. In effect, we are allowing the algorithm to undo its decisions by sending flow back in the opposite direction. This is best seen by example. Starting from our original graph and selecting the augmenting path s, a, d, t, we obtain the graphs in Figure 9.46.

Notice that in the residual graph, there are edges in both directions between a and d. Either one more unit of flow can be pushed from a to d, or up to three units can be pushed back—we can undo flow. Now the algorithm finds the augmenting path $s, b, d, a, c, t,$ of flow 2. By pushing two units of flow from d to a, the algorithm takes two units of flow away from the edge (a, d) and is essentially changing its mind. Figure 9.47 shows the new graphs.

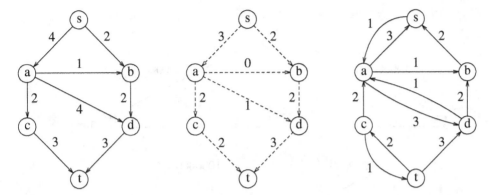

Figure 9.47 Graphs after two units of flow added along s, b, d, a, c, t using correct algorithm

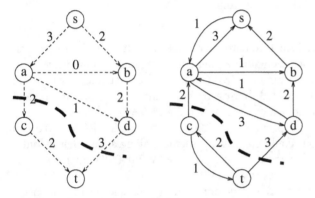

Figure 9.48 The vertices reachable from s in the residual graph form one side of a cut; the unreachables form the other side of the cut.

There is no augmenting path in this graph, so the algorithm terminates. Note that the same result would occur if at Figure 9.46, the augmenting path s, a, c, t was chosen which allows one unit of flow, because then a subsequent augmenting path could be found.

It is easy to see that *if* the algorithm terminates, then it must terminate with a maximum flow. Termination implies that there is no path from s to t in the residual graph. So cut the residual graph, putting the vertices reachable from s on one side, and the unreachables (which include t) on the other side. Figure 9.48 shows the cut. Clearly any edges in the original graph G that cross the cut must be saturated; otherwise, there would be residual flow remaining on one of the edges, which would then imply an edge that crosses the cut (in the wrong disallowed direction) in G_r. But that means that the flow in G is exactly equal to the capacity of a cut in G; hence we have a maximum flow.

If the edge costs in the graph are integers, then the algorithm *must* terminate; each augmentation adds a unit of flow, so we eventually reach the maximum flow, though there

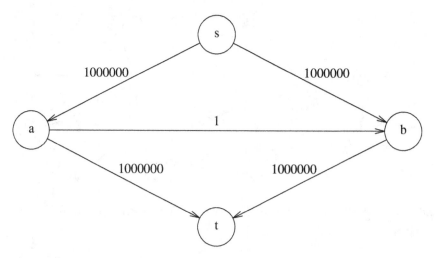

Figure 9.49   The classic bad case for augmenting

is no guarantee that this will be efficient. In particular, if the capacities are all integers and the maximum flow is f, then, since each augmenting path increases the flow value by at least $1, f$ stages suffice, and the total running time is $O(f \cdot |E|)$, since an augmenting path can be found in $O(|E|)$ time by an unweighted shortest-path algorithm. The classic example of why this is a bad running time is shown by the graph in Figure 9.49.

The maximum flow is seen by inspection to be 2,000,000 by sending 1,000,000 down each side. Random augmentations could continually augment along a path that includes the edge connected by a and b. If this were to occur repeatedly, 2,000,000 augmentations would be required, when we could get by with only 2.

A simple method to get around this problem is always to choose the augmenting path that allows the largest increase in flow. Finding such a path is similar to solving a weighted shortest-path problem and a single-line modification to Dijkstra's algorithm will do the trick. If cap_{max} is the maximum edge capacity, then one can show that $O(|E| \log cap_{max})$ augmentations will suffice to find the maximum flow. In this case, since $O(|E| \log |V|)$ time is used for each calculation of an augmenting path, a total bound of $O(|E|^2 \log |V| \log cap_{max})$ is obtained. If the capacities are all small integers, this reduces to $O(|E|^2 \log |V|)$.

Another way to choose augmenting paths is always to take the path with the least number of edges, with the plausible expectation that by choosing a path in this manner, it is less likely that a small, flow-restricting edge will turn up on the path. With this rule, each augmenting step computes the shortest unweighted path from s to t in the residual graph, so assume that each vertex in the graph maintains d_v, representing the shortest-path distance from s to v in the residual graph. Each augmenting step can add new edges into the residual graph, but it is clear that no d_v can decrease, because an edge is added in the opposite direction of an existing shortest path.

Each augmenting step saturates at least one edge. Suppose edge (u, v) is saturated; at that point, u had distance d_u and v had distance $d_v = d_u + 1$; then (u, v) was removed from the residual graph, and edge (v, u) was added. (u, v) cannot reappear in the residual graph again, unless and until (v, u) appears in a future augmenting path. But if it does, then the distance to u at that point must be $d_v + 1$, which would be two higher than at the time (u, v) was previously removed.

This means that each time (u, v) reappears, u's distance goes up by 2. This means that any edge can reappear at most $|V|/2$ times. Each augmentation causes some edge to reappear so the number of augmentations is $O(|E||V|)$. Each step takes $O(|E|)$, due to the unweighted shortest-path calculation, yielding an $O(|E|^2|V|)$ bound on the running time.

Further data structure improvements are possible to this algorithm, and there are several, more complicated, algorithms. A long history of improved bounds has lowered the current best-known bound for this problem. Although no $O(|E||V|)$ algorithm has been reported yet, algorithms with $O(|E||V| \log(|V|^2/|E|))$ and $O(|E||V| + |V|^{2+\varepsilon})$ bounds have been discovered (see the references). There are also a host of very good bounds for special cases. For instance, $O(|E||V|^{1/2})$ time finds a maximum flow in a graph, having the property that all vertices except the source and sink have either a single incoming edge of capacity 1 or a single outgoing edge of capacity 1. These graphs occur in many applications.

The analyses required to produce these bounds are rather intricate, and it is not clear how the worst-case results relate to the running times encountered in practice. A related, even more difficult problem is the **min-cost flow** problem. Each edge has not only a capacity, but also a cost per unit of flow. The problem is to find, among all maximum flows, the one flow of minimum cost. Both of these problems are being actively researched.

9.5 Minimum Spanning Tree

The next problem we will consider is that of finding a **minimum spanning tree** in an undirected graph. The problem makes sense for directed graphs but appears to be more difficult. Informally, a minimum spanning tree of an undirected graph G is a tree formed from graph edges that connects all the vertices of G at lowest total cost. A minimum spanning tree exists if and only if G is connected. Although a robust algorithm should report the case that G is unconnected, we will assume that G is connected and leave the issue of robustness as an exercise to the reader.

In Figure 9.50 the second graph is a minimum spanning tree of the first (it happens to be unique, but this is unusual). Notice that the number of edges in the minimum spanning tree is $|V| - 1$. The minimum spanning tree is a *tree* because it is acyclic, it is *spanning* because it covers every vertex, and it is *minimum* for the obvious reason. If we need to wire a house with a minimum of cable (assuming no other electrical constraints), then a minimum spanning tree problem needs to be solved.

For any spanning tree T, if an edge e that is not in T is added, a cycle is created. The removal of any edge on the cycle reinstates the spanning tree property. The cost of the spanning tree is lowered if e has lower cost than the edge that was removed. If, as a spanning tree is created, the edge that is added is the one of minimum cost that avoids creation of a cycle, then the cost of the resulting spanning tree cannot be improved, because any replacement edge would have cost at least as much as an edge already in the spanning tree.

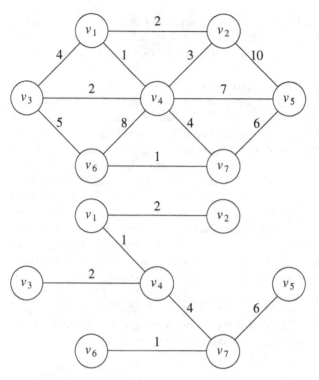

Figure 9.50   A graph G and its minimum spanning tree

This shows that greed works for the minimum spanning tree problem. The two algorithms we present differ in how a minimum edge is selected.

9.5.1   Prim's Algorithm

One way to compute a minimum spanning tree is to grow the tree in successive stages. In each stage, one node is picked as the root, and we add an edge, and thus an associated vertex, to the tree.

At any point in the algorithm, we can see that we have a set of vertices that have already been included in the tree; the rest of the vertices have not. The algorithm then finds, at each stage, a new vertex to add to the tree by choosing the edge (u, v) such that the cost of (u, v) is the smallest among all edges where u is in the tree and v is not. Figure 9.51 shows how this algorithm would build the minimum spanning tree, starting from v_1. Initially, v_1 is in the tree as a root with no edges. Each step adds one edge and one vertex to the tree.

We can see that Prim's algorithm is essentially identical to Dijkstra's algorithm for shortest paths. As before, for each vertex we keep values d_v and p_v and an indication of whether it is *known* or *unknown*. d_v is the weight of the shortest edge connecting v to a *known* vertex, and p_v, as before, is the last vertex to cause a change in d_v. The rest of the algorithm is exactly the same, with the exception that since the definition of d_v is different, so is the

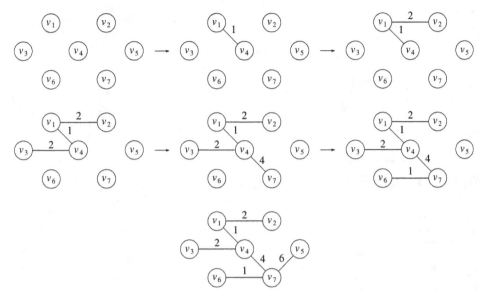

Figure 9.51 Prim's algorithm after each stage

v	known	d_v	p_v
v_1	F	0	0
v_2	F	∞	0
v_3	F	∞	0
v_4	F	∞	0
v_5	F	∞	0
v_6	F	∞	0
v_7	F	∞	0

Figure 9.52 Initial configuration of table used in Prim's algorithm

update rule. For this problem, the update rule is even simpler than before: After a vertex v is selected, for each *unknown* w adjacent to v, $d_w = \min(d_w, c_{w,v})$.

The initial configuration of the table is shown in Figure 9.52. v_1 is selected, and v_2, v_3, and v_4 are updated. The table resulting from this is shown in Figure 9.53. The next vertex selected is v_4. Every vertex is adjacent to v_4. v_1 is not examined, because it is *known*. v_2 is unchanged, because it has $d_v = 2$ and the edge cost from v_4 to v_2 is 3; all the rest are updated. Figure 9.54 shows the resulting table. The next vertex chosen is v_2 (arbitrarily breaking a tie). This does not affect any distances. Then v_3 is chosen, which affects the distance in v_6, producing Figure 9.55. Figure 9.56 results from the selection of v_7, which forces v_6 and v_5 to be adjusted. v_6 and then v_5 are selected, completing the algorithm.

v	known	d_v	p_v
v_1	T	0	0
v_2	F	2	v_1
v_3	F	4	v_1
v_4	F	1	v_1
v_5	F	∞	0
v_6	F	∞	0
v_7	F	∞	0

Figure 9.53 The table after v_1 is declared *known*

v	known	d_v	p_v
v_1	T	0	0
v_2	F	2	v_1
v_3	F	2	v_4
v_4	T	1	v_1
v_5	F	7	v_4
v_6	F	8	v_4
v_7	F	4	v_4

Figure 9.54 The table after v_4 is declared *known*

v	known	d_v	p_v
v_1	T	0	0
v_2	T	2	v_1
v_3	T	2	v_4
v_4	T	1	v_1
v_5	F	7	v_4
v_6	F	5	v_3
v_7	F	4	v_4

Figure 9.55 The table after v_2 and then v_3 are declared *known*

The final table is shown in Figure 9.57. The edges in the spanning tree can be read from the table: (v_2, v_1), (v_3, v_4), (v_4, v_1), (v_5, v_7), (v_6, v_7), (v_7, v_4). The total cost is 16.

The entire implementation of this algorithm is virtually identical to that of Dijkstra's algorithm, and everything that was said about the analysis of Dijkstra's algorithm applies here. Be aware that Prim's algorithm runs on undirected graphs, so when coding it,

v	known	d_v	p_v
v_1	T	0	0
v_2	T	2	v_1
v_3	T	2	v_4
v_4	T	1	v_1
v_5	F	6	v_7
v_6	F	1	v_7
v_7	T	4	v_4

Figure 9.56   The table after v_7 is declared *known*

v	known	d_v	p_v
v_1	T	0	0
v_2	T	2	v_1
v_3	T	2	v_4
v_4	T	1	v_1
v_5	T	6	v_7
v_6	T	1	v_7
v_7	T	4	v_4

Figure 9.57   The table after v_6 and v_5 are selected (Prim's algorithm terminates)

remember to put every edge in two adjacency lists. The running time is $O(|V|^2)$ without heaps, which is optimal for dense graphs, and $O(|E| \log |V|)$ using binary heaps, which is good for sparse graphs.

9.5.2   Kruskal's Algorithm

A second greedy strategy is continually to select the edges in order of smallest weight and accept an edge if it does not cause a cycle. The action of the algorithm on the graph in the preceding example is shown in Figure 9.58.

Formally, Kruskal's algorithm maintains a forest—a collection of trees. Initially, there are $|V|$ single-node trees. Adding an edge merges two trees into one. When the algorithm terminates, there is only one tree, and this is the minimum spanning tree. Figure 9.59 shows the order in which edges are added to the forest.

The algorithm terminates when enough edges are accepted. It turns out to be simple to decide whether edge (u, v) should be accepted or rejected. The appropriate data structure is the union/find algorithm from Chapter 8.

The invariant we will use is that at any point in the process, two vertices belong to the same set if and only if they are connected in the current spanning forest. Thus, each

Edge	Weight	Action
(v_1, v_4)	1	Accepted
(v_6, v_7)	1	Accepted
(v_1, v_2)	2	Accepted
(v_3, v_4)	2	Accepted
(v_2, v_4)	3	Rejected
(v_1, v_3)	4	Rejected
(v_4, v_7)	4	Accepted
(v_3, v_6)	5	Rejected
(v_5, v_7)	6	Accepted

Figure 9.58   Action of Kruskal's algorithm on G

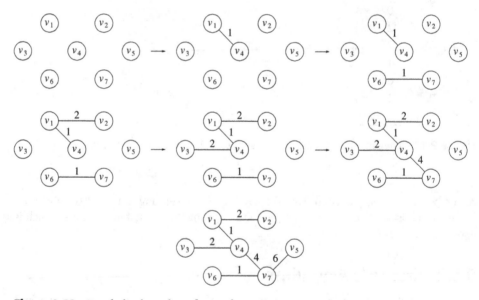

Figure 9.59   Kruskal's algorithm after each stage

vertex is initially in its own set. If u and v are in the same set, the edge is rejected, because since they are already connected, adding (u, v) would form a cycle. Otherwise, the edge is accepted, and a `union` is performed on the two sets containing u and v. It is easy to see that this maintains the set invariant, because once the edge (u, v) is added to the spanning forest, if w was connected to u and x was connected to v, then x and w must now be connected, and thus belong in the same set.

The edges could be sorted to facilitate the selection, but building a heap in linear time is a much better idea. Then `deleteMins` give the edges to be tested in order. Typically, only a small fraction of the edges need to be tested before the algorithm can terminate, although

```
ArrayList<Edge> kruskal( List<Edge> edges, int numVertices )
{
    DisjSets ds = new DisjSets( numVertices );
    PriorityQueue<Edge> pq = new PriorityQueue<>( edges );
    List<Edge> mst = new ArrayList<>( );

    while( mst.size( ) != numVertices - 1 )
    {
        Edge e = pq.deleteMin( );          // Edge e = (u, v)
        SetType uset = ds.find( e.getu( ) );
        SetType vset = ds.find( e.getv( ) );

        if( uset != vset )
        {
            // Accept the edge
            mst.add( e );
            ds.union( uset, vset );
        }
    }
    return mst;
}
```

Figure 9.60 Pseudocode for Kruskal's algorithm

it is always possible that all the edges must be tried. For instance, if there was an extra vertex v_8 and edge (v_5, v_8) of cost 100, all the edges would have to be examined. Method kruskal in Figure 9.60 finds a minimum spanning tree.

The worst-case running time of this algorithm is $O(|E| \log |E|)$, which is dominated by the heap operations. Notice that since $|E| = O(|V|^2)$, this running time is actually $O(|E| \log |V|)$. In practice, the algorithm is much faster than this time bound would indicate.

9.6 Applications of Depth-First Search

Depth-first search is a generalization of preorder traversal. Starting at some vertex, v, we process v and then recursively traverse all vertices adjacent to v. If this process is performed on a tree, then all tree vertices are systematically visited in a total of $O(|E|)$ time, since $|E| = \Theta(|V|)$. If we perform this process on an arbitrary graph, we need to be careful to avoid cycles. To do this, when we visit a vertex v, we *mark* it visited, since now we have been there, and recursively call depth-first search on all adjacent vertices that are not already marked. We implicitly assume that for undirected graphs every edge (v, w) appears twice in the adjacency lists: once as (v, w) and once as (w, v). The procedure in Figure 9.61 performs a depth-first search (and does absolutely nothing else) and is a template for the general style.

```
void dfs( Vertex v )
{
    v.visited = true;
    for each Vertex w adjacent to v
        if( !w.visited )
            dfs( w );
}
```

Figure 9.61   Template for depth-first search (pseudocode)

For each vertex, the field visited is initialized to false. By recursively calling the procedures only on nodes that have not been visited, we guarantee that we do not loop indefinitely. If the graph is undirected and not connected, or directed and not strongly connected, this strategy might fail to visit some nodes. We then search for an unmarked node, apply a depth-first traversal there, and continue this process until there are no unmarked nodes.[2] Because this strategy guarantees that each edge is encountered only once, the total time to perform the traversal is $O(|E| + |V|)$, as long as adjacency lists are used.

9.6.1   Undirected Graphs

An undirected graph is connected if and only if a depth-first search starting from any node visits every node. Because this test is so easy to apply, we will assume that the graphs we deal with are connected. If they are not, then we can find all the connected components and apply our algorithm on each of these in turn.

As an example of depth-first search, suppose in the graph of Figure 9.62 we start at vertex A. Then we mark A as visited and call dfs(B) recursively. dfs(B) marks B as visited and calls dfs(C) recursively. dfs(C) marks C as visited and calls dfs(D) recursively. dfs(D) sees both A and B, but both of these are marked, so no recursive calls are made. dfs(D) also sees that C is adjacent but marked, so no recursive call is made there, and dfs(D) returns back to dfs(C). dfs(C) sees B adjacent, ignores it, finds a previously unseen vertex E adjacent, and thus calls dfs(E). dfs(E) marks E, ignores A and C, and returns to dfs(C). dfs(C) returns to dfs(B). dfs(B) ignores both A and D and returns. dfs(A) ignores both D and E and returns. (We have actually touched every edge twice, once as (v, w) and again as (w, v), but this is really once per adjacency list entry.)

We graphically illustrate these steps with a **depth-first spanning tree**. The root of the tree is A, the first vertex visited. Each edge (v, w) in the graph is present in the tree. If, when we process (v, w), we find that w is unmarked, or if, when we process (w, v), we find that v is unmarked, we indicate this with a tree edge. If, when we process (v, w), we find that w is already marked, and when processing (w, v), we find that v is already marked, we draw

---

[2] An efficient way of implementing this is to begin the depth-first search at v_1. If we need to restart the depth-first search, we examine the sequence $v_k, v_{k+1}, \ldots$ for an unmarked vertex, where v_{k-1} is the vertex where the last depth-first search was started. This guarantees that throughout the algorithm, only $O(|V|)$ is spent looking for vertices where new depth-first search trees can be started.

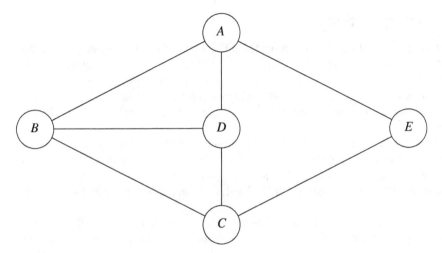

Figure 9.62 An undirected graph

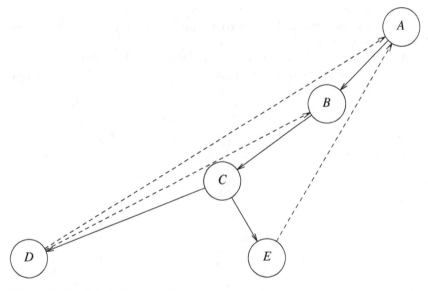

Figure 9.63 Depth-first search of previous graph

a dashed line, which we will call a **back edge**, to indicate that this "edge" is not really part of the tree. The depth-first search of the graph in Figure 9.62 is shown in Figure 9.63.

The tree will simulate the traversal we performed. A preorder numbering of the tree, using only tree edges, tells us the order in which the vertices were marked. If the graph is not connected, then processing all nodes (and edges) requires several calls to dfs, and each generates a tree. This entire collection is a **depth-first spanning forest**.

9.6.2 Biconnectivity

A connected undirected graph is **biconnected** if there are no vertices whose removal disconnects the rest of the graph. The graph in the example above is biconnected. If the nodes are computers and the edges are links, then if any computer goes down, network mail is unaffected, except, of course, at the down computer. Similarly, if a mass transit system is biconnected, users always have an alternate route should some terminal be disrupted.

If a graph is not biconnected, the vertices whose removal would disconnect the graph are known as **articulation points**. These nodes are critical in many applications. The graph in Figure 9.64 is not biconnected: C and D are articulation points. The removal of C would disconnect G, and the removal of D would disconnect E and F, from the rest of the graph.

Depth-first search provides a linear-time algorithm to find all articulation points in a connected graph. First, starting at any vertex, we perform a depth-first search and number the nodes as they are visited. For each vertex v, we call this preorder number $Num(v)$. Then, for every vertex v in the depth-first search spanning tree, we compute the lowest-numbered vertex, which we call $Low(v)$, that is reachable from v by taking zero or more tree edges and then possibly one back edge (in that order). The depth-first search tree in Figure 9.65 shows the preorder number first, and then the lowest-numbered vertex reachable under the rule described above.

The lowest-numbered vertex reachable by A, B, and C is vertex 1 (A), because they can all take tree edges to D and then one back edge back to A. We can efficiently compute Low

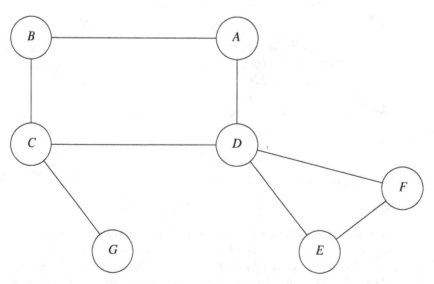

Figure 9.64   A graph with articulation points C and D

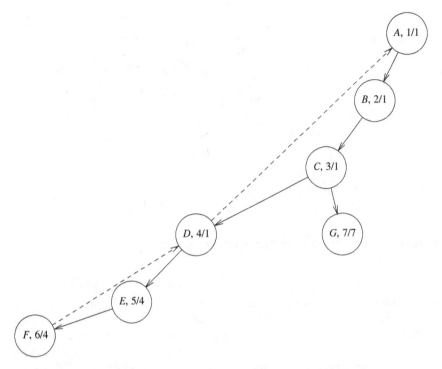

Figure 9.65 Depth-first tree for previous graph, with *Num* and *Low*

by performing a postorder traversal of the depth-first spanning tree. By the definition of *Low*, *Low*(*v*) is the minimum of

1. *Num*(*v*)
2. the lowest *Num*(*w*) among all back edges (*v*, *w*)
3. the lowest *Low*(*w*) among all tree edges (*v*, *w*)

The first condition is the option of taking no edges, the second way is to choose no tree edges and a back edge, and the third way is to choose some tree edges and possibly a back edge. This third method is succinctly described with a recursive call. Since we need to evaluate *Low* for all the children of *v* before we can evaluate *Low*(*v*), this is a postorder traversal. For any edge (*v*, *w*), we can tell whether it is a tree edge or a back edge merely by checking *Num*(*v*) and *Num*(*w*). Thus, it is easy to compute *Low*(*v*): We merely scan down *v*'s adjacency list, apply the proper rule, and keep track of the minimum. Doing all the computation takes $O(|E| + |V|)$ time.

All that is left to do is to use this information to find articulation points. The root is an articulation point if and only if it has more than one child, because if it has two children, removing the root disconnects nodes in different subtrees, and if it has only one child,

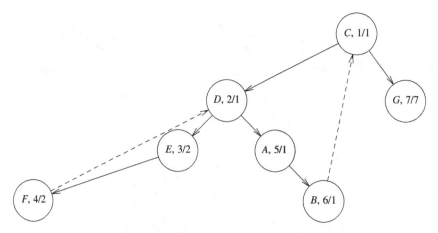

Figure 9.66   Depth-first tree that results if depth-first search starts at *C*

removing the root merely disconnects the root. Any other vertex *v* is an articulation point if and only if *v* has some child *w* such that $Low(w) \geq Num(v)$. Notice that this condition is always satisfied at the root; hence the need for a special test.

The *if* part of the proof is clear when we examine the articulation points that the algorithm determines, namely, *C* and *D*. *D* has a child *E*, and $Low(E) \geq Num(D)$, since both are 4. Thus, there is only one way for *E* to get to any node above *D*, and that is by going through *D*. Similarly, *C* is an articulation point, because $Low(G) \geq Num(C)$. To prove that this algorithm is correct, one must show that the *only if* part of the assertion is true (that is, this finds *all* articulation points). We leave this as an exercise. As a second example, we show (Figure 9.66) the result of applying this algorithm on the same graph, starting the depth-first search at *C*.

We close by giving pseudocode to implement this algorithm. We will assume that Vertex contains the data fields visited (initialized to false), num, low, and parent. We will also keep a (Graph) class variable called counter, which is initialized to 1, to assign the preorder traversal numbers, num. We also leave out the easily implemented test for the root.

As we have already stated, this algorithm can be implemented by performing a preorder traversal to compute *Num* and then a postorder traversal to compute *Low*. A third traversal can be used to check which vertices satisfy the articulation point criteria. Performing three traversals, however, would be a waste. The first pass is shown in Figure 9.67.

The second and third passes, which are postorder traversals, can be implemented by the code in Figure 9.68. The last if statement handles a special case. If *w* is adjacent to *v*, then the recursive call to *w* will find *v* adjacent to *w*. This is not a back edge, only an edge that has already been considered and needs to be ignored. Otherwise, the procedure computes the minimum of the various low and num entries, as specified by the algorithm.

There is no rule that a traversal must be either preorder or postorder. It is possible to do processing both before and after the recursive calls. The procedure in Figure 9.69 combines the two routines assignNum and assignLow in a straightforward manner to produce the procedure findArt.

```
// Assign Num and compute parents
void assignNum( Vertex v )
{
    v.num = counter++;
    v.visited = true;
    for each Vertex w adjacent to v
        if( !w.visited )
        {
            w.parent = v;
            assignNum( w );
        }
}
```

Figure 9.67   Routine to assign *Num* to vertices (pseudocode)

```
// Assign low; also check for articulation points.
void assignLow( Vertex v )
{
    v.low = v.num;  // Rule 1
    for each Vertex w adjacent to v
    {
        if( w.num > v.num )  // Forward edge
        {
            assignLow( w );
            if( w.low >= v.num )
                System.out.println( v + " is an articulation point" );
            v.low = min( v.low, w.low );  // Rule 3
        }
        else
        if( v.parent != w )  // Back edge
            v.low = min( v.low, w.num );  // Rule 2
    }
}
```

Figure 9.68   Pseudocode to compute *Low* and to test for articulation points (test for the root is omitted)

9.6.3 Euler Circuits

Consider the three figures in Figure 9.70. A popular puzzle is to reconstruct these figures using a pen, drawing each line exactly once. The pen may not be lifted from the paper while the drawing is being performed. As an extra challenge, make the pen finish at the same point at which it started. This puzzle has a surprisingly simple solution. Stop reading if you would like to try to solve it.

```
void findArt( Vertex v )
{
    v.visited = true;
    v.low = v.num = counter++;  // Rule 1
    for each Vertex w adjacent to v
    {
        if( !w.visited )  // Forward edge
        {
            w.parent = v;
            findArt( w );
            if( w.low >= v.num )
                System.out.println( v + " is an articulation point" );
            v.low = min( v.low, w.low );  // Rule 3
        }
        else
        if( v.parent != w )  // Back edge
            v.low = min( v.low, w.num );  // Rule 2
    }
}
```

Figure 9.69 Testing for articulation points in one depth-first search (test for the root is omitted) (pseudocode)

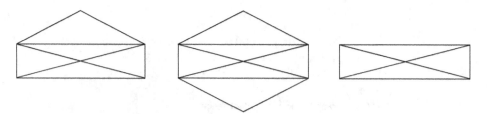

Figure 9.70 Three drawings

The first figure can be drawn only if the starting point is the lower left- or right-hand corner, and it is not possible to finish at the starting point. The second figure is easily drawn with the finishing point the same as the starting point, but the third figure cannot be drawn at all within the parameters of the puzzle.

We can convert this problem to a graph theory problem by assigning a vertex to each intersection. Then the edges can be assigned in the natural manner, as in Figure 9.71.

After this conversion is performed, we must find a path in the graph that visits every edge exactly once. If we are to solve the "extra challenge," then we must find a cycle that visits every edge exactly once. This graph problem was solved in 1736 by Euler and marked the beginning of graph theory. The problem is thus commonly referred to as an **Euler path** (sometimes **Euler tour**) or **Euler circuit problem**, depending on the specific problem

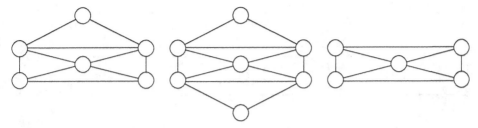

Figure 9.71   Conversion of puzzle to graph

statement. The Euler tour and Euler circuit problems, though slightly different, have the same basic solution. Thus, we will consider the Euler circuit problem in this section.

The first observation that can be made is that an Euler circuit, which must end on its starting vertex, is possible only if the graph is connected and each vertex has an even degree (number of edges). This is because, on the Euler circuit, a vertex is entered and then left. If any vertex v has odd degree, then eventually we will reach the point where only one edge into v is unvisited, and taking it will strand us at v. If exactly two vertices have odd degree, an Euler tour, which must visit every edge but need not return to its starting vertex, is still possible if we start at one of the odd-degree vertices and finish at the other. If more than two vertices have odd degree, then an Euler tour is not possible.

The observations of the preceding paragraph provide us with a necessary condition for the existence of an Euler circuit. It does not, however, tell us that all connected graphs that satisfy this property must have an Euler circuit, nor does it give us guidance on how to find one. It turns out that the necessary condition is also sufficient. That is, any connected graph, all of whose vertices have even degree, must have an Euler circuit. Furthermore, a circuit can be found in linear time.

We can assume that we know that an Euler circuit exists, since we can test the necessary and sufficient condition in linear time. Then the basic algorithm is to perform a depth-first search. There are a surprisingly large number of "obvious" solutions that do not work. Some of these are presented in the exercises.

The main problem is that we might visit a portion of the graph and return to the starting point prematurely. If all the edges coming out of the start vertex have been used up, then part of the graph is untraversed. The easiest way to fix this is to find the first vertex on this path that has an untraversed edge, and perform another depth-first search. This will give another circuit, which can be spliced into the original. This is continued until all edges have been traversed.

As an example, consider the graph in Figure 9.72. It is easily seen that this graph has an Euler circuit. Suppose we start at vertex 5, and traverse the circuit 5, 4, 10, 5. Then we are stuck, and most of the graph is still untraversed. The situation is shown in Figure 9.73.

We then continue from vertex 4, which still has unexplored edges. A depth-first search might come up with the path 4, 1, 3, 7, 4, 11, 10, 7, 9, 3, 4. If we splice this path into the previous path of 5, 4, 10, 5, then we get a new path of 5, 4, 1, 3, 7, 4, 11, 10, 7, 9, 3, 4, 10, 5.

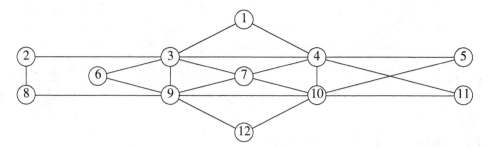

Figure 9.72   Graph for Euler circuit problem

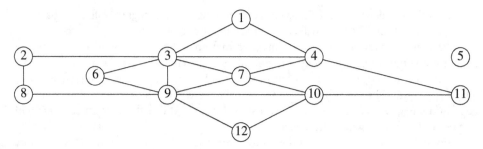

Figure 9.73   Graph remaining after 5, 4, 10, 5

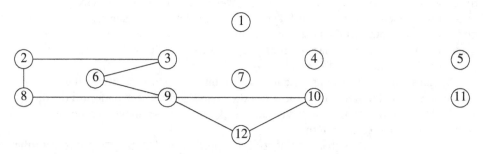

Figure 9.74   Graph after the path 5, 4, 1, 3, 7, 4, 11, 10, 7, 9, 3, 4, 10, 5

The graph that remains after this is shown in Figure 9.74. Notice that in this graph all the vertices must have even degree, so we are guaranteed to find a cycle to add. The remaining graph might not be connected, but this is not important. The next vertex on the path that has untraversed edges is vertex 3. A possible circuit would then be 3, 2, 8, 9, 6, 3. When spliced in, this gives the path 5, 4, 1, 3, 2, 8, 9, 6, 3, 7, 4, 11, 10, 7, 9, 3, 4, 10, 5.

The graph that remains is in Figure 9.75. On this path, the next vertex with an untraversed edge is 9, and the algorithm finds the circuit 9, 12, 10, 9. When this is added to the current path, a circuit of 5, 4, 1, 3, 2, 8, 9, 12, 10, 9, 6, 3, 7, 4, 11, 10, 7, 9, 3, 4, 10, 5 is obtained. As all the edges are traversed, the algorithm terminates with an Euler circuit.

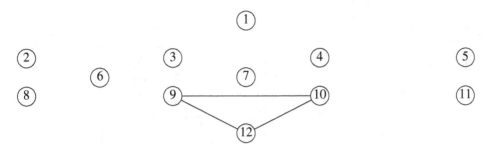

Figure 9.75   Graph remaining after the path 5, 4, 1, 3, 2, 8, 9, 6, 3, 7, 4, 11, 10, 7, 9, 3, 4, 10, 5

To make this algorithm efficient, we must use appropriate data structures. We will sketch some of the ideas, leaving the implementation as an exercise. To make splicing simple, the path should be maintained as a linked list. To avoid repetitious scanning of adjacency lists, we must maintain, for each adjacency list, the last edge scanned. When a path is spliced in, the search for a new vertex from which to perform the next depth-first search must begin at the start of the splice point. This guarantees that the total work performed on the vertex search phase is $O(|E|)$ during the entire life of the algorithm. With the appropriate data structures, the running time of the algorithm is $O(|E| + |V|)$.

A very similar problem is to find a simple cycle, in an undirected graph, that visits every vertex. This is known as the **Hamiltonian cycle problem**. Although it seems almost identical to the Euler circuit problem, no efficient algorithm for it is known. We shall see this problem again in Section 9.7.

9.6.4 Directed Graphs

Using the same strategy as with undirected graphs, directed graphs can be traversed in linear time, using depth-first search. If the graph is not strongly connected, a depth-first search starting at some node might not visit all nodes. In this case we repeatedly perform depth-first searches, starting at some unmarked node, until all vertices have been visited. As an example, consider the directed graph in Figure 9.76.

We arbitrarily start the depth-first search at vertex B. This visits vertices B, C, A, D, E, and F. We then restart at some unvisited vertex. Arbitrarily, we start at H, which visits J and I. Finally, we start at G, which is the last vertex that needs to be visited. The corresponding depth-first search tree is shown in Figure 9.77.

The dashed arrows in the depth-first spanning forest are edges (v, w) for which w was already marked at the time of consideration. In undirected graphs, these are always back edges, but, as we can see, there are three types of edges that do not lead to new vertices. First, there are **back edges**, such as (A, B) and (I, H). There are also **forward edges**, such as (C, D) and (C, E), that lead from a tree node to a descendant. Finally, there are **cross edges**, such as (F, C) and (G, F), which connect two tree nodes that are not directly related. Depth-first search forests are generally drawn with children and new trees added to the forest from

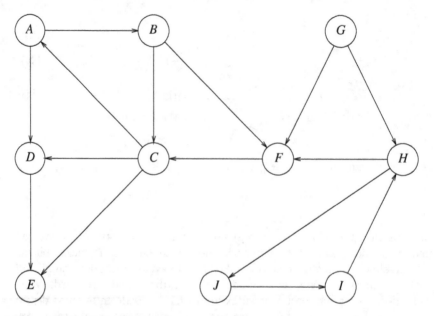

Figure 9.76   A directed graph

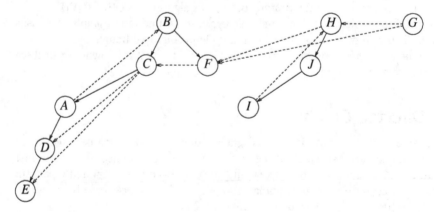

Figure 9.77   Depth-first search of previous graph

left to right. In a depth-first search of a directed graph drawn in this manner, cross edges always go from right to left.

Some algorithms that use depth-first search need to distinguish between the three types of nontree edges. This is easy to check as the depth-first search is being performed, and it is left as an exercise.

One use of depth-first search is to test whether or not a directed graph is acyclic. The rule is that a directed graph is acyclic if and only if it has no back edges. (The graph above has back edges, and thus is not acyclic.) The reader may remember that a topological sort can also be used to determine whether a graph is acyclic. Another way to perform topological sorting is to assign the vertices topological numbers $N, N - 1, \ldots, 1$ by postorder traversal of the depth-first spanning forest. As long as the graph is acyclic, this ordering will be consistent.

9.6.5 Finding Strong Components

By performing two depth-first searches, we can test whether a directed graph is strongly connected, and if it is not, we can actually produce the subsets of vertices that are strongly connected to themselves. This can also be done in only one depth-first search, but the method used here is much simpler to understand.

First, a depth-first search is performed on the input graph G. The vertices of G are numbered by a postorder traversal of the depth-first spanning forest, and then all edges in G are reversed, forming G_r. The graph in Figure 9.78 represents G_r for the graph G shown in Figure 9.76; the vertices are shown with their numbers.

The algorithm is completed by performing a depth-first search on G_r, always starting a new depth-first search at the highest-numbered vertex. Thus, we begin the depth-first search of G_r at vertex G, which is numbered 10. This leads nowhere, so the next search is started at H. This call visits I and J. The next call starts at B and visits A, C, and F. The next calls after this are dfs(D) and finally dfs(E). The resulting depth-first spanning forest is shown in Figure 9.79.

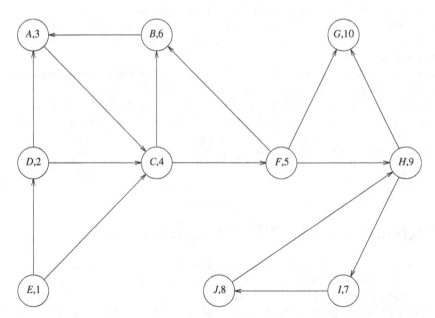

Figure 9.78 G_r numbered by postorder traversal of G (from Figure 9.76)

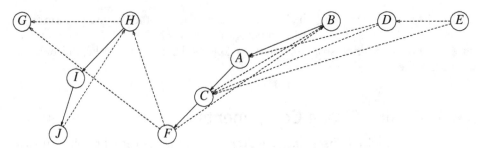

Figure 9.79 Depth-first search of G_r—strong components are $\{G\}$, $\{H, I, J\}$, $\{B, A, C, F\}$, $\{D\}$, $\{E\}$

Each of the trees (this is easier to see if you completely ignore all nontree edges) in this depth-first spanning forest forms a strongly connected component. Thus, for our example, the strongly connected components are $\{G\}$, $\{H, I, J\}$, $\{B, A, C, F\}$, $\{D\}$, and $\{E\}$.

To see why this algorithm works, first note that if two vertices v and w are in the same strongly connected component, then there are paths from v to w and from w to v in the original graph G, and hence also in G_r. Now, if two vertices v and w are not in the same depth-first spanning tree of G_r, clearly they cannot be in the same strongly connected component.

To prove that this algorithm works, we must show that if two vertices v and w are in the same depth-first spanning tree of G_r, there must be paths from v to w and from w to v. Equivalently, we can show that if x is the root of the depth-first spanning tree of G_r containing v, then there is a path from x to v and from v to x. Applying the same logic to w would then give a path from x to w and from w to x. These paths would imply paths from v to w and w to v (going through x).

Since v is a descendant of x in G_r's depth-first spanning tree, there is a path from x to v in G_r and thus a path from v to x in G. Furthermore, since x is the root, x has the higher postorder number from the first depth-first search. Therefore, during the first depth-first search, all the work processing v was completed before the work at x was completed. Since there is a path from v to x, it follows that v must be a descendant of x in the spanning tree for G—otherwise v would finish *after* x. This implies a path from x to v in G and completes the proof.

9.7 Introduction to NP-Completeness

In this chapter, we have seen solutions to a wide variety of graph theory problems. All these problems have polynomial running times, and with the exception of the network flow problem, the running time is either linear or only slightly more than linear ($O(|E| \log |E|)$). We have also mentioned, in passing, that for some problems certain variations seem harder than the original.

Recall that the Euler circuit problem, which finds a path that touches every edge exactly once, is solvable in linear time. The Hamiltonian cycle problem asks for a simple cycle that contains every vertex. No linear algorithm is known for this problem.

The single-source unweighted shortest-path problem for directed graphs is also solvable in linear time. No linear-time algorithm is known for the corresponding longest-simple-path problem.

The situation for these problem variations is actually much worse than we have described. Not only are no linear algorithms known for these variations, but there are no known algorithms that are guaranteed to run in polynomial time. The best known algorithms for these problems could take exponential time on some inputs.

In this section we will take a brief look at this problem. This topic is rather complex, so we will only take a quick and informal look at it. Because of this, the discussion may be (necessarily) somewhat imprecise in places.

We will see that there are a host of important problems that are roughly equivalent in complexity. These problems form a class called the *NP-complete* problems. The exact complexity of these *NP*-complete problems has yet to be determined and remains the foremost open problem in theoretical computer science. Either all these problems have polynomial-time solutions or none of them do.

9.7.1 Easy vs. Hard

When classifying problems, the first step is to examine the boundaries. We have already seen that many problems can be solved in linear time. We have also seen some $O(\log N)$ running times, but these either assume some preprocessing (such as input already being read or a data structure already being built) or occur on arithmetic examples. For instance, the *gcd* algorithm, when applied on two numbers M and N, takes $O(\log N)$ time. Since the numbers consist of $\log M$ and $\log N$ bits respectively, the *gcd* algorithm is really taking time that is linear in the *amount* or *size* of input. Thus, when we measure running time, we will be concerned with the running time as a function of the amount of input. Generally, we cannot expect better than linear running time.

At the other end of the spectrum lie some truly hard problems. These problems are so hard that they are *impossible*. This does not mean the typical exasperated moan, which means that it would take a genius to solve the problem. Just as real numbers are not sufficient to express a solution to $x^2 < 0$, one can prove that computers cannot solve every problem that happens to come along. These "impossible" problems are called **undecidable problems**.

One particular undecidable problem is the **halting problem**. Is it possible to have your Java compiler have an extra feature that not only detects syntax errors, but also all infinite loops? This seems like a hard problem, but one might expect that if some very clever programmers spent enough time on it, they could produce this enhancement.

The intuitive reason that this problem is undecidable is that such a program might have a hard time checking itself. For this reason, these problems are sometimes called **recursively undecidable.**

If an infinite loop–checking program could be written, surely it could be used to check itself. We could then produce a program called *LOOP. LOOP* takes as input a program

P and runs *P* on itself. It prints out the phrase *YES* if *P* loops when run on itself. If *P* terminates when run on itself, a natural thing to do would be to print out *NO*. Instead of doing that, we will have *LOOP* go into an infinite loop.

What happens when *LOOP* is given itself as input? Either *LOOP* halts, or it does not halt. The problem is that both these possibilities lead to contradictions, in much the same way as does the phrase "This sentence is a lie."

By our definition, *LOOP*(*P*) goes into an infinite loop if *P*(*P*) terminates. Suppose that when *P* = *LOOP*, *P*(*P*) terminates. Then, according to the *LOOP* program, *LOOP*(*P*) is obligated to go into an infinite loop. Thus, we must have *LOOP*(*LOOP*) terminating *and* entering an infinite loop, which is clearly not possible. On the other hand, suppose that when *P* = *LOOP*, *P*(*P*) enters an infinite loop. Then *LOOP*(*P*) must terminate, and we arrive at the same set of contradictions. Thus, we see that the program *LOOP* cannot possibly exist.

9.7.2 The Class NP

A few steps down from the horrors of undecidable problems lies the class *NP*. *NP* stands for **nondeterministic polynomial-time**. A deterministic machine, at each point in time, is executing an instruction. Depending on the instruction, it then goes to some next instruction, which is unique. A nondeterministic machine has a choice of next steps. It is free to choose any that it wishes, and if one of these steps leads to a solution, it will always choose the correct one. A nondeterministic machine thus has the power of extremely good (optimal) guessing. This probably seems like a ridiculous model, since nobody could possibly build a nondeterministic computer, and because it would seem to be an incredible upgrade to your standard computer (every problem might now seem trivial). We will see that nondeterminism is a very useful theoretical construct. Furthermore, nondeterminism is not as powerful as one might think. For instance, undecidable problems are still undecidable, even if nondeterminism is allowed.

A simple way to check if a problem is in *NP* is to phrase the problem as a yes/no question. The problem is in *NP* if, in polynomial time, we can prove that any "yes" instance is correct. We do not have to worry about "no" instances, since the program always makes the right choice. Thus, for the Hamiltonian cycle problem, a "yes" instance would be any simple circuit in the graph that includes all the vertices. This is in *NP*, since, given the path, it is a simple matter to check that it is really a Hamiltonian cycle. Appropriately phrased questions, such as "Is there a simple path of length > *K*?" can also easily be checked and are in *NP*. Any path that satisfies this property can be checked trivially.

The class *NP* includes all problems that have polynomial-time solutions, since obviously the solution provides a check. One would expect that since it is so much easier to check an answer than to come up with one from scratch, there would be problems in *NP* that do not have polynomial-time solutions. To date no such problem has been found, so it is entirely possible, though not considered likely by experts, that nondeterminism is not such an important improvement. The problem is that proving exponential lower bounds is an extremely difficult task. The information theory bound technique, which we used to show that sorting requires $\Omega(N \log N)$ comparisons, does not seem to be adequate for the task, because the decision trees are not nearly large enough.

Notice also that not all decidable problems are in *NP*. Consider the problem of determining whether a graph *does not* have a Hamiltonian cycle. To prove that a graph has a Hamiltonian cycle is a relatively simple matter—we just need to exhibit one. Nobody knows how to show, in polynomial time, that a graph does not have a Hamiltonian cycle. It seems that one must enumerate all the cycles and check them one by one. Thus the Non–Hamiltonian cycle problem is not known to be in *NP*.

9.7.3 NP-Complete Problems

Among all the problems known to be in *NP*, there is a subset, known as the **NP-complete problems**, which contains the hardest. An *NP*-complete problem has the property that any problem in *NP* can be **polynomially reduced** to it.

A problem P_1 can be reduced to P_2 as follows: Provide a mapping so that any instance of P_1 can be transformed to an instance of P_2. Solve P_2, and then map the answer back to the original. As an example, numbers are entered into a pocket calculator in decimal. The decimal numbers are converted to binary, and all calculations are performed in binary. Then the final answer is converted back to decimal for display. For P_1 to be polynomially reducible to P_2, all the work associated with the transformations must be performed in polynomial time.

The reason that *NP*-complete problems are the hardest *NP* problems is that a problem that is *NP*-complete can essentially be used as a subroutine for *any* problem in *NP*, with only a polynomial amount of overhead. Thus, if any *NP*-complete problem has a polynomial-time solution, then *every* problem in *NP* must have a polynomial-time solution. This makes the *NP*-complete problems the hardest of all *NP* problems.

Suppose we have an *NP*-complete problem P_1. Suppose P_2 is known to be in *NP*. Suppose further that P_1 polynomially reduces to P_2, so that we can solve P_1 by using P_2 with only a polynomial time penalty. Since P_1 is *NP*-complete, every problem in *NP* polynomially reduces to P_1. By applying the closure property of polynomials, we see that every problem in *NP* is polynomially reducible to P_2: We reduce the problem to P_1 and then reduce P_1 to P_2. Thus, P_2 is *NP*-complete.

As an example, suppose that we already know that the Hamiltonian cycle problem is *NP*-complete. The **traveling salesman problem** is as follows.

Traveling Salesman Problem.
Given a complete graph $G = (V, E)$, with edge costs, and an integer K, is there a simple cycle that visits all vertices and has total cost $\leq K$?

The problem is different from the Hamiltonian cycle problem, because all $|V|(|V|-1)/2$ edges are present and the graph is weighted. This problem has many important applications. For instance, printed circuit boards need to have holes punched so that chips, resistors, and other electronic components can be placed. This is done mechanically. Punching the hole is a quick operation; the time-consuming step is positioning the hole puncher. The time required for positioning depends on the distance traveled from hole to hole. Since we would like to punch every hole (and then return to the start for the next

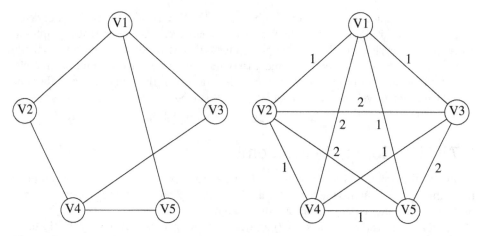

Figure 9.80   Hamiltonian cycle problem transformed to traveling salesman problem

board), and minimize the total amount of time spent traveling, what we have is a traveling salesman problem.

The traveling salesman problem is *NP*-complete. It is easy to see that a solution can be checked in polynomial time, so it is certainly in *NP*. To show that it is *NP*-complete, we polynomially reduce the Hamiltonian cycle problem to it. To do this we construct a new graph G'. G' has the same vertices as G. For G', each edge (v, w) has a weight of 1 if $(v, w) \in G$, and 2 otherwise. We choose $K = |V|$. See Figure 9.80.

It is easy to verify that G has a Hamiltonian cycle if and only if G' has a traveling salesman tour of total weight $|V|$.

There is now a long list of problems known to be *NP*-complete. To prove that some new problem is *NP*-complete, it must be shown to be in *NP*, and then an appropriate *NP*-complete problem must be transformed into it. Although the transformation to a traveling salesman problem was rather straightforward, most transformations are actually quite involved and require some tricky constructions. Generally, several different *NP*-complete problems are considered before the problem that actually provides the reduction. As we are only interested in the general ideas, we will not show any more transformations; the interested reader can consult the references.

The alert reader may be wondering how the first *NP*-complete problem was actually proven to be *NP*-complete. Since proving that a problem is *NP*-complete requires transforming it from another *NP*-complete problem, there must be some *NP*-complete problem for which this strategy will not work. The first problem that was proven to be *NP*-complete was the satisfiability problem. The **satisfiability problem** takes as input a Boolean expression and asks whether the expression has an assignment to the variables that gives a value of **true**.

Satisfiability is certainly in *NP*, since it is easy to evaluate a Boolean expression and check whether the result is true. In 1971, Cook showed that satisfiability was *NP*-complete by directly proving that all problems that are in *NP* could be transformed to satisfiability. To do this, he used the one known fact about every problem in *NP*: Every problem in *NP* can be solved in polynomial time by a nondeterministic computer. The formal model for a computer is known as a **Turing machine**. Cook showed how the actions of this machine could be simulated by an extremely complicated and long, but still polynomial, Boolean formula. This Boolean formula would be true if and only if the program which was being run by the Turing machine produced a "yes" answer for its input.

Once satisfiability was shown to be *NP*-complete, a host of new *NP*-complete problems, including some of the most classic problems, were also shown to be *NP*-complete.

In addition to the satisfiability, Hamiltonian circuit, traveling salesman, and longest-path problems, which we have already examined, some of the more well-known *NP*-complete problems which we have not discussed are *bin packing, knapsack, graph coloring,* and *clique.* The list is quite extensive and includes problems from operating systems (scheduling and security), database systems, operations research, logic, and especially graph theory.

Summary

In this chapter we have seen how graphs can be used to model many real-life problems. Many of the graphs that occur are typically very sparse, so it is important to pay attention to the data structures that are used to implement them.

We have also seen a class of problems that do not seem to have efficient solutions. In Chapter 10, some techniques for dealing with these problems will be discussed.

Exercises

9.1   Find a topological ordering for the graph in Figure 9.81.

9.2   If a stack is used instead of a queue for the topological sort algorithm in Section 9.2, does a different ordering result? Why might one data structure give a "better" answer?

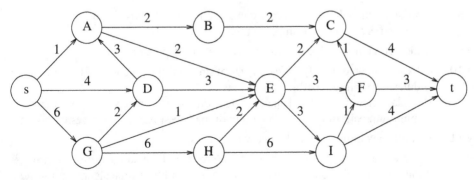

Figure 9.81   Graph used in Exercises 9.1 and 9.11

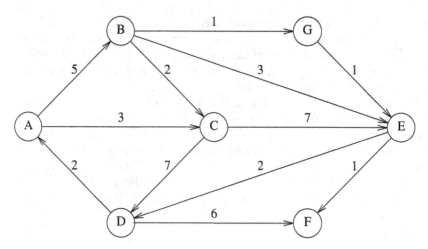

Figure 9.82   Graph used in Exercise 9.5

9.3   Write a program to perform a topological sort on a graph.

9.4   An adjacency matrix requires $O(|V|^2)$ merely to initialize using a standard double loop. Propose a method that stores a graph in an adjacency matrix (so that testing for the existence of an edge is $O(1)$) but avoids the quadratic running time.

9.5   a. Find the shortest path from A to all other vertices for the graph in Figure 9.82.
b. Find the shortest unweighted path from B to all other vertices for the graph in Figure 9.82.

9.6   What is the worst-case running time of Dijkstra's algorithm when implemented with d-heaps (Section 6.5)?

9.7   a. Give an example where Dijkstra's algorithm gives the wrong answer in the presence of a negative edge but no negative-cost cycle.
**b. Show that the weighted shortest-path algorithm suggested in Section 9.3.3 works if there are negative-weight edges, but no negative-cost cycles, and that the running time of this algorithm is $O(|E| \cdot |V|)$.

*9.8   Suppose all the edge weights in a graph are integers between 1 and $|E|$. How fast can Dijkstra's algorithm be implemented?

9.9   Write a program to solve the single-source shortest-path problem.

9.10   a. Explain how to modify Dijkstra's algorithm to produce a count of the number of different minimum paths from v to w.
b. Explain how to modify Dijkstra's algorithm so that if there is more than one minimum path from v to w, a path with the fewest number of edges is chosen.

9.11   Find the maximum flow in the network of Figure 9.81.

9.12   Suppose that $G = (V, E)$ is a tree, s is the root, and we add a vertex t and edges of infinite capacity from all leaves in G to t. Give a linear-time algorithm to find a maximum flow from s to t.

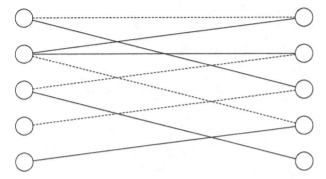

Figure 9.83 A bipartite graph

9.13 A bipartite graph, $G = (V, E)$, is a graph such that V can be partitioned into two subsets V_1 and V_2 and no edge has both its vertices in the same subset.
   a. Give a linear algorithm to determine whether a graph is bipartite.
   b. The bipartite matching problem is to find the largest subset E' of E such that no vertex is included in more than one edge. A matching of four edges (indicated by dashed edges) is shown in Figure 9.83. There is a matching of five edges, which is maximum.
   Show how the bipartite matching problem can be used to solve the following problem: We have a set of instructors, a set of courses, and a list of courses that each instructor is qualified to teach. If no instructor is required to teach more than one course, and only one instructor may teach a given course, what is the maximum number of courses that can be offered?
   c. Show that the network flow problem can be used to solve the bipartite matching problem.
   d. What is the time complexity of your solution to part (b)?

*9.14 a. Give an algorithm to find an augmenting path that permits the maximum flow.
   b. Let f be the amount of flow remaining in the residual graph. Show that the augmenting path produced by the algorithm in part (a) admits a path of capacity $f/|E|$.
   c. Show that after $|E|$ consecutive iterations, the total flow remaining in the residual graph is reduced from f to at most f/e, where $e \approx 2.71828$.
   d. Show that $|E| \ln f$ iterations suffice to produce the maximum flow.

9.15 a. Find a minimum spanning tree for the graph in Figure 9.84 using both Prim's and Kruskal's algorithms.
   b. Is this minimum spanning tree unique? Why?

9.16 Does either Prim's or Kruskal's algorithm work if there are negative edge weights?

9.17 Show that a graph of V vertices can have V^{V-2} minimum spanning trees.

9.18 Write a program to implement Kruskal's algorithm.

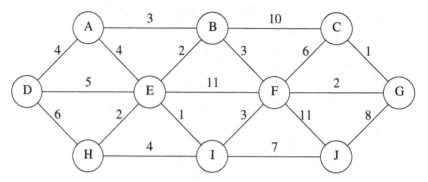

Figure 9.84   Graph used in Exercise 9.15

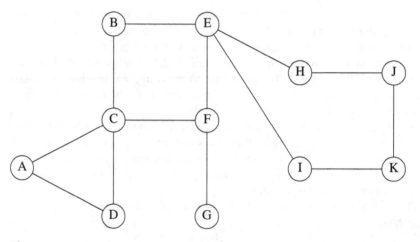

Figure 9.85   Graph used in Exercise 9.21

9.19   If all the edges in a graph have weights between 1 and $|E|$, how fast can the minimum spanning tree be computed?

9.20   Give an algorithm to find a *maximum* spanning tree. Is this harder than finding a minimum spanning tree?

9.21   Find all the articulation points in the graph in Figure 9.85. Show the depth-first spanning tree and the values of *Num* and *Low* for each vertex.

9.22   Prove that the algorithm to find articulation points works.

9.23   a. Give an algorithm to find the minimum number of edges that need to be removed from an undirected graph so that the resulting graph is acyclic.
    *b. Show that this problem is *NP*-complete for directed graphs.

9.24   Prove that in a depth-first spanning forest of a directed graph, all cross edges go from right to left.

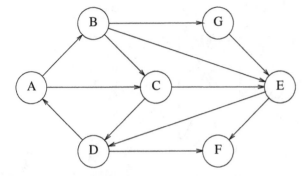

Figure 9.86 Graph used in Exercise 9.26

9.25 Give an algorithm to decide whether an edge (v, w) in a depth-first spanning forest of a directed graph is a tree, back, cross, or forward edge.

9.26 Find the strongly connected components in the graph of Figure 9.86.

9.27 Write a program to find the strongly connected components in a digraph.

⋆9.28 Give an algorithm that finds the strongly connected components in only one depth-first search. Use an algorithm similar to the biconnectivity algorithm.

9.29 The *biconnected components* of a graph G is a partition of the edges into sets such that the graph formed by each set of edges is biconnected. Modify the algorithm in Figure 9.69 to find the biconnected components instead of the articulation points.

9.30 Suppose we perform a breadth-first search of an undirected graph and build a breadth-first spanning tree. Show that all edges in the tree are either tree edges or cross edges.

9.31 Give an algorithm to find in an undirected (connected) graph a path that goes through every edge exactly once in each direction.

9.32 a. Write a program to find an Euler circuit in a graph if one exists.
  b. Write a program to find an Euler tour in a graph if one exists.

9.33 An Euler circuit in a directed graph is a cycle in which every edge is visited exactly once.
  ⋆a. Prove that a directed graph has an Euler circuit if and only if it is strongly connected and every vertex has equal indegree and outdegree.
  ⋆b. Give a linear-time algorithm to find an Euler circuit in a directed graph where one exists.

9.34 a. Consider the following solution to the Euler circuit problem: Assume that the graph is biconnected. Perform a depth-first search, taking back edges only as a last resort. If the graph is not biconnected, apply the algorithm recursively on the biconnected components. Does this algorithm work?
  b. Suppose that when taking back edges, we take the back edge to the nearest ancestor. Does the algorithm work?

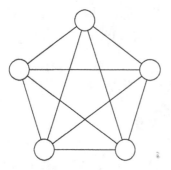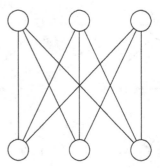

Figure 9.87   Graph used in Exercise 9.35

9.35 A planar graph is a graph that can be drawn in a plane without any two edges intersecting.
  *a. Show that neither of the graphs in Figure 9.87 is planar.
  b. Show that in a planar graph, there must exist some vertex which is connected to no more than five nodes.
  **c. Show that in a planar graph, $|E| \le 3|V| - 6$.

9.36 A *multigraph* is a graph in which multiple edges are allowed between pairs of vertices. Which of the algorithms in this chapter work without modification for multigraphs? What modifications need to be done for the others?

*9.37 Let $G = (V, E)$ be an undirected graph. Use depth-first search to design a linear algorithm to convert each edge in G to a directed edge such that the resulting graph is strongly connected, or determine that this is not possible.

9.38 You are given a set of N sticks, which are lying on top of each other in some configuration. Each stick is specified by its two endpoints; each endpoint is an ordered triple giving its x, y, and z coordinates; no stick is vertical. A stick may be picked up only if there is no stick on top of it.
  a. Explain how to write a routine that takes two sticks a and b and reports whether a is above, below, or unrelated to b. (This has nothing to do with graph theory.)
  b. Give an algorithm that determines whether it is possible to pick up all the sticks, and if so, provides a sequence of stick pickups that accomplishes this.

9.39 A graph is k-colorable if each vertex can be given one of k colors, and no edge connects identically colored vertices. Give a linear-time algorithm to test a graph for two-colorability. Assume graphs are stored in adjacency list format; you must specify any additional data structures that are needed.

9.40 Give a polynomial-time algorithm that finds $\lceil V/2 \rceil$ vertices that collectively cover at least three-fourths (3/4) of the edges in an arbitrary undirected graph.

9.41 Show how to modify the topological sort algorithm so that if the graph is not acyclic, the algorithm will print out some cycle. You may not use depth-first search.

9.42   Let G be a directed graph with N vertices. A vertex s is called a **sink** if, for every v in V such that $s \neq v$, there is an edge (v, s), and there are no edges of the form (s, v). Give an $O(N)$ algorithm to determine whether or not G has a sink, assuming that G is given by its $N \times N$ adjacency matrix.

9.43   When a vertex and its incident edges are removed from a tree, a collection of sub-trees remains. Give a linear-time algorithm that finds a vertex whose removal from an N vertex tree leaves no subtree with more than $N/2$ vertices.

9.44   Give a linear-time algorithm to determine the longest unweighted path in an acyclic undirected graph (that is, a tree).

9.45   Consider an N-by-N grid in which some squares are occupied by black circles. Two squares belong to the same group if they share a common edge. In Figure 9.88, there is one group of four occupied squares, three groups of two occupied squares, and two individual occupied squares. Assume that the grid is represented by a two-dimensional array. Write a program that does the following:
a.   Computes the size of a group when a square in the group is given.
b.   Computes the number of different groups.
c.   Lists all groups.

9.46   Section 8.7 described the generating of mazes. Suppose we want to output the path in the maze. Assume that the maze is represented as a matrix; each cell in the matrix stores information about what walls are present (or absent).
a.   Write a program that computes enough information to output a path in the maze. Give output in the form SEN... (representing go south, then east, then north, etc.).
b.   Write a program that draws the maze and, at the press of a button, draws the path.

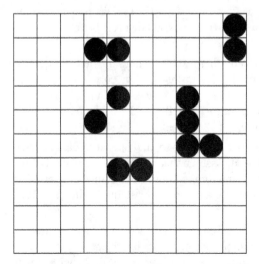

Figure 9.88   Grid for Exercise 9.45

9.47   Suppose that walls in the maze can be knocked down, with a penalty of *P* squares. *P* is specified as a parameter to the algorithm. (If the penalty is 0, then the problem is trivial.) Describe an algorithm to solve this version of the problem. What is the running time of your algorithm?

9.48   Suppose that the maze may or may not have a solution.
   a. Describe a linear-time algorithm that determines the minimum number of walls that need to be knocked down to create a solution. (*Hint:* Use a double-ended queue.)
   b. Describe an algorithm (not necessarily linear-time) that finds a shortest path after knocking down the minimum number of walls. Note that the solution to part (a) would give no information about which walls would be the best to knock down. (*Hint:* Use Exercise 9.47.)

9.49   Write a program to compute word ladders where single-character substitutions have a cost of 1, and single-character additions or deletions have a cost of $p > 0$, specified by the user. As mentioned at the end of Section 9.3.6, this is essentially a weighted shortest-path problem.

Explain how each of the following problems (Exercises 9.50–9.53) can be solved by applying a shortest-path algorithm. Then design a mechanism for representing an input, and write a program that solves the problem.

9.50   The input is a list of league game scores (and there are no ties). If all teams have at least one win and a loss, we can generally prove, by a silly transitivity argument, that any team is better than any other. For instance, in the six-team league where everyone plays three games, suppose we have the following results: *A* beat *B* and *C*; *B* beat *C* and *F*; *C* beat *D*; *D* beat *E*; *E* beat *A*; *F* beat *D* and *E*. Then we can prove that *A* is better than *F*, because *A* beat *B*, who in turn beat *F*. Similarly, we can prove that *F* is better than *A* because *F* beat *E* and *E* beat *A*. Given a list of game scores and two teams *X* and *Y*, either find a proof (if one exists) that *X* is better than *Y*, or indicate that no proof of this form can be found.

9.51   The input is a collection of currencies and their exchange rates. Is there a sequence of exchanges that makes money instantly? For instance, if the currencies are *X*, *Y*, and *Z* and the exchange rate is 1 *X* equals 2 *Y*s, 1 *Y* equals 2 *Z*s, and 1 *X* equals 3 *Z*s, then 300 *Z*s will buy 100 *X*s, which in turn will buy 200 *Y*s, which in turn will buy 400 *Z*s. We have thus made a profit of 33 percent.

9.52   A student needs to take a certain number of courses to graduate, and these courses have prerequisites that must be followed. Assume that all courses are offered every semester and that the student can take an unlimited number of courses. Given a list of courses and their prerequisites, compute a schedule that requires the minimum number of semesters.

9.53   The object of the *Kevin Bacon Game* is to link a movie actor to Kevin Bacon via shared movie roles. The minimum number of links is an actor's *Bacon number.* For instance, Tom Hanks has a Bacon number of 1; he was in *Apollo 13* with Kevin Bacon. Sally Field has a Bacon number of 2, because she was in *Forrest Gump* with

Tom Hanks, who was in *Apollo 13* with Kevin Bacon. Almost all well-known actors have a Bacon number of 1 or 2. Assume that you have a comprehensive list of actors, with roles,[3] and do the following:

a. Explain how to find an actor's Bacon number.

b. Explain how to find the actor with the highest Bacon number.

c. Explain how to find the minimum number of links between two arbitrary actors.

9.54 The *clique* problem can be stated as follows: Given an undirected graph $G = (V, E)$ and an integer K, does G contain a complete subgraph of at least K vertices?

The *vertex cover* problem can be stated as follows: Given an undirected graph $G = (V, E)$ and an integer K, does G contain a subset $V' \subset V$ such that $|V'| \leq K$ and every edge in G has a vertex in V'? Show that the clique problem is polynomially reducible to vertex cover.

9.55 Assume that the Hamiltonian cycle problem is NP-complete for undirected graphs.

a. Prove that the Hamiltonian cycle problem is NP-complete for directed graphs.

b. Prove that the unweighted simple longest-path problem is NP-complete for directed graphs.

9.56 The *baseball card collector* problem is as follows: Given packets $P_1, P_2, \ldots, P_M$, each of which contains a subset of the year's baseball cards, and an integer K, is it possible to collect all the baseball cards by choosing $\leq K$ packets? Show that the baseball card collector problem is NP-complete.

References

Good graph theory textbooks include [9], [14], [24], and [39]. More advanced topics, including the more careful attention to running times, are covered in [41], [43], and [50].

Use of adjacency lists was advocated in [26]. The topological sort algorithm is from [31], as described in [36]. Dijkstra's algorithm appeared in [10]. The improvements using *d*-heaps and Fibonacci heaps are described in [30] and [16], respectively. The shortest-path algorithm with negative edge weights is due to Bellman [3]; Tarjan [50] describes a more efficient way to guarantee termination.

Ford and Fulkerson's seminal work on network flow is [15]. The idea of augmenting along shortest paths or on paths admitting the largest flow increase is from [13]. Other approaches to the problem can be found in [11], [34], [23], [7], [35], and [22]. An algorithm for the min-cost flow problem can be found in [20].

An early minimum spanning tree algorithm can be found in [4]. Prim's algorithm is from [44]; Kruskal's algorithm appears in [37]. Two $O(|E| \log \log |V|)$ algorithms are [6] and [51]. The theoretically best-known algorithms appear in [16], [18], [32] and [5].

---

[3] For instance, see the *Internet Movie Database* files `actors.list.gz` and `actresses.list.gz` at `ftp://ftp.fu-berlin.de/pub/misc/movies/database`.

An empirical study of these algorithms suggests that Prim's algorithm, implemented with `decreaseKey`, is best in practice on most graphs [42].

The algorithm for biconnectivity is from [46]. The first linear-time strong components algorithm (Exercise 9.28) appears in the same paper. The algorithm presented in the text is due to Kosaraju (unpublished) and Sharir [45]. Other applications of depth-first search appear in [27], [28], [47], and [48] (as mentioned in Chapter 8, the results in [47] and [48] have been improved, but the basic algorithm is unchanged).

The classic reference work for the theory of NP-complete problems is [21]. Additional material can be found in [1]. The NP-completeness of satisfiability is shown in [8]. The other seminal paper is [33], which showed the NP-completeness of 21 problems. An excellent survey of complexity theory is [49]. An approximation algorithm for the traveling salesman problem, which generally gives nearly optimal results, can be found in [40].

A solution to Exercise 9.8 can be found in [2]. Solutions to the bipartite matching problem in Exercise 9.13 can be found in [25] and [38]. The problem can be generalized by adding weights to the edges and removing the restriction that the graph is bipartite. Efficient solutions for the unweighted matching problem for general graphs are quite complex. Details can be found in [12], [17], and [19].

Exercise 9.35 deals with planar graphs, which commonly arise in practice. Planar graphs are very sparse, and many difficult problems are easier on planar graphs. An example is the graph isomorphism problem, which is solvable in linear time for planar graphs [29]. No polynomial time algorithm is known for general graphs.

1. A. V. Aho, J. E. Hopcroft, and J. D. Ullman, *The Design and Analysis of Computer Algorithms,* Addison-Wesley, Reading, Mass., 1974.

2. R. K. Ahuja, K. Melhorn, J. B. Orlin, and R. E. Tarjan, "Faster Algorithms for the Shortest Path Problem," *Journal of the ACM,* 37 (1990), 213–223.

3. R. E. Bellman, "On a Routing Problem," *Quarterly of Applied Mathematics,* 16 (1958), 87–90.

4. O. Borůvka, "Ojistém problému minimálním (On a Minimal Problem)," *Práca Moravské Přirodo-vědecké Společnosti,* 3 (1926), 37–58.

5. B. Chazelle, "A Minimum Spanning Tree Algorithm with Inverse-Ackermann Type Complexity," *Journal of the ACM,* 47 (2000), 1028–1047.

6. D. Cheriton and R. E. Tarjan, "Finding Minimum Spanning Trees," *SIAM Journal on Computing,* 5 (1976), 724–742.

7. J. Cheriyan and T. Hagerup, "A Randomized Maximum-Flow Algorithm," *SIAM Journal on Computing,* 24 (1995), 203–226.

8. S. Cook, "The Complexity of Theorem Proving Procedures," *Proceedings of the Third Annual ACM Symposium on Theory of Computing* (1971), 151–158.

9. N. Deo, *Graph Theory with Applications to Engineering and Computer Science,* Prentice Hall, Englewood Cliffs, N.J., 1974.

10. E. W. Dijkstra, "A Note on Two Problems in Connexion with Graphs," *Numerische Mathematik,* 1 (1959), 269–271.

11. E. A. Dinic, "Algorithm for Solution of a Problem of Maximum Flow in Networks with Power Estimation," *Soviet Mathematics Doklady,* 11 (1970), 1277–1280.

12. J. Edmonds, "Paths, Trees, and Flowers," *Canadian Journal of Mathematics,* 17 (1965), 449–467.

13. J. Edmonds and R. M. Karp, "Theoretical Improvements in Algorithmic Efficiency for Network Flow Problems," *Journal of the ACM,* 19 (1972), 248–264.

14. S. Even, *Graph Algorithms,* Computer Science Press, Potomac, Md., 1979.

15. L. R. Ford, Jr., and D. R. Fulkerson, *Flows in Networks,* Princeton University Press, Princeton, N.J., 1962.

16. M. L. Fredman and R. E. Tarjan, "Fibonacci Heaps and Their Uses in Improved Network Optimization Algorithms," *Journal of the ACM,* 34 (1987), 596–615.

17. H. N. Gabow, "Data Structures for Weighted Matching and Nearest Common Ancestors with Linking," *Proceedings of First Annual ACM-SIAM Symposium on Discrete Algorithms* (1990), 434–443.

18. H. N. Gabow, Z. Galil, T. H. Spencer, and R. E. Tarjan, "Efficient Algorithms for Finding Minimum Spanning Trees on Directed and Undirected Graphs," *Combinatorica,* 6 (1986), 109–122.

19. Z. Galil, "Efficient Algorithms for Finding Maximum Matchings in Graphs," *ACM Computing Surveys,* 18 (1986), 23–38.

20. Z. Galil and E. Tardos, "An $O(n^2(m + n \log n) \log n)$ Min-Cost Flow Algorithm," *Journal of the ACM,* 35 (1988), 374–386.

21. M. R. Garey and D. S. Johnson, *Computers and Intractability: A Guide to the Theory of NP-Completeness,* Freeman, San Francisco, 1979.

22. A. V. Goldberg and S. Rao, "Beyond the Flow Decomposition Barrier," *Journal of the ACM,* 45 (1998), 783–797.

23. A. V. Goldberg and R. E. Tarjan, "A New Approach to the Maximum-Flow Problem," *Journal of the ACM,* 35 (1988), 921–940.

24. F. Harary, *Graph Theory,* Addison-Wesley, Reading, Mass., 1969.

25. J. E. Hopcroft and R. M. Karp, "An $n^{5/2}$ Algorithm for Maximum Matchings in Bipartite Graphs," *SIAM Journal on Computing,* 2 (1973), 225–231.

26. J. E. Hopcroft and R. E. Tarjan, "Algorithm 447: Efficient Algorithms for Graph Manipulation," *Communications of the ACM,* 16 (1973), 372–378.

27. J. E. Hopcroft and R. E. Tarjan, "Dividing a Graph into Triconnected Components," *SIAM Journal on Computing,* 2 (1973), 135–158.

28. J. E. Hopcroft and R. E. Tarjan, "Efficient Planarity Testing," *Journal of the ACM,* 21 (1974), 549–568.

29. J. E. Hopcroft and J. K. Wong, "Linear Time Algorithm for Isomorphism of Planar Graphs," *Proceedings of the Sixth Annual ACM Symposium on Theory of Computing* (1974), 172–184.

30. D. B. Johnson, "Efficient Algorithms for Shortest Paths in Sparse Networks," *Journal of the ACM,* 24 (1977), 1–13.

31. A. B. Kahn, "Topological Sorting of Large Networks," *Communications of the ACM,* 5 (1962), 558–562.

32. D. R. Karger, P. N. Klein, and R. E. Tarjan, "A Randomized Linear-Time Algorithm to Find Minimum Spanning Trees," *Journal of the ACM,* 42 (1995), 321–328.

33. R. M. Karp, "Reducibility among Combinatorial Problems," *Complexity of Computer Computations* (eds. R. E. Miller and J. W. Thatcher), Plenum Press, New York, 1972, 85–103.

34. A. V. Karzanov, "Determining the Maximal Flow in a Network by the Method of Preflows," *Soviet Mathematics Doklady,* 15 (1974), 434–437.

35. V. King, S. Rao, and R. E. Tarjan, "A Faster Deterministic Maximum Flow Algorithm," *Journal of Algorithms,* 17 (1994), 447–474.

36. D. E. Knuth, *The Art of Computer Programming, Vol. 1: Fundamental Algorithms,* 3d ed., Addison-Wesley, Reading, Mass., 1997.

37. J. B. Kruskal, Jr., "On the Shortest Spanning Subtree of a Graph and the Traveling Salesman Problem," *Proceedings of the American Mathematical Society,* 7 (1956), 48–50.

38. H. W. Kuhn, "The Hungarian Method for the Assignment Problem," *Naval Research Logistics Quarterly,* 2 (1955), 83–97.

39. E. L. Lawler, *Combinatorial Optimization: Networks and Matroids,* Holt, Reinhart and Winston, New York, 1976.

40. S. Lin and B. W. Kernighan, "An Effective Heuristic Algorithm for the Traveling Salesman Problem," *Operations Research,* 21 (1973), 498–516.

41. K. Melhorn, *Data Structures and Algorithms 2: Graph Algorithms and NP-completeness,* Springer-Verlag, Berlin, 1984.

42. B. M. E. Moret and H. D. Shapiro, "An Empirical Analysis of Algorithms for Constructing a Minimum Spanning Tree," *Proceedings of the Second Workshop on Algorithms and Data Structures* (1991), 400–411.

43. C. H. Papadimitriou and K. Steiglitz, *Combinatorial Optimization: Algorithms and Complexity,* Prentice Hall, Englewood Cliffs, N.J., 1982.

44. R. C. Prim, "Shortest Connection Networks and Some Generalizations," *Bell System Technical Journal,* 36 (1957), 1389–1401.

45. M. Sharir, "A Strong-Connectivity Algorithm and Its Application in Data Flow Analysis," *Computers and Mathematics with Applications,* 7 (1981), 67–72.

46. R. E. Tarjan, "Depth First Search and Linear Graph Algorithms," *SIAM Journal on Computing,* 1 (1972), 146–160.

47. R. E. Tarjan, "Testing Flow Graph Reducibility," *Journal of Computer and System Sciences,* 9 (1974), 355–365.

48. R. E. Tarjan, "Finding Dominators in Directed Graphs," *SIAM Journal on Computing,* 3 (1974), 62–89.

49. R. E. Tarjan, "Complexity of Combinatorial Algorithms," *SIAM Review,* 20 (1978), 457–491.

50. R. E. Tarjan, *Data Structures and Network Algorithms,* Society for Industrial and Applied Mathematics, Philadelphia, 1983.

51. A. C. Yao, "An $O(|E| \log \log |V|)$ Algorithm for Finding Minimum Spanning Trees," *Information Processing Letters,* 4 (1975), 21–23.

CHAPTER 10

Algorithm Design Techniques

So far, we have been concerned with the efficient implementation of algorithms. We have seen that when an algorithm is given, the actual data structures need not be specified. It is up to the programmer to choose the appropriate data structure in order to make the running time as small as possible.

In this chapter, we switch our attention from the *implementation* of algorithms to the *design* of algorithms. Most of the algorithms that we have seen so far are straightforward and simple. Chapter 9 contains some algorithms that are much more subtle, and some require an argument (in some cases lengthy) to show that they are indeed correct. In this chapter, we will focus on five of the common types of algorithms used to solve problems. For many problems, it is quite likely that at least one of these methods will work. Specifically, for each type of algorithm we will

- See the general approach.
- Look at several examples (the exercises at the end of the chapter provide many more examples).
- Discuss, in general terms, the time and space complexity, where appropriate.

10.1 Greedy Algorithms

The first type of algorithm we will examine is the **greedy algorithm.** We have already seen three greedy algorithms in Chapter 9: Dijkstra's, Prim's, and Kruskal's algorithms. Greedy algorithms work in phases. In each phase, a decision is made that appears to be good, without regard for future consequences. Generally, this means that some *local optimum* is chosen. This "take what you can get now" strategy is the source of the name for this class of algorithms. When the algorithm terminates, we hope that the local optimum is equal to the **global optimum.** If this is the case, then the algorithm is correct; otherwise, the algorithm has produced a suboptimal solution. If the absolute best answer is not required, then simple greedy algorithms are sometimes used to generate approximate answers, rather than using the more complicated algorithms generally required to generate an exact answer.

There are several real-life examples of greedy algorithms. The most obvious is the coin-changing problem. To make change in U.S. currency, we repeatedly dispense the largest denomination. Thus, to give out seventeen dollars and sixty-one cents in change, we give out a ten-dollar bill, a five-dollar bill, two one-dollar bills, two quarters, one dime, and one penny. By doing this, we are guaranteed to minimize the number of bills and coins. This algorithm does not work in all monetary systems, but fortunately, we can prove that it does work in the American monetary system. Indeed, it works even if two-dollar bills and fifty-cent pieces are allowed.

Traffic problems provide an example where making locally optimal choices does not always work. For example, during certain rush hour times in Miami, it is best to stay off the prime streets even if they look empty, because traffic will come to a standstill a mile down the road, and you will be stuck. Even more shocking, it is better in some cases to make a temporary detour in the direction opposite your destination in order to avoid all traffic bottlenecks.

In the remainder of this section, we will look at several applications that use greedy algorithms. The first application is a simple scheduling problem. Virtually all scheduling problems are either *NP*-complete (or of similar difficult complexity) or are solvable by a greedy algorithm. The second application deals with file compression and is one of the earliest results in computer science. Finally, we will look at an example of a greedy approximation algorithm.

10.1.1 A Simple Scheduling Problem

We are given jobs $j_1, j_2, \ldots, j_N$, all with known running times $t_1, t_2, \ldots, t_N$, respectively. We have a single processor. What is the best way to schedule these jobs in order to minimize the average completion time? In this entire section, we will assume **nonpreemptive scheduling**: Once a job is started, it must run to completion.

As an example, suppose we have the four jobs and associated running times shown in Figure 10.1. One possible schedule is shown in Figure 10.2. Because j_1 finishes in 15 (time units), j_2 in 23, j_3 in 26, and j_4 in 36, the average completion time is 25. A better schedule, which yields a mean completion time of 17.75, is shown in Figure 10.3.

The schedule given in Figure 10.3 is arranged by shortest job first. We can show that this will always yield an optimal schedule. Let the jobs in the schedule be $j_{i_1}, j_{i_2}, \ldots, j_{i_N}$. The first job finishes in time t_{i_1}. The second job finishes after $t_{i_1} + t_{i_2}$, and the third job finishes after $t_{i_1} + t_{i_2} + t_{i_3}$. From this, we see that the total cost, C, of the schedule is

$$C = \sum_{k=1}^{N} (N - k + 1) t_{i_k} \tag{10.1}$$

$$C = (N + 1) \sum_{k=1}^{N} t_{i_k} - \sum_{k=1}^{N} k \cdot t_{i_k} \tag{10.2}$$

Notice that in Equation (10.2), the first sum is independent of the job ordering, so only the second sum affects the total cost. Suppose that in an ordering there exists some $x > y$ such that $t_{i_x} < t_{i_y}$. Then a calculation shows that by swapping j_{i_x} and j_{i_y}, the second sum increases, decreasing the total cost. Thus, any schedule of jobs in which the times are

Job	Time
j_1	15
j_2	8
j_3	3
j_4	10

Figure 10.1   Jobs and times

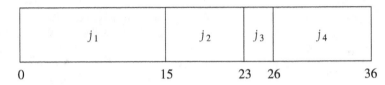

Figure 10.2   Schedule #1

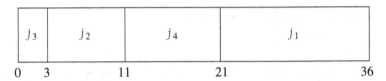

Figure 10.3   Schedule #2 (optimal)

not monotonically nondecreasing must be suboptimal. The only schedules left are those in which the jobs are arranged by smallest running time first, breaking ties arbitrarily.

This result indicates the reason the operating system scheduler generally gives precedence to shorter jobs.

The Multiprocessor Case

We can extend this problem to the case of several processors. Again we have jobs $j_1, j_2, \ldots, j_N$, with associated running times $t_1, t_2, \ldots, t_N$, and a number P of processors. We will assume without loss of generality that the jobs are ordered, shortest running time first. As an example, suppose $P = 3$, and the jobs are as shown in Figure 10.4.

Figure 10.5 shows an optimal arrangement to minimize mean completion time. Jobs j_1, j_4, and j_7 are run on Processor 1. Processor 2 handles j_2, j_5, and j_8, and Processor 3 runs the remaining jobs. The total time to completion is 165, for an average of $\frac{165}{9} = 18.33$.

The algorithm to solve the multiprocessor case is to start jobs in order, cycling through processors. It is not hard to show that no other ordering can do better, although if the number of processors P evenly divides the number of jobs N, there are many optimal orderings. This is obtained by, for each $0 \leq i < N/P$, placing each of the jobs j_{iP+1} through $j_{(i+1)P}$ on a different processor. In our case, Figure 10.6 shows a second optimal solution.

Job	Time
j_1	3
j_2	5
j_3	6
j_4	10
j_5	11
j_6	14
j_7	15
j_8	18
j_9	20

Figure 10.4 Jobs and times

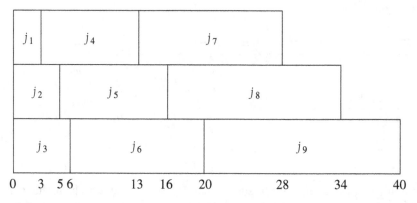

Figure 10.5 An optimal solution for the multiprocessor case

Even if P does not divide N exactly, there can still be many optimal solutions, even if all the job times are distinct. We leave further investigation of this as an exercise.

Minimizing the Final Completion Time

We close this section by considering a very similar problem. Suppose we are only concerned with when the last job finishes. In our two examples above, these completion times are 40 and 38, respectively. Figure 10.7 shows that the minimum final completion time is 34, and this clearly cannot be improved, because every processor is always busy.

Although this schedule does not have minimum mean completion time, it has merit in that the completion time of the entire sequence is earlier. If the same user owns all these jobs, then this is the preferable method of scheduling. Although these problems are very similar, this new problem turns out to be *NP*-complete; it is just another way of phrasing the knapsack or bin-packing problems, which we will encounter later in this section. Thus,

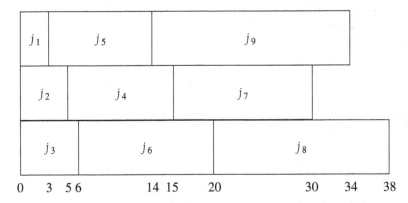

Figure 10.6   A second optimal solution for the multiprocessor case

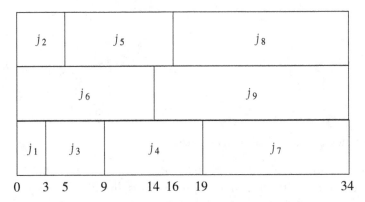

Figure 10.7   Minimizing the final completion time

minimizing the final completion time is apparently much harder than minimizing the mean completion time.

10.1.2 Huffman Codes

In this section, we consider a second application of greedy algorithms, known as **file compression**.

The normal ASCII character set consists of roughly 100 "printable" characters. In order to distinguish these characters, $\lceil \log 100 \rceil = 7$ bits are required. Seven bits allow the representation of 128 characters, so the ASCII character set adds some other "nonprintable" characters. An eighth bit is added as a parity check. The important point, however, is that if the size of the character set is C, then $\lceil \log C \rceil$ bits are needed in a standard encoding.

Suppose we have a file that contains only the characters *a, e, i, s, t,* plus blank spaces and *newlines*. Suppose further, that the file has ten *a*'s, fifteen *e*'s, twelve *i*'s, three *s*'s, four *t*'s,

Character	Code	Frequency	Total Bits
a	000	10	30
e	001	15	45
i	010	12	36
s	011	3	9
t	100	4	12
space	101	13	39
newline	110	1	3
Total			174

Figure 10.8   Using a standard coding scheme

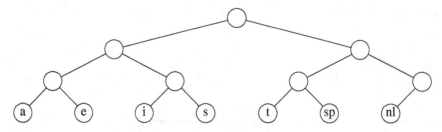

Figure 10.9   Representation of the original code in a tree

thirteen blanks, and one *newline*. As the table in Figure 10.8 shows, this file requires 174 bits to represent, since there are 58 characters and each character requires three bits.

In real life, files can be quite large. Many of the very large files are output of some program and there is usually a big disparity between the most frequent and least frequent characters. For instance, many large data files have an inordinately large amount of digits, blanks, and *newlines*, but few *q*'s and *x*'s. We might be interested in reducing the file size in the case where we are transmitting it over a slow phone line. Also, since on virtually every machine, disk space is precious, one might wonder if it would be possible to provide a better code and reduce the total number of bits required.

The answer is that this is possible, and a simple strategy achieves 25 percent savings on typical large files and as much as 50 to 60 percent savings on many large data files. The general strategy is to allow the code length to vary from character to character and to ensure that the frequently occurring characters have short codes. Notice that if all the characters occur with the same frequency, then there are not likely to be any savings.

The binary code that represents the alphabet can be represented by the binary tree shown in Figure 10.9.

The tree in Figure 10.9 has data only at the leaves. The representation of each character can be found by starting at the root and recording the path, using a 0 to indicate the left branch and a 1 to indicate the right branch. For instance, *s* is reached by going left, then

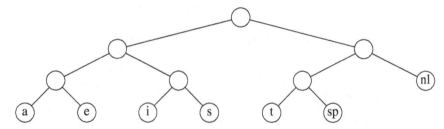

Figure 10.10 A slightly better tree

right, and finally right. This is encoded as 011. This data structure is sometimes referred to as a **trie**. If character c_i is at depth d_i and occurs f_i times, then the *cost* of the code is equal to $\sum d_i f_i$.

A better code than the one given in Figure 10.9 can be obtained by noticing that the *newline* is an only child. By placing the *newline* symbol one level higher at its parent, we obtain the new tree in Figure 10.10. This new tree has cost of 173, but is still far from optimal.

Notice that the tree in Figure 10.10 is a **full tree**: All nodes either are leaves or have two children. An optimal code will always have this property, since otherwise, as we have already seen, nodes with only one child could move up a level.

If the characters are placed only at the leaves, any sequence of bits can always be decoded unambiguously. For instance, suppose 0100111100010110001000111 is the encoded string. 0 is not a character code, 01 is not a character code, but 010 represents i, so the first character is i. Then 011 follows, giving an s. Then 11 follows, which is a *newline*. The remainder of the code is a, *space, t, i, e,* and *newline*. Thus, it does not matter if the character codes are different lengths, as long as no character code is a prefix of another character code. Such an encoding is known as a **prefix code**. Conversely, if a character is contained in a nonleaf node, it is no longer possible to guarantee that the decoding will be unambiguous.

Putting these facts together, we see that our basic problem is to find the full binary tree of minimum total cost (as defined above), where all characters are contained in the leaves. The tree in Figure 10.11 shows the optimal tree for our sample alphabet. As can be seen in Figure 10.12, this code uses only 146 bits.

Notice that there are many optimal codes. These can be obtained by swapping children in the encoding tree. The main unresolved question, then, is how the coding tree is constructed. The algorithm to do this was given by Huffman in 1952. Thus, this coding system is commonly referred to as a Huffman code.

Huffman's Algorithm

Throughout this section we will assume that the number of characters is C. Huffman's algorithm can be described as follows: We maintain a forest of trees. The *weight* of a tree is equal to the sum of the frequencies of its leaves. $C - 1$ times, select the two trees, T_1 and T_2, of smallest weight, breaking ties arbitrarily, and form a new tree with subtrees T_1 and T_2.

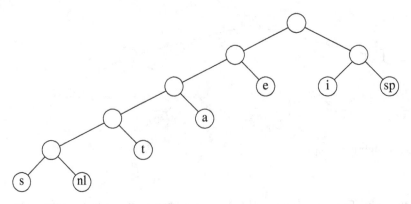

Figure 10.11 Optimal prefix code

Character	Code	Frequency	Total Bits
a	001	10	30
e	01	15	30
i	10	12	24
s	00000	3	15
t	0001	4	16
space	11	13	26
newline	00001	1	5
Total			146

Figure 10.12 Optimal prefix code

Figure 10.13 Initial stage of Huffman's algorithm

At the beginning of the algorithm, there are C single-node trees—one for each character. At the end of the algorithm there is one tree, and this is the optimal Huffman coding tree.

A worked example will make the operation of the algorithm clear. Figure 10.13 shows the initial forest; the weight of each tree is shown in small type at the root. The two trees of lowest weight are merged together, creating the forest shown in Figure 10.14. We will name the new root $T1$, so that future merges can be stated unambiguously. We have made s the left child arbitrarily; any tiebreaking procedure can be used. The total weight of the new tree is just the sum of the weights of the old trees, and can thus be easily computed.

Figure 10.14   Huffman's algorithm after the first merge

Figure 10.15   Huffman's algorithm after the second merge

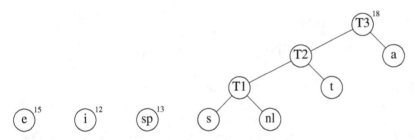

Figure 10.16   Huffman's algorithm after the third merge

It is also a simple matter to create the new tree, since we merely need to get a new node, set the left and right links, and record the weight.

Now there are six trees, and we again select the two trees of smallest weight. These happen to be $T1$ and t, which are then merged into a new tree with root $T2$ and weight 8. This is shown in Figure 10.15. The third step merges $T2$ and a, creating $T3$, with weight $10 + 8 = 18$. Figure 10.16 shows the result of this operation.

After the third merge is completed, the two trees of lowest weight are the single-node trees representing i and the blank space. Figure 10.17 shows how these trees are merged into the new tree with root $T4$. The fifth step is to merge the trees with roots e and $T3$, since these trees have the two smallest weights. The result of this step is shown in Figure 10.18.

Finally, the optimal tree, which was shown in Figure 10.11, is obtained by merging the two remaining trees. Figure 10.19 shows this optimal tree, with root $T6$.

We will sketch the ideas involved in proving that Huffman's algorithm yields an optimal code; we will leave the details as an exercise. First, it is not hard to show by contradiction that the tree must be full, since we have already seen how a tree that is not full is improved.

Next, we must show that the two least frequent characters α and β must be the two deepest nodes (although other nodes may be as deep). Again, this is easy to show by

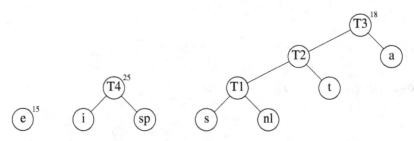

Figure 10.17   Huffman's algorithm after the fourth merge

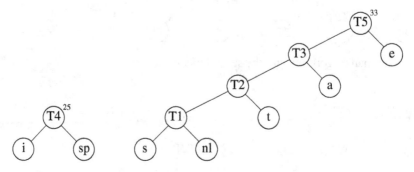

Figure 10.18   Huffman's algorithm after the fifth merge

contradiction, since if either α or β is not a deepest node, then there must be some γ that is (recall that the tree is full). If α is less frequent than γ, then we can improve the cost by swapping them in the tree.

We can then argue that the characters in any two nodes at the same depth can be swapped without affecting optimality. This shows that an optimal tree can always be found that contains the two least frequent symbols as siblings; thus the first step is not a mistake.

The proof can be completed by using an induction argument. As trees are merged, we consider the new character set to be the characters in the roots. Thus, in our example, after four merges, we can view the character set as consisting of e and the metacharacters $T3$ and $T4$. This is probably the trickiest part of the proof; you are urged to fill in all of the details.

The reason that this is a greedy algorithm is that at each stage we perform a merge without regard to global considerations. We merely select the two smallest trees.

If we maintain the trees in a priority queue, ordered by weight, then the running time is $O(C \log C)$, since there will be one buildHeap, $2C - 2$ deleteMins, and $C - 2$ inserts, on a priority queue that never has more than C elements. A simple implementation of the priority queue, using a list, would give an $O(C^2)$ algorithm. The choice of priority queue implementation depends on how large C is. In the typical case of an ASCII character set, C is small enough that the quadratic running time is acceptable. In such an application, virtually all the running time will be spent on the disk I/O required to read the input file and write out the compressed version.

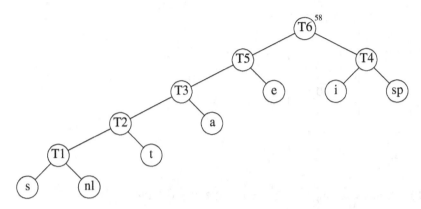

Figure 10.19 Huffman's algorithm after the final merge

There are two details that must be considered. First, the encoding information must be transmitted at the start of the compressed file, since otherwise it will be impossible to decode. There are several ways of doing this; see Exercise 10.4. For small files, the cost of transmitting this table will override any possible savings in compression, and the result will probably be file expansion. Of course, this can be detected and the original left intact. For large files, the size of the table is not significant.

The second problem is that as described, this is a two-pass algorithm. The first pass collects the frequency data and the second pass does the encoding. This is obviously not a desirable property for a program dealing with large files. Some alternatives are described in the references.

10.1.3 Approximate Bin Packing

In this section, we will consider some algorithms to solve the **bin-packing** problem. These algorithms will run quickly but will not necessarily produce optimal solutions. We will prove, however, that the solutions that are produced are not too far from optimal.

We are given N items of sizes $s_1, s_2, \ldots, s_N$. All sizes satisfy $0 < s_i \le 1$. The problem is to pack these items in the fewest number of bins, given that each bin has unit capacity. As an example, Figure 10.20 shows an optimal packing for an item list with sizes 0.2, 0.5, 0.4, 0.7, 0.1, 0.3, 0.8.

There are two versions of the bin-packing problem. The first version is **online bin packing**. In this version, each item must be placed in a bin before the next item can be processed. The second version is the **off-line bin-packing problem**. In an off-line algorithm, we do not need to do anything until all the input has been read. The distinction between online and off-line algorithms was discussed in Section 8.2.

Online Algorithms

The first issue to consider is whether or not an online algorithm can actually always give an optimal answer, even if it is allowed unlimited computation. Remember that even

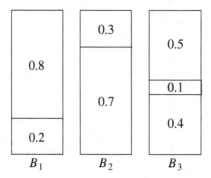

Figure 10.20 Optimal packing for $0.2, 0.5, 0.4, 0.7, 0.1, 0.3, 0.8$

though unlimited computation is allowed, an online algorithm must place an item before processing the next item and cannot change its decision.

To show that an online algorithm cannot always give an optimal solution, we will give it particularly difficult data to work on. Consider an input sequence I_1 of M small items of weight $\frac{1}{2} - \epsilon$ followed by M large items of weight $\frac{1}{2} + \epsilon$, $0 < \epsilon < 0.01$. It is clear that these items can be packed in M bins if we place one small item and one large item in each bin. Suppose there were an optimal online algorithm A that could perform this packing. Consider the operation of algorithm A on the sequence I_2, consisting of only M small items of weight $\frac{1}{2} - \epsilon$. I_2 can be packed in $\lceil M/2 \rceil$ bins. However, A will place each item in a separate bin, since A must yield the same results on I_2 as it does for the first half of I_1, and the first half of I_1 is exactly the same input as I_2. This means that A will use twice as many bins as is optimal for I_2. What we have proven is that there is no optimal algorithm for online bin packing.

What the argument above shows is that an online algorithm never knows when the input might end, so any performance guarantees it provides must hold at every instant throughout the algorithm. If we follow the foregoing strategy, we can prove the following.

Theorem 10.1.
There are inputs that force any online bin-packing algorithm to use at least $\frac{4}{3}$ the optimal number of bins.

Proof.
Suppose otherwise, and suppose for simplicity that M is even. Consider any online algorithm A running on the input sequence I_1, above. Recall that this sequence consists of M small items followed by M large items. Let us consider what the algorithm A has done after processing the Mth item. Suppose A has already used b bins. At this point in the algorithm, the optimal number of bins is $M/2$, because we can place two elements in each bin. Thus we know that $2b/M < \frac{4}{3}$, by our assumption of a better-than-$\frac{4}{3}$ performance guarantee.

Now consider the performance of algorithm A after all items have been packed. All bins created after the bth bin must contain exactly one item, since all small items are placed in the first b bins, and two large items will not fit in a bin. Since the

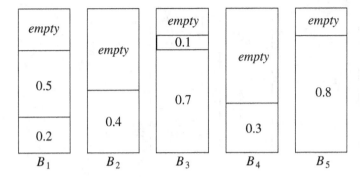

Figure 10.21 Next fit for 0.2, 0.5, 0.4, 0.7, 0.1, 0.3, 0.8

first b bins can have at most two items each, and the remaining bins have one item each, we see that packing $2M$ items will require at least $2M - b$ bins. Since the $2M$ items can be optimally packed using M bins, our performance guarantee assures us that $(2M - b)/M < \frac{4}{3}$.

The first inequality implies that $b/M < \frac{2}{3}$, and the second inequality implies that $b/M > \frac{2}{3}$, which is a contradiction. Thus, no online algorithm can guarantee that it will produce a packing with less than $\frac{4}{3}$ the optimal number of bins.

There are three simple algorithms that guarantee that the number of bins used is no more than twice optimal. There are also quite a few more complicated algorithms with better guarantees.

Next Fit

Probably the simplest algorithm is **next fit**. When processing any item, we check to see whether it fits in the same bin as the last item. If it does, it is placed there; otherwise, a new bin is created. This algorithm is incredibly simple to implement and runs in linear time. Figure 10.21 shows the packing produced for the same input as Figure 10.20.

Not only is next fit simple to program, its worst-case behavior is also easy to analyze.

Theorem 10.2.

Let M be the optimal number of bins required to pack a list I of items. Then next fit never uses more than $2M$ bins. There exist sequences such that next fit uses $2M - 2$ bins.

Proof.

Consider any adjacent bins B_j and B_{j+1}. The sum of the sizes of all items in B_j and B_{j+1} must be larger than 1, since otherwise all of these items would have been placed in B_j. If we apply this result to all pairs of adjacent bins, we see that at most half of the space is wasted. Thus next fit uses at most twice the optimal number of bins.

To see that this ratio, 2, is tight, suppose that the N items have size $s_i = 0.5$ if i is odd and $s_i = 2/N$ if i is even. Assume N is divisible by 4. The optimal packing, shown in Figure 10.22, consists of $N/4$ bins, each containing 2 elements of size 0.5, and one

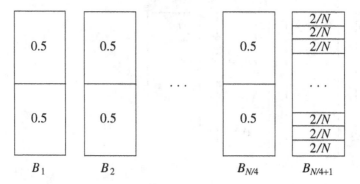

Figure 10.22   Optimal packing for 0.5, 2/N, 0.5, 2/N, 0.5, 2/N, ...

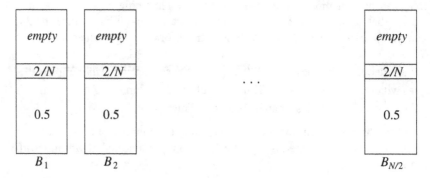

Figure 10.23   Next fit packing for 0.5, 2/N, 0.5, 2/N, 0.5, 2/N, ...

bin containing the $N/2$ elements of size $2/N$, for a total of $(N/4) + 1$. Figure 10.23 shows that next fit uses $N/2$ bins. Thus, next fit can be forced to use almost twice as many bins as optimal.

First Fit

Although next fit has a reasonable performance guarantee, it performs poorly in practice, because it creates new bins when it does not need to. In the sample run, it could have placed the item of size 0.3 in either B_1 or B_2, rather than create a new bin.

The **first fit** strategy is to scan the bins in order and place the new item in the first bin that is large enough to hold it. Thus, a new bin is created only when the results of previous placements have left no other alternative. Figure 10.24 shows the packing that results from first fit on our standard input.

A simple method of implementing first fit would process each item by scanning down the list of bins sequentially. This would take $O(N^2)$. It is possible to implement first fit to run in $O(N \log N)$; we leave this as an exercise.

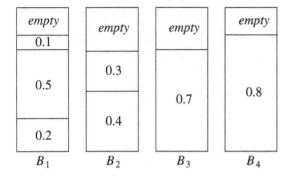

Figure 10.24 First fit for 0.2, 0.5, 0.4, 0.7, 0.1, 0.3, 0.8

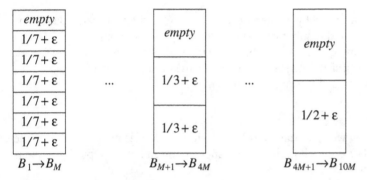

Figure 10.25 A case where first fit uses 10M bins instead of 6M

A moment's thought will convince you that at any point, at most one bin can be more than half empty, since if a second bin were also half empty, its contents would fit into the first bin. Thus, we can immediately conclude that first fit guarantees a solution with at most twice the optimal number of bins.

On the other hand, the bad case that we used in the proof of next fit's performance bound does not apply for first fit. Thus, one might wonder if a better bound can be proven. The answer is yes, but the proof is complicated.

Theorem 10.3.

Let M be the optimal number of bins required to pack a list I of items. Then first fit never uses more than $\frac{17}{10}M + \frac{7}{10}$ bins. There exist sequences such that first fit uses $\frac{17}{10}(M - 1)$ bins.

Proof.

See the references at the end of the chapter.

An example where first fit does almost as poorly as the previous theorem would indicate is shown in Figure 10.25. The input consists of 6M items of size $\frac{1}{7} + \epsilon$, followed by

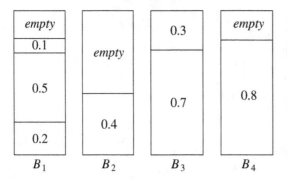

Figure 10.26 Best fit for 0.2, 0.5, 0.4, 0.7, 0.1, 0.3, 0.8

6M items of size $\frac{1}{3} + \epsilon$, followed by 6M items of size $\frac{1}{2} + \epsilon$. One simple packing places one item of each size in a bin and requires 6M bins. First fit requires 10M bins.

When first fit is run on a large number of items with sizes uniformly distributed between 0 and 1, empirical results show that first fit uses roughly 2 percent more bins than optimal. In many cases, this is quite acceptable.

Best Fit

The third online strategy we will examine is **best fit**. Instead of placing a new item in the first spot that is found, it is placed in the tightest spot among all bins. A typical packing is shown in Figure 10.26.

Notice that the item of size 0.3 is placed in B_3, where it fits perfectly, instead of B_2. One might expect that since we are now making a more educated choice of bins, the performance guarantee would improve. This is not the case, because the generic bad cases are the same. Best fit is never more than roughly 1.7 times as bad as optimal, and there are inputs for which it (nearly) achieves this bound. Nevertheless, best fit is also simple to code, especially if an $O(N \log N)$ algorithm is required, and it does perform better for random inputs.

Off-line Algorithms

If we are allowed to view the entire item list before producing an answer, then we should expect to do better. Indeed, since we can eventually find the optimal packing by exhaustive search, we already have a theoretical improvement over the online case.

The major problem with all the online algorithms is that it is hard to pack the large items, especially when they occur late in the input. The natural way around this is to sort the items, placing the largest items first. We can then apply first fit or best fit, yielding the algorithms **first fit decreasing** and **best fit decreasing**, respectively. Figure 10.27 shows that in our case this yields an optimal solution (although, of course, this is not true in general).

In this section, we will deal with first fit decreasing. The results for best fit decreasing are almost identical. Since it is possible that the item sizes are not distinct, some authors

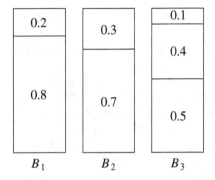

Figure 10.27 First fit for 0.8, 0.7, 0.5, 0.4, 0.3, 0.2, 0.1

prefer to call the algorithm *first fit nonincreasing*. We will stay with the original name. We will also assume, without loss of generality, that input sizes are already sorted.

The first remark we can make is that the bad case, which showed first fit using $10M$ bins instead of $6M$ bins, does not apply when the items are sorted. We will show that if an optimal packing uses M bins, then first fit decreasing never uses more than $(4M + 1)/3$ bins.

The result depends on two observations. First, all the items with weight larger than $\frac{1}{3}$ will be placed in the first M bins. This implies that all the items in the extra bins have weight at most $\frac{1}{3}$. The second observation is that the number of items in the extra bins can be at most $M - 1$. Combining these two results, we find that at most $\lceil (M - 1)/3 \rceil$ extra bins can be required. We now prove these two observations.

Lemma 10.1.

Let the N items have (sorted in decreasing order) input sizes $s_1, s_2, \ldots, s_N$, respectively, and suppose that the optimal packing is M bins. Then all items that first fit decreasing places in extra bins have size at most $\frac{1}{3}$.

Proof.

Suppose the ith item is the first placed in bin $M + 1$. We need to show that $s_i \le \frac{1}{3}$. We will prove this by contradiction. Assume $s_i > \frac{1}{3}$.

It follows that $s_1, s_2, \ldots, s_{i-1} > \frac{1}{3}$, since the sizes are arranged in sorted order. From this it follows that all bins $B_1, B_2, \ldots, B_M$ have at most two items each.

Consider the state of the system after the $(i - 1)$st item is placed in a bin, but before the ith item is placed. We now want to show that (under the assumption that $s_i > \frac{1}{3}$) the first M bins are arranged as follows: First there are some bins with exactly one element, and then the remaining bins have two elements.

Suppose there were two bins B_x and B_y, such that $1 \le x < y \le M$, B_x has two items, and B_y has one item. Let x_1 and x_2 be the two items in B_x, and let y_1 be the item in B_y. $x_1 \ge y_1$, since x_1 was placed in the earlier bin. $x_2 \ge s_i$, by similar reasoning. Thus, $x_1 + x_2 \ge y_1 + s_i$. This implies that s_i could be placed in B_y. By our assumption this is not possible. Thus, if $s_i > \frac{1}{3}$, then, at the time that we try to process s_i, the first

M bins are arranged such that the first j have one element and the next $M - j$ have two elements.

To prove the lemma we will show that there is no way to place all the items in M bins, which contradicts the premise of the lemma.

Clearly, no two items $s_1, s_2, \ldots, s_j$ can be placed in one bin, by any algorithm, since if they could, first fit would have done so too. We also know that first fit has not placed any of the items of size $s_{j+1}, s_{j+2}, \ldots, s_i$ into the first j bins, so none of them fit. Thus, in any packing, specifically the optimal packing, there must be j bins that do not contain these items. It follows that the items of size $s_{j+1}, s_{j+2}, \ldots, s_{i-1}$ must be contained in some set of $M - j$ bins, and from previous considerations, the total number of such items is $2(M - j)$.[1]

The proof is completed by noting that if $s_i > \frac{1}{3}$, there is no way for s_i to be placed in one of these M bins. Clearly, it cannot go in one of the j bins, since if it could, then first fit would have done so too. To place it in one of the remaining $M - j$ bins requires distributing $2(M - j) + 1$ items into the $M - j$ bins. Thus, some bin would have to have three items, each of which is larger than $\frac{1}{3}$, a clear impossibility.

This contradicts the fact that all the sizes can be placed in M bins, so the original assumption must be incorrect. Thus, $s_i \leq \frac{1}{3}$.

Lemma 10.2.
The number of objects placed in extra bins is at most $M - 1$.

Proof.
Assume that there are at least M objects placed in extra bins. We know that $\sum_{i=1}^{N} s_i \leq M$, since all the objects fit in M bins. Suppose that B_j is filled with W_j total weight for $1 \leq j \leq M$. Suppose the first M extra objects have sizes $x_1, x_2, \ldots, x_M$. Then, since the items in the first M bins plus the first M extra items are a subset of all the items, it follows that

$$\sum_{i=1}^{N} s_i \geq \sum_{j=1}^{M} W_j + \sum_{j=1}^{M} x_j \geq \sum_{j=1}^{M} (W_j + x_j)$$

Now $W_j + x_j > 1$, since otherwise the item corresponding to x_j would have been placed in B_j. Thus

$$\sum_{i=1}^{N} s_i > \sum_{j=1}^{M} 1 > M$$

But this is impossible if the N items can be packed in M bins. Thus, there can be at most $M - 1$ extra items.

---

[1] Recall that first fit packed these elements into $M - j$ bins and placed two items in each bin. Thus, there are $2(M - j)$ items.

Theorem 10.4.

Let M be the optimal number of bins required to pack a list I of items. Then first fit decreasing never uses more than $(4M + 1)/3$ bins.

Proof.

There are at most $M - 1$ extra items, of size at most $\frac{1}{3}$. Thus, there can be at most $\lceil(M - 1)/3\rceil$ extra bins. The total number of bins used by first fit decreasing is thus at most $\lceil(4M - 1)/3\rceil \leq (4M + 1)/3$.

It is possible to prove a much tighter bound for both first fit decreasing and next fit decreasing.

Theorem 10.5.

Let M be the optimal number of bins required to pack a list I of items. Then first fit decreasing never uses more than $\frac{11}{9}M + \frac{6}{9}$ bins. There exist sequences such that first fit decreasing uses $\frac{11}{9}M + \frac{6}{9}$ bins.

Proof.

The upper bound requires a very complicated analysis. The lower bound is exhibited by a sequence consisting of $6k + 4$ elements of size $\frac{1}{2} + \epsilon$, followed by $6k + 4$ elements of size $\frac{1}{4} + 2\epsilon$, followed by $6k + 4$ elements of size $\frac{1}{4} + \epsilon$, followed by $12k + 8$ elements of size $\frac{1}{4} - 2\epsilon$. Figure 10.28 shows that the optimal packing requires $9k + 6$ bins, but first fit decreasing uses $11k + 8$ bins. Set $M = 9k + 6$, and the result follows.

In practice, first fit decreasing performs extremely well. If sizes are chosen uniformly over the unit interval, then the expected number of extra bins is $\Theta(\sqrt{M})$. Bin packing is a fine example of how simple greedy heuristics can give good results.

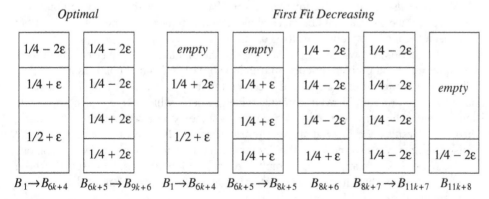

Figure 10.28 Example where first fit decreasing uses $11k + 8$ bins, but only $9k + 6$ bins are required

10.2 Divide and Conquer

Another common technique used to design algorithms is **divide and conquer**. Divide-and conquer-algorithms consist of two parts:

Divide: Smaller problems are solved recursively (except, of course, base cases).

Conquer: The solution to the original problem is then formed from the solutions to the subproblems.

Traditionally, routines in which the text contains at least two recursive calls are called divide-and-conquer algorithms, while routines whose text contains only one recursive call are not. We generally insist that the subproblems be disjoint (that is, essentially nonoverlapping). Let us review some of the recursive algorithms that have been covered in this text.

We have already seen several divide-and-conquer algorithms. In Section 2.4.3, we saw an $O(N \log N)$ solution to the maximum subsequence sum problem. In Chapter 4, we saw linear-time tree traversal strategies. In Chapter 7, we saw the classic examples of divide and conquer, namely mergesort and quicksort, which have $O(N \log N)$ worst-case and average-case bounds, respectively.

We have also seen several examples of recursive algorithms that probably do not classify as divide and conquer, but merely reduce to a single simpler case. In Section 1.3, we saw a simple routine to print a number. In Chapter 2, we used recursion to perform efficient exponentiation. In Chapter 4, we examined simple search routines for binary search trees. In Section 6.6, we saw simple recursion used to merge leftist heaps. In Section 7.7, an algorithm was given for selection that takes linear average time. The disjoint set find operation was written recursively in Chapter 8. Chapter 9 showed routines to recover the shortest path in Dijkstra's algorithm and other procedures to perform depth-first search in graphs. None of these algorithms are really divide-and-conquer algorithms, because only one recursive call is performed.

We have also seen, in Section 2.4, a very bad recursive routine to compute the Fibonacci numbers. This could be called a divide-and-conquer algorithm, but it is terribly inefficient, because the problem really is not divided at all.

In this section, we will see more examples of the divide-and-conquer paradigm. Our first application is a problem in **computational geometry**. Given N points in a plane, we will show that the closest pair of points can be found in $O(N \log N)$ time. The exercises describe some other problems in computational geometry which can be solved by divide and conquer. The remainder of the section shows some extremely interesting, but mostly theoretical, results. We provide an algorithm that solves the selection problem in $O(N)$ worst-case time. We also show that 2 N-bit numbers can be multiplied in $o(N^2)$ operations and that two $N \times N$ matrices can be multiplied in $o(N^3)$ operations. Unfortunately, even though these algorithms have better worst-case bounds than the conventional algorithms, none are practical except for very large inputs.

10.2.1   Running Time of Divide-and-Conquer Algorithms

All the efficient divide-and-conquer algorithms we will see divide the problems into sub-problems, each of which is some fraction of the original problem, and then perform some additional work to compute the final answer. As an example, we have seen that merge-sort operates on two problems, each of which is half the size of the original, and then uses $O(N)$ additional work. This yields the running time equation (with appropriate initial conditions)

$$T(N) = 2T(N/2) + O(N)$$

We saw in Chapter 7 that the solution to this equation is $O(N \log N)$. The following theorem can be used to determine the running time of most divide-and-conquer algorithms.

Theorem 10.6.
The solution to the equation $T(N) = aT(N/b) + \Theta(N^k)$, where $a \geq 1$ and $b > 1$, is

$$T(N) = \begin{cases} O(N^{\log_b a}) & \text{if } a > b^k \\ O(N^k \log N) & \text{if } a = b^k \\ O(N^k) & \text{if } a < b^k \end{cases}$$

Proof.
Following the analysis of mergesort in Chapter 7, we will assume that N is a power of b; thus, let $N = b^m$. Then $N/b = b^{m-1}$ and $N^k = (b^m)^k = b^{mk} = b^{km} = (b^k)^m$. Let us assume $T(1) = 1$, and ignore the constant factor in $\Theta(N^k)$. Then we have

$$T(b^m) = aT(b^{m-1}) + (b^k)^m$$

If we divide through by a^m, we obtain the equation

$$\frac{T(b^m)}{a^m} = \frac{T(b^{m-1})}{a^{m-1}} + \left\{ \frac{b^k}{a} \right\}^m \tag{10.3}$$

We can apply this equation for other values of m, obtaining

$$\frac{T(b^{m-1})}{a^{m-1}} = \frac{T(b^{m-2})}{a^{m-2}} + \left\{ \frac{b^k}{a} \right\}^{m-1} \tag{10.4}$$

$$\frac{T(b^{m-2})}{a^{m-2}} = \frac{T(b^{m-3})}{a^{m-3}} + \left\{ \frac{b^k}{a} \right\}^{m-2} \tag{10.5}$$

$$\cdots$$

$$\frac{T(b^1)}{a^1} = \frac{T(b^0)}{a^0} + \left\{ \frac{b^k}{a} \right\}^1 \tag{10.6}$$

We use our standard trick of adding up the telescoping Equations (10.3) through (10.6). Virtually all the terms on the left cancel the leading terms on the right, yielding

$$\frac{T(b^m)}{a^m} = 1 + \sum_{i=1}^{m} \left\{ \frac{b^k}{a} \right\}^i \tag{10.7}$$

$$= \sum_{i=0}^{m} \left\{ \frac{b^k}{a} \right\}^i \tag{10.8}$$

Thus

$$T(N) = T(b^m) = a^m \sum_{i=0}^{m} \left\{ \frac{b^k}{a} \right\}^i \tag{10.9}$$

If $a > b^k$, then the sum is a geometric series with ratio smaller than 1. Since the sum of infinite series would converge to a constant, this finite sum is also bounded by a constant, and thus Equation (10.10) applies:

$$T(N) = O(a^m) = O(a^{\log_b N}) = O(N^{\log_b a}) \tag{10.10}$$

If $a = b^k$, then each term in the sum is 1. Since the sum contains $1 + \log_b N$ terms and $a = b^k$ implies that $\log_b a = k$,

$$T(N) = O(a^m \log_b N) = O(N^{\log_b a} \log_b N) = O(N^k \log_b N)$$
$$= O(N^k \log N) \tag{10.11}$$

Finally, if $a < b^k$, then the terms in the geometric series are larger than 1, and the second formula in Section 1.2.3 applies. We obtain

$$T(N) = a^m \frac{(b^k/a)^{m+1} - 1}{(b^k/a) - 1} = O(a^m (b^k/a)^m) = O((b^k)^m) = O(N^k) \tag{10.12}$$

proving the last case of the theorem.

As an example, mergesort has $a = b = 2$ and $k = 1$. The second case applies, giving the answer $O(N \log N)$. If we solve three problems, each of which is half the original size, and combine the solutions with $O(N)$ additional work, then $a = 3, b = 2$, and $k = 1$. Case 1 applies here, giving a bound of $O(N^{\log_2 3}) = O(N^{1.59})$. An algorithm that solved three half-sized problems, but required $O(N^2)$ work to merge the solution, would have an $O(N^2)$ running time, since the third case would apply.

There are two important cases that are not covered by Theorem 10.6. We state two more theorems, leaving the proofs as exercises. Theorem 10.7 generalizes the previous theorem.

Theorem 10.7.
The solution to the equation $T(N) = aT(N/b) + \Theta(N^k \log^p N)$, where $a \geq 1, b > 1$, and $p \geq 0$ is

$$T(N) = \begin{cases} O(N^{\log_b a}) & \text{if } a > b^k \\ O(N^k \log^{p+1} N) & \text{if } a = b^k \\ O(N^k \log^p N) & \text{if } a < b^k \end{cases}$$

Theorem 10.8.

If $\sum_{i=1}^{k} \alpha_i < 1$, then the solution to the equation $T(N) = \sum_{i=1}^{k} T(\alpha_i N) + O(N)$ is $T(N) = O(N)$.

10.2.2 Closest-Points Problem

The input to our first problem is a list P of points in a plane. If $p_1 = (x_1, y_1)$ and $p_2 = (x_2, y_2)$, then the Euclidean distance between p_1 and p_2 is $[(x_1 - x_2)^2 + (y_1 - y_2)^2]^{1/2}$. We are required to find the closest pair of points. It is possible that two points have the same position; in that case that pair is the closest, with distance zero.

If there are N points, then there are $N(N - 1)/2$ pairs of distances. We can check all of these, obtaining a very short program, but at the expense of an $O(N^2)$ algorithm. Since this approach is just an exhaustive search, we should expect to do better.

Let us assume that the points have been sorted by x coordinate. At worst, this adds $O(N \log N)$ to the final time bound. Since we will show an $O(N \log N)$ bound for the entire algorithm, this sort is essentially free, from a complexity standpoint.

Figure 10.29 shows a small sample point set P. Since the points are sorted by x coordinate, we can draw an imaginary vertical line that partitions the point set into two halves, P_L and P_R. This is certainly simple to do. Now we have almost exactly the same situation as we saw in the maximum subsequence sum problem in Section 2.4.3. Either the closest points are both in P_L, or they are both in P_R, or one is in P_L and the other is in P_R. Let us call these distances d_L, d_R, and d_C. Figure 10.30 shows the partition of the point set and these three distances.

Figure 10.29 A small point set

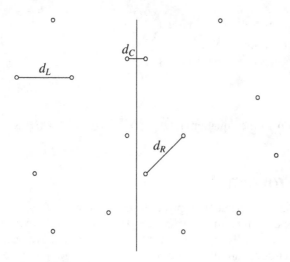

Figure 10.30　P partitioned into P_L and P_R; shortest distances are shown

We can compute d_L and d_R recursively. The problem, then, is to compute d_C. Since we would like an $O(N \log N)$ solution, we must be able to compute d_C with only $O(N)$ additional work. We have already seen that if a procedure consists of two half-sized recursive calls and $O(N)$ additional work, then the total time will be $O(N \log N)$.

Let $\delta = \min(d_L, d_R)$. The first observation is that we only need to compute d_C if d_C improves on δ. If d_C is such a distance, then the two points that define d_C must be within δ of the dividing line; we will refer to this area as a **strip**. As shown in Figure 10.31, this observation limits the number of points that need to be considered (in our case, $\delta = d_R$).

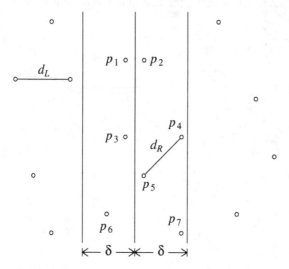

Figure 10.31　Two-lane strip, containing all points considered for d_C strip

```
// Points are all in the strip

for( i = 0; i < numPointsInStrip; i++ )
    for( j = i + 1; j < numPointsInStrip; j++ )
        if( dist(p_i, p_j) < δ )
            δ = dist(p_i, p_j);
```

Figure 10.32 Brute-force calculation of $\min(\delta, d_C)$

There are two strategies that can be tried to compute d_C. For large point sets that are uniformly distributed, the number of points that are expected to be in the strip is very small. Indeed, it is easy to argue that only $O(\sqrt{N})$ points are in the strip on average. Thus, we could perform a brute-force calculation on these points in $O(N)$ time. The pseudocode in Figure 10.32 implements this strategy, assuming the Java convention that the points are indexed starting at 0.

In the worst case, all the points could be in the strip, so this strategy does not always work in linear time. We can improve this algorithm with the following observation: The y coordinates of the two points that define d_C can differ by at most δ. Otherwise, $d_C > \delta$. Suppose that the points in the strip are sorted by their y coordinates. Therefore, if p_i and p_j's y coordinates differ by more than δ, then we can proceed to p_{i+1}. This simple modification is implemented in Figure 10.33.

This extra test has a significant effect on the running time, because for each p_i only a few points p_j are examined before p_i's and p_j's y coordinates differ by more than δ and force an exit from the inner **for** loop. Figure 10.34 shows, for instance, that for point p_3, only the two points p_4 and p_5 lie in the strip within δ vertical distance.

In the worst case, for any point p_i, at most 7 points p_j are considered. This is because these points must lie either in the δ by δ square in the left half of the strip or in the δ by δ square in the right half of the strip. On the other hand, all the points in each δ by δ square are separated by at least δ. In the worst case, each square contains four points, one at each corner. One of these points is p_i, leaving at most seven points to be considered. This worst-case situation is shown in Figure 10.35. Notice that even though p_{L2} and p_{R1} have the same coordinates, they could be different points. For the actual analysis, it is only

```
// Points are all in the strip and sorted by y-coordinate

for( i = 0; i < numPointsInStrip; i++ )
    for( j = i + 1; j < numPointsInStrip; j++ )
        if( p_i and p_j's y-coordinates differ by more than δ )
            break;        // Go to next p_i.
        else
        if( dist(p_i, p_j) < δ )
            δ = dist(p_i, p_j);
```

Figure 10.33 Refined calculation of $\min(\delta, d_C)$

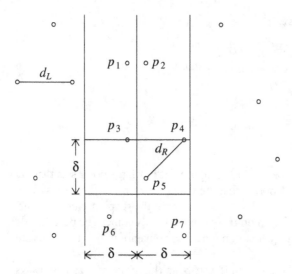

Figure 10.34   Only p_4 and p_5 are considered in the second for loop

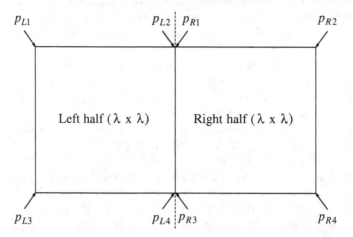

Figure 10.35   At most eight points fit in the rectangle; there are two coordinates shared by two points each

important that the number of points in the λ by 2λ rectangle be $O(1)$, and this much is certainly clear.

Because at most seven points are considered for each p_i, the time to compute a d_C that is better than δ is $O(N)$. Thus, we appear to have an $O(N \log N)$ solution to the closest-points problem, based on the two half-sized recursive calls plus the linear extra work to combine the two results. However, we do not quite have an $O(N \log N)$ solution yet.

The problem is that we have assumed that a list of points sorted by y coordinate is available. If we perform this sort for each recursive call, then we have $O(N \log N)$ extra work:

This gives an $O(N \log^2 N)$ algorithm. This is not all that bad, especially when compared to the brute force $O(N^2)$. However, it is not hard to reduce the work for each recursive call to $O(N)$, thus ensuring an $O(N \log N)$ algorithm.

We will maintain two lists. One is the point list sorted by x coordinate, and the other is the point list sorted by y coordinate. We will call these lists P and Q, respectively. These can be obtained by a preprocessing sorting step at cost $O(N \log N)$ and thus does not affect the time bound. P_L and Q_L are the lists passed to the left-half recursive call, and P_R and Q_R are the lists passed to the right-half recursive call. We have already seen that P is easily split in the middle. Once the dividing line is known, we step through Q sequentially, placing each element in Q_L or Q_R as appropriate. It is easy to see that Q_L and Q_R will be automatically sorted by y coordinate. When the recursive calls return, we scan through the Q list and discard all the points whose x coordinates are not within the strip. Then Q contains only points in the strip, and these points are guaranteed to be sorted by their y coordinates.

This strategy ensures that the entire algorithm is $O(N \log N)$, because only $O(N)$ extra work is performed.

10.2.3 The Selection Problem

The **selection problem** requires us to find the kth smallest element in a collection S of N elements. Of particular interest is the special case of finding the median. This occurs when $k = \lceil N/2 \rceil$.

In Chapters 1, 6, and 7 we have seen several solutions to the selection problem. The solution in Chapter 7 uses a variation of quicksort and runs in $O(N)$ average time. Indeed, it is described in Hoare's original paper on quicksort.

Although this algorithm runs in linear average time, it has a worst case of $O(N^2)$. Selection can easily be solved in $O(N \log N)$ worst-case time by sorting the elements, but for a long time it was unknown whether or not selection could be accomplished in $O(N)$ worst-case time. The *quickselect* algorithm outlined in Section 7.7.6 is quite efficient in practice, so this was mostly a question of theoretical interest.

Recall that the basic algorithm is a simple recursive strategy. Assuming that N is larger than the cutoff point where elements are simply sorted, an element v, known as the pivot, is chosen. The remaining elements are placed into two sets, S_1 and S_2. S_1 contains elements that are guaranteed to be no larger than v, and S_2 contains elements that are no smaller than v. Finally, if $k \le |S_1|$, then the kth smallest element in S can be found by recursively computing the kth smallest element in S_1. If $k = |S_1| + 1$, then the pivot is the kth smallest element. Otherwise, the kth smallest element in S is the $(k - |S_1| - 1)$st smallest element in S_2. The main difference between this algorithm and quicksort is that there is only one subproblem to solve instead of two.

In order to obtain a linear algorithm, we must ensure that the subproblem is only a fraction of the original and not merely only a few elements smaller than the original. Of course, we can always find such an element if we are willing to spend some time to do so. The difficult problem is that we cannot spend too much time finding the pivot.

For quicksort, we saw that a good choice for pivot was to pick three elements and use their median. This gives some expectation that the pivot is not too bad but does not provide a guarantee. We could choose 21 elements at random, sort them in constant time, use the 11th largest as pivot, and get a pivot that is even more likely to be good. However, if these 21 elements were the 21 largest, then the pivot would still be poor. Extending this, we could use up to $O(N/\log N)$ elements, sort them using heapsort in $O(N)$ total time, and be almost certain, from a statistical point of view, of obtaining a good pivot. In the worst case, however, this does not work because we might select the $O(N/\log N)$ largest elements, and then the pivot would be the $[N - O(N/\log N)]$th largest element, which is not a constant fraction of N.

The basic idea is still useful. Indeed, we will see that we can use it to improve the expected number of comparisons that quickselect makes. To get a good worst case, however, the key idea is to use one more level of indirection. Instead of finding the median from a sample of random elements, we will find the median from a **sample of medians**.

The basic pivot selection algorithm is as follows:

1. Arrange the N elements into $\lfloor N/5 \rfloor$ groups of five elements, ignoring the (at most four) extra elements.
2. Find the median of each group. This gives a list M of $\lfloor N/5 \rfloor$ medians.
3. Find the median of M. Return this as the pivot, v.

We will use the term **median-of-median-of-five partitioning** to describe the quick-select algorithm that uses the pivot selection rule given above. We will now show that median-of-median-of-five partitioning guarantees that each recursive subproblem is at most roughly 70 percent as large as the original. We will also show that the pivot can be computed quickly enough to guarantee an $O(N)$ running time for the entire selection algorithm.

Let us assume for the moment that N is divisible by 5, so there are no extra elements. Suppose also that $N/5$ is odd, so that the set M contains an odd number of elements. This provides some symmetry, as we shall see. We are thus assuming, for convenience, that N is of the form $10k + 5$. We will also assume that all the elements are distinct. The actual algorithm must make sure to handle the case where this is not true. Figure 10.36 shows how the pivot might be chosen when $N = 45$.

In Figure 10.36, v represents the element which is selected by the algorithm as pivot. Since v is the median of nine elements, and we are assuming that all elements are distinct, there must be four medians that are larger than v and four that are smaller. We denote these by L and S, respectively. Consider a group of five elements with a large median (type L). The median of the group is smaller than two elements in the group and larger than two elements in the group. We will let H represent the *huge* elements. These are elements that are known to be larger than a large median. Similarly, T represents the *tiny* elements, which are smaller than a small median. There are 10 elements of type H: Two are in each of the groups with an L type median, and two elements are in the same group as v. Similarly, there are 10 elements of type T.

Elements of type L or H are guaranteed to be larger than v, and elements of type S or T are guaranteed to be smaller than v. There are thus guaranteed to be 14 large and 14 small elements in our problem. Therefore, a recursive call could be on at most $45 - 14 - 1 = 30$ elements.

Sorted groups of five elements

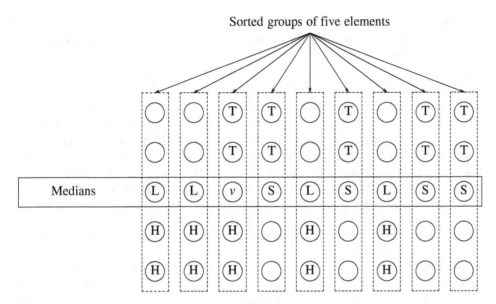

Figure 10.36   How the pivot is chosen

Let us extend this analysis to general N of the form $10k + 5$. In this case, there are k elements of type L and k elements of type S. There are $2k + 2$ elements of type H, and also $2k + 2$ elements of type T. Thus, there are $3k + 2$ elements that are guaranteed to be larger than v and $3k + 2$ elements that are guaranteed to be smaller. Thus, in this case, the recursive call can contain at most $7k + 2 < 0.7N$ elements. If N is not of the form $10k + 5$, similar arguments can be made without affecting the basic result.

It remains to bound the running time to obtain the pivot element. There are two basic steps. We can find the median of five elements in constant time. For instance, it is not hard to sort five elements in eight comparisons. We must do this $\lfloor N/5 \rfloor$ times, so this step takes $O(N)$ time. We must then compute the median of a group of $\lfloor N/5 \rfloor$ elements. The obvious way to do this is to sort the group and return the element in the middle. But this takes $O(\lfloor N/5 \rfloor \log \lfloor N/5 \rfloor) = O(N \log N)$ time, so this does not work. The solution is to call the selection algorithm recursively on the $\lfloor N/5 \rfloor$ elements.

This completes the description of the basic algorithm. There are still some details that need to be filled in if an actual implementation is desired. For instance, duplicates must be handled correctly, and the algorithm needs a cutoff large enough to ensure that the recursive calls make progress. There is quite a large amount of overhead involved, and this algorithm is not practical at all, so we will not describe any more of the details that need to be considered. Even so, from a theoretical standpoint, the algorithm is a major breakthrough, because, as the following theorem shows, the running time is linear in the worst case.

Theorem 10.9.
The running time of quickselect using median-of-median-of-five partitioning is $O(N)$.

Proof.

The algorithm consists of two recursive calls of size $0.7N$ and $0.2N$, plus linear extra work. By Theorem 10.8, the running time is linear.

Reducing the Average Number of Comparisons

Divide and conquer can also be used to reduce the expected number of comparisons required by the selection algorithm. Let us look at a concrete example. Suppose we have a group S of 1,000 numbers and are looking for the 100th smallest number, which we will call X. We choose a subset S' of S consisting of 100 numbers. We would expect that the value of X is similar in size to the 10th smallest number in S'. More specifically, the fifth smallest number in S' is almost certainly less than X, and the 15th smallest number in S' is almost certainly greater than X.

More generally, a sample S' of s elements is chosen from the N elements. Let δ be some number, which we will choose later so as to minimize the average number of comparisons used by the procedure. We find the $(v_1 = ks/N - \delta)$th and $(v_2 = ks/N + \delta)$th smallest elements in S'. Almost certainly, the kth smallest element in S will fall between v_1 and v_2, so we are left with a selection problem on 2δ elements. With low probability, the kth smallest element does not fall in this range, and we have considerable work to do. However, with a good choice of s and δ, we can ensure, by the laws of probability, that the second case does not adversely affect the total work.

If an analysis is performed, we find that if $s = N^{2/3} \log^{1/3} N$ and $\delta = N^{1/3} \log^{2/3} N$, then the expected number of comparisons is $N + k + O(N^{2/3} \log^{1/3} N)$, which is optimal except for the low-order term. (If $k > N/2$, we can consider the symmetric problem of finding the $(N - k)$th largest element.)

Most of the analysis is easy to do. The last term represents the cost of performing the two selections to determine v_1 and v_2. The average cost of the partitioning, assuming a reasonably clever strategy, is equal to N plus the expected rank of v_2 in S, which is $N + k + O(N\delta/s)$. If the kth element winds up in S', the cost of finishing the algorithm is equal to the cost of selection on S', namely, $O(s)$. If the kth smallest element doesn't wind up in S', the cost is $O(N)$. However, s and δ have been chosen to guarantee that this happens with very low probability $o(1/N)$, so the expected cost of this possibility is $o(1)$, which is a term that goes to zero as N gets large. An exact calculation is left as Exercise 10.22.

This analysis shows that finding the median requires about $1.5N$ comparisons on average. Of course, this algorithm requires some floating-point arithmetic to compute s, which can slow down the algorithm on some machines. Even so, experiments have shown that if correctly implemented, this algorithm compares favorably with the quickselect implementation in Chapter 7.

10.2.4 Theoretical Improvements for Arithmetic Problems

In this section we describe a divide-and-conquer algorithm that multiplies two N-digit numbers. Our previous model of computation assumed that multiplication was done in constant time, because the numbers were small. For large numbers, this assumption is no longer valid. If we measure multiplication in terms of the size of numbers being multiplied, then the natural multiplication algorithm takes quadratic time. The divide-and-conquer algorithm runs in subquadratic time. We also present the classic divide-and-conquer algorithm that multiplies two N by N matrices in subcubic time.

Multiplying Integers

Suppose we want to multiply two N-digit numbers X and Y. If exactly one of X and Y is negative, then the answer is negative; otherwise it is positive. Thus, we can perform this check and then assume that $X, Y \geq 0$. The algorithm that almost everyone uses when multiplying by hand requires $\Theta(N^2)$ operations, because each digit in X is multiplied by each digit in Y.

If $X = 61{,}438{,}521$ and $Y = 94{,}736{,}407$, $XY = 5{,}820{,}464{,}730{,}934{,}047$. Let us break X and Y into two halves, consisting of the most significant and least significant digits, respectively. Then $X_L = 6{,}143$, $X_R = 8{,}521$, $Y_L = 9{,}473$, and $Y_R = 6{,}407$. We also have $X = X_L 10^4 + X_R$ and $Y = Y_L 10^4 + Y_R$. It follows that

$$XY = X_L Y_L 10^8 + (X_L Y_R + X_R Y_L)10^4 + X_R Y_R$$

Notice that this equation consists of four multiplications, $X_L Y_L$, $X_L Y_R$, $X_R Y_L$, and $X_R Y_R$, which are each half the size of the original problem ($N/2$ digits). The multiplications by 10^8 and 10^4 amount to the placing of zeros. This and the subsequent additions add only $O(N)$ additional work. If we perform these four multiplications recursively using this algorithm, stopping at an appropriate base case, then we obtain the recurrence

$$T(N) = 4T(N/2) + O(N)$$

From Theorem 10.6, we see that $T(N) = O(N^2)$, so, unfortunately, we have not improved the algorithm. To achieve a subquadratic algorithm, we must use less than four recursive calls. The key observation is that

$$X_L Y_R + X_R Y_L = (X_L - X_R)(Y_R - Y_L) + X_L Y_L + X_R Y_R$$

Thus, instead of using two multiplications to compute the coefficient of 10^4, we can use one multiplication, plus the result of two multiplications that have already been performed. Figure 10.37 shows how only three recursive subproblems need to be solved.

It is easy to see that now the recurrence equation satisfies

$$T(N) = 3T(N/2) + O(N)$$

and so we obtain $T(N) = O(N^{\log_2 3}) = O(N^{1.59})$. To complete the algorithm, we must have a base case, which can be solved without recursion.

When both numbers are one-digit, we can do the multiplication by table lookup. If one number has zero digits, then we return zero. In practice, if we were to use this algorithm, we would choose the base case to be that which is most convenient for the machine.

Although this algorithm has better asymptotic performance than the standard quadratic algorithm, it is rarely used, because for small N the overhead is significant, and for larger N there are even better algorithms. These algorithms also make extensive use of divide and conquer.

Function	Value	Computational Complexity
X_L	6,143	Given
X_R	8,521	Given
Y_L	9,473	Given
Y_R	6,407	Given
$D_1 = X_L - X_R$	$-2,378$	$O(N)$
$D_2 = Y_R - Y_L$	$-3,066$	$O(N)$
$X_L Y_L$	58,192,639	$T(N/2)$
$X_R Y_R$	54,594,047	$T(N/2)$
$D_1 D_2$	7,290,948	$T(N/2)$
$D_3 = D_1 D_2 + X_L Y_L + X_R Y_R$	120,077,634	$O(N)$
$X_R Y_R$	54,594,047	Computed above
$D_3 10^4$	1,200,776,340,000	$O(N)$
$X_L Y_L 10^8$	5,819,263,900,000,000	$O(N)$
$X_L Y_L 10^8 + D_3 10^4 + X_R Y_R$	5,820,464,730,934,047	$O(N)$

Figure 10.37   The divide-and-conquer algorithm in action

Matrix Multiplication

A fundamental numerical problem is the multiplication of two matrices. Figure 10.38 gives a simple $O(N^3)$ algorithm to compute $\mathbf{C} = \mathbf{AB}$, where $\mathbf{A}$, $\mathbf{B}$, and $\mathbf{C}$ are $N \times N$ matrices. The algorithm follows directly from the definition of matrix multiplication. To compute $C_{i,j}$, we compute the dot product of the ith row in $\mathbf{A}$ with the jth column in $\mathbf{B}$. As usual, arrays begin at index 0.

For a long time it was assumed that $\Omega(N^3)$ was required for matrix multiplication. However, in the late sixties Strassen showed how to break the $\Omega(N^3)$ barrier. The basic idea of Strassen's algorithm is to divide each matrix into four quadrants, as shown in Figure 10.39. Then it is easy to show that

$$C_{1,1} = A_{1,1}B_{1,1} + A_{1,2}B_{2,1}$$
$$C_{1,2} = A_{1,1}B_{1,2} + A_{1,2}B_{2,2}$$
$$C_{2,1} = A_{2,1}B_{1,1} + A_{2,2}B_{2,1}$$
$$C_{2,2} = A_{2,1}B_{1,2} + A_{2,2}B_{2,2}$$

As an example, to perform the multiplication $\mathbf{AB}$

$$\mathbf{AB} = \begin{bmatrix} 3 & 4 & 1 & 6 \\ 1 & 2 & 5 & 7 \\ 5 & 1 & 2 & 9 \\ 4 & 3 & 5 & 6 \end{bmatrix} \begin{bmatrix} 5 & 6 & 9 & 3 \\ 4 & 5 & 3 & 1 \\ 1 & 1 & 8 & 4 \\ 3 & 1 & 4 & 1 \end{bmatrix}$$

```
1        /**
2         * Standard matrix multiplication.
3         * Arrays start at 0.
4         * Assumes a and b are square.
5         */
6        public static int [ ][ ] multiply( int [ ][ ] a, int [ ][ ] b )
7        {
8            int n = a.length;
9            int [ ][ ] c = new int[ n ][ n ];
10
11           for( int i = 0; i < n; i++ )     // Initialization
12               for( int j = 0; j < n; j++ )
13                   c[ i ][ j ] = 0;
14
15           for( int i = 0; i < n; i++ )
16               for( int j = 0; j < n; j++ )
17                   for( int k = 0; k < n; k++ )
18                       c[ i ][ j ] += a[ i ][ k ] * b[ k ][ j ];
19
20           return c;
21       }
```

Figure 10.38   Simple $O(N^3)$ matrix multiplication

$$\begin{bmatrix} A_{1,1} & A_{1,2} \\ A_{2,1} & A_{2,2} \end{bmatrix} \begin{bmatrix} B_{1,1} & B_{1,2} \\ B_{2,1} & B_{2,2} \end{bmatrix} = \begin{bmatrix} C_{1,1} & C_{1,2} \\ C_{2,1} & C_{2,2} \end{bmatrix}$$

Figure 10.39   Decomposing $\mathbf{AB} = \mathbf{C}$ into four quadrants

we define the following eight $N/2$ by $N/2$ matrices:

$$A_{1,1} = \begin{bmatrix} 3 & 4 \\ 1 & 2 \end{bmatrix} \quad A_{1,2} = \begin{bmatrix} 1 & 6 \\ 5 & 7 \end{bmatrix} \quad B_{1,1} = \begin{bmatrix} 5 & 6 \\ 4 & 5 \end{bmatrix} \quad B_{1,2} = \begin{bmatrix} 9 & 3 \\ 3 & 1 \end{bmatrix}$$

$$A_{2,1} = \begin{bmatrix} 5 & 1 \\ 4 & 3 \end{bmatrix} \quad A_{2,2} = \begin{bmatrix} 2 & 9 \\ 5 & 6 \end{bmatrix} \quad B_{2,1} = \begin{bmatrix} 1 & 1 \\ 3 & 1 \end{bmatrix} \quad B_{2,2} = \begin{bmatrix} 8 & 4 \\ 4 & 1 \end{bmatrix}$$

We could then perform eight $N/2$ by $N/2$ matrix multiplications and four $N/2$ by $N/2$ matrix additions. The matrix additions take $O(N^2)$ time. If the matrix multiplications are done recursively, then the running time satisfies

$$T(N) = 8T(N/2) + O(N^2)$$

From Theorem 10.6, we see that $T(N) = O(N^3)$, so we do not have an improvement. As we saw with integer multiplication, we must reduce the number of subproblems below 8.

Strassen used a strategy similar to the integer multiplication divide-and-conquer algorithm and showed how to use only seven recursive calls by carefully arranging the computations. The seven multiplications are

$$M_1 = (A_{1,2} - A_{2,2})(B_{2,1} + B_{2,2})$$
$$M_2 = (A_{1,1} + A_{2,2})(B_{1,1} + B_{2,2})$$
$$M_3 = (A_{1,1} - A_{2,1})(B_{1,1} + B_{1,2})$$
$$M_4 = (A_{1,1} + A_{1,2})B_{2,2}$$
$$M_5 = A_{1,1}(B_{1,2} - B_{2,2})$$
$$M_6 = A_{2,2}(B_{2,1} - B_{1,1})$$
$$M_7 = (A_{2,1} + A_{2,2})B_{1,1}$$

Once the multiplications are performed, the final answer can be obtained with eight more additions.

$$C_{1,1} = M_1 + M_2 - M_4 + M_6$$
$$C_{1,2} = M_4 + M_5$$
$$C_{2,1} = M_6 + M_7$$
$$C_{2,2} = M_2 - M_3 + M_5 - M_7$$

It is straightforward to verify that this tricky ordering produces the desired values. The running time now satisfies the recurrence

$$T(N) = 7T(N/2) + O(N^2)$$

The solution of this recurrence is $T(N) = O(N^{\log_2 7}) = O(N^{2.81})$.

As usual, there are details to consider, such as the case when N is not a power of 2, but these are basically minor nuisances. Strassen's algorithm is worse than the straight-forward algorithm until N is fairly large. It does not generalize for the case where the matrices are sparse (contain many zero entries), and it does not easily parallelize. When run with floating-point entries, it is less stable numerically than the classic algorithm. Thus, it has only limited applicability. Nevertheless, it represents an important theoretical milestone and certainly shows that in computer science, as in many other fields, even though a problem seems to have an intrinsic complexity, nothing is certain until proven.

10.3 Dynamic Programming

In the previous section, we saw that a problem that can be mathematically expressed recursively can also be expressed as a recursive algorithm, in many cases yielding a significant performance improvement over a more naïve exhaustive search.

Any recursive mathematical formula could be directly translated to a recursive algorithm, but the underlying reality is that often the compiler will not do justice to the recursive algorithm, and an inefficient program results. When we suspect that this is likely to be the case, we must provide a little more help to the compiler, by rewriting the recursive algorithm as a nonrecursive algorithm that systematically records the answers to the subproblems in a table. One technique that makes use of this approach is known as **dynamic programming**.

10.3.1 Using a Table Instead of Recursion

In Chapter 2, we saw that the natural recursive program to compute the Fibonacci numbers is very inefficient. Recall that the program shown in Figure 10.40 has a running time $T(N)$ that satisfies $T(N) \geq T(N-1) + T(N-2)$. Since $T(N)$ satisfies the same recurrence relation as the Fibonacci numbers and has the same initial conditions, $T(N)$ in fact grows at the same rate as the Fibonacci numbers and is thus exponential.

On the other hand, since to compute F_N, all that is needed is F_{N-1} and F_{N-2}, we only need to record the two most recently computed Fibonacci numbers. This yields the $O(N)$ algorithm in Figure 10.41.

The reason that the recursive algorithm is so slow is because of the algorithm used to simulate recursion. To compute F_N, there is one call to F_{N-1} and F_{N-2}. However, since F_{N-1} recursively makes a call to F_{N-2} and F_{N-3}, there are actually two separate calls to compute F_{N-2}. If one traces out the entire algorithm, then we can see that F_{N-3} is computed three times, F_{N-4} is computed five times, F_{N-5} is computed eight times, and so on. As Figure 10.42 shows, the growth of redundant calculations is explosive. If the compiler's recursion simulation algorithm were able to keep a list of all precomputed values and not make a recursive call for an already solved subproblem, then this exponential explosion would be avoided. This is why the program in Figure 10.41 is so much more efficient.

As a second example, we saw in Chapter 7 how to solve the recurrence $C(N) = (2/N)\sum_{i=0}^{N-1} C(i) + N$, with $C(0) = 1$. Suppose that we want to check, numerically,

```
1      /**
2       * Compute Fibonacci numbers as described in Chapter 1.
3       */
4      public static int fib( int n )
5      {
6          if( n <= 1 )
7              return 1;
8          else
9              return fib( n - 1 ) + fib( n - 2 );
10     }
```

Figure 10.40 Inefficient algorithm to compute Fibonacci numbers

```
1      /**
2       * Compute Fibonacci numbers as described in Chapter 1.
3       */
4      public static int fibonacci( int n )
5      {
6          if( n <= 1 )
7              return 1;
8
9          int last = 1;
10         int nextToLast = 1;
11         int answer = 1;
12
13         for( int i = 2; i <= n; i++ )
14         {
15             answer = last + nextToLast;
16             nextToLast = last;
17             last = answer;
18         }
19         return answer;
20     }
```

Figure 10.41   Linear algorithm to compute Fibonacci numbers

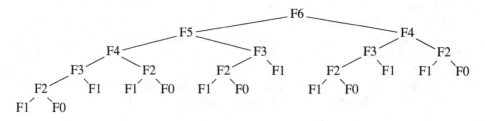

Figure 10.42   Trace of the recursive calculation of Fibonacci numbers

whether the solution we obtained is correct. We could then write the simple program in Figure 10.43 to evaluate the recursion.

Once again, the recursive calls duplicate work. In this case, the running time $T(N)$ satisfies $T(N) = \sum_{i=0}^{N-1} T(i) + N$, because, as shown in Figure 10.44, there is one (direct) recursive call of each size from 0 to $N - 1$, plus $O(N)$ additional work (where else have we seen the tree shown in Figure 10.44?). Solving for $T(N)$, we find that it grows exponentially. By using a table, we obtain the program in Figure 10.45. This program avoids the redundant recursive calls and runs in $O(N^2)$. It is not a perfect program; as an exercise, you should make the simple change that reduces its running time to $O(N)$.

```
1      public static double eval( int n )
2      {
3          if( n == 0 )
4              return 1.0;
5          else
6          {
7              double sum = 0.0;
8              for( int i = 0; i < n; i++ )
9                  sum += eval( i );
10             return 2.0 * sum / n + n;
11         }
12     }
```

Figure 10.43 Recursive method to evaluate $C(N) = 2/N \sum_{i=0}^{N-1} C(i) + N$

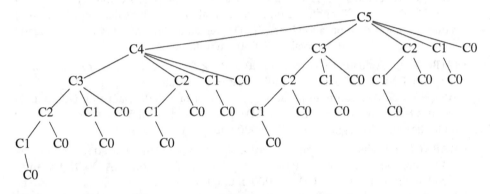

Figure 10.44 Trace of the recursive calculation in eval

```
1      public static double eval( int n )
2      {
3          double [ ] c = new double [ n + 1 ];
4
5          c[ 0 ] = 1.0;
6          for( int i = 1; i <= n; i++ )
7          {
8              double sum = 0.0;
9              for( int j = 0; j < i; j++ )
10                 sum += c[ j ];
11             c[ i ] =  2.0 * sum / i + i;
12         }
13
14         return c[ n ];
15     }
```

Figure 10.45 Evaluating $C(N) = 2/N \sum_{i=0}^{N-1} C(i) + N$ with a table

10.3.2 Ordering Matrix Multiplications

Suppose we are given four matrices, **A**, **B**, **C**, and **D**, of dimensions $\mathbf{A} = 50 \times 10$, $\mathbf{B} = 10 \times 40$, $\mathbf{C} = 40 \times 30$, and $\mathbf{D} = 30 \times 5$. Although matrix multiplication is not commutative, it is associative, which means that the matrix product **ABCD** can be parenthesized, and thus evaluated, in any order. The obvious way to multiply two matrices of dimensions $p \times q$ and $q \times r$, respectively, uses pqr scalar multiplications. (Using a theoretically superior algorithm such as Strassen's algorithm does not significantly alter the problem we will consider, so we will assume this performance bound.) What is the best way to perform the three matrix multiplications required to compute **ABCD**?

In the case of four matrices, it is simple to solve the problem by exhaustive search, since there are only five ways to order the multiplications. We evaluate each case below:

- **(A((BC)D))**: Evaluating **BC** requires $10 \times 40 \times 30 = 12{,}000$ multiplications. Evaluating **(BC)D** requires the 12,000 multiplications to compute **BC**, plus an additional $10 \times 30 \times 5 = 1{,}500$ multiplications, for a total of 13,500. Evaluating **(A((BC)D))** requires 13,500 multiplications for **(BC)D**, plus an additional $50 \times 10 \times 5 = 2{,}500$ multiplications, for a grand total of 16,000 multiplications.

- **(A(B(CD)))**: Evaluating **CD** requires $40 \times 30 \times 5 = 6{,}000$ multiplications. Evaluating **B(CD)** requires the 6,000 multiplications to compute **CD**, plus an additional $10 \times 40 \times 5 = 2{,}000$ multiplications, for a total of 8,000. Evaluating **(A(B(CD)))** requires 8,000 multiplications for **B(CD)**, plus an additional $50 \times 10 \times 5 = 2{,}500$ multiplications, for a grand total of 10,500 multiplications.

- **((AB)(CD))**: Evaluating **CD** requires $40 \times 30 \times 5 = 6{,}000$ multiplications. Evaluating **AB** requires $50 \times 10 \times 40 = 20{,}000$ multiplications. Evaluating **((AB)(CD))** requires 6,000 multiplications for **CD**, 20,000 multiplications for **AB**, plus an additional $50 \times 40 \times 5 = 10{,}000$ multiplications for a grand total of 36,000 multiplications.

- **(((AB)C)D)**: Evaluating **AB** requires $50 \times 10 \times 40 = 20{,}000$ multiplications. Evaluating **(AB)C** requires the 20,000 multiplications to compute **AB**, plus an additional $50 \times 40 \times 30 = 60{,}000$ multiplications, for a total of 80,000. Evaluating **(((AB)C)D)** requires 80,000 multiplications for **(AB)C**, plus an additional $50 \times 30 \times 5 = 7{,}500$ multiplications, for a grand total of 87,500 multiplications.

- **((A(BC))D)**: Evaluating **BC** requires $10 \times 40 \times 30 = 12{,}000$ multiplications. Evaluating **A(BC)** requires the 12,000 multiplications to compute **BC**, plus an additional $50 \times 10 \times 30 = 15{,}000$ multiplications, for a total of 27,000. Evaluating **((A(BC))D)** requires 27,000 multiplications for **A(BC)**, plus an additional $50 \times 30 \times 5 = 7{,}500$ multiplications, for a grand total of 34,500 multiplications.

The calculations show that the best ordering uses roughly one-ninth the number of multiplications as the worst ordering. Thus, it might be worthwhile to perform a few calculations to determine the optimal ordering. Unfortunately, none of the obvious greedy strategies seems to work. Moreover, the number of possible orderings grows quickly. Suppose we define $T(N)$ to be this number. Then $T(1) = T(2) = 1$, $T(3) = 2$, and $T(4) = 5$, as we have seen. In general,

$$T(N) = \sum_{i=1}^{N-1} T(i)T(N-i)$$

To see this, suppose that the matrices are $A_1, A_2, \ldots, A_N$, and the last multiplication performed is $(A_1 A_2 \cdots A_i)(A_{i+1} A_{i+2} \cdots A_N)$. Then there are $T(i)$ ways to compute $(A_1 A_2 \cdots A_i)$ and $T(N-i)$ ways to compute $(A_{i+1} A_{i+2} \cdots A_N)$. Thus, there are $T(i)T(N-i)$ ways to compute $(A_1 A_2 \cdots A_i)(A_{i+1} A_{i+2} \cdots A_N)$ for each possible i.

The solution of this recurrence is the well-known **Catalan numbers**, which grow exponentially. Thus, for large N, an exhaustive search through all possible orderings is useless. Nevertheless, this counting argument provides a basis for a solution that is substantially better than exponential. Let c_i be the number of columns in matrix A_i for $1 \leq i \leq N$. Then A_i has c_{i-1} rows, since otherwise the multiplications are not valid. We will define c_0 to be the number of rows in the first matrix, A_1.

Suppose $m_{Left,Right}$ is the number of multiplications required to multiply $A_{Left} A_{Left+1} \cdots A_{Right-1} A_{Right}$. For consistency, $m_{Left,Left} = 0$. Suppose the last multiplication is $(A_{Left} \cdots A_i)(A_{i+1} \cdots A_{Right})$, where $Left \leq i < Right$. Then the number of multiplications used is $m_{Left,i} + m_{i+1,Right} + c_{Left-1} c_i c_{Right}$. These three terms represent the multiplications required to compute $(A_{Left} \cdots A_i)$, $(A_{i+1} \cdots A_{Right})$, and their product, respectively.

If we define $M_{Left,Right}$ to be the number of multiplications required in an *optimal* ordering, then, if $Left < Right$,

$$M_{Left,Right} = \min_{Left \leq i < Right} \{M_{Left,i} + M_{i+1,Right} + c_{Left-1} c_i c_{Right}\}$$

This equation implies that if we have an optimal multiplication arrangement of $A_{Left} \cdots A_{Right}$, the subproblems $A_{Left} \cdots A_i$ and $A_{i+1} \cdots A_{Right}$ cannot be performed suboptimally. This should be clear, since otherwise we could improve the entire result by replacing the suboptimal computation by an optimal computation.

The formula translates directly to a recursive program, but, as we have seen in the last section, such a program would be blatantly inefficient. However, since there are only approximately $N^2/2$ values of $M_{Left,Right}$ that ever need to be computed, it is clear that a table can be used to store these values. Further examination shows that if $Right - Left = k$, then the only values $M_{x,y}$ that are needed in the computation of $M_{Left,Right}$ satisfy $y - x < k$. This tells us the order in which we need to compute the table.

If we want to print out the actual ordering of the multiplications in addition to the final answer $M_{1,N}$, then we can use the ideas from the shortest-path algorithms in Chapter 9. Whenever we make a change to $M_{Left,Right}$, we record the value of i that is responsible. This gives the simple program shown in Figure 10.46.

Although the emphasis of this chapter is not coding, it is worth noting that many programmers tend to shorten variable names to a single letter. c, i, and k are used as single-letter variables because this agrees with the names we have used in the description of the algorithm, which is very mathematical. However, it is generally best to avoid l as a variable name, because l looks too much like 1 and can make for very difficult debugging if you make a transcription error.

```
1      /**
2       * Compute optimal ordering of matrix multiplication.
3       * c contains the number of columns for each of the n matrices.
4       * c[ 0 ] is the number of rows in matrix 1.
5       * The minimum number of multiplications is left in m[ 1 ][ n ].
6       * Actual ordering is computed via another procedure using lastChange.
7       * m and lastChange are indexed starting at 1, instead of 0.
8       * Note: Entries below main diagonals of m and lastChange
9       * are meaningless and uninitialized.
10      */
11     public static void optMatrix( int [ ] c, long [ ][ ] m, int [ ][ ] lastChange )
12     {
13         int n = c.length - 1;
14
15         for( int left = 1; left <= n; left++ )
16             m[ left ][ left ] = 0;
17         for( int k = 1; k < n; k++ )    // k is right - left
18             for( int left = 1; left <= n - k; left++ )
19             {
20                 // For each position
21                 int right = left + k;
22                 m[ left ][ right ] = INFINITY;
23                 for( int i = left; i < right; i++ )
24                 {
25                     long thisCost = m[ left ][  i ] + m[ i + 1 ][ right ]
26                         + c[ left - 1 ] * c[ i ] * c[ right ];
27
28                     if( thisCost < m[ left ][ right ] )  // Update min
29                     {
30                         m[ left ][ right ] = thisCost;
31                         lastChange[ left ][ right ] = i;
32                     }
33             }
34         }
35     }
```

Figure 10.46   Program to find optimal ordering of matrix multiplications

Returning to the algorithmic issues, this program contains a triply nested loop and is easily seen to run in $O(N^3)$ time. The references describe a faster algorithm, but since the time to perform the actual matrix multiplication is still likely to be much larger than the time to compute the optimal ordering, this algorithm is still quite practical.

10.3.3 Optimal Binary Search Tree

Our second dynamic programming example considers the following input: We are given a list of words, $w_1, w_2, \ldots, w_N$, and *fixed* probabilities $p_1, p_2, \ldots, p_N$ of their occurrence. The problem is to arrange these words in a binary search tree in a way that minimizes the expected total access time. In a binary search tree, the number of comparisons needed to access an element at depth d is $d + 1$, so if w_i is placed at depth d_i, then we want to minimize $\sum_{i=1}^{N} p_i(1 + d_i)$.

As an example, Figure 10.47 shows seven words along with their probability of occurrence in some context. Figure 10.48 shows three possible binary search trees. Their searching costs are shown in Figure 10.49.

The first tree was formed using a greedy strategy. The word with the highest probability of being accessed was placed at the root. The left and right subtrees were then formed recursively. The second tree is the perfectly balanced search tree. Neither of these trees is optimal, as demonstrated by the existence of the third tree. From this we can see that neither of the obvious solutions works.

This is initially surprising, since the problem appears to be very similar to the construction of a Huffman encoding tree, which, as we have already seen, can be solved by a greedy algorithm. Construction of an optimal binary search tree is harder, because the data are not constrained to appear only at the leaves, and also because the tree must satisfy the binary search tree property.

Word	Probability
a	0.22
am	0.18
and	0.20
egg	0.05
if	0.25
the	0.02
two	0.08

Figure 10.47 Sample input for optimal binary search tree problem

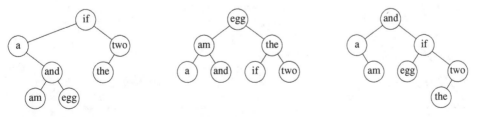

Figure 10.48 Three possible binary search trees for data in previous table

	Input	Tree #1		Tree #2		Tree #3	
Word	Probability	Access Cost		Access Cost		Access Cost	
w_i	p_i	Once	Sequence	Once	Sequence	Once	Sequence
a	0.22	2	0.44	3	0.66	2	0.44
am	0.18	4	0.72	2	0.36	3	0.54
and	0.20	3	0.60	3	0.60	1	0.20
egg	0.05	4	0.20	1	0.05	3	0.15
if	0.25	1	0.25	3	0.75	2	0.50
the	0.02	3	0.06	2	0.04	4	0.08
two	0.08	2	0.16	3	0.24	3	0.24
Totals	1.00		2.43		2.70		2.15

Figure 10.49 Comparison of the three binary search trees

A dynamic programming solution follows from two observations. Once again, suppose we are trying to place the (sorted) words $w_{Left}, w_{Left+1}, \ldots, w_{Right-1}, w_{Right}$ into a binary search tree. Suppose the optimal binary search tree has w_i as the root, where $Left \le i \le Right$. Then the left subtree must contain $w_{Left}, \ldots, w_{i-1}$, and the right subtree must contain $w_{i+1}, \ldots, w_{Right}$ (by the binary search tree property). Further, both of these subtrees must also be optimal, since otherwise they could be replaced by optimal subtrees, which would give a better solution for $w_{Left}, \ldots, w_{Right}$. Thus, we can write a formula for the cost $C_{Left,Right}$ of an optimal binary search tree. Figure 10.50 may be helpful.

If $Left > Right$, then the cost of the tree is 0; this is the null case, which we always have for binary search trees. Otherwise, the root costs p_i. The left subtree has a cost of $C_{Left,i-1}$, relative to its root, and the right subtree has a cost of $C_{i+1,Right}$ relative to its root. As Figure 10.50 shows, each node in these subtrees is one level deeper from w_i than from their respective roots, so we must add $\sum_{j=Left}^{i-1} p_j$ and $\sum_{j=i+1}^{Right} p_j$. This gives the formula

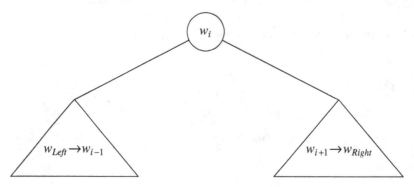

Figure 10.50 Structure of an optimal binary search tree

$$C_{Left,Right} = \min_{Left \le i \le Right} \left\{ p_i + C_{Left,i-1} + C_{i+1,Right} + \sum_{j=Left}^{i-1} p_j + \sum_{j=i+1}^{Right} p_j \right\}$$

$$= \min_{Left \le i \le Right} \left\{ C_{Left,i-1} + C_{i+1,Right} + \sum_{j=Left}^{Right} p_j \right\}$$

From this equation, it is straightforward to write a program to compute the cost of the optimal binary search tree. As usual, the actual search tree can be maintained by saving the value of i that minimizes $C_{Left,Right}$. The standard recursive routine can be used to print the actual tree.

Figure 10.51 shows the table that will be produced by the algorithm. For each subrange of words, the cost and root of the optimal binary search tree are maintained. The bottommost entry computes the optimal binary search tree for the entire set of words in the input. The optimal tree is the third tree shown in Figure 10.48.

The precise computation for the optimal binary search tree for a particular subrange, namely, am..if, is shown in Figure 10.52. It is obtained by computing the minimum-cost tree obtained by placing am, and, egg, and if at the root. For instance, when and is placed at the root, the left subtree contains am..am (of cost 0.18, via previous calculation), the right subtree contains egg..if (of cost 0.35), and $p_{am} + p_{and} + p_{egg} + p_{if} = 0.68$, for a total cost of 1.21.

	Left=1		Left=2		Left=3		Left=4		Left=5		Left=6		Left=7	
Iteration=1	a..a		am..am		and..and		egg..egg		if..if		the..the		two..two	
	.22	a	.18	am	.20	and	.05	egg	.25	if	.02	the	.08	two
Iteration=2	a..am		am..and		and..egg		egg..if		if..the		the..two			
	.58	a	.56	and	.30	and	.35	if	.29	if	.12	two		
Iteration=3	a..and		am..egg		and..if		egg..the		if..two					
	1.02	am	.66	and	.80	if	.39	if	.47	if				
Iteration=4	a..egg		am..if		and..the		egg..two							
	1.17	am	1.21	and	.84	if	.57	if						
Iteration=5	a..if		am..the		and..two									
	1.83	and	1.27	and	1.02	if								
Iteration=6	a..the		am..two											
	1.89	and	1.53	and										
Iteration=7	a..two													
	2.15	and												

Figure 10.51 Computation of the optimal binary search tree for sample input

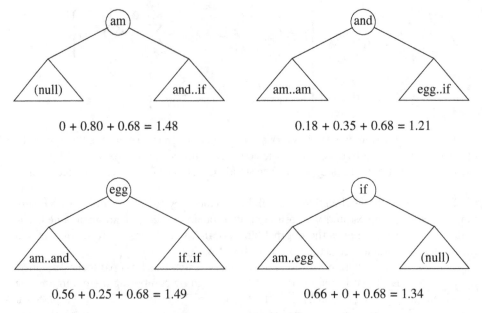

Figure 10.52   Computation of table entry (1.21, *and*) for *am..if*

The running time of this algorithm is $O(N^3)$, because when it is implemented, we obtain a triple loop. An $O(N^2)$ algorithm for the problem is sketched in the exercises.

10.3.4 All-Pairs Shortest Path

Our third and final dynamic programming application is an algorithm to compute shortest weighted paths between every pair of points in a directed graph $G = (V, E)$. In Chapter 9, we saw an algorithm for the *single-source shortest-path* problem, which finds the shortest path from some arbitrary vertex s to all others. That algorithm (Dijkstra's) runs in $O(|V|^2)$ time on dense graphs, but substantially faster on sparse graphs. We will give a short algorithm to solve the all-pairs problem for dense graphs. The running time of the algorithm is $O(|V|^3)$, which is not an asymptotic improvement over $|V|$ iterations of Dijkstra's algorithm but could be faster on a very dense graph, because its loops are tighter. The algorithm also performs correctly if there are negative edge costs, but no negative-cost cycles; Dijkstra's algorithm fails in this case.

Let us recall the important details of Dijkstra's algorithm (the reader may wish to review Section 9.3). Dijkstra's algorithm starts at a vertex s and works in stages. Each vertex in the graph is eventually selected as an intermediate vertex. If the current selected vertex is v, then for each $w \in V$, we set $d_w = \min(d_w, d_v + c_{v,w})$. This formula says that the best distance to w (from s) is either the previously known distance to w from s, or the result of going from s to v (optimally) and then directly from v to w.

Dijkstra's algorithm provides the idea for the dynamic programming algorithm: we select the vertices in sequential order. We will define $D_{k,i,j}$ to be the weight of the shortest path from v_i to v_j that uses only $v_1, v_2, \ldots, v_k$ as intermediates. By this definition, $D_{0,i,j} = c_{i,j}$, where $c_{i,j}$ is ∞ if (v_i, v_j) is not an edge in the graph. Also, by definition, $D_{|V|,i,j}$ is the shortest path from v_i to v_j in the graph.

As Figure 10.53 shows, when $k > 0$ we can write a simple formula for $D_{k,i,j}$. The shortest path from v_i to v_j that uses only $v_1, v_2, \ldots, v_k$ as intermediates is the shortest path that either does not use v_k as an intermediate at all, or consists of the merging of the two paths $v_i \rightarrow v_k$ and $v_k \rightarrow v_j$, each of which uses only the first $k-1$ vertices as intermediates. This leads to the formula

```
1    /**
2     * Compute all-shortest paths.
3     * a[ ][ ] contains the adjacency matrix with
4     * a[ i ][ i ] presumed to be zero.
5     * d[ ] contains the values of the shortest path.
6     * Vertices are numbered starting at 0; all arrays
7     * have equal dimension. A negative cycle exists if
8     * d[ i ][ i ] is set to a negative value.
9     * Actual path can be computed using path[ ][ ].
10    * NOT_A_VERTEX is -1
11    */
12   public static void allPairs( int [ ][ ] a, int [ ][ ] d, int [ ][ ] path )
13   {
14       int n = a.length;
15
16       // Initialize d and path
17       for( int i = 0; i < n; i++ )
18           for( int j = 0; j < n; j++ )
19           {
20               d[ i ][ j ] = a[ i ][ j ];
21               path[ i ][ j ] = NOT_A_VERTEX;
22           }
23
24       for( int k = 0; k < n; k++ )
25           // Consider each vertex as an intermediate
26           for( int i = 0; i < n; i++ )
27               for( int j = 0; j < n; j++ )
28                   if( d[ i ][ k ] + d[ k ][ j ] < d[ i ][ j ] )
29                   {
30                       // Update shortest path
31                       d[ i ][ j ] = d[ i ][ k ] + d[ k ][ j ];
32                       path[ i ][ j ] = k;
33                   }
34   }
```

Figure 10.53 All-pairs shortest path

$$D_{k,i,j} = \min\left\{D_{k-1,i,j}, D_{k-1,i,k} + D_{k-1,k,j}\right\}$$

The time requirement is once again $O(|V|^3)$. Unlike the two previous dynamic programming examples, this time bound has not been substantially lowered by another approach.

Because the kth stage depends only on the $(k - 1)$th stage, it appears that only two $|V| \times |V|$ matrices need to be maintained. However, using k as an *intermediate* vertex on a path that starts or finishes with k does not improve the result unless there is a negative cycle. Thus, only one matrix is necessary, because $D_{k-1,i,k} = D_{k,i,k}$ and $D_{k-1,k,j} = D_{k,k,j}$, which implies that none of the terms on the right change values and need to be saved. This observation leads to the simple program in Figure 10.53, which numbers vertices starting at zero to conform with Java's conventions.

On a complete graph, where every pair of vertices is connected (in both directions), this algorithm is almost certain to be faster than $|V|$ iterations of Dijkstra's algorithm, because the loops are so tight. Lines 17 through 22 can be executed in parallel, as can lines 26 through 33. Thus, this algorithm seems to be well suited for parallel computation.

Dynamic programming is a powerful algorithm design technique, which provides a starting point for a solution. It is essentially the divide-and-conquer paradigm of solving simpler problems first, with the important difference being that the simpler problems are not a clear division of the original. Because subproblems are repeatedly solved, it is important to record their solutions in a table rather than recompute them. In some cases, the solution can be improved (although it is certainly not always obvious and frequently difficult), and in other cases, the dynamic programming technique is the best approach known.

In some sense, if you have seen one dynamic programming problem, you have seen them all. More examples of dynamic programming can be found in the exercises and references.

10.4 Randomized Algorithms

Suppose you are a professor who is giving weekly programming assignments. You want to make sure that the students are doing their own programs or, at the very least, understand the code they are submitting. One solution is to give a quiz on the day that each program is due. On the other hand, these quizzes take time out of class, so it might only be practical to do this for roughly half of the programs. Your problem is to decide when to give the quizzes.

Of course, if the quizzes are announced in advance, that could be interpreted as an implicit license to cheat for the 50 percent of the programs that will not get a quiz. One could adopt the unannounced strategy of giving quizzes on alternate programs, but students would figure out the strategy before too long. Another possibility is to give quizzes on what seem like the important programs, but this would likely lead to similar quiz patterns from semester to semester. Student grapevines being what they are, this strategy would probably be worthless after a semester.

One method that seems to eliminate these problems is to use a coin. A quiz is made for every program (making quizzes is not nearly as time-consuming as grading them), and at the start of class, the professor will flip a coin to decide whether the quiz is to be given. This way, it is impossible to know before class whether or not the quiz will occur, and these patterns do not repeat from semester to semester. Thus, the students will have to expect that a quiz will occur with 50 percent probability, regardless of previous quiz patterns. The disadvantage is that it is possible that there is no quiz for an entire semester. This is not a likely occurrence, unless the coin is suspect. Each semester, the expected number of quizzes is half the number of programs, and with high probability, the number of quizzes will not deviate much from this.

This example illustrates what we call **randomized algorithms**. At least once during the algorithm, a random number is used to make a decision. The running time of the algorithm depends not only on the particular input, but also on the random numbers that occur.

The worst-case running time of a randomized algorithm is often the same as the worst-case running time of the nonrandomized algorithm. The important difference is that a good randomized algorithm has no bad inputs, but only bad random numbers (relative to the particular input). This may seem like only a philosophical difference, but actually it is quite important, as the following example shows.

Consider two variants of quicksort. Variant A uses the first element as pivot, while variant B uses a randomly chosen element as pivot. In both cases, the worst-case running time is $\Theta(N^2)$, because it is possible at each step that the largest element is chosen as pivot. The difference between these worst cases is that there is a particular input that can always be presented to variant A to cause the bad running time. Variant A will run in $\Theta(N^2)$ time every single time it is given an already sorted list. If variant B is presented with the same input twice, it will have two different running times, depending on what random numbers occur.

Throughout the text, in our calculations of running times, we have assumed that all inputs are equally likely. This is not true, because nearly sorted input, for instance, occurs much more often than is statistically expected, and this causes problems, particularly for quicksort and binary search trees. By using a randomized algorithm, the particular input is no longer important. The random numbers are important, and we can get an *expected* running time, where we now average over all possible random numbers instead of over all possible inputs. Using quicksort with a random pivot gives an $O(N \log N)$-expected-time algorithm. This means that for any input, including already-sorted input, the running time is expected to be $O(N \log N)$, based on the statistics of random numbers. An expected running-time bound is somewhat stronger than an average-case bound but, of course, is weaker than the corresponding worst-case bound. On the other hand, as we saw in the selection problem, solutions that obtain the worst-case bound are frequently not as practical as their average-case counterparts. Randomized algorithms usually are.

Randomized algorithms were used implicitly in perfect and universal hashing (Sections 5.7 and 5.8). In this section we will examine two additional uses of randomization. First, we will see a novel scheme for supporting the binary search tree operations in $O(\log N)$ expected time. Once again, this means that there are no bad inputs, just bad random numbers. From a theoretical point of view, this is not terribly exciting, since balanced search trees achieve this bound in the worst case. Nevertheless, the use of randomization leads to relatively simple algorithms for searching, inserting, and especially deleting.

Our second application is a randomized algorithm to test the primality of large numbers. The algorithm we present runs quickly but occasionally makes an error. The probability of error can, however, be made negligibly small.

10.4.1   Random Number Generators

Since our algorithms require random numbers, we must have a method to generate them. Actually, true randomness is virtually impossible to do on a computer, since these numbers will depend on the algorithm and thus cannot possibly be random. Generally, it suffices to produce **pseudorandom** numbers, which are numbers that appear to be random. Random numbers have many known statistical properties; pseudorandom numbers satisfy most of these properties. Surprisingly, this too is much easier said than done.

Suppose we only need to flip a coin; thus, we must generate a 0 (for heads) or 1 (for tails) randomly. One way to do this is to examine the system clock. The clock might record time as an integer that counts the number of seconds since some starting time. We could then use the lowest bit. The problem is that this does not work well if a sequence of random numbers is needed. One second is a long time, and the clock might not change at all while the program is running. Even if the time was recorded in units of microseconds, if the program was running by itself the sequence of numbers that would be generated would be far from random, since the time between calls to the generator would be essentially identical on every program invocation. We see, then, that what is really needed is a **sequence** of random numbers.[2] These numbers should appear independent. If a coin is flipped and heads appears, the next coin flip should still be equally likely to come up heads or tails.

The simplest method to generate random numbers is the **linear congruential generator**, which was first described by Lehmer in 1951. Numbers $x_1, x_2, \ldots$ are generated satisfying

$$x_{i+1} = A x_i \bmod M$$

To start the sequence, some value of x_0 must be given. This value is known as the **seed**. If $x_0 = 0$, then the sequence is far from random, but if A and M are correctly chosen, then any other $1 \leq x_0 < M$ is equally valid. If M is prime, then x_i is never 0. As an example, if $M = 11, A = 7$, and $x_0 = 1$, then the numbers generated are

$$7, 5, 2, 3, 10, 4, 6, 9, 8, 1, 7, 5, 2, \ldots$$

Notice that after $M - 1 = 10$ numbers, the sequence repeats. Thus, this sequence has a period of $M - 1$, which is as large as possible (by the pigeonhole principle). If M is prime, there are always choices of A that give a full period of $M - 1$. Some choices of A do not; if $A = 5$ and $x_0 = 1$, the sequence has a short period of 5.

---

[2] We will use random in place of pseudorandom in the rest of this section.

$$5, 3, 4, 9, 1, 5, 3, 4, \ldots$$

If M is chosen to be a large, 31-bit prime, the period should be significantly large for most applications. Lehmer suggested the use of the 31-bit prime $M = 2^{31} - 1 = 2,147,483,647$. For this prime, $A = 48,271$ is one of the many values that gives a full-period generator. Its use has been well studied and is recommended by experts in the field. We will see later that with random number generators, tinkering usually means breaking, so one is well advised to stick with this formula until told otherwise.[3]

This seems like a simple routine to implement. Generally, a class variable is used to hold the current value in the sequence of x's. When debugging a program that uses random numbers, it is probably best to set $x_0 = 1$, so that the same random sequence occurs all the time. When the program seems to work, either the system clock can be used or the user can be asked to input a value for the seed.

It is also common to return a random real number in the open interval $(0, 1)$ (0 and 1 are not possible values); this can be done by dividing by M. From this, a random number in any closed interval $[\alpha, \beta]$ can be computed by normalizing. This yields the "obvious" class in Figure 10.54 which, unfortunately, is erroneous.

The problem with this class is that the multiplication could overflow; although this is not an error, it affects the result and thus the pseudorandomness. Even though we could use 64-bit `long`s, this would slow down the computation. Schrage gave a procedure in which all the calculations can be done on a 32-bit machine without overflow. We compute the quotient and remainder of M/A and define these as Q and R, respectively. In our case, $Q = 44,488, R = 3,399$, and $R < Q$. We have

$$x_{i+1} = A x_i \bmod M = A x_i - M \left\lfloor \frac{A x_i}{M} \right\rfloor$$

$$= A x_i - M \left\lfloor \frac{x_i}{Q} \right\rfloor + M \left\lfloor \frac{x_i}{Q} \right\rfloor - M \left\lfloor \frac{A x_i}{M} \right\rfloor$$

$$= A x_i - M \left\lfloor \frac{x_i}{Q} \right\rfloor + M \left(\left\lfloor \frac{x_i}{Q} \right\rfloor - \left\lfloor \frac{A x_i}{M} \right\rfloor \right)$$

Since $x_i = Q \lfloor \frac{x_i}{Q} \rfloor + x_i \bmod Q$, we can replace the leading $A x_i$ and obtain

$$x_{i+1} = A \left(Q \left\lfloor \frac{x_i}{Q} \right\rfloor + x_i \bmod Q \right) - M \left\lfloor \frac{x_i}{Q} \right\rfloor + M \left(\left\lfloor \frac{x_i}{Q} \right\rfloor - \left\lfloor \frac{A x_i}{M} \right\rfloor \right)$$

$$= (AQ - M) \left\lfloor \frac{x_i}{Q} \right\rfloor + A(x_i \bmod Q) + M \left(\left\lfloor \frac{x_i}{Q} \right\rfloor - \left\lfloor \frac{A x_i}{M} \right\rfloor \right)$$

---

[3] For instance, it seems that

$$x_{i+1} = (48,271 x_i + 1) \bmod(2^{31} - 1)$$

would somehow be even more random. This illustrates how fragile these generators are.

$$[48,271(179,424,105) + 1] \bmod(2^{31} - 1) = 179,424,105$$

so if the seed is 179,424,105, the generator gets stuck in a cycle of period 1.

```
1    public class Random
2    {
3        private static final int A = 48271;
4        private static final int M = 2147483647;
5
6        public Random( )
7        {
8            state = System.currentTimeMillis( ) % Integer.MAX_VALUE ;
9        }
10
11       /**
12        * Return a pseudorandom int, and change the
13        * internal state. DOES NOT WORK.
14        * @return the pseudorandom int.
15        */
16       public int randomIntWRONG( )
17       {
18           return state = ( A * state ) % M;
19       }
20
21       /**
22        * Return a pseudorandom double in the open range 0..1
23        * and change the internal state.
24        * @return the pseudorandom double.
25        */
26       public double random0_1( )
27       {
28           return (double) randomInt( ) / M;
29       }
30
31       private int state;
32   }
```

Figure 10.54 Random number generator that does not work

Since $M = AQ + R$, it follows that $AQ - M = -R$. Thus, we obtain

$$x_{i+1} = A(x_i \bmod Q) - R\left\lfloor \frac{x_i}{Q} \right\rfloor + M\left(\left\lfloor \frac{x_i}{Q} \right\rfloor - \left\lfloor \frac{A x_i}{M} \right\rfloor\right)$$

The term $\delta(x_i) = \lfloor \frac{x_i}{Q} \rfloor - \lfloor \frac{A x_i}{M} \rfloor$ is either 0 or 1, because both terms are integers and their difference lies between 0 and 1. Thus, we have

$$x_{i+1} = A(x_i \bmod Q) - R\left\lfloor \frac{x_i}{Q} \right\rfloor + M\delta(x_i)$$

```
1   public class Random
2   {
3       private static final int A = 48271;
4       private static final int M = 2147483647;
5       private static final int Q = M / A;
6       private static final int R = M % A;
7
8       /**
9        * Return a pseudorandom int, and change the internal state.
10       * @return the pseudorandom int.
11       */
12      public int randomInt( )
13      {
14          int tmpState = A * ( state % Q ) - R * ( state / Q );
15
16          if( tmpState >= 0 )
17              state = tmpState;
18          else
19              state = tmpState + M;
20
21          return state;
22      }
23
24      // Remainder of this class is the same as Figure 10.54
```

Figure 10.55   Random number generator that does not overflow

A quick check shows that because $R < Q$, all the remaining terms can be calculated without overflow (this is one of the reasons for choosing $A = 48,271$). Furthermore, $\delta(x_i) = 1$ only if the remaining terms evaluate to less than zero. Thus $\delta(x_i)$ does not need to be explicitly computed but can be determined by a simple test. This leads to the revisions in Figure 10.55.

One might be tempted to assume that all machines have a random number generator at least as good as the one in Figure 10.55 in their standard library. Sadly, this is not true. Many libraries have generators based on the function

$$x_{i+1} = (A x_i + C) \bmod 2^B$$

where B is chosen to match the number of bits in the machine's integer, and C is odd. Unfortunately, these generators always produce values of x_i that alternate between even and odd—hardly a desirable property. Indeed, the lower k bits cycle with period 2^k (at best). Many other random number generators have much smaller cycles than the one provided in Figure 10.55. These are not suitable for the case where long sequences of random numbers are needed. The Java library and the UNIX drand48 function use a generator of this form. However, they use a 48-bit linear congruential generator and return only the

high 32 bits, thus avoiding the cycling problem in the low-order bits. The constants are
$A = 25{,}214{,}903{,}917$, $B = 48$, and $C = 13$.

Because Java provides 64-bit longs, implementing a basic 48-bit random number generator in standard Java can be illustrated in only a page of code. It is somewhat slower than the 31-bit random number generator, but not much so, and yields a significantly longer period. Figure 10.56 shows a respectable implementation of this random number generator.

Lines 7–10 show the basic constants of the random number generator. Because M is a power of 2, we can use bitwise operators. $M = 2^B$ can be computed by a bit shift, and instead of using the modulus operator %, we can use a bitwise and operator. This is because MASK=M-1 consists of the low 48 bits all set to 1, and a bitwise and operator with MASK thus has the effect of yielding a 48-bit result.

The next routine returns a specified number (at most 32) of random bits from the computed state, using the high-order bits which are more random than the lower bits. Line 34 is a direct application of the previously stated linear congruential formula, and line 36 is a bitwise shift (zero-filled in the high bits to avoid negative numbers). randomInt obtains 32 bits, while random0_1 obtains 53 bits (representing the mantissa; the other 11 bits of a double represent the exponent) in two separate calls.

The 48-bit random number generator (and even the 31-bit generator) is quite adequate for many applications, simple to implement in 64-bit arithmetic, and uses little space. However, linear congruential generators are unsuitable for some applications, such as cryptography or in simulations that require large numbers of highly independent and uncorrelated random numbers. In those cases, the Java class java.security.SecureRandom should be used.

10.4.2　Skip Lists

Our first use of randomization is a data structure that supports both searching and insertion in $O(\log N)$ expected time. As mentioned in the introduction to this section, this means that the running time for each operation on *any input sequence* has expected value $O(\log N)$, where the expectation is based on the random number generator. It is possible to add deletion and all the operations that involve ordering and obtain expected time bounds that match the average time bounds of binary search trees.

The simplest possible data structure to support searching is the linked list. Figure 10.57 shows a simple linked list. The time to perform a search is proportional to the number of nodes that have to be examined, which is at most N.

Figure 10.58 shows a linked list in which every other node has an additional link to the node two ahead of it in the list. Because of this, at most $\lceil N/2 \rceil + 1$ nodes are examined in the worst case.

We can extend this idea and obtain Figure 10.59. Here, every fourth node has a link to the node four ahead. Only $\lceil N/4 \rceil + 2$ nodes are examined.

The limiting case of this argument is shown in Figure 10.60. Every 2^ith node has a link to the node 2^i ahead of it. The total number of links has only doubled, but now at most $\lceil \log N \rceil$ nodes are examined during a search. It is not hard to see that the total time spent for a search is $O(\log N)$, because the search consists of either advancing to a new node or

```
1     /**
2      * Random number class, using a 48-bit
3      * linear congruential generator.
4      */
5     public class Random48
6     {
7         private static final long A = 25_214_903_917L;
8         private static final long B = 48;
9         private static final long C = 11;
10        private static final long M = (1L<<B);
11        private static final long MASK = M-1;
12
13        public Random48( )
14          { state = System.nanoTime( ) & MASK; }
15
16        public int randomInt( )
17          { return next( 32 ); }
18
19        public double random0_1( )
20          { return ( ( (long) ( next( 26 ) ) << 27 ) + next( 27 ) / (double) ( 1L << 53 ); }
21
22        /**
23         * Return specified number of random bits
24         * @param bits number of bits to return
25         * @return specified random bits
26         * @throws IllegalArgumentException if bits is more than 32
27         */
28        private int next( int bits )
29        {
30            if( bits <= 0 || bits > 32 )
31                throw new IllegalArgumentException( );
32
33            state = ( A * state + C ) & MASK;
34
35            return (int) ( state >>> ( B - bits ) );
36        }
37
38        private long state;
39    }
```

Figure 10.56 48-bit random number generator

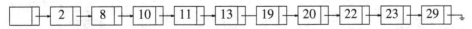

Figure 10.57 Simple linked list

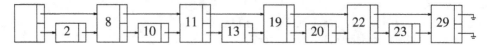

Figure 10.58   Linked list with links to two cells ahead

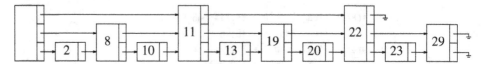

Figure 10.59   Linked list with links to four cells ahead

dropping to a lower link in the same node. Each of these steps consumes at most $O(\log N)$ total time during a search. Notice that the search in this data structure is essentially a binary search.

The problem with this data structure is that it is much too rigid to allow efficient insertion. The key to making this data structure usable is to relax the structure conditions slightly. We define a *level k node* to be a node that has k links. As Figure 10.60 shows, the ith link in any level k node ($k \geq i$) links to the next node with at least i levels. This is an easy property to maintain; however, Figure 10.60 shows a more restrictive property than this. We thus drop the restriction that the ith link links to the node 2^i ahead, and we replace it with the less restrictive condition above.

When it comes time to insert a new element, we allocate a new node for it. We must at this point decide what level the node should be. Examining Figure 10.60, we find that roughly half the nodes are level 1 nodes, roughly a quarter are level 2, and, in general, approximately $1/2^i$ nodes are level i. We choose the level of the node randomly, in accordance with this probability distribution. The easiest way to do this is to flip a coin until a head occurs and use the total number of flips as the node level. Figure 10.61 shows a typical skip list.

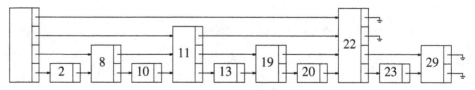

Figure 10.60   Linked list with links to 2^i cells ahead

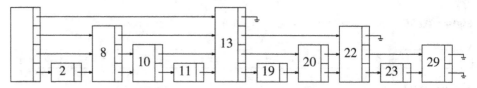

Figure 10.61   A skip list

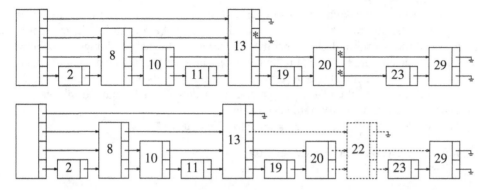

Figure 10.62 Before and after an insertion

Given this, the skip list algorithms are simple to describe. To perform a search, we start at the highest link at the header. We traverse along this level until we find that the next node is larger than the one we are looking for (or null). When this occurs, we go to the next lower level and continue the strategy. When progress is stopped at level 1, either we are in front of the node we are looking for, or it is not in the list. To perform an insert, we proceed as in a search, and keep track of each point where we switch to a lower level. The new node, whose level is determined randomly, is then spliced into the list. This operation is shown in Figure 10.62.

A cursory analysis shows that since the expected number of nodes at each level is unchanged from the original (nonrandomized) algorithm, the total amount of work that is expected to be performed traversing to nodes on the same level is unchanged. This tells us that these operations have $O(\log N)$ *expected* costs. Of course, a more formal proof is required, but it is not much different from this.

Skip lists are similar to hash tables, in that they require an estimate of the number of elements that will be in the list (so that the number of levels can be determined). If an estimate is not available, we can assume a large number or use a technique similar to rehashing. Experiments have shown that skip lists are as efficient as many balanced search tree implementations and are certainly much simpler to implement in many languages. Skip lists also have efficient concurrent implementations, unlike balanced binary search trees. Hence they are provided in the Java library class `java.util.concurrent.ConcurrentSkipList`.

10.4.3 Primality Testing

In this section we examine the problem of determining whether or not a large number is prime. As was mentioned at the end of Chapter 2, some cryptography schemes depend on the difficulty of factoring a large, 400-digit number into two 200-digit primes. In order to implement this scheme, we need a method of generating these two primes. If d is the number of digits in N, the obvious method of testing for the divisibility by odd numbers from 3 to $\sqrt{N}$ requires roughly $\frac{1}{2}\sqrt{N}$ divisions, which is about $10^{d/2}$ and is completely impractical for 200-digit numbers.

In this section, we will give a polynomial-time algorithm that can test for primality. If the algorithm declares that the number is not prime, we can be certain that the number is not prime. If the algorithm declares that the number is prime, then, with high probability but not 100 percent certainty, the number is prime. The error probability does not depend on the particular number that is being tested but instead depends on random choices made by the algorithm. Thus, this algorithm occasionally makes a mistake, but we will see that the error ratio can be made arbitrarily negligible.

The key to the algorithm is a well-known theorem due to Fermat.

Theorem 10.10. *(Fermat's Lesser Theorem)*
If P is prime, and $0 < A < P$, then $A^{P-1} \equiv 1 \pmod{P}$.

Proof.
A proof of this theorem can be found in any textbook on number theory.

For instance, since 67 is prime, $2^{66} \equiv 1 \pmod{67}$. This suggests an algorithm to test whether a number N is prime. Merely check whether $2^{N-1} \equiv 1 \pmod{N}$. If $2^{N-1} \not\equiv 1 \pmod{N}$, then we can be certain that N is not prime. On the other hand, if the equality holds, then N is probably prime. For instance, the smallest N that satisfies $2^{N-1} \equiv 1 \pmod{N}$ but is not prime is $N = 341$.

This algorithm will occasionally make errors, but the problem is that it will always make the same errors. Put another way, there is a fixed set of N for which it does not work. We can attempt to randomize the algorithm as follows: Pick $1 < A < N - 1$ at random. If $A^{N-1} \equiv 1 \pmod{N}$, declare that N is probably prime, otherwise declare that N is definitely not prime. If $N = 341$, and $A = 3$, we find that $3^{340} \equiv 56 \pmod{341}$. Thus, if the algorithm happens to choose $A = 3$, it will get the correct answer for $N = 341$.

Although this seems to work, there are numbers that fool even this algorithm for most choices of A. One such set of numbers is known as the **Carmichael numbers**. These are not prime but satisfy $A^{N-1} \equiv 1 \pmod{N}$ for *all* $0 < A < N$ that are relatively prime to N. The smallest such number is 561. Thus, we need an additional test to improve the chances of not making an error.

In Chapter 7, we proved a theorem related to quadratic probing. A special case of this theorem is the following:

Theorem 10.11.
If P is prime and $0 < X < P$, the only solutions to $X^2 \equiv 1 \pmod{P}$ are $X = 1, P - 1$.

Proof.
$X^2 \equiv 1 \pmod{P}$ implies that $X^2 - 1 \equiv 0 \pmod{P}$. This implies $(X - 1)(X + 1) \equiv 0 \pmod{P}$. Since P is prime, $0 < X < P$, and P must divide either $(X - 1)$ or $(X + 1)$, the theorem follows.

Therefore, if at any point in the computation of $A^{N-1} \pmod{N}$ we discover a violation of this theorem, we can conclude that N is definitely not prime. If we use pow, from Section 2.4.4, we see that there will be several opportunities to apply this test. We modify this routine to perform operations mod N, and apply the test of Theorem 10.11. This strategy is implemented in the pseudocode shown in Figure 10.63.

```java
1      /**
2       * Method that implements the basic primality test. If witness does not return 1,
3       * n is definitely composite. Do this by computing a^i (mod n) and looking for
4       * nontrivial square roots of 1 along the way.
5       */
6      private static long witness( long a, long i, long n )
7      {
8          if( i == 0 )
9              return 1;
10
11         long x = witness( a, i / 2, n );
12         if( x == 0 )     // If n is recursively composite, stop
13             return 0;
14
15         // n is not prime if we find a nontrivial square root of 1
16         long y = ( x * x ) % n;
17         if( y == 1 && x != 1 && x != n - 1 )
18             return 0;
19
20         if( i % 2 != 0 )
21             y = ( a * y ) % n;
22
23         return y;
24     }
25
26     /**
27      * The number of witnesses queried in randomized primality test.
28      */
29     public static final int TRIALS = 5;
30
31     /**
32      * Randomized primality test.
33      * Adjust TRIALS to increase confidence level.
34      * @param n the number to test.
35      * @return if false, n is definitely not prime.
36      *      If true, n is probably prime.
37      */
38     public static boolean isPrime( long n )
39     {
40         Random r = new Random( );
41
42         for( int counter = 0; counter < TRIALS; counter++ )
43             if( witness( r.randomLong( 2, n - 2 ), n - 1, n ) != 1 )
44                 return false;
45
46         return true;
47     }
```

Figure 10.63 A probabilistic primality testing algorithm

Recall that if `witness` returns anything but 1, it has *proven* that N cannot be prime. The proof is nonconstructive, because it gives no method of actually finding the factors. It has been shown that for any (sufficiently large) N, at most $(N - 9)/4$ values of A fool this algorithm. Thus, if A is chosen at random, and the algorithm answers that N is (probably) prime, then the algorithm is correct at least 75 percent of the time. Suppose `witness` is run 50 times. The probability that the algorithm is fooled once is at most $\frac{1}{4}$. Thus, the probability that 50 independent random trials fool the algorithm is never more than $1/4^{50} = 2^{-100}$. This is actually a very conservative estimate, which holds for only a few choices of N. Even so, one is more likely to see a hardware error than an incorrect claim of primality.

Randomized algorithms for primality testing are important because they have long been significantly faster than the best nonrandomized algorithms, and although the randomized algorithm can occasionally produce a false positive, the chances of this happening can be made small enough to be negligible.

For many years, it was suspected that it was possible to test definitively the primality of a d-digit number in time polynomial in d, but no such algorithm was known. Recently, however, deterministic polynomial time algorithms for primality testing have been discovered. While these algorithms are tremendously exciting theoretical results, they are not yet competitive with the randomized algorithms. The end of chapter references provide more information.

10.5 Backtracking Algorithms

The last algorithm design technique we will examine is **backtracking**. In many cases, a backtracking algorithm amounts to a clever implementation of exhaustive search, with generally unfavorable performance. This is not always the case, however, and even so, in some cases, the savings over a brute-force exhaustive search can be significant. Performance is, of course, relative: an $O(N^2)$ algorithm for sorting is pretty bad, but an $O(N^5)$ algorithm for the traveling salesman (or any NP-complete) problem would be a landmark result.

A practical example of a backtracking algorithm is the problem of arranging furniture in a new house. There are many possibilities to try, but typically only a few are actually considered. Starting with no arrangement, each piece of furniture is placed in some part of the room. If all the furniture is placed and the owner is happy, then the algorithm terminates. If we reach a point where all subsequent placement of furniture is undesirable, we have to undo the last step and try an alternative. Of course, this might force another undo, and so forth. If we find that we undo all possible first steps, then there is no placement of furniture that is satisfactory. Otherwise, we eventually terminate with a satisfactory arrangement. Notice that although this algorithm is essentially brute force, it does not try all possibilities directly. For instance, arrangements that consider placing the sofa in the kitchen are never tried. Many other bad arrangements are discarded early, because an undesirable subset of the arrangement is detected. The elimination of a large group of possibilities in one step is known as **pruning**.

We will see two examples of backtracking algorithms. The first is a problem in computational geometry. Our second example shows how computers select moves in games, such as chess and checkers.

10.5.1   The Turnpike Reconstruction Problem

Suppose we are given N points, $p_1, p_2, \ldots, p_N$, located on the x-axis. x_i is the x coordinate of p_i. Let us further assume that $x_1 = 0$ and the points are given from left to right. These N points determine $N(N-1)/2$ (not necessarily unique) distances $d_1, d_2, \ldots, d_N$ between every pair of points of the form $|x_i - x_j|$ ($i \neq j$). It is clear that if we are given the set of points, it is easy to *construct* the set of distances in $O(N^2)$ time. This set will not be sorted, but if we are willing to settle for an $O(N^2 \log N)$ time bound, the distances can be sorted, too. The **turnpike reconstruction problem** is to reconstruct a point set from the distances. This finds applications in physics and molecular biology (see the references for pointers to more specific information). The name derives from the analogy of points to turnpike exits on East Coast highways. Just as factoring seems harder than multiplication, the reconstruction problem seems harder than the construction problem. Nobody has been able to give an algorithm that is guaranteed to work in polynomial time. The algorithm that we will present generally runs in $O(N^2 \log N)$ but can take exponential time in the worst case.

Of course, given one solution to the problem, an infinite number of others can be constructed by adding an offset to all the points. This is why we insist that the first point is anchored at 0 and that the point set that constitutes a solution is output in nondecreasing order.

Let D be the set of distances, and assume that $|D| = M = N(N-1)/2$. As an example, suppose that

$$D = \{1, 2, 2, 2, 3, 3, 3, 4, 5, 5, 5, 6, 7, 8, 10\}$$

Since $|D| = 15$, we know that $N = 6$. We start the algorithm by setting $x_1 = 0$. Clearly, $x_6 = 10$, since 10 is the largest element in D. We remove 10 from D. The points that we have placed and the remaining distances are as shown in the following figure.

$x_1 = 0 \qquad\qquad\qquad\qquad\qquad\qquad x_6 = 10$
$D = \{1, 2, 2, 2, 3, 3, 3, 4, 5, 5, 5, 6, 7, 8\}$

The largest remaining distance is 8, which means that either $x_2 = 2$ or $x_5 = 8$. By symmetry, we can conclude that the choice is unimportant, since either both choices lead to solutions (which are mirror images of each other), or neither do, so we can set $x_5 = 8$ without affecting the solution. We then remove the distances $x_6 - x_5 = 2$ and $x_5 - x_1 = 8$ from D, obtaining

$x_1 = 0 \qquad\qquad\qquad\qquad\qquad x_5 = 8 \qquad x_6 = 10$
$D = \{1, 2, 2, 3, 3, 3, 4, 5, 5, 5, 6, 7\}$

The next step is not obvious. Since 7 is the largest value in D, either $x_4 = 7$ or $x_2 = 3$. If $x_4 = 7$, then the distances $x_6 - 7 = 3$ and $x_5 - 7 = 1$ must also be present in D. A quick check shows that indeed they are. On the other hand, if we set $x_2 = 3$, then $3 - x_1 = 3$ and $x_5 - 3 = 5$ must be present in D. These distances are also in D, so we have no guidance on which choice to make. Thus, we try one and see if it leads to a solution. If it turns out that it does not, we can come back and try the other. Trying the first choice, we set $x_4 = 7$, which leaves

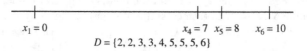

$$D = \{2, 2, 3, 3, 4, 5, 5, 5, 6\}$$

At this point, we have $x_1 = 0$, $x_4 = 7$, $x_5 = 8$, and $x_6 = 10$. Now the largest distance is 6, so either $x_3 = 6$ or $x_2 = 4$. But if $x_3 = 6$, then $x_4 - x_3 = 1$, which is impossible, since 1 is no longer in D. On the other hand, if $x_2 = 4$ then $x_2 - x_0 = 4$, and $x_5 - x_2 = 4$. This is also impossible, since 4 only appears once in D. Thus, this line of reasoning leaves no solution, so we backtrack.

Since $x_4 = 7$ failed to produce a solution, we try $x_2 = 3$. If this also fails, we give up and report no solution. We now have

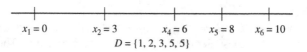

$$D = \{1, 2, 2, 3, 3, 4, 5, 5, 6\}$$

Once again, we have to choose between $x_4 = 6$ and $x_3 = 4$. $x_3 = 4$ is impossible, because D only has one occurrence of 4, and two would be implied by this choice. $x_4 = 6$ is possible, so we obtain

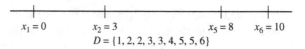

$$D = \{1, 2, 3, 5, 5\}$$

The only remaining choice is to assign $x_3 = 5$; this works because it leaves D empty, and so we have a solution.

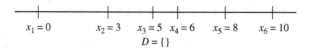

$$D = \{\}$$

Figure 10.64 shows a decision tree representing the actions taken to arrive at the solution. Instead of labeling the branches, we have placed the labels in the branches' destination nodes. A node with an asterisk indicates that the points chosen are inconsistent with the given distances; nodes with two asterisks have only impossible nodes as children, and thus represent an incorrect path.

The pseudocode to implement this algorithm is mostly straightforward. The driving routine, turnpike, is shown in Figure 10.65. It receives the point array x (which need not

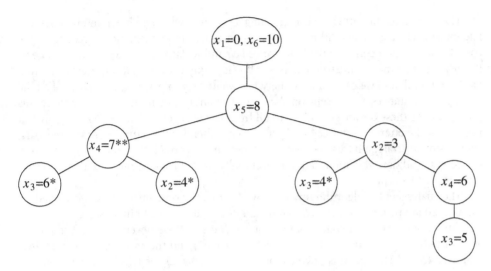

Figure 10.64   Decision tree for the worked turnpike reconstruction example

```
boolean turnpike( int [ ] x, DistSet d, int n )
{
    x[ 1 ] = 0;
    x[ n ] = d.deleteMax( );
    x[ n - 1 ] = d.deleteMax( );
    if( x[ n ] - x[ n - 1 ] ∈ d )
    {
        d.remove( x[ n ] - x[ n - 1 ] );
        return place( x, d, n, 2, n - 2 );
    }
    else
        return false;
}
```

Figure 10.65   Turnpike reconstruction algorithm: driver routine (pseudocode)

be initialized) and the distance set D and N.[4] If a solution is discovered, then **true** will be returned, the answer will be placed in x, and D will be empty. Otherwise, **false** will be returned, x will be undefined, and the distance set D will be untouched. The routine sets x_1, x_{N-1}, and x_N, as described above, alters D, and calls the backtracking algorithm place to place the other points. We presume that a check has already been made to ensure that $|D| = N(N-1)/2$.

---

[4] We have used one-letter variable names, which is generally poor style, for consistency with the worked example. We also, for simplicity, do not give the type of variables. Finally, we index arrays starting at 1, instead of 0.

The more difficult part is the backtracking algorithm, which is shown in Figure 10.66. Like most backtracking algorithms, the most convenient implementation is recursive. We pass the same arguments plus the boundaries *Left* and *Right*; $x_{Left}, \ldots, x_{Right}$ are the x coordinates of points that we are trying to place. If D is empty (or *Left* > *Right*), then a solution has been found, and we can return. Otherwise, we first try to place $x_{Right} = D_{max}$. If all the appropriate distances are present (in the correct quantity), then we tentatively place this point, remove these distances, and try to fill from *Left* to *Right* − 1. If the distances are not present, or the attempt to fill *Left* to *Right* − 1 fails, then we try setting $x_{Left} = x_N - d_{max}$, using a similar strategy. If this does not work, then there is no solution; otherwise a solution has been found, and this information is eventually passed back to turnpike by the return statement and x array.

The analysis of the algorithm involves two factors. Suppose lines 9 through 11 and 18 through 20 are never executed. We can maintain D as a balanced binary search (or splay) tree (this would require a code modification, of course). If we never backtrack, there are at most $O(N^2)$ operations involving D, such as deletion and the contains implied at lines 4 and 12 to 13. This claim is obvious for deletions, since D has $O(N^2)$ elements and no element is ever reinserted. Each call to place uses at most 2N contains, and since place never backtracks in this analysis, there can be at most $2N^2$ contains operations. Thus, if there is no backtracking, the running time is $O(N^2 \log N)$.

Of course, backtracking happens, and if it happens repeatedly, then the performance of the algorithm is affected. This can be forced to happen by construction of a pathological case. Experiments have shown that if the points have integer coordinates distributed uniformly and randomly from $[0, D_{max}]$, where $D_{max} = \Theta(N^2)$, then, almost certainly, at most one backtrack is performed during the entire algorithm.

10.5.2 Games

As our last application, we will consider the strategy that a computer might use to play a strategic game, such as checkers or chess. We will use, as an example, the much simpler game of tic-tac-toe, because it makes the points easier to illustrate.

Tic-tac-toe is a draw if both sides play optimally. By performing a careful case-by-case analysis, it is not a difficult matter to construct an algorithm that never loses and always wins when presented the opportunity. This can be done, because certain positions are known traps and can be handled by a lookup table. Other strategies, such as taking the center square when it is available, make the analysis simpler. If this is done, then by using a table we can always choose a move based only on the current position. Of course, this strategy requires the programmer, and not the computer, to do most of the thinking.

Minimax Strategy

The more general strategy is to use an evaluation function to quantify the "goodness" of a position. A position that is a win for a computer might get the value of +1; a draw could get 0; and a position that the computer has lost would get a −1. A position for which this assignment can be determined by examining the board is known as a **terminal position**.

```
      /**
       * Backtracking algorithm to place the points x[left] ... x[right].
       * x[1]...x[left-1] and x[right+1]...x[n] already tentatively placed.
       * If place returns true, then x[left]...x[right] will have values.
       */
      boolean place( int [ ] x, DistSet d, int n, int left, int right )
      {
          int dmax;
          boolean found = false;

 1        if( d.isEmpty( ) )
 2            return true;

 3        dmax = d.findMax( );

          // Check if setting x[right] = dmax is feasible.
 4        if( | x[j] - dmax | ∈ d for all 1≤j<left and right<j≤n )
          {
 5            x[right] = dmax;                    // Try x[right]=dmax
 6            for( 1≤j<left, right<j≤n )
 7                d.remove( | x[j] - dmax | );
 8            found = place( x, d, n, left, right-1 );

 9            if( !found )      // Backtrack
10                for( 1≤j<left, right<j≤n ) // Undo the deletion
11                    d.insert( | x[j] - dmax | );
          }
          // If first attempt failed, try to see if setting
          // x[left]=x[n]-dmax is feasible.
12        if( !found && ( | x[n] - dmax - x[j] | ∈ d
13                            for all 1≤j<left and right<j≤n ) )
          {
14            x[left] = x[n] - dmax;      // Same logic as before
15            for( 1≤j<left, right<j≤n )
16                d.remove( | x[n] - dmax - x[j] | );
17            found = place( x, d, n, left+1, right );

18            if( !found )      // Backtrack
19                for( 1≤j<left, right<j≤n ) // Undo the deletion
20                    d.insert( | x[n] - dmax - x[j] | );
          }
21        return found;
      }
```

Figure 10.66 Turnpike reconstruction algorithm: backtracking steps (pseudocode)

If a position is not terminal, the value of the position is determined by recursively assuming optimal play by both sides. This is known as a **minimax strategy**, because one player (the human) is trying to minimize the value of the position, while the other player (the computer) is trying to maximize it.

A **successor position** of P is any position P_s that is reachable from P by playing one move. If the computer is to move when in some position P, it recursively evaluates the value of all the successor positions. The computer chooses the move with the largest value; this is the value of P. To evaluate any successor position P_s, all of P_s's successors are recursively evaluated, and the smallest value is chosen. This smallest value represents the most favorable reply for the human player.

The code in Figure 10.67 makes the computer's strategy more clear. Lines 22 through 25 evaluate immediate wins or draws. If neither of these cases apply, then the position is nonterminal. Recalling that `value` should contain the maximum of all possible successor positions, line 28 initializes it to the smallest possible value, and the loop in lines 29 through 42 searches for improvements. Each successor position is recursively evaluated in turn by lines 32 through 34. This is recursive, because, as we will see, `findHumanMove` calls `findCompMove`. If the human's response to a move leaves the computer with a more favorable position than that obtained with the previously best computer move, then the `value` and `bestMove` are updated. Figure 10.68 shows the method for the human's move selection. The logic is virtually identical, except that the human player chooses the move that leads to the lowest-valued position. Indeed, it is not difficult to combine these two procedures into one by passing an extra variable, which indicates whose turn it is to move. This does make the code somewhat less readable, so we have stayed with separate routines.

Since these routines must pass back both the value of the position and the best move, we pass these two variables in a `MoveInfo` object.

We leave supporting routines as an exercise. The most costly computation is the case where the computer is asked to pick the opening move. Since at this stage the game is a forced draw, the computer selects square 1.[5] A total of 97,162 positions were examined, and the calculation took a few seconds. No attempt was made to optimize the code. When the computer moves second, the number of positions examined is 5,185 if the human selects the center square, 9,761 when a corner square is selected, and 13,233 when a noncorner edge square is selected.

For more complex games, such as checkers and chess, it is obviously infeasible to search all the way to the terminal nodes.[6] In this case, we have to stop the search after a certain depth of recursion is reached. The nodes where the recursion is stopped become terminal nodes. These terminal nodes are evaluated with a function that estimates the

---

[5] We numbered the squares starting from the top left and moving right. However, this is only important for the supporting routines.

[6] It is estimated that if this search were conducted for chess, at least 10^{100} positions would be examined for the first move. Even if the improvements described later in this section were incorporated, this number could not be reduced to a practical level.

```java
1   public class MoveInfo
2   {
3       public int move;
4       public int value;
5
6       public MoveInfo( int m, int v )
7         { move = m; value = v; }
8   }
9
10      /**
11       * Recursive method to find best move for computer.
12       * MoveInfo.move returns a number from 1-9 indicating square.
13       * Possible evaluations satisfy COMP_LOSS < DRAW < COMP_WIN.
14       * Complementary method findHumanMove is Figure 10.68.
15       */
16      public MoveInfo findCompMove( )
17      {
18          int i, responseValue;
19          int value, bestMove = 1;
20          MoveInfo quickWinInfo;
21
22          if( fullBoard( ) )
23              value = DRAW;
24          else if( ( quickWinInfo = immediateCompWin( ) ) != null )
25              return quickWinInfo;
26          else
27          {
28              value = COMP_LOSS;
29              for( i = 1; i <= 9; i++ )  // Try each square
30                  if( isEmpty( i ) )
31                  {
32                      place( i, COMP );
33                      responseValue = findHumanMove( ).value;
34                      unplace( i );  // Restore board
35
36                      if( responseValue > value )
37                      {
38                          // Update best move
39                          value = responseValue;
40                          bestMove = i;
41                      }
42                  }
43          }
44
45          return new MoveInfo( bestMove, value );
46      }
```

Figure 10.67 Minimax tic-tac-toe algorithm: computer selection

```
1      public MoveInfo findHumanMove( )
2      {
3          int i, responseValue;
4          int value, bestMove = 1;
5          MoveInfo quickWinInfo;
6
7          if( fullBoard( ) )
8              value = DRAW;
9          else
10         if( ( quickWinInfo = immediateHumanWin( ) ) != null )
11             return quickWinInfo;
12         else
13         {
14             value = COMP_WIN;
15             for( i = 1; i <= 9; i++ )  // Try each square
16             {
17                 if( isEmpty( i ) )
18                 {
19                     place( i, HUMAN );
20                     responseValue = findCompMove( ).value;
21                     unplace( i );  // Restore board
22
23                     if( responseValue < value )
24                     {
25                         // Update best move
26                         value = responseValue;
27                         bestMove = i;
28                     }
29                 }
30             }
31         }
32
33         return new MoveInfo( bestMove, value );
34     }
```

Figure 10.68   Minimax tic-tac-toe algorithm: human selection

value of the position. For instance, in a chess program, the evaluation function measures such variables as the relative amount and strength of pieces and positional factors. The evaluation function is crucial for success, because the computer's move selection is based on maximizing this function. The best computer chess programs have surprisingly sophisticated evaluation functions.

Nevertheless, for computer chess, the single most important factor seems to be number of moves of look-ahead the program is capable of. This is sometimes known as **ply**; it is

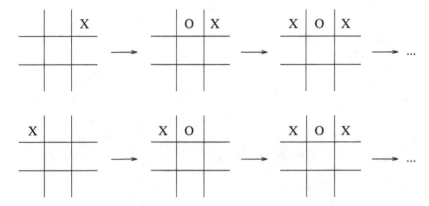

Figure 10.69 Two searches that arrive at identical position

equal to the depth of the recursion. To implement this, an extra parameter is given to the search routines.

The basic method to increase the look-ahead factor in game programs is to come up with methods that evaluate fewer nodes without losing any information. One method which we have already seen is to use a table to keep track of all positions that have been evaluated. For instance, in the course of searching for the first move, the program will examine the positions in Figure 10.69. If the values of the positions are saved, the second occurrence of a position need not be recomputed; it essentially becomes a terminal position. The data structure that records this is known as a **transposition table**; it is almost always implemented by hashing. In many cases, this can save considerable computation. For instance, in a chess endgame, where there are relatively few pieces, the time savings can allow a search to go several levels deeper.

α–β Pruning

Probably the most significant improvement one can obtain in general is known as **α–β pruning**. Figure 10.70 shows the trace of the recursive calls used to evaluate some hypothetical position in a hypothetical game. This is commonly referred to as a **game tree**. (We have avoided the use of this term until now, because it is somewhat misleading: No tree is actually constructed by the algorithm. The game tree is just an abstract concept.) The value of the game tree is 44.

Figure 10.71 shows the evaluation of the same game tree, with several (but not all possible) unevaluated nodes. Almost half of the terminal nodes have not been checked. We show that evaluating them would not change the value at the root.

First, consider node D. Figure 10.72 shows the information that has been gathered when it is time to evaluate D. At this point, we are still in findHumanMove and are contemplating a call to findCompMove on D. However, we already know that findHumanMove will return at most 40, since it is a *min* node. On the other hand, its *max* node parent has already found a sequence that guarantees 44. Nothing that D does can possibly increase this value. Therefore, D does not need to be evaluated. This pruning

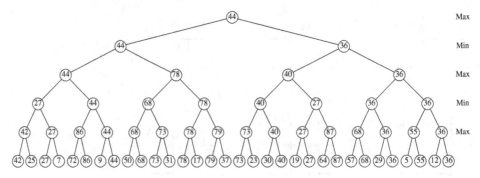

Figure 10.70   A hypothetical game tree

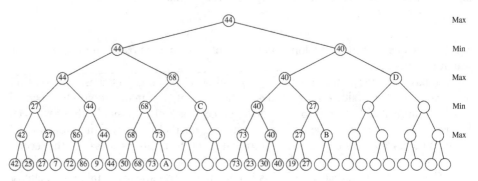

Figure 10.71   A pruned game tree

of the tree is known as α pruning. An identical situation occurs at node *B*. To implement α pruning, findCompMove passes its tentative maximum (α) to findHumanMove. If the tentative minimum of findHumanMove falls below this value, then findHumanMove returns immediately.

A similar thing happens at nodes *A* and *C*. This time, we are in the middle of a findCompMove and are about to make a call to findHumanMove to evaluate *C*. Figure 10.73 shows the situation that is encountered at node *C*. However, the findHumanMove, at the *min* level, which has called findCompMove, has already determined that it can force a value of at most 44 (recall that low values are good for the human side). Since findCompMove has a tentative maximum of 68, nothing that *C* does will affect the result at the *min* level. Therefore, *C* should not be evaluated. This type of pruning is known as β pruning; it is the symmetric version of α pruning. When both techniques are combined, we have α–β pruning.

Implementing α–β pruning requires surprisingly little code. Figure 10.74 shows half of the α–β pruning scheme (minus type declarations); you should have no trouble coding the other half.

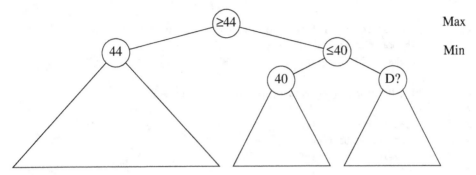

Figure 10.72   The node marked ? is unimportant

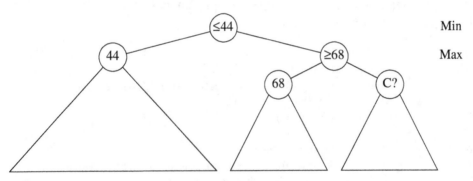

Figure 10.73   The node marked ? is unimportant

To take full advantage of α–β pruning, game programs usually try to apply the evaluation function to nonterminal nodes in an attempt to place the best moves early in the search. The result is even more pruning than one would expect from a random ordering of the nodes. Other techniques, such as searching deeper in more active lines of play, are also employed.

In practice, α–β pruning limits the searching to only $O(\sqrt{N})$ nodes, where N is the size of the full game tree. This is a huge savings and means that searches using α–β pruning can go twice as deep as compared to an unpruned tree. Our tic-tac-toe example is not ideal, because there are so many identical values, but even so, the initial search of 97,162 nodes is reduced to 4,493 nodes. (These counts include nonterminal nodes.)

In many games, computers are among the best players in the world. The techniques used are very interesting and can be applied to more serious problems. More details can be found in the references.

```
1      /**
2       * Same as before, but perform alpha-beta pruning.
3       * The main routine should make the call with
4       * alpha = COMP_LOSS and beta = COMP_WIN.
5       */
6      public MoveInfo findCompMove( int alpha, int beta )
7      {
8          int i, responseValue;
9          int value, bestMove = 1;
10         MoveInfo quickWinInfo;
11
12         if( fullBoard( ) )
13             value = DRAW;
14         else
15         if( ( quickWinInfo = immediateCompWin( ) ) != null )
16             return quickWinInfo;
17         else
18         {
19             value = alpha;
20             for( i = 1; i <= 9 && value < beta; i++ )  // Try each square
21             {
22                 if( isEmpty( i ) )
23                 {
24                     place( i, COMP );
25                     responseValue = findHumanMove( value, beta ).value;
26                     unplace( i );  // Restore board
27
28                     if( responseValue > value )
29                     {
30                             // Update best move
31                         value = responseValue;
32                         bestMove = i;
33                     }
34                 }
35             }
36         }
37
38         return new MoveInfo( bestMove, value );
39     }
```

Figure 10.74 Minimax tic-tac-toe algorithm with α–β pruning: computer selection

Summary

This chapter illustrates five of the most common techniques found in algorithm design. When confronted with a problem, it is worthwhile to see if any of these methods apply. A proper choice of algorithm, combined with judicious use of data structures, can often lead quickly to efficient solutions.

Exercises

10.1 Show that the greedy algorithm to minimize the mean completion time for multiprocessor job scheduling works.

10.2 The input is a set of jobs $j_1, j_2, \ldots, j_N$, each of which takes one time unit to complete. Each job j_i earns d_i dollars if it is completed by the time limit t_i, but no money if completed after the time limit.
  a. Give an $O(N^2)$ greedy algorithm to solve the problem.
  **b. Modify your algorithm to obtain an $O(N \log N)$ time bound. (*Hint:* The time bound is due entirely to sorting the jobs by money. The rest of the algorithm can be implemented, using the disjoint set data structure, in $o(N \log N)$.)

10.3 A file contains only colons, spaces, newlines, commas, and digits in the following frequency: colon (100), space (605), newline (100), comma (705), 0 (431), 1 (242), 2 (176), 3 (59), 4 (185), 5 (250), 6 (174), 7 (199), 8 (205), 9 (217). Construct the Huffman code.

10.4 Part of the encoded file must be a header indicating the Huffman code. Give a method for constructing the header of size at most $O(N)$ (in addition to the symbols), where N is the number of symbols.

10.5 Complete the proof that Huffman's algorithm generates an optimal prefix code.

10.6 Show that if the symbols are sorted by frequency, Huffman's algorithm can be implemented in linear time.

10.7 Write a program to implement file compression (and uncompression) using Huffman's algorithm.

*10.8 Show that any online bin-packing algorithm can be forced to use at least $\frac{3}{2}$ the optimal number of bins, by considering the following sequence of items: N items of size $\frac{1}{6} - 2\epsilon$, N items of size $\frac{1}{3} + \epsilon$, N items of size $\frac{1}{2} + \epsilon$.

10.9 Give a simple analysis to show the performance bound for first fit decreasing bin packing when
  a. The smallest item size is larger than $\frac{1}{3}$.
  *b. The smallest item size is larger than $\frac{1}{4}$.
  *c. The smallest item size is smaller than $\frac{2}{11}$.

10.10 Explain how to implement first fit and best fit in $O(N \log N)$ time.

10.11 Show the operation of all the bin-packing strategies discussed in Section 10.1.3 on the input 0.42, 0.25, 0.27, 0.07, 0.72, 0.86, 0.09, 0.44, 0.50, 0.68, 0.73, 0.31, 0.78, 0.17, 0.79, 0.37, 0.73, 0.23, 0.30.

10.12 Write a program that compares the performance (both in time and number of bins used) of the various bin-packing heuristics.

10.13 Prove Theorem 10.7.

10.14 Prove Theorem 10.8.

*10.15 N points are placed in a unit square. Show that the distance between the closest pair is $O(N^{-1/2})$.

*10.16 Argue that for the closest-points algorithm, the average number of points in the strip is $O(\sqrt{N})$. (*Hint:* Use the result of the previous exercise.)

10.17 Write a program to implement the closest-pair algorithm.

10.18 What is the asymptotic running time of quickselect, using a median-of-median-of-three partitioning strategy?

10.19 Show that quickselect with median-of-median-of-seven partitioning is linear. Why is median-of-median-of-seven partitioning not used in the proof?

10.20 Implement the quickselect algorithm in Chapter 7, quickselect using median-of-median-of-five partitioning, and the sampling algorithm at the end of Section 10.2.3. Compare the running times.

10.21 Much of the information used to compute the median-of-median-of-five is thrown away. Show how the number of comparisons can be reduced by more careful use of the information.

*10.22 Complete the analysis of the sampling algorithm described at the end of Section 10.2.3, and explain how the values of δ and s are chosen.

10.23 Show how the recursive multiplication algorithm computes XY, where $X = 1234$ and $Y = 4321$. Include all recursive computations.

10.24 Show how to multiply two complex numbers $X = a + bi$ and $Y = c + di$ using only three multiplications.

10.25 a. Show that

$$X_L Y_R + X_R Y_L = (X_L + X_R)(Y_L + Y_R) - X_L Y_L - X_R Y_R$$

b. This gives an $O(N^{1.59})$ algorithm to multiply N-bit numbers. Compare this method to the solution in the text.

10.26 *a. Show how to multiply two numbers by solving five problems that are roughly one-third of the original size.

   **b. Generalize this problem to obtain an $O(N^{1+\epsilon})$ algorithm for any constant $\epsilon > 0$.

   c. Is the algorithm in part (b) better than $O(N \log N)$?

10.27 Why is it important that Strassen's algorithm does not use commutativity in the multiplication of 2×2 matrices?

10.28   Two 70×70 matrices can be multiplied using 143,640 multiplications. Show how this can be used to improve the bound given by Strassen's algorithm.

10.29   What is the optimal way to compute $A_1A_2A_3A_4A_5A_6$, where the dimensions of the matrices are $A_1 : 10 \times 20$, $A_2 : 20 \times 1$, $A_3 : 1 \times 40$, $A_4 : 40 \times 5$, $A_5 : 5 \times 30$, $A_6 : 30 \times 15$?

10.30   Show that none of the following greedy algorithms for chained matrix multiplication work. At each step
a. Compute the cheapest multiplication.
b. Compute the most expensive multiplication.
c. Compute the multiplication between the two matrices M_i and M_{i+1}, such that the number of columns in M_i is minimized (breaking ties by one of the rules above).

10.31   Write a program to compute the best ordering of matrix multiplication. Include the routine to print out the actual ordering.

10.32   Show the optimal binary search tree for the following words, where the frequency of occurrence is in parentheses: a (0.18), and (0.19), I (0.23), it (0.21), or (0.19).

*10.33   Extend the optimal binary search tree algorithm to allow for unsuccessful searches. In this case, q_j, for $1 \le j < N$, is the probability that a search is performed for any word W satisfying $w_j < W < w_{j+1}$. q_0 is the probability of performing a search for $W < w_1$, and q_N is the probability of performing a search for $W > w_N$. Notice that $\sum_{i=1}^{N} p_i + \sum_{j=0}^{N} q_j = 1$.

*10.34   Suppose $C_{i,i} = 0$ and that otherwise

$$C_{i,j} = W_{i,j} + \min_{i < k \le j} (C_{i,k-1} + C_{k,j})$$

Suppose that $\mathbf{W}$ satisfies the *quadrangle inequality*, namely, for all $i \le i' \le j \le j'$,

$$W_{i,j} + W_{i',j'} \le W_{i',j} + W_{i,j'}$$

Suppose further, that $\mathbf{W}$ is *monotone*: If $i \le i'$ and $j \le j'$, then $W_{i,j} \le W_{i',j'}$.
a. Prove that $\mathbf{C}$ satisfies the quadrangle inequality.
b. Let $R_{i,j}$ be the largest k that achieves the minimum $C_{i,k-1} + C_{k,j}$. (That is, in case of ties, choose the largest k.) Prove that

$$R_{i,j} \le R_{i,j+1} \le R_{i+1,j+1}$$

c. Show that $\mathbf{R}$ is nondecreasing along each row and column.
d. Use this to show that all entries in $\mathbf{C}$ can be computed in $O(N^2)$ time.
e. Which of the dynamic programming algorithms can be solved in $O(N^2)$ using these techniques?

10.35   Write a routine to reconstruct the shortest paths from the algorithm in Section 10.3.4.

10.36   The binomial coefficients $C(N, k)$ can be defined recursively as follows: $C(N,0) = 1$, $C(N, N) = 1$, and, for $0 < k < N$, $C(N, k) = C(N - 1, k) + C(N - 1, k - 1)$.

```
1        CoinSide flip( )
2        {
3            if( ( random( ) % 2 ) == 0 )
4                return HEADS;
5            else
6                return TAILS;
7        }
```

Figure 10.75 Questionable coin flipper

Write a method and give an analysis of the running time to compute the binomial coefficients as follows:

a. recursively

b. using dynamic programming

10.37 Write the routines to perform insertion, deletion, and searching in skip lists.

10.38 Give a formal proof that the expected time for the skip list operations is $O(\log N)$.

10.39 Figure 10.75 shows a routine to flip a coin, assuming that random returns an integer (which is prevalent in many systems). What is the expected performance of the skip list algorithms if the random number generator uses a modulus of the form $M = 2^B$ (which is unfortunately prevalent on many systems)?

10.40 a. Use the exponentiation algorithm to prove that $2^{340} \equiv 1 \pmod{341}$.

b. Show how the randomized primality test works for $N = 561$ with several choices of A.

10.41 Implement the turnpike reconstruction algorithm.

10.42 Two point sets are **homometric** if they yield the same distance set and are not rotations of each other. The following distance set gives two distinct point sets: $\{ 1, 2, 3, 4, 5, 6, 7, 8, 9, 10, 11, 12, 13, 16, 17 \}$. Find the two point sets.

10.43 Extend the reconstruction algorithm to find *all* homometric point sets given a distance set.

10.44 Show the result of α–β pruning of the tree in Figure 10.76.

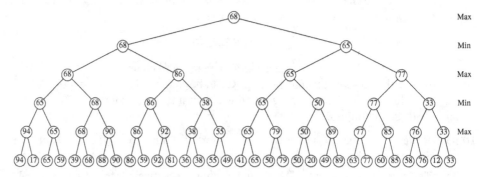

Figure 10.76 Game tree, which can be pruned

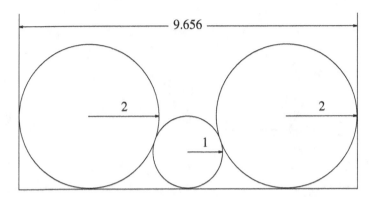

Figure 10.77 Sample for circle packing problem

10.45 a. Does the code in Figure 10.74 implement α pruning or β pruning?
b. Implement the complementary routine.

10.46 Write the remaining procedures for tic-tac-toe.

10.47 The **one-dimensional circle packing problem** is as follows: You have N circles of radii $r_1, r_2, \ldots, r_N$. These circles are packed in a box such that each circle is tangent to the bottom of the box and are arranged in the original order. The problem is to find the width of the minimum-sized box. Figure 10.77 shows an example with circles of radii 2, 1, 2 respectively. The minimum-sized box has width $4 + 4\sqrt{2}$.

*10.48 Suppose that the edges in an undirected graph G satisfy the triangle inequality: $c_{u,v} + c_{v,w} \geq c_{u,w}$. Show how to compute a traveling salesman tour of cost at most twice optimal. (*Hint:* Construct a minimum spanning tree.)

*10.49 You are a tournament director and need to arrange a round robin tournament among $N = 2^k$ players. In this tournament, everyone plays exactly one game each day; after $N - 1$ days, a match has occurred between every pair of players. Give a recursive algorithm to do this.

*10.50 a. Prove that in a round robin tournament it is always possible to arrange the players in an order $p_{i_1}, p_{i_2}, \ldots, p_{i_N}$ such that for all $1 \leq j < N$, p_{i_j} has won the match against $p_{i_{j+1}}$.
b. Give an $O(N \log N)$ algorithm to find one such arrangement. Your algorithm may serve as a proof for part (a).

*10.51 We are given a set $P = p_1, p_2, \ldots, p_N$ of N points in a plane. A **Voronoi diagram** is a partition of the plane into N regions R_i such that all points in R_i are closer to p_i than any other point in P. Figure 10.78 shows a sample Voronoi diagram for seven (nicely arranged) points. Give an $O(N \log N)$ algorithm to construct the Voronoi diagram.

*10.52 A **convex polygon** is a polygon with the property that any line segment whose endpoints are on the polygon lies entirely within the polygon. The **convex hull**

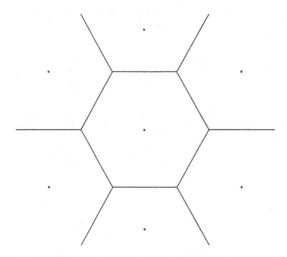

Figure 10.78 Voronoi diagram

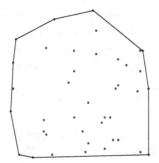

Figure 10.79 Example of a convex hull

problem consists of finding the smallest (area) convex polygon that encloses a set of points in the plane. Figure 10.79 shows the convex hull for a set of 40 points. Give an $O(N \log N)$ algorithm to find the convex hull.

*10.53 Consider the problem of right-justifying a paragraph. The paragraph contains a sequence of words $w_1, w_2, \ldots, w_N$ of length $a_1, a_2, \ldots, a_N$, which we wish to break into lines of length L. Words are separated by blanks whose ideal length is b (millimeters), but blanks can stretch or shrink as necessary (but must be >0), so that a line $w_i w_{i+1} \ldots w_j$ has length exactly L. However, for each blank b' we charge $|b' - b|$ *ugliness points*. The exception to this is the last line, for which we charge only if $b' < b$ (in other words, we charge only for shrinking), since the last line does not need to be justified. Thus, if b_i is the length of the blank between a_i and a_{i+1}, then the *ugliness* of setting any line (but the last) $w_i w_{i+1} \ldots w_j$ for $j > i$ is $\sum_{k=i}^{j-1} |b_k - b| = (j - i)|b' - b|$, where b' is the average size of a blank on this line. This is true of the last line only if $b' < b$, otherwise the last line is not ugly at all.

a. Give a dynamic programming algorithm to find the least ugly setting of $w_1, w_2, \ldots, w_N$ into lines of length L. (*Hint:* For $i = N, N - 1, \ldots, 1$, compute the best way to set $w_i, w_{i+1}, \ldots, w_N$.)

b. Give the time and space complexities for your algorithm (as a function of the number of words, N).

c. Consider the special case where we are using a fixed-width font, and assume the optimal value of b is 1 (space). In this case, no shrinking of blanks is allowed, since the next smallest blank space would be 0. Give a linear-time algorithm to generate the least ugly setting for this case.

*10.54 The *longest increasing subsequence* problem is as follows: Given numbers $a_1, a_2, \ldots, a_N$, find the maximum value of k such that $a_{i_1} < a_{i_2} < \cdots < a_{i_k}$, and $i_1 < i_2 < \cdots < i_k$. As an example, if the input is 3, 1, 4, 1, 5, 9, 2, 6, 5, the maximum increasing subsequence has length four (1, 4, 5, 9 among others). Give an $O(N^2)$ algorithm to solve the longest increasing subsequence problem.

*10.55 The *longest common subsequence* problem is as follows: Given two sequences $A = a_1, a_2, \ldots, a_M$, and $B = b_1, b_2, \ldots, b_N$, find the length, k, of the longest sequence $C = c_1, c_2, \ldots, c_k$ such that C is a subsequence (not necessarily continguous) of both A and B. As an example, if

$$A = \text{d,y,n,a,m,i,c}$$

and

$$B = \text{p,r,o,g,r,a,m,m,i,n,g,}$$

then the longest common subsequence is a,m,i and has length 3. Give an algorithm to solve the longest common subsequence problem. Your algorithm should run in $O(MN)$ time.

*10.56 The *pattern-matching problem* is as follows: Given a string S of text, and a pattern P, find the first occurrence of P in S. *Approximate pattern matching* allows k mismatches of three types:

1. A character can be in S that is not in P.
2. A character can be in P that is not in S.
3. P and S can differ in a position.

As an example, if we are searching for the pattern "textbook" with at most three mismatches in the string "data structures txtborpk", we find a match (insert an e, change an r to an o, delete a p). Give an $O(MN)$ algorithm to solve the approximate string matching problem, where $M = |P|$ and $N = |S|$.

*10.57 One form of the *knapsack problem* is as follows: We are given a set of integers $A = a_1, a_2, \ldots, a_N$ and an integer K. Is there a subset of A whose sum is exactly K?

a. Give an algorithm that solves the knapsack problem in $O(NK)$ time.

b. Why does this not show that $P = NP$?

*10.58 You are given a currency system with coins of (decreasing) value $c_1, c_2, \ldots, c_N$ cents.
   a. Give an algorithm that computes the minimum number of coins required to give K cents in change.
   b. Give an algorithm that computes the number of different ways to give K cents in change.

*10.59 Consider the problem of placing eight queens on an (eight-by-eight) chess board. Two queens are said to attack each other if they are on the same row, column, or (not necessarily main) diagonal.
   a. Give a randomized algorithm to place eight nonattacking queens on the board.
   b. Give a backtracking algorithm to solve the same problem.
   c. Implement both algorithms and compare the running time.

*10.60 In the game of chess, a knight in row R and column C may move to row $1 \le R' \le B$ and column $1 \le C' \le B$ (where B is the size of the board) provided that either

$$|R - R'| = 2 \quad \text{and} \quad |C - C'| = 1$$

$$or$$

$$|R - R'| = 1 \quad \text{and} \quad |C - C'| = 2$$

A *knight's tour* is a sequence of moves that visits all squares exactly once before returning to the starting point.
   a. If B is odd, show that a knight's tour cannot exist.
   b. Give a backtracking algorithm to find a knight's tour.

10.61 Consider the recursive algorithm in Figure 10.80 for finding the shortest weighted path in an acyclic graph, from s to t.

```
Distance shortest( s,t )
{
    Distance dₜ, tmp;

    if( s == t )
        return 0;

    dₜ = ∞;
    for each Vertex v adjacent to s
    {
        tmp = shortest(v,t );
        if( c_{s,v} + tmp < dₜ)
            dₜ = c_{s,v} + tmp;
    }
    return dₜ;
}
```

Figure 10.80 Recursive shortest-path algorithm pseudocode

  a. Why does this algorithm not work for general graphs?
  b. Prove that this algorithm terminates for acyclic graphs.
  c. What is the worst-case running time of the algorithm?

10.62 Let A be an N-by-N matrix of zeros and ones. A submatrix S of A is any group of *contiguous* entries that forms a square.
  a. Design an $O(N^2)$ algorithm that determines the size of the largest submatrix of ones in A. For instance, in the matrix that follows, the largest submatrix is a 4-by-4 square.

```
10111000
00010100
00111000
00111010
00111111
01011110
01011110
00011110
```

  **b. Repeat part (a) if S is allowed to be a rectangle instead of a square. *Largest* is measured by area.

10.63 Even if the computer has a move that gives an immediate win, it may not make it if it detects another move that is also guaranteed to win. Some early chess programs were problematic in that they would get into a repetition of position when a forced win was detected, thereby allowing the opponent to claim a draw. In tic-tac-toe, this is not a problem, because the program eventually will win. Modify the tic-tac-toe algorithm so that when a winning position is found, the move that leads to the shortest win is always taken. You can do this by adding `9-depth` to `COMP_WIN` so that the quickest win gives the highest value.

10.64 Write a program to play 5-by-5 tic-tac-toe, where 4 in a row wins. Can you search to terminal nodes?

10.65 The game of Boggle consists of a grid of letters and a word list. The object is to find words in the grid that are subject to the constraint that two adjacent letters must be adjacent in the grid, and each item in the grid can be used, at most, once per word. Write a program to play Boggle.

10.66 Write a program to play MAXIT. The board is represented as an N-by-N grid of numbers randomly placed at the start of the game. One position is designated as the initial current position. Two players alternate turns. At each turn, a player must select a grid element in the current row or column. The value of the selected position is added to the player's score, and that position becomes the current position and cannot be selected again. Players alternate until all grid elements in the current row and column are already selected, at which point the game ends and the player with the higher score wins.

10.67 Othello played on a 6-by-6 board is a forced win for black. Prove this by writing a program. What is the final score if play on both sides is optimal?

References

The original paper on Huffman codes is [25]. Variations on the algorithm are discussed in [33], [34], and [37]. Another popular compression scheme is Ziv–Lempel encoding [67], [68]. Here the codes have a fixed length but represent strings instead of characters. [9] and [39] are good surveys of the common compression schemes.

The analysis of bin-packing heuristics first appeared in Johnson's Ph.D. thesis and was published in [26]. Improvements in the additive constants of the bounds for first fit and first fit decreasing were given in [63] and [16], respectively. The improved lower bound for online bin packing given in Exercise 10.8 is from [64]; this result has been improved further in [41] and [61]. [54] describes another approach to online bin packing.

Theorem 10.7 is from [8]. The closest points algorithm appeared in [56]. [58] and describes the turnpike reconstruction problem and its applications. The exponential worst-case input was given by [66]. Books on computational geometry include [17], [50], [45], and [46]. [2] contains the lecture notes for a computational geometry course taught at MIT; it includes an extensive bibliography.

The linear-time selection algorithm appeared in [10]. The best bound for selecting the median is currently $\sim 2.95N$ comparisons [15]. [20] discusses the sampling approach that finds the median in $1.5N$ expected comparisons. The $O(N^{1.59})$ multiplication is from [27]. Generalizations are discussed in [11] and [29]. Strassen's algorithm appears in the short paper [59]. The paper states the results and not much else. Pan [47] gives several divide-and-conquer algorithms, including the one in Exercise 10.28. The best-known bound is $O(N^{2.376})$, which is due to Coppersmith and Winograd [14].

The classic references on dynamic programming are the books [6] and [7]. The matrix ordering problem was first studied in [22]. It was shown in [24] that the problem can be solved in $O(N \log N)$ time.

An $O(N^2)$ algorithm was provided for the construction of optimal binary search trees by Knuth [30]. The all-pairs shortest-path algorithm is from Floyd [19]. A theoretically better $O(N^3 (\log \log N / \log N)^{1/3})$ algorithm is given by Fredman [21], but not surprisingly, it is not practical. A slightly improved bound (with 1/2 instead of 1/3) is given in [60], and the current best bound is $O(N^3 \sqrt{\log \log N / \log N})$ [69]; see also [4] for related results. For undirected graphs, the all-pairs problem can be solved in $O(|E||V| \log \alpha(|E|, |V|))$, where α was previously seen in the union/find analysis in Chapter 8 [49]. Under certain conditions, the running time of dynamic programs can automatically be improved by a factor of N or more. This is discussed in Exercise 10.34, [18], and [65].

The discussion of random number generators is based on [48]. Park and Miller attribute the portable implementation to Schrage [57]. Skip lists are discussed by Pugh in [51]. An alternative, namely the *treap,* is discussed in Chapter 12. The randomized primality-testing algorithm is due to Miller [42] and Rabin [53]. The theorem that at most $(N - 9)/4$ values of A fool the algorithm is from Monier [43]. In 2002, an $O(d^{12})$ deterministic polynomial-time primality testing algorithm was discovered [3], and subsequently an improved algorithm with running time $O(d^6)$ was found [40]. However, these algorithms are slower than the randomized algorithm. Other randomized algorithms are discussed in [52]. More examples of randomization techniques can be found in [24], [28], and [44].

More information on α–β pruning can be found in [1], [31], and [34]. The top programs that play chess, checkers, Othello, and backgammon all achieved world class status in the 1990s. The world's leading checkers program, Chinook, has improved to the point that in 2007, it probably could not lose a game [55]. [38] describes an Othello program. The paper appears in a special issue on computer games (mostly chess); this issue is a gold mine of ideas. One of the papers describes the use of dynamic programming to solve chess endgames completely when only a few pieces are left on the board. Related research resulted in the change in 1989 (later revoked in 1992) of the 50-move rule in certain cases.

Exercise 10.42 is solved in [9]. Determining whether a homometric point set with no duplicate distances exists for $N > 6$ is open. Christofides [13] gives a solution to Exercise 10.48, and also an algorithm that generates a tour at most $\frac{3}{2}$ optimal. Exercise 10.53 is discussed in [32]. Exercise 10.56 is solved in [62]. An $O(kN)$ algorithm is given in [35]. Exercise 10.58 is discussed in [12], but do not be misled by the title of the paper.

1. B. Abramson, "Control Strategies for Two-Player Games," *ACM Computing Surveys,* 21 (1989), 137–161.

2. A. Aggarwal and J. Wein, *Computational Geometry: Lecture Notes for 18.409,* MIT Laboratory for Computer Science, 1988.

3. M. Agrawal, N. Kayal, and N. Saxena, "Primes in P (preprint)" (2002) (see http://www.cse.iitk.ac.in/news/primality.pdf).

4. N. Alon, Z. Galil, and O. Margalit, "On the Exponent of the All-Pairs Shortest Path Problem," *Proceedings of the Thirty-Second Annual Symposium on the Foundations of Computer Science* (1991), 569–575.

5. T. Bell, I. H. Witten, and J. G. Cleary, "Modeling for Text Compression," *ACM Computing Surveys,* 21 (1989), 557–591.

6. R. E. Bellman, *Dynamic Programming,* Princeton University Press, Princeton, N. J., 1957.

7. R. E. Bellman and S. E. Dreyfus, *Applied Dynamic Programming,* Princeton University Press, Princeton, N.J., 1962.

8. J. L. Bentley, D. Haken, and J. B. Saxe, "A General Method for Solving Divide-and-Conquer Recurrences," *SIGACT News,* 12 (1980), 36–44.

9. G. S. Bloom, "A Counterexample to the Theorem of Piccard," *Journal of Combinatorial Theory A* (1977), 378–379.

10. M. Blum, R. W. Floyd, V. R. Pratt, R. L. Rivest, and R. E. Tarjan, "Time Bounds for Selection," *Journal of Computer and System Sciences,* 7 (1973), 448–461.

11. A. Borodin and J. I. Munro, *The Computational Complexity of Algebraic and Numerical Problems,* American Elsevier, New York, 1975.

12. L. Chang and J. Korsh, "Canonical Coin Changing and Greedy Solutions," *Journal of the ACM,* 23 (1976), 418–422.

13. N. Christofides, "Worst-case Analysis of a New Heuristic for the Traveling Salesman Problem," *Management Science Research Report #388,* Carnegie-Mellon University, Pittsburgh, Pa., 1976.

14. D. Coppersmith and S. Winograd, "Matrix Multiplication via Arithmetic Progressions," *Proceedings of the Nineteenth Annual ACM Symposium on the Theory of Computing* (1987), 1–6.

15. D. Dor and U. Zwick, "Selecting the Median," *SIAM Journal on Computing*, 28 (1999), 1722–1758.

16. G. Dosa, "The Tight Bound of First Fit Decreasing Bin-Packing Algorithm Is FFD (I)=(11/9)OPT(I)+6/9," *Combinatorics, Algorithms, Probabilistic and Experimental Methodologies* (ESCAPE 2007), (2007), 1–11.

17. H. Edelsbrunner, *Algorithms in Combinatorial Geometry*, Springer-Verlag, Berlin, 1987.

18. D. Eppstein, Z. Galil, and R. Giancarlo, "Speeding up Dynamic Programming," *Proceedings of the Twenty-ninth Annual IEEE Symposium on the Foundations of Computer Science* (1988), 488–495.

19. R. W. Floyd, "Algorithm 97: Shortest Path," *Communications of the ACM*, 5 (1962), 345.

20. R. W. Floyd and R. L. Rivest, "Expected Time Bounds for Selection," *Communications of the ACM*, 18 (1975), 165–172.

21. M. L. Fredman, "New Bounds on the Complexity of the Shortest Path Problem," *SIAM Journal on Computing*, 5 (1976), 83–89.

22. S. Godbole, "On Efficient Computation of Matrix Chain Products," *IEEE Transactions on Computers*, 9 (1973), 864–866.

23. R. Gupta, S. A. Smolka, and S. Bhaskar, "On Randomization in Sequential and Distributed Algorithms," *ACM Computing Surveys*, 26 (1994), 7–86.

24. T. C. Hu and M. R. Shing, "Computations of Matrix Chain Products, Part I," *SIAM Journal on Computing*, 11 (1982), 362–373.

25. D. A. Huffman, "A Method for the Construction of Minimum Redundancy Codes," *Proceedings of the IRE*, 40 (1952), 1098–1101.

26. D. S. Johnson, A. Demers, J. D. Ullman, M. R. Garey, and R. L. Graham, "Worst-case Performance Bounds for Simple One-Dimensional Packing Algorithms," *SIAM Journal on Computing*, 3 (1974), 299–325.

27. A. Karatsuba and Y. Ofman, "Multiplication of Multi-digit Numbers on Automata," *Doklady Akademii Nauk SSSR*, 145 (1962), 293–294.

28. D. R. Karger, "Random Sampling in Graph Optimization Problems," Ph. D. thesis, Stanford University, 1995.

29. D. E. Knuth, *The Art of Computer Programming, Vol 2: Seminumerical Algorithms*, 3d ed., Addison-Wesley, Reading, Mass., 1998.

30. D. E. Knuth, "Optimum Binary Search Trees," *Acta Informatica*, 1 (1971), 14–25.

31. D. E. Knuth, "An Analysis of Alpha-Beta Cutoffs," *Artificial Intelligence*, 6 (1975), 293–326.

32. D. E. Knuth, *TEX and Metafont, New Directions in Typesetting*, Digital Press, Bedford, Mass., 1981.

33. D. E. Knuth, "Dynamic Huffman Coding," *Journal of Algorithms*, 6 (1985), 163–180.

34. D. E. Knuth and R. W. Moore, "Estimating the Efficiency of Backtrack Programs," *Mathematics of Computation*, 29 (1975), 121–136.

35. G. M. Landau and U. Vishkin, "Introducing Efficient Parallelism into Approximate String Matching and a New Serial Algorithm," *Proceedings of the Eighteenth Annual ACM Symposium on Theory of Computing* (1986), 220–230.

36. L. L. Larmore, "Height-Restricted Optimal Binary Trees," *SIAM Journal on Computing*, 16 (1987), 1115–1123.

37. L. L. Larmore and D. S. Hirschberg, "A Fast Algorithm for Optimal Length-Limited Huffman Codes," *Journal of the ACM,* 37 (1990), 464–473.

38. K. Lee and S. Mahajan, "The Development of a World Class Othello Program," *Artificial Intelligence,* 43 (1990), 21–36.

39. D. A. Lelewer and D. S. Hirschberg, "Data Compression," *ACM Computing Surveys,* 19 (1987), 261–296.

40. H. W. Lenstra, Jr. and C. Pomerance, "Primality Testing with Gaussian Periods," manuscript (2003).

41. F. M. Liang, "A Lower Bound for On-line Bin Packing," *Information Processing Letters,* 10 (1980), 76–79.

42. G. L. Miller, "Riemann's Hypothesis and Tests for Primality," *Journal of Computer and System Sciences,* 13 (1976), 300–317.

43. L. Monier, "Evaluation and Comparison of Two Efficient Probabilistic Primality Testing Algorithms," *Theoretical Computer Science,* 12 (1980), 97–108.

44. R. Motwani and P. Raghavan, *Randomized Algorithms,* Cambridge University Press, New York, 1995.

45. K. Mulmuley, *Computational Geometry: An Introduction Through Randomized Algorithms,* Prentice Hall, Englewood Cliffs, N.J., 1994.

46. J. O'Rourke, *Computational Geometry in C,* Cambridge University Press, New York, 1994.

47. V. Pan, "Strassen's Algorithm Is Not Optimal," *Proceedings of the Nineteenth Annual IEEE Symposium on the Foundations of Computer Science* (1978), 166–176.

48. S. K. Park and K. W. Miller, "Random Number Generators: Good Ones Are Hard to Find," *Communications of the ACM,* 31 (1988), 1192–1201. (See also *Technical Correspondence,* in 36 (1993) 105–110.)

49. S. Pettie and V. Ramachandran, "A Shortest Path Algorithm for Undirected Graphs," *SIAM Journal on Computing,* 34 (2005), 1398–1431.

50. F. P. Preparata and M. I. Shamos, *Computational Geometry: An Introduction,* Springer-Verlag, New York, 1985.

51. W. Pugh, "Skip Lists: A Probabilistic Alternative to Balanced Trees," *Communications of the ACM,* 33 (1990), 668–676.

52. M. O. Rabin, "Probabilistic Algorithms," in *Algorithms and Complexity, Recent Results and New Directions* (J. F. Traub, ed.), Academic Press, New York, 1976, 21–39.

53. M. O. Rabin, "Probabilistic Algorithms for Testing Primality," *Journal of Number Theory,* 12 (1980), 128–138.

54. P. Ramanan, D. J. Brown, C. C. Lee, and D. T. Lee, "On-line Bin Packing in Linear Time," *Journal of Algorithms,* 10 (1989), 305–326.

55. J. Schaeffer, N. Burch, Y. Björnsson, A. Kishimoto, M. Müller, R. Lake, P. Lu, and S. Sutphen, "Checkers is Solved," *Science,* 317 (2007), 1518–1522.

56. M. I. Shamos and D. Hoey, "Closest-Point Problems," *Proceedings of the Sixteenth Annual IEEE Symposium on the Foundations of Computer Science* (1975), 151–162.

57. L. Schrage, "A More Portable FORTRAN Random Number Generator," *ACM Transactions on Mathematics Software,* 5 (1979), 132–138.

58. S. S. Skiena, W. D. Smith, and P. Lemke, "Reconstructing Sets from Interpoint Distances," *Proceedings of the Sixth Annual ACM Symposium on Computational Geometry* (1990), 332–339.

59. V. Strassen, "Gaussian Elimination Is Not Optimal," *Numerische Mathematik,* 13 (1969), 354–356.

60. T. Takaoka, "A New Upper Bound on the Complexity of the All-Pairs Shortest Path Problem," *Information Processing Letters,* 43 (1992), 195–199.

61. A. van Vliet, "An Improved Lower Bound for On-Line Bin Packing Algorithms," *Information Processing Letters,* 43 (1992), 277–284.

62. R. A. Wagner and M. J. Fischer, "The String-to-String Correction Problem," *Journal of the ACM,* 21 (1974), 168–173.

63. B. Xia and Z. Tan, "Tighter Bounds of the First Fit Algorithm for the Bin-packing Problem," *Discrete Applied Mathematics,* (2010), 1668–1675.

64. A. C. Yao, "New Algorithms for Bin Packing," *Journal of the ACM,* 27 (1980), 207–227.

65. F. F. Yao, "Efficient Dynamic Programming Using Quadrangle Inequalities," *Proceedings of the Twelfth Annual ACM Symposium on the Theory of Computing* (1980), 429–435.

66. Z. Zhang, "An Exponential Example for a Partial Digest Mapping Algorithm," *Journal of Computational Molecular Biology,* 1 (1994), 235–239.

67. J. Ziv and A. Lempel, "A Universal Algorithm for Sequential Data Compression," *IEEE Transactions on Information Theory* IT23 (1977), 337–343.

68. J. Ziv and A. Lempel, "Compression of Individual Sequences via Variable-rate Coding," *IEEE Transactions on Information Theory* IT24 (1978), 530–536.

69. U. Zwick, "A Slightly Improved Sub-cubic Algorithm for the All Pairs Shortest Paths Problem with Real Edge Lengths," *Proceedings of the 15th International Symposium on Algorithms and Computation* (2004), 921–932.

CHAPTER 11

Amortized Analysis

In this chapter, we will analyze the running times for several of the advanced data structures that have been presented in Chapters 4 and 6. In particular, we will consider the worst-case running time for any sequence of M operations. This contrasts with the more typical analysis, in which a worst-case bound is given for any *single* operation.

As an example, we have seen that AVL trees support the standard tree operations in $O(\log N)$ worst-case time per operation. AVL trees are somewhat complicated to implement, not only because there are a host of cases, but also because height balance information must be maintained and updated correctly. The reason that AVL trees are used is that a sequence of $\Theta(N)$ operations on an unbalanced search tree could require $\Theta(N^2)$ time, which would be expensive. For search trees, the $O(N)$ worst-case running time of an operation is not the real problem. The major problem is that this could happen repeatedly. Splay trees offer a pleasant alternative. Although any operation can still require $\Theta(N)$ time, this degenerate behavior cannot occur repeatedly, and we can prove that any sequence of M operations takes $O(M \log N)$ worst-case time (total). Thus, in the long run this data structure behaves as though each operation takes $O(\log N)$. We call this an **amortized time bound**.

Amortized bounds are weaker than the corresponding worst-case bounds, because there is no guarantee for any single operation. Since this is generally not important, we are willing to sacrifice the bound on a single operation, if we can retain the same bound for the sequence of operations and at the same time simplify the data structure. Amortized bounds are stronger than the equivalent average-case bound. For instance, binary search trees have $O(\log N)$ average time per operation, but it is still possible for a sequence of M operations to take $O(MN)$ time.

Because deriving an amortized bound requires us to look at an entire sequence of operations instead of just one, we expect that the analysis will be more tricky. We will see that this expectation is generally realized.

In this chapter we shall

- Analyze the binomial queue operations.
- Analyze skew heaps.
- Introduce and analyze the Fibonacci heap.
- Analyze splay trees.

11.1   An Unrelated Puzzle

Consider the following puzzle: Two kittens are placed on opposite ends of a football field, 100 yards apart. They walk toward each other at the speed of 10 yards per minute. At the same time, their mother is at one end of the field. She can run at 100 yards per minute. The mother runs from one kitten to the other, making turns with no loss of speed, until the kittens (and thus the mother) meet at midfield. How far does the mother run?

It is not hard to solve this puzzle with a brute-force calculation. We leave the details to you, but one expects that this calculation will involve computing the sum of an infinite geometric series. Although this straightforward calculation will lead to an answer, it turns out that a much simpler solution can be arrived at by introducing an extra variable, namely, time.

Because the kittens are 100 yards apart and approach each other at a combined velocity of 20 yards per minute, it takes them five minutes to get to midfield. Since the mother runs 100 yards per minute, her total is 500 yards.

This puzzle illustrates the point that sometimes it is easier to solve a problem indirectly than directly. The amortized analyses that we will perform will use this idea. We will introduce an extra variable, known as the **potential**, to allow us to prove results that seem very difficult to establish otherwise.

11.2   Binomial Queues

The first data structure we will look at is the binomial queue of Chapter 6, which we now review briefly. Recall that a **binomial tree** B_0 is a one-node tree, and for $k > 0$, the binomial tree B_k is built by melding two binomial trees B_{k-1} together. Binomial trees B_0 through B_4 are shown in Figure 11.1.

The **rank** of a node in a binomial tree is equal to the number of children; in particular, the rank of the root of B_k is k. A **binomial queue** is a collection of heap-ordered binomial trees, in which there can be at most one binomial tree B_k for any k. Two binomial queues, H_1 and H_2, are shown in Figure 11.2.

The most important operation is merge. To merge two binomial queues, an operation similar to addition of binary integers is performed: At any stage we may have zero, one, two, or possibly three B_k trees, depending on whether or not the two priority queues contain a B_k tree and whether or not a B_k tree is carried over from the previous step. If there is zero or one B_k tree, it is placed as a tree in the resultant binomial queue. If there are two B_k trees, they are melded into a B_{k+1} tree and carried over; if there are three B_k trees, one is placed as a tree in the binomial queue and the other two are melded and carried over. The result of merging H_1 and H_2 is shown in Figure 11.3.

Insertion is performed by creating a one-node binomial queue and performing a merge. The time to do this is $M + 1$, where M represents the smallest type of binomial tree B_M not present in the binomial queue. Thus, insertion into a binomial queue that has a B_0 tree but no B_1 tree requires two steps. Deletion of the minimum is accomplished by removing the

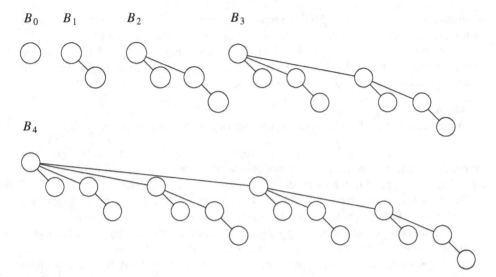

Figure 11.1 Binomial trees B_0, B_1, B_2, B_3, and B_4

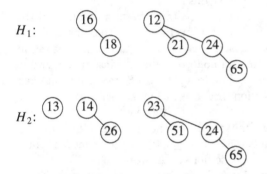

Figure 11.2 Two binomial queues H_1 and H_2

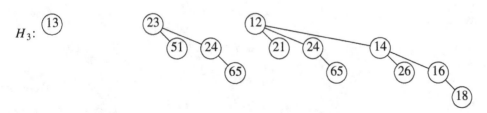

Figure 11.3 Binomial queue H_3: the result of merging H_1 and H_2

minimum and splitting the original binomial queue into two binomial queues, which are then merged. A less terse explanation of these operations is given in Chapter 6.

We consider a very simple problem first. Suppose we want to build a binomial queue of N elements. We know that building a binary heap of N elements can be done in $O(N)$, so we expect a similar bound for binomial queues.

Claim.
A binomial queue of N elements can be built by N successive insertions in $O(N)$ time.

The claim, if true, would give an extremely simple algorithm. Since the worst-case time for each insertion is $O(\log N)$, it is not obvious that the claim is true. Recall that if this algorithm were applied to binary heaps, the running time would be $O(N \log N)$.

To prove the claim, we could do a direct calculation. To measure the running time, we define the cost of each insertion to be one time unit plus an extra unit for each linking step. Summing this cost over all insertions gives the total running time. This total is N units plus the total number of linking steps. The 1st, 3rd, 5th, and all odd-numbered steps require no linking steps, since there is no B_0 present at the time of insertion. Thus, half the insertions require no linking steps. A quarter of the insertions require only one linking step (2nd, 6th, 10th, and so on). An eighth requires two, and so on. We could add this all up and bound the number of linking steps by N, proving the claim. This brute-force calculation will not help when we try to analyze a sequence of operations that include more than just insertions, so we will use another approach to prove this result.

Consider the result of an insertion. If there is no B_0 tree present at the time of the insertion, then the insertion costs a total of one unit, using the same accounting as above. The result of the insertion is that there is now a B_0 tree, and thus we have added one tree to the forest of binomial trees. If there is a B_0 tree but no B_1 tree, then the insertion costs two units. The new forest will have a B_1 tree but will no longer have a B_0 tree, so the number of trees in the forest is unchanged. An insertion that costs three units will create a B_2 tree but destroy a B_0 and B_1 tree, yielding a net loss of one tree in the forest. In fact, it is easy to see that, in general, an insertion that costs c units results in a net increase of $2 - c$ trees in the forest, because a B_{c-1} tree is created but all B_i trees $0 \le i < c - 1$ are removed. Thus, expensive insertions remove trees, while cheap insertions create trees.

Let C_i be the cost of the ith insertion. Let T_i be the number of trees *after* the ith insertion. $T_0 = 0$ is the number of trees initially. Then we have the invariant

$$C_i + (T_i - T_{i-1}) = 2 \tag{11.1}$$

We then have

$$C_1 + (T_1 - T_0) = 2$$
$$C_2 + (T_2 - T_1) = 2$$

$$\vdots$$

$$C_{N-1} + (T_{N-1} - T_{N-2}) = 2$$
$$C_N + (T_N - T_{N-1}) = 2$$

If we add all these equations, most of the T_i terms cancel, leaving

$$\sum_{i=1}^{N} C_i + T_N - T_0 = 2N$$

or equivalently,

$$\sum_{i=1}^{N} C_i = 2N - (T_N - T_0)$$

Recall that $T_0 = 0$ and T_N, the number of trees after the N insertions, is certainly not negative, so $(T_N - T_0)$ is not negative. Thus

$$\sum_{i=1}^{N} C_i \leq 2N$$

which proves the claim.

During the `buildBinomialQueue` routine, each insertion had a worst-case time of $O(\log N)$, but since the entire routine used at most $2N$ units of time, the insertions behaved as though each used no more than two units each.

This example illustrates the general technique we will use. The state of the data structure at any time is given by a function known as the *potential*. The potential function is not maintained by the program but rather is an accounting device that will help with the analysis. When operations take less time than we have allocated for them, the unused time is "saved" in the form of a higher potential. In our example, the potential of the data structure is simply the number of trees. In the analysis above, when we have insertions that use only one unit instead of the two units that are allocated, the extra unit is saved for later by an increase in potential. When operations occur that exceed the allotted time, then the excess time is accounted for by a decrease in potential. One may view the potential as representing a savings account. If an operation uses less than its allotted time, the difference is saved for use later on by more expensive operations. Figure 11.4 shows the cumulative running time used by `buildBinomialQueue` over a sequence of insertions. Observe that the running time never exceeds $2N$ and that the potential in the binomial queue after any insertion measures the amount of savings.

Once a potential function is chosen, we write the main equation:

$$T_{\text{actual}} + \Delta Potential = T_{\text{amortized}} \tag{11.2}$$

T_{actual}, the *actual time* of an operation, represents the exact (observed) amount of time required to execute a particular operation. In a binary search tree, for example, the actual time to perform a `contains(x)` is 1 plus the depth of the node containing x. If we sum the basic equation over the entire sequence, and if the final potential is at least as large as the initial potential, then the amortized time is an upper bound on the actual time used during the execution of the sequence. Notice that while T_{actual} varies from operation to operation, $T_{\text{amortized}}$ is stable.

Picking a potential function that proves a meaningful bound is a very tricky task; there is no one method that is used. Generally, many potential functions are tried before the one

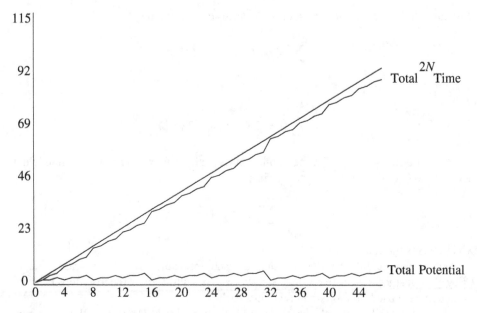

Figure 11.4 A sequence of N inserts

that works is found. Nevertheless, the discussion above suggests a few rules, which tell us the properties that good potential functions have. The potential function should

- Always assume its minimum at the start of the sequence. A popular method of choosing potential functions is to ensure that the potential function is initially 0, and always nonnegative. All the examples that we will encounter use this strategy.

- Cancel a term in the actual time. In our case, if the actual cost was c, then the potential change was $2 - c$. When these are added, an amortized cost of 2 is obtained. This is shown in Figure 11.5.

We can now perform a complete analysis of binomial queue operations.

Theorem 11.1.
The amortized running times of insert, deleteMin, and merge are $O(1)$, $O(\log N)$, and $O(\log N)$, respectively, for binomial queues.

Proof.
The potential function is the number of trees. The initial potential is 0, and the potential is always nonnegative, so the amortized time is an upper bound on the actual time. The analysis for insert follows from the argument above. For merge, assume the two queues have N_1 and N_2 nodes with T_1 and T_2 trees, respectively. Let $N = N_1 + N_2$. The actual time to perform the merge is $O(\log(N_1) + \log(N_2)) = O(\log N)$. After the merge, there can be at most $\log N$ trees, so the potential can increase by at most $O(\log N)$. This gives an amortized bound of $O(\log N)$. The deleteMin bound follows in a similar manner.

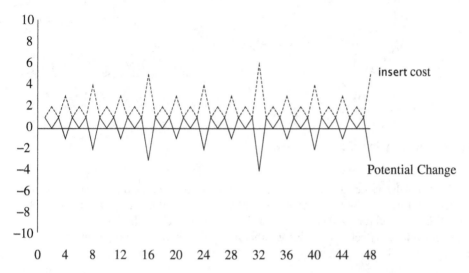

Figure 11.5 The insertion cost and potential change for each operation in a sequence

11.3 Skew Heaps

The analysis of binomial queues is a fairly easy example of an amortized analysis. We now look at skew heaps. As is common with many of our examples, once the right potential function is found, the analysis is easy. The difficult part is choosing a meaningful potential function.

Recall that for skew heaps, the key operation is merging. To merge two skew heaps, we merge their right paths and make this the new left path. For each node on the new path, except the last, the old left subtree is attached as the right subtree. The last node on the new left path is known to not have a right subtree, so it is silly to give it one. The bound does not depend on this exception, and if the routine is coded recursively, this is what will happen naturally. Figure 11.6 shows the result of merging two skew heaps.

Suppose we have two heaps, H_1 and H_2, and there are r_1 and r_2 nodes on their respective right paths. Then the actual time to perform the merge is proportional to $r_1 + r_2$, so we will drop the Big-Oh notation and charge one unit of time for each node on the paths. Since the heaps have no structure, it is possible that all the nodes in both heaps lie on the right path, and this would give a $\Theta(N)$ worst-case bound to merge the heaps (Exercise 11.3 asks you to construct an example). We will show that the amortized time to merge two skew heaps is $O(\log N)$.

What is needed is some sort of a potential function that captures the effect of skew heap operations. Recall that the effect of a merge is that every node on the right path is moved to the left path, and its old left child becomes the new right child. One idea might be to classify each node as a right node or left node, depending on whether or not it is a right child, and use the number of right nodes as a potential function. Although the potential is

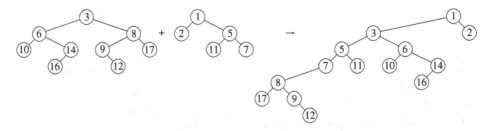

Figure 11.6   Merging of two skew heaps

initially 0 and always nonnegative, the problem is that the potential does not decrease after a merge and thus does not adequately reflect the savings in the data structure. The result is that this potential function cannot be used to prove the desired bound.

A similar idea is to classify nodes as either heavy or light, depending on whether or not the right subtree of any node has more nodes than the left subtree.

Definition 11.1.

A node p is **heavy** if the number of descendants of p's right subtree is at least half of the number of descendants of p, and **light** otherwise. Note that the number of descendants of a node includes the node itself.

As an example, Figure 11.7 shows a skew heap. The nodes with values 15, 3, 6, 12, and 7 are heavy, and all other nodes are light.

The potential function we will use is the number of heavy nodes in the (collection of) heaps. This seems like a good choice, because a long right path will contain an inordinate

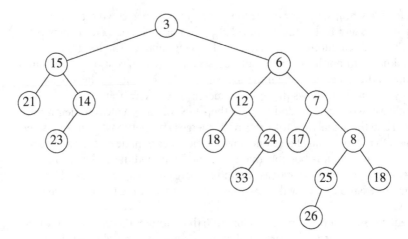

Figure 11.7   Skew heap—heavy nodes are 3, 6, 7, 12, and 15

number of heavy nodes. Because nodes on this path have their children swapped, these nodes will be converted to light nodes as a result of the merge.

Theorem 11.2.
The amortized time to merge two skew heaps is $O(\log N)$.

Proof.
Let H_1 and H_2 be the two heaps, with N_1 and N_2 nodes respectively. Suppose the right path of H_1 has l_1 light nodes and h_1 heavy nodes, for a total of $l_1 + h_1$. Likewise, H_2 has l_2 light and h_2 heavy nodes on its right path, for a total of $l_2 + h_2$ nodes.

If we adopt the convention that the cost of merging two skew heaps is the total number of nodes on their right paths, then the actual time to perform the merge is $l_1 + l_2 + h_1 + h_2$. Now the only nodes whose heavy/light status can change are nodes that are initially on the right path (and wind up on the left path), since no other nodes have their subtrees altered. This is shown by the example in Figure 11.8.

If a heavy node is initially on the right path, then after the merge it must become a light node. The other nodes that were on the right path were light and may or may not become heavy, but since we are proving an upper bound, we will have to assume the worst, which is that they become heavy and increase the potential. Then the net change in the number of heavy nodes is at most $l_1 + l_2 - h_1 - h_2$. Adding the actual time and the potential change [Equation (11.2)] gives an amortized bound of $2(l_1 + l_2)$.

Now we must show that $l_1 + l_2 = O(\log N)$. Since l_1 and l_2 are the number of light nodes on the original right paths, and the right subtree of a light node is less than half the size of the tree rooted at the light node, it follows directly that the number of light nodes on the right path is at most $\log N_1 + \log N_2$, which is $O(\log N)$.

The proof is completed by noting that the initial potential is 0 and that the potential is always nonnegative. It is important to verify this, since otherwise the amortized time does not bound the actual time and is meaningless.

Since the `insert` and `deleteMin` operations are basically just `merges`, they also have $O(\log N)$ amortized bounds.

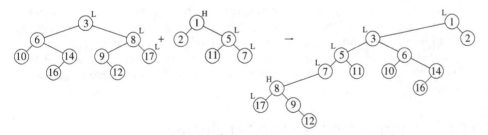

Figure 11.8   Change in heavy/light status after a merge

11.4   Fibonacci Heaps

In Section 9.3.2, we showed how to use priority queues to improve on the naïve $O(|V|^2)$ running time of Dijkstra's shortest-path algorithm. The important observation was that the running time was dominated by $|E|$ decreaseKey operations and $|V|$ insert and deleteMin operations. These operations take place on a set of size at most $|V|$. By using a binary heap, all these operations take $O(\log |V|)$ time, so the resulting bound for Dijkstra's algorithm can be reduced to $O(|E| \log |V|)$.

In order to lower this time bound, the time required to perform the decreaseKey operation must be improved. d-heaps, which were described in Section 6.5, give an $O(\log_d |V|)$ time bound for the decreaseKey operation as well as for insert, but an $O(d \log_d |V|)$ bound for deleteMin. By choosing d to balance the costs of $|E|$ decreaseKey operations with $|V|$ deleteMin operations, and remembering that d must always be at least 2, we see that a good choice for d is

$$d = \max(2, \lfloor |E|/|V| \rfloor)$$

This improves the time bound for Dijkstra's algorithm to

$$O(|E| \log_{(2+\lfloor |E|/|V| \rfloor)} |V|)$$

The **Fibonacci heap** is a data structure that supports all the basic heap operations in $O(1)$ amortized time, with the exception of deleteMin and delete, which take $O(\log N)$ amortized time. It immediately follows that the heap operations in Dijkstra's algorithm will require a total of $O(|E| + |V| \log |V|)$ time.

Fibonacci heaps[1] generalize binomial queues by adding two new concepts:

A different implementation of decreaseKey: The method we have seen before is to percolate the element up toward the root. It does not seem reasonable to expect an $O(1)$ amortized bound for this strategy, so a new method is needed.

Lazy merging: Two heaps are merged only when it is required to do so. This is similar to lazy deletion. For lazy merging, merges are cheap, but because lazy merging does not actually combine trees, the deleteMin operation could encounter lots of trees, making that operation expensive. Any one deleteMin could take linear time, but it is always possible to charge the time to previous merge operations. In particular, an expensive deleteMin must have been preceded by a large number of unduly cheap merges, which were able to store up extra potential.

11.4.1   Cutting Nodes in Leftist Heaps

In binary heaps, the decreaseKey operation is implemented by lowering the value at a node and then percolating it up toward the root until heap order is established. In the worst

---

[1] The name comes from a property of this data structure, which we will prove later in the section.

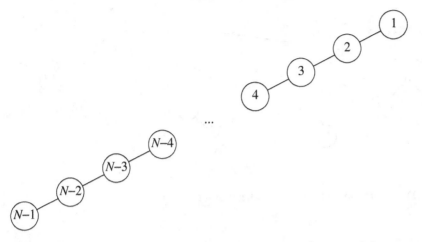

Figure 11.9 Decreasing $N - 1$ to 0 via percolate up would take $\Theta(N)$ time

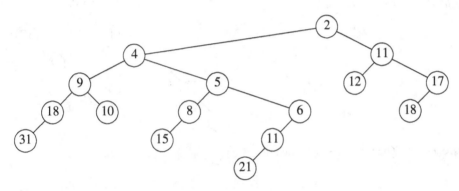

Figure 11.10 Sample leftist heap H

case, this can take $O(\log N)$ time, which is the length of the longest path toward the root in a balanced tree.

This strategy does not work if the tree that represents the priority queue does not have $O(\log N)$ depth. As an example, if this strategy is applied to leftist heaps, then the decreaseKey operation could take $\Theta(N)$ time, as the example in Figure 11.9 shows.

We see that for leftist heaps, another strategy is needed for the decreaseKey operation. Our example will be the leftist heap in Figure 11.10. Suppose we want to decrease the key with value 9 down to 0. If we make the change, we find that we have created a violation of heap order, which is indicated by a dashed line in Figure 11.11.

We do not want to percolate the 0 to the root, because, as we have seen, there are cases where this could be expensive. The solution is to **cut** the heap along the dashed line, thus creating two trees, and then merge the two trees back into one. Let X be the node to which the decreaseKey operation is being applied, and let P be its parent. After the cut, we have

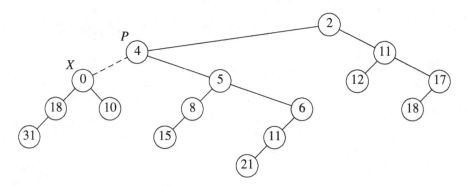

Figure 11.11   Decreasing 9 to 0 creates a heap order violation

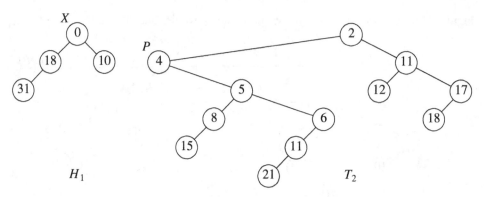

Figure 11.12   The two trees after the cut

two trees, namely, H_1 with root X, and T_2, which is the original tree with H_1 removed. The situation is shown in Figure 11.12.

If these two trees were both leftist heaps, then they could be merged in $O(\log N)$ time, and we would be done. It is easy to see that H_1 is a leftist heap, since none of its nodes have had any changes in their descendants. Thus, since all of its nodes originally satisfied the leftist property, they still must.

Nevertheless, it seems that this scheme will not work, because T_2 is not necessarily leftist. However, it is easy to reinstate the leftist heap property by using two observations:

- Only nodes on the path from P to the root of T_2 can be in violation of the leftist heap property; these can be fixed by swapping children.
- Since the maximum right path length has at most $\lfloor \log(N+1) \rfloor$ nodes, we only need to check the first $\lfloor \log(N+1) \rfloor$ nodes on the path from P to the root of T_2. Figure 11.13 shows H_1 and T_2 after T_2 is converted to a leftist heap.

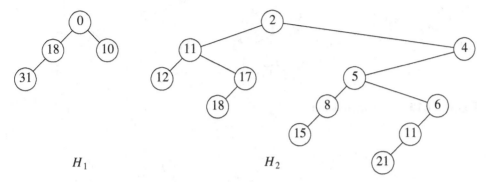

H_1 H_2

Figure 11.13 T_2 converted to the leftist heap H_2

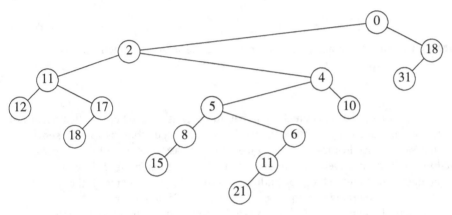

Figure 11.14 decreaseKey$(X, 9)$ completed by merging H_1 and H_2

Because we can convert T_2 to the leftist heap H_2 in $O(\log N)$ steps, and then merge H_1 and H_2, we have an $O(\log N)$ algorithm for performing the decreaseKey operation in leftist heaps. The heap that results in our example is shown in Figure 11.14.

11.4.2 Lazy Merging for Binomial Queues

The second idea that is used by Fibonacci heaps is **lazy merging**. We will apply this idea to binomial queues and show that the amortized time to perform a merge operation (as well as insertion, which is a special case) is $O(1)$. The amortized time for deleteMin will still be $O(\log N)$.

The idea is as follows: To merge two binomial queues, merely concatenate the two lists of binomial trees, creating a new binomial queue. This new queue may have several trees of the same size, so it violates the binomial queue property. We will call this a **lazy binomial queue** in order to maintain consistency. This is a fast operation that always takes constant (worst-case) time. As before, an insertion is done by creating a one-node binomial queue and merging. The difference is that the merge is lazy.

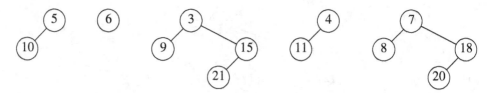

Figure 11.15   Lazy binomial queue

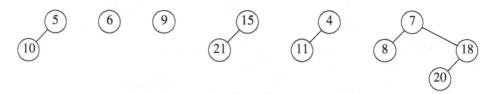

Figure 11.16   Lazy binomial queue after removing the smallest element (3)

The deleteMin operation is much more painful, because it is where we finally convert the lazy binomial queue back into a standard binomial queue, but, as we will show, it is still $O(\log N)$ amortized time—but not $O(\log N)$ worst-case time, as before. To perform a deleteMin, we find (and eventually return) the minimum element. As before, we delete it from the queue, making each of its children new trees. We then merge all the trees into a binomial queue by merging two equal-sized trees until it is no longer possible.

As an example, Figure 11.15 shows a lazy binomial queue. In a lazy binomial queue, there can be more than one tree of the same size. To perform the deleteMin, we remove the smallest element, as before, and obtain the tree in Figure 11.16.

We now have to merge all the trees and obtain a standard binomial queue. A standard binomial queue has at most one tree of each rank. In order to do this efficiently, we must be able to perform the merge in time proportional to the number of trees present (T) (or $\log N$, whichever is larger). To do this, we form an array of lists, $L_0, L_1, \ldots, L_{R_{max}+1}$, where R_{max} is the rank of the largest tree. Each list L_R contains all of the trees of rank R. The procedure in Figure 11.17 is then applied.

```
1     for( R = 0; R <= ⌊log N⌋; R++ )
2         while( |L_R| >= 2 )
3         {
4             Remove two trees from L_R;
5             Merge the two trees into a new tree;
6             Add the new tree to L_R+1;
7         }
```

Figure 11.17   Procedure to reinstate a binomial queue

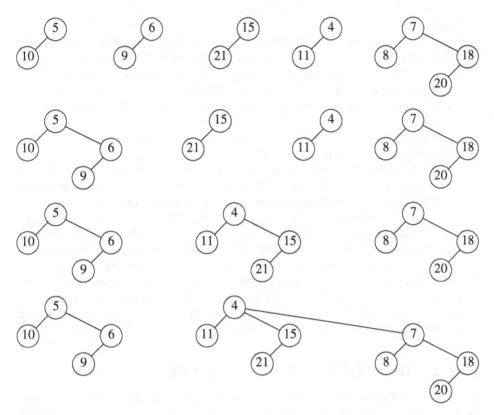

Figure 11.18 Combining the binomial trees into a binomial queue

Each time through the loop, at lines 4 through 6, the total number of trees is reduced by 1. This means that this part of the code, which takes constant time per execution, can only be performed $T - 1$ times, where T is the number of trees. The for loop counters and tests at the end of the while loop take $O(\log N)$ time, so the running time is $O(T + \log N)$, as required. Figure 11.18 shows the execution of this algorithm on the previous collection of binomial trees.

Amortized Analysis of Lazy Binomial Queues

To carry out the amortized analysis of lazy binomial queues, we will use the same potential function that was used for standard binomial queues. Thus, the potential of a lazy binomial queue is the number of trees.

Theorem 11.3.
The amortized running times of merge and insert are both $O(1)$ for lazy binomial queues. The amortized running time of deleteMin is $O(\log N)$.

Proof.

The potential function is the number of trees in the collection of binomial queues. The initial potential is 0, and the potential is always nonnegative. Thus, over a sequence of operations, the total amortized time is an upper bound on the total actual time.

For the `merge` operation, the actual time is constant, and the number of trees in the collection of binomial queues is unchanged, so, by Equation (11.2), the amortized time is $O(1)$.

For the `insert` operation, the actual time is constant, and the number of trees can increase by at most 1, so the amortized time is $O(1)$.

The `deleteMin` operation is more complicated. Let R be the rank of the tree that contains the minimum element, and let T be the number of trees. Thus, the potential at the start of the `deleteMin` operation is T. To perform a `deleteMin`, the children of the smallest node are split off into separate trees. This creates $T + R$ trees, which must be merged into a standard binomial queue. The actual time to perform this is $T + R + \log N$, if we ignore the constant in the Big-Oh notation, by the argument above.[2] Once this is done, there can be at most $\log N$ trees remaining, so the potential function can increase by at most $(\log N) - T$. Adding the actual time and the change in potential gives an amortized bound of $2 \log N + R$. Since all the trees are binomial trees, we know that $R \leq \log N$. Thus we arrive at an $O(\log N)$ amortized time bound for the `deleteMin` operation.

11.4.3 The Fibonacci Heap Operations

As we mentioned before, the Fibonacci heap combines the leftist heap `decreaseKey` operation with the lazy binomial queue `merge` operation. Unfortunately, we cannot use both operations without a slight modification. The problem is that if arbitrary cuts are made in the binomial trees, the resulting forest will no longer be a collection of binomial trees. Because of this, it will no longer be true that the rank of every tree is at most $\lfloor \log N \rfloor$. Since the amortized bound for `deleteMin` in lazy binomial queues was shown to be $2 \log N + R$, we need $R = O(\log N)$ for the `deleteMin` bound to hold.

In order to ensure that $R = O(\log N)$, we apply the following rules to all nonroot nodes:

- Mark a (nonroot) node the first time that it loses a child (because of a cut).
- If a marked node loses another child, then cut it from its parent. This node now becomes the root of a separate tree and is no longer marked. This is called a **cascading cut**, because several of these could occur in one `decreaseKey` operation.

Figure 11.19 shows one tree in a Fibonacci heap prior to a `decreaseKey` operation. When the node with key 39 is changed to 12, the heap order is violated. Therefore, the node is cut from its parent, becoming the root of a new tree. Since the node containing 33 is marked, this is its second lost child, and thus it is cut from its parent (10). Now 10 has

---

[2] We can do this because we can place the constant implied by the Big-Oh notation in the potential function and still get the cancellation of terms, which is needed in the proof.

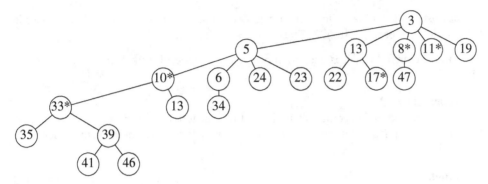

Figure 11.19 A tree in the Fibonacci heap prior to decreasing 39 to 12

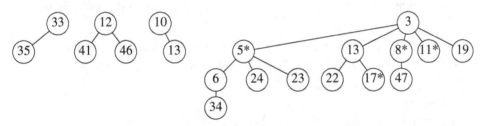

Figure 11.20 The resulting segment of the Fibonacci heap after the `decreaseKey` operation

lost its second child, so it is cut from 5. The process stops here, since 5 was unmarked. The node 5 is now marked. The result is shown in Figure 11.20.

Notice that 10 and 33, which used to be marked nodes, are no longer marked, because they are now root nodes. This will be a crucial observation in our proof of the time bound.

11.4.4 Proof of the Time Bound

Recall that the reason for marking nodes is that we needed to bound the rank (number of children) R of any node. We will now show that any node with N descendants has rank $O(\log N)$.

Lemma 11.1.
Let X be any node in a Fibonacci heap. Let c_i be the ith oldest child of X. Then the rank of c_i is at least $i - 2$.

Proof.
At the time when c_i was linked to X, X already had (older) children $c_1, c_2, \ldots, c_{i-1}$. Thus, X had at least $i - 1$ children when it linked to c_i. Since nodes are linked only if they have the same rank, it follows that at the time that c_i was linked to X, c_i had

at least $i - 1$ children. Since that time, it could have lost at most one child, or else it would have been cut from X. Thus, c_i has at least $i - 2$ children.

From Lemma 11.1, it is easy to show that any node of rank R must have a lot of descendants.

Lemma 11.2.

Let F_k be the Fibonacci numbers defined (in Section 1.2) by $F_0 = 1$, $F_1 = 1$, and $F_k = F_{k-1} + F_{k-2}$. Any node of rank $R \geq 1$ has at least F_{R+1} descendants (including itself).

Proof.

Let S_R be the smallest tree of rank R. Clearly, $S_0 = 1$ and $S_1 = 2$. By Lemma 11.1, a tree of rank R must have subtrees of rank at least $R - 2, R - 3, \ldots, 1$, and 0, plus another subtree, which has at least one node. Along with the root of S_R itself, this gives a minimum value for $S_{R>1}$ of $S_R = 2 + \sum_{i=0}^{R-2} S_i$. It is easy to show that $S_R = F_{R+1}$ (Exercise 1.11(a)).

Because it is well known that the Fibonacci numbers grow exponentially, it immediately follows that any node with s descendants has rank at most $O(\log s)$. Thus, we have

Lemma 11.3.

The rank of any node in a Fibonacci heap is $O(\log N)$.

Proof.

Immediate from the discussion above.

If all we were concerned about were the time bounds for the merge, insert, and deleteMin operations, then we could stop here and prove the desired amortized time bounds. Of course, the whole point of Fibonacci heaps is to obtain an $O(1)$ time bound for decreaseKey as well.

The actual time required for a decreaseKey operation is 1 plus the number of cascading cuts that are performed during the operation. Since the number of cascading cuts could be much more than $O(1)$, we will need to pay for this with a loss in potential. If we look at Figure 11.20, we see that the number of trees actually increases with each cascading cut, so we will have to enhance the potential function to include something that decreases during cascading cuts. Notice that we cannot just throw out the number of trees from the potential function, since then we will not be able to prove the time bound for the merge operation. Looking at Figure 11.20 again, we see that a cascading cut causes a decrease in the number of marked nodes, because each node that is the victim of a cascading cut becomes an unmarked root. Since each cascading cut costs 1 unit of actual time and increases the tree potential by 1, we will count each marked node as two units of potential. This way, we have a chance of canceling out the number of cascading cuts.

Theorem 11.4.

The amortized time bounds for Fibonacci heaps are $O(1)$ for insert, merge, and decreaseKey and $O(\log N)$ for deleteMin.

Proof.

The potential is the number of trees in the collection of Fibonacci heaps plus twice the number of marked nodes. As usual, the initial potential is 0 and is always nonnegative. Thus, over a sequence of operations, the total amortized time is an upper bound on the total actual time.

For the `merge` operation, the actual time is constant, and the number of trees and marked nodes is unchanged, so, by Equation (11.2), the amortized time is $O(1)$.

For the `insert` operation, the actual time is constant, the number of trees increases by 1, and the number of marked nodes is unchanged. Thus, the potential increases by at most 1, so the amortized time is $O(1)$.

For the `deleteMin` operation, let R be the rank of the tree that contains the minimum element, and let T be the number of trees before the operation. To perform a `deleteMin`, we once again split the children of a tree, creating an additional R new trees. Notice that, although this can remove marked nodes (by making them unmarked roots), this cannot create any additional marked nodes. These R new trees, along with the other T trees, must now be merged, at a cost of $T + R + \log N = T + O(\log N)$, by Lemma 11.3. Since there can be at most $O(\log N)$ trees, and the number of marked nodes cannot increase, the potential change is at most $O(\log N) - T$. Adding the actual time and potential change gives the $O(\log N)$ amortized bound for `deleteMin`.

Finally, for the `decreaseKey` operation, let C be the number of cascading cuts. The actual cost of a `decreaseKey` is $C + 1$, which is the total number of cuts performed. The first (noncascading) cut creates a new tree and thus increases the potential by 1. Each cascading cut creates a new tree but converts a marked node to an unmarked (root) node, for a net loss of one unit per cascading cut. The last cut also can convert an unmarked node (in Figure 11.20 it is node 5) into a marked node, thus increasing the potential by 2. The total change in potential is thus at most $3 - C$. Adding the actual time and the potential change gives a total of 4, which is $O(1)$.

11.5 Splay Trees

As a final example, we analyze the running time of splay trees. Recall, from Chapter 4, that after an access of some item X is performed, a splaying step moves X to the root by a series of three operations: *zig, zig-zag,* and *zig-zig.* These tree rotations are shown in Figure 11.21. We adopt the convention that if a tree rotation is being performed at node X, then prior to the rotation P is its parent and (if X is not a child of the root) G is its grandparent.

Recall that the time required for any tree operation on node X is proportional to the number of nodes on the path from the root to X. If we count each *zig* operation as one rotation and each *zig-zig* or *zig-zag* as two rotations, then the cost of any access is equal to 1 plus the number of rotations.

In order to show an $O(\log N)$ amortized bound for the splaying step, we need a potential function that can increase by at most $O(\log N)$ over the entire splaying step but that will also cancel out the number of rotations performed during the step. It is not at all easy to find a potential function that satisfies these criteria. A simple first guess at a potential

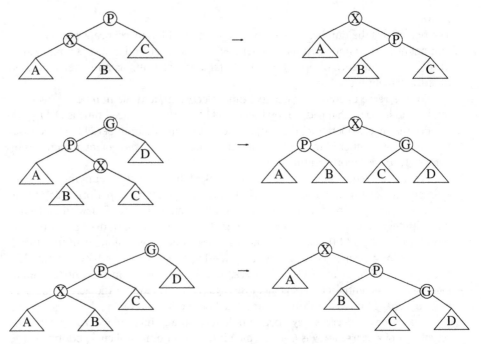

Figure 11.21   *zig, zig-zag,* and *zig-zig* operations; each has a symmetric case (not shown)

function might be the sum of the depths of all the nodes in the tree. This does not work, because the potential can increase by $\Theta(N)$ during an access. A canonical example of this occurs when elements are inserted in sequential order.

A potential function Φ that does work is defined as

$$\Phi(T) = \sum_{i \in T} \log S(i)$$

where $S(i)$ represents the number of descendants of i (including i itself). The potential function is the sum, over all nodes i in the tree T, of the logarithm of $S(i)$.

To simplify the notation, we will define

$$R(i) = \log S(i)$$

This makes

$$\Phi(T) = \sum_{i \in T} R(i)$$

$R(i)$ represents the *rank* of node i. The terminology is similar to what we used in the analysis of the disjoint set algorithm, binomial queues, and Fibonacci heaps. In all these data structures, the meaning of *rank* is somewhat different, but the rank is generally meant to be on the order (magnitude) of the logarithm of the size of the tree. For a tree T with N

nodes, the rank of the root is simply $R(T) = \log N$. Using the sum of ranks as a potential function is similar to using the sum of heights as a potential function. The important difference is that while a rotation can change the heights of many nodes in the tree, only X, P, and G can have their ranks changed.

Before proving the main theorem, we need the following lemma.

Lemma 11.4.
If $a + b \leq c$, and a and b are both positive integers, then

$$\log a + \log b \leq 2 \log c - 2$$

Proof.
By the arithmetic-geometric mean inequality,

$$\sqrt{ab} \leq (a + b)/2$$

Thus

$$\sqrt{ab} \leq c/2$$

Squaring both sides gives

$$ab \leq c^2/4$$

Taking logarithms of both sides proves the lemma.

With the preliminaries taken care of, we are ready to prove the main theorem.

Theorem 11.5.
The amortized time to splay a tree with root T at node X is at most $3(R(T) - R(X)) + 1 = O(\log N)$.

Proof.
The potential function is the sum of the ranks of the nodes in T.

If X is the root of T, then there are no rotations, so there is no potential change. The actual time is 1 to access the node; thus, the amortized time is 1 and the theorem is true. Thus, we may assume that there is at least one rotation.

For any splaying step, let $R_i(X)$ and $S_i(X)$ be the rank and size of X before the step, and let $R_f(X)$ and $S_f(X)$ be the rank and size of X immediately after the splaying step. We will show that the amortized time required for a zig is at most $3(R_f(X) - R_i(X)) + 1$ and that the amortized time for either a zig-zag or zig-zig is at most $3(R_f(X) - R_i(X))$. We will show that when we add over all steps, the sum telescopes to the desired time bound.

Zig step: For the zig step, the actual time is 1 (for the single rotation), and the potential change is $R_f(X) + R_f(P) - R_i(X) - R_i(P)$. Notice that the potential change is easy to compute, because only X's and P's trees change size. Thus, using AT to represent amortized time,

$$AT_{\text{zig}} = 1 + R_f(X) + R_f(P) - R_i(X) - R_i(P)$$

From Figure 11.21 we see that $S_i(P) \geq S_f(P)$; thus, it follows that $R_i(P) \geq R_f(P)$. Thus,

$$AT_{\text{zig}} \leq 1 + R_f(X) - R_i(X)$$

Since $S_f(X) \geq S_i(X)$, it follows that $R_f(X) - R_i(X) \geq 0$, so we may increase the right side, obtaining

$$AT_{\text{zig}} \leq 1 + 3(R_f(X) - R_i(X))$$

Zig-zag step: For the zig-zag case, the actual cost is 2, and the potential change is $R_f(X) + R_f(P) + R_f(G) - R_i(X) - R_i(P) - R_i(G)$. This gives an amortized time bound of

$$AT_{\text{zig-zag}} = 2 + R_f(X) + R_f(P) + R_f(G) - R_i(X) - R_i(P) - R_i(G)$$

From Figure 11.21 we see that $S_f(X) = S_i(G)$, so their ranks must be equal. Thus, we obtain

$$AT_{\text{zig-zag}} = 2 + R_f(P) + R_f(G) - R_i(X) - R_i(P)$$

We also see that $S_i(P) \geq S_i(X)$. Consequently, $R_i(X) \leq R_i(P)$. Making this substitution gives

$$AT_{\text{zig-zag}} \leq 2 + R_f(P) + R_f(G) - 2R_i(X)$$

From Figure 11.21 we see that $S_f(P) + S_f(G) \leq S_f(X)$. If we apply Lemma 11.4, we obtain

$$\log S_f(P) + \log S_f(G) \leq 2 \log S_f(X) - 2$$

By the definition of *rank*, this becomes

$$R_f(P) + R_f(G) \leq 2R_f(X) - 2$$

Substituting this, we obtain

$$AT_{\text{zig-zag}} \leq 2R_f(X) - 2R_i(X)$$
$$\leq 2(R_f(X) - R_i(X))$$

Since $R_f(X) \geq R_i(X)$, we obtain

$$AT_{\text{zig-zag}} \leq 3(R_f(X) - R_i(X))$$

Zig-zig step: The third case is the zig-zig. The proof of this case is very similar to the zig-zag case. The important inequalities are $R_f(X) = R_i(G)$, $R_f(X) \geq R_f(P)$, $R_i(X) \leq R_i(P)$, and $S_i(X) + S_f(G) \leq S_f(X)$. We leave the details as Exercise 11.8.

The amortized cost of an entire splay is the sum of the amortized costs of each splay step. Figure 11.22 shows the steps that are performed in a splay at node 2. Let $R_1(2)$, $R_2(2)$, $R_3(2)$, and $R_4(2)$ be the rank of node 2 in each of the four trees. The cost of the first step, which is a zig-zag, is at most $3(R_2(2) - R_1(2))$. The cost of the second step, which is a zig-zig, is $3(R_3(2) - R_2(2))$. The last step is a zig and has cost no larger than $3(R_4(2) - R_3(2)) + 1$. The total cost thus telescopes to $3(R_4(2) - R_1(2)) + 1$.

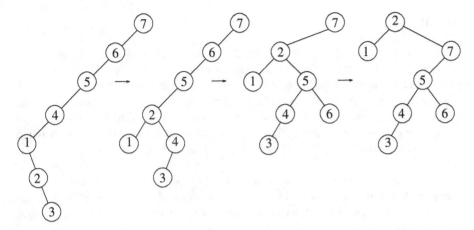

Figure 11.22 The splaying steps involved in splaying at node 2

In general, by adding up the amortized costs of all the rotations, of which at most one can be a *zig*, we see that the total amortized cost to splay at node X is at most $3(R_f(X) - R_i(X)) + 1$, where $R_i(X)$ is the rank of X before the first splaying step and $R_f(X)$ is the rank of X after the last splaying step. Since the last splaying step leaves X at the root, we obtain an amortized bound of $3(R(T) - R_i(X)) + 1$, which is $O(\log N)$.

Because every operation on a splay tree requires a splay, the amortized cost of any operation is within a constant factor of the amortized cost of a splay. Thus, all splay tree access operations take $O(\log N)$ amortized time. To show that insertions and deletions take $O(\log N)$, amortized time, potential changes that occur either prior to or after the splaying step should be accounted for.

In the case of insertion, assume we are inserting into an $N - 1$ node tree. Thus, after the insertion, we have an N-node tree, and the splaying bound applies. However, the insertion at the leaf node adds potential prior to the splay to each node on the path from the leaf node to the root. Let $n_1, n_2, \ldots, n_k$ be the nodes on the path prior to the insertion of the leaf (n_k is the root), and assume they have sizes $s_1, s_2, \ldots, s_k$. After the insertions, the sizes are $s_1 + 1, s_2 + 1, \ldots, s_k + 1$. (The leaf will contribute 0 to the potential so we can ignore it.) Note that (excluding the root node) $s_j + 1 \leq s_{j+1}$, so the new rank of n_j is no more than the old rank of n_{j+1}. Thus, the increase of ranks, which is the maximum increase in potential that results from adding a new leaf, is limited by the new rank of the root, which is $O(\log N)$.

A deletion consists of a nonsplaying step that attaches one tree to another. This does increase the rank of one node, but that is limited by $\log N$ (and is compensated by the removal of a node, which at the time was a root). Thus the splaying costs accurately bound the cost of a deletion.

By using a more general potential function, it is possible to show that splay trees have several remarkable properties. This is discussed in more detail in the exercises.

Summary

In this chapter, we have seen how an amortized analysis can be used to apportion charges among operations. To perform the analysis, we invent a fictitious potential function. The potential function measures the state of the system. A high-potential data structure is volatile, having been built on relatively cheap operations. When the expensive bill comes for an operation, it is paid for by the savings of previous operations. One can view potential as standing for *potential for disaster,* in that very expensive operations can occur only when the data structure has a high potential and has used considerably less time than has been allocated.

Low potential in a data structure means that the cost of each operation has been roughly equal to the amount allocated for it. Negative potential means debt; more time has been spent than has been allocated, so the allocated (or amortized) time is not a meaningful bound.

As expressed by Equation (11.2), the amortized time for an operation is equal to the sum of the actual time and potential change. Taken over an entire sequence of operations, the amortized time for the sequence is equal to the total sequence time plus the net change in potential. As long as this net change is positive, then the amortized bound provides an *upper bound* for the actual time spent and is meaningful.

The keys to choosing a potential function are to guarantee that the minimum potential occurs at the beginning of the algorithm, and to have the potential increase for cheap operations and decrease for expensive operations. It is important that the excess or saved time be measured by an opposite change in potential. Unfortunately, this is sometimes easier said than done.

Exercises

11.1 When do M consecutive insertions into a binomial queue take less than $2M$ time units?

11.2 Suppose a binomial queue of $N = 2^k - 1$ elements is built. Alternately perform M insert and deleteMin pairs. Clearly, each operation takes $O(\log N)$ time. Why does this not contradict the amortized bound of $O(1)$ for insertion?

*11.3 Show that the amortized bound of $O(\log N)$ for the skew heap operations described in the text cannot be converted to a worst-case bound, by giving a sequence of operations that lead to a merge requiring $\Theta(N)$ time.

*11.4 Show how to merge two skew heaps with one top-down pass and reduce the merge cost to $O(1)$ amortized time.

11.5 Extend skew heaps to support the decreaseKey operation in $O(\log N)$ amortized time.

11.6 Implement Fibonacci heaps and compare their performance with that of binary heaps when used in Dijkstra's algorithm.

11.7 A standard implementation of Fibonacci heaps requires four links per node (parent, child, and two siblings). Show how to reduce the number of links, at the cost of at most a constant factor in the running time.

11.8 Show that the amortized time of a *zig-zig* splay is at most $3(R_f(X) - R_i(X))$.

11.9 By changing the potential function, it is possible to prove different bounds for splaying. Let the *weight function* $W(i)$ be some function assigned to each node in the tree, and let $S(i)$ be the sum of the weights of all the nodes in the subtree rooted at i, including i itself. The special case $W(i) = 1$ for all nodes corresponds to the function used in the proof of the splaying bound. Let N be the number of nodes in the tree, and let M be the number of accesses. Prove the following two theorems:
   a. The total access time is $O(M + (M + N) \log N)$.
   *b. If q_i is the number of times that item i is accessed, and $q_i > 0$ for all i, then the total access time is
   $$O\left(M + \sum_{i=1}^{N} q_i \log(M/q_i)\right)$$

11.10 a. Show how to implement the `merge` operation on splay trees so that any sequence of $N-1$ `merges` starting from N single-element trees takes $O(N \log^2 N)$ time.
   *b. Improve the bound to $O(N \log N)$.

11.11 In Chapter 5, we described *rehashing*: When a table becomes more than half full, a new table twice as large is constructed, and the entire old table is rehashed. Give a formal amortized analysis, with potential function, to show that the amortized cost of an insertion is still $O(1)$.

11.12 What is the maximum depth of a Fibonacci heap?

11.13 A *deque* with *heap order* is a data structure consisting of a list of items, on which the following operations are possible:
   `push(x)`: Insert item x on the front end of the deque.
   `pop()`: Remove the front item from the deque and return it.
   `inject(x)`: Insert item x on the rear end of the deque.
   `eject()`: Remove the rear item from the deque and return it.
   `findMin()`: Return the smallest item from the deque (breaking ties arbitrarily).
   a. Describe how to support these operations in constant amortized time per operation.
   **b. Describe how to support these operations in constant worst-case time per operation.

11.14 Show that the binomial queues actually support merging in $O(1)$ amortized time. Define the potential of a binomial queue to be the number of trees plus the rank of the largest tree.

11.15   Suppose that in an attempt to save time, we splay on every second tree operation. Does the amortized cost remain logarithmic?

11.16   Using the potential function in the proof of the splay tree bound, what is the maximum and minimum potential of a splay tree? By how much can the potential function decrease in one splay? By how much can the potential function increase in one splay? You may give Big-Oh answers.

11.17   As a result of a splay, most of the nodes on the access path are moved halfway towards the root, while a couple of nodes on the path move down one level. This suggests using the sum over all nodes of the logarithm of each node's depth as a potential function.
   a. What is the maximum value of the potential function?
   b. What is the minimum value of the potential function?
   c. The difference in the answers to parts (a) and (b) gives some indication that this potential function isn't too good. Show that a splaying operation could increase the potential by $\Theta(N/\log N)$.

References

An excellent survey of amortized analysis is provided in [10].

Most of the references below duplicate citations in earlier chapters. We cite them again for convenience and completeness. Binomial queues were first described in [11] and analyzed in [1]. Solutions to Exercises 11.3 and 11.4 appear in [9]. Fibonacci heaps are described in [3]. Exercise 11.9(a) shows that splay trees are optimal, to within a constant factor of the best static search trees. Exercise 11.9(b) shows that splay trees are optimal, to within a constant factor of the best optimal search trees. These, as well as two other strong results, are proved in the original splay tree paper [7].

Amortization is used in [2] to merge a balanced search tree efficiently. The `merge` operation for splay trees is described in [6]. A solution to Exercise 11.13 can be found in [4]. Exercise 11.14 is from [5].

Amortized analysis is used in [8] to design an online algorithm that processes a series of queries in time only a constant factor larger than any off-line algorithm in its class.

1. M. R. Brown, "Implementation and Analysis of Binomial Queue Algorithms," *SIAM Journal on Computing,* 7 (1978), 298–319.

2. M. R. Brown and R. E. Tarjan, "Design and Analysis of a Data Structure for Representing Sorted Lists," *SIAM Journal on Computing,* 9 (1980), 594–614.

3. M. L. Fredman and R. E. Tarjan, "Fibonacci Heaps and Their Uses in Improved Network Optimization Algorithms," *Journal of the ACM,* 34 (1987), 596–615.

4. H. Gajewska and R. E. Tarjan, "Deques with Heap Order," *Information Processing Letters,* 22 (1986), 197–200.

5. C. M. Khoong and H. W. Leong, "Double-Ended Binomial Queues," *Proceedings of the Fourth Annual International Symposium on Algorithms and Computation* (1993), 128–137.

6. G. Port and A. Moffat, "A Fast Algorithm for Melding Splay Trees," *Proceedings of First Workshop on Algorithms and Data Structures* (1989), 450–459.

7. D. D. Sleator and R. E. Tarjan, "Self-adjusting Binary Search Trees," *Journal of the ACM,* 32 (1985), 652–686.

8. D. D. Sleator and R. E. Tarjan, "Amortized Efficiency of List Update and Paging Rules," *Communications of the ACM,* 28 (1985), 202–208.

9. D. D. Sleator and R. E. Tarjan, "Self-adjusting Heaps," *SIAM Journal on Computing,* 15 (1986), 52–69.

10. R. E. Tarjan, "Amortized Computational Complexity," *SIAM Journal on Algebraic and Discrete Methods,* 6 (1985), 306–318.

11. J. Vuillemin, "A Data Structure for Manipulating Priority Queues," *Communications of the ACM,* 21 (1978), 309–314.

CHAPTER 12

Advanced Data Structures and Implementation

In this chapter, we discuss six data structures with an emphasis on practicality. We begin by examining alternatives to the AVL tree discussed in Chapter 4. These include an optimized version of the splay tree, the red-black tree, and the treap. We also examine the *suffix tree*, which allows searching for a pattern in a large text.

We then examine a data structure that can be used for multidimensional data. In this case, each item may have several keys. The *k*-d tree allows searching relative to any key.

Finally, we examine the pairing heap, which seems to be the most practical alternative to the Fibonacci heap.

Recurring themes include

- Nonrecursive, top-down (instead of bottom-up) search tree implementations when appropriate.
- Detailed, optimized implementations that make use of, among other things, sentinel nodes.

12.1 Top-Down Splay Trees

In Chapter 4, we discussed the basic splay tree operation. When an item X is inserted as a leaf, a series of tree rotations, known as a *splay*, makes X the new root of the tree. A splay is also performed during searches, and if an item is not found, a splay is performed on the last node on the access path. In Chapter 11, we showed that the *amortized* cost of a splay tree operation is $O(\log N)$.

A direct implementation of this strategy requires a traversal from the root down the tree, and then a bottom-up traversal to implement the splaying step. This can be done either by maintaining parent links, or by storing the access path on a stack. Unfortunately, both methods require a substantial amount of overhead, and both must handle many special cases. In this section, we show how to perform rotations on the initial access path. The result is a procedure that is faster in practice, uses only $O(1)$ extra space, but retains the $O(\log N)$ amortized time bound.

Figure 12.1 shows the rotations for the *zig*, *zig-zig*, and *zig-zag* cases. (As is customary, three symmetric rotations are omitted.) At any point in the access, we have a current node

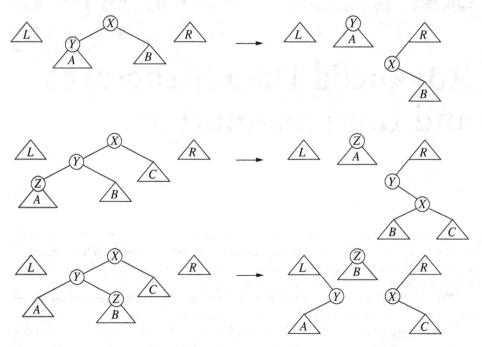

Figure 12.1   Top-down splay rotations: *zig, zig-zig,* and *zig-zag*

X that is the root of its subtree; this is represented in our diagrams as the "middle" tree.[1] Tree L stores nodes in the tree T that are less than X, but not in X's subtree; similarly tree R stores nodes in the tree T that are larger than X, but not in X's subtree. Initially, X is the root of T, and L and R are empty.

If the rotation should be a *zig*, then the tree rooted at Y becomes the new root of the middle tree. X and subtree B are attached as a left child of the smallest item in R; X's left child is logically made null.[2] As a result, X is the new smallest item in R. Note carefully that Y does not have to be a leaf for the *zig* case to apply. If we are searching for an item that is smaller than Y, and Y has no left child (but does have a right child), then the *zig* case will apply.

For the *zig-zig* case, we have a similar dissection. The crucial point is that a rotation between X and Y is performed. The *zig-zag* case brings the bottom node Z to the top in the middle tree and attaches subtrees X and Y to R and L, respectively. Note that Y is attached to, and then becomes, the largest item in L.

The *zig-zag* step can be simplified somewhat because no rotations are performed. Instead of making Z the root of the middle tree, we make Y the root. This is shown in Figure 12.2. This simplifies the coding because the action for the *zig-zag* case becomes

---

[1] For simplicity we don't distinguish between a "node" and the item in the node.

[2] In the code, the smallest node in R does not have a null left link because there is no need for it. This means that printTree(r) will include some items that logically are not in R.

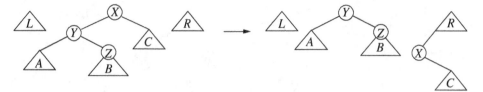

Figure 12.2   Simplified top-down *zig-zag*

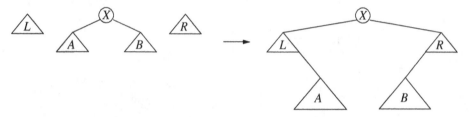

Figure 12.3   Final arrangement for top-down splaying

identical to the *zig* case. This would seem advantageous because testing for a host of cases is time-consuming. The disadvantage is that by descending only one level, we have more iterations in the splaying procedure.

Once we have performed the final splaying step, Figure 12.3 shows how *L*, *R*, and the middle tree are arranged to form a single tree. Note carefully that the result is different from bottom-up splaying. The crucial fact is that the $O(\log N)$ amortized bound is preserved (Exercise 12.1).

An example of the top-down splaying algorithm is shown in Figure 12.4. We attempt to access 19 in the tree. The first step is a *zig-zag*. In accordance with (a symmetric version of) Figure 12.2, we bring the subtree rooted at 25 to the root of the middle tree and attach 12 and its left subtree to *L*.

Next we have a *zig-zig*: 15 is elevated to the root of the middle tree, and a rotation between 20 and 25 is performed, with the resulting subtree being attached to *R*. The search for 19 then results in a terminal *zig*. The middle tree's new root is 18, and 15 and its left subtree are attached as a right child of *L*'s largest node. The reassembly, in accordance with Figure 12.3, terminates the splay step.

We will use a header with left and right links to eventually reference the roots of the left and right trees. Since these trees are initially empty, a header is used to correspond to the min or max node of the right or left tree, respectively, in this initial state. This way the code can avoid checking for empty trees. The first time the left tree becomes nonempty, the right link will get initialized and will not change in the future; thus it will contain the root of the left tree at the end of the top-down search. Similarly, the left link will eventually contain the root of the right tree.

The SplayTree class, whose skeleton is shown in Figure 12.5, includes a constructor that is used to allocate the nullNode sentinel. We use the sentinel nullNode to represent logically a null reference. We will repeatedly use this technique to simplify the code (and consequently make the code somewhat faster). Figure 12.6 (shown on page 546) gives the

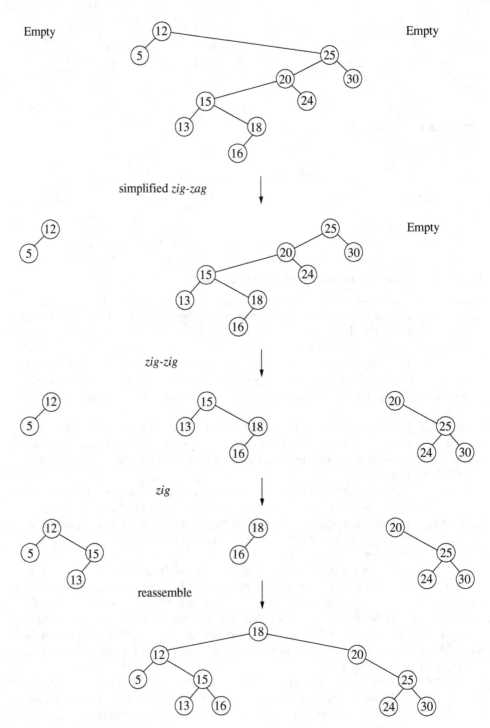

Figure 12.4 Steps in top-down splay (access 19 in top tree)

```
1    public class SplayTree<AnyType extends Comparable<? super AnyType>>
2    {
3        public SplayTree( )
4        {
5            nullNode = new BinaryNode<>( null );
6            nullNode.left = nullNode.right = nullNode;
7            root = nullNode;
8        }
9
10       private BinaryNode<AnyType> splay( AnyType x, BinaryNode<AnyType> t )
11         { /* Figure 12.6 */ }
12       public void insert( AnyType x )
13         { /* Figure 12.7 */ }
14       public void remove( AnyType x )
15         { /* Figure 12.8 */ }
16
17       public AnyType findMin( )
18         { /* See online code */ }
19       public AnyType findMax( )
20         { /* See online code */ }
21       public boolean contains( AnyType x )
22         { /* See online code */ }
23       public void makeEmpty( )
24         { root = nullNode; }
25       public boolean isEmpty( )
26         { return root == nullNode; }
27
28       // Basic node stored in unbalanced binary search trees
29       private static class BinaryNode<AnyType>
30         { /* Same as in Figure 4.16 */ }
31
32       private BinaryNode<AnyType> root;
33       private BinaryNode<AnyType> nullNode;
34       private BinaryNode<AnyType> header = new BinaryNode<>( null ); // For splay
35       private BinaryNode<AnyType> newNode = null;   // Used between different inserts
36
37       private BinaryNode<AnyType> rotateWithLeftChild( BinaryNode<AnyType> k2 )
38         { /* See online code */ }
39       private BinaryNode<AnyType> rotateWithRightChild( BinaryNode<AnyType> k1 )
40         { /* See online code */ }
41   }
```

Figure 12.5 Splay trees: class skeleton

```
1      /**
2       * Internal method to perform a top-down splay.
3       * The last accessed node becomes the new root.
4       * @param x the target item to splay around.
5       * @param t the root of the subtree to splay.
6       * @return the subtree after the splay.
7       */
8      private BinaryNode<AnyType> splay( AnyType x, BinaryNode<AnyType> t )
9      {
10         BinaryNode<AnyType> leftTreeMax, rightTreeMin;
11
12         header.left = header.right = nullNode;
13         leftTreeMax = rightTreeMin = header;
14
15         nullNode.element = x;    // Guarantee a match
16
17         for( ; ; )
18             if( x.compareTo( t.element ) < 0 )
19             {
20                 if( x.compareTo( t.left.element ) < 0 )
21                     t = rotateWithLeftChild( t );
22                 if( t.left == nullNode )
23                     break;
24                 // Link Right
25                 rightTreeMin.left = t;
26                 rightTreeMin = t;
27                 t = t.left;
28             }
29             else if( x.compareTo( t.element ) > 0 )
30             {
31                 if( x.compareTo( t.right.element ) > 0 )
32                     t = rotateWithRightChild( t );
33                 if( t.right == nullNode )
34                     break;
35                 // Link Left
36                 leftTreeMax.right = t;
37                 leftTreeMax = t;
38                 t = t.right;
39             }
40             else
41                 break;
42
43         leftTreeMax.right = t.left;
44         rightTreeMin.left = t.right;
45         t.left = header.right;
46         t.right = header.left;
47         return t;
48     }
```

Figure 12.6 Top-down splaying method

```
 1        /**
 2         * Insert into the tree.
 3         * @param x the item to insert.
 4         */
 5        public void insert( AnyType x )
 6        {
 7            if( newNode == null )
 8                newNode = new BinaryNode<>( null );
 9            newNode.element = x;
10
11            if( root == nullNode )
12            {
13                newNode.left = newNode.right = nullNode;
14                root = newNode;
15            }
16            else
17            {
18                root = splay( x, root );
19                if( x.compareTo( root.element ) < 0 )
20                {
21                    newNode.left = root.left;
22                    newNode.right = root;
23                    root.left = nullNode;
24                    root = newNode;
25                }
26                else
27                if( x.compareTo( root.element ) > 0 )
28                {
29                    newNode.right = root.right;
30                    newNode.left = root;
31                    root.right = nullNode;
32                    root = newNode;
33                }
34                else
35                    return;    // No duplicates
36            }
37            newNode = null;    // So next insert will call new
38        }
```

Figure 12.7   Top-down splay tree insert

code for the splaying procedure. The header node allows us to be certain that we can attach X to the largest node in R without having to worry that R might be empty (and similarly for the symmetric case dealing with L).

As we mentioned above, before the reassembly at the end of the splay, header.left and header.right reference the roots of R and L, respectively (this is not a typo—follow the links). Except for this detail, the code is relatively straightforward.

Figure 12.7 shows the method to insert an item into a tree. A new node is allocated (if necessary), and if the tree is empty, a one-node tree is created. Otherwise we splay root around the inserted value x. If the data in the new root is equal to x, we have a duplicate; instead of reinserting x, we preserve newNode for a future insertion and return immediately. If the new root contains a value larger than x, then the new root and its right subtree become a right subtree of newNode, and root's left subtree becomes the left subtree of newNode. Similar logic applies if root's new root contains a value smaller than x. In either case, newNode becomes the new root.

In Chapter 4, we showed that deletion in splay trees is easy, because a splay will place the target of the deletion at the root. We close by showing the deletion routine in Figure 12.8. It is indeed rare that a deletion procedure is shorter than the corresponding insertion procedure.

```
1      /**
2       * Remove from the tree.
3       * @param x the item to remove.
4       */
5      public void remove( AnyType x )
6      {
7          BinaryNode<AnyType> newTree;
8
9              // If x is found, it will be at the root
10         root = splay( x, root );
11         if( root.element.compareTo( x ) != 0 )
12             return;   // Item not found; do nothing
13
14         if( root.left == nullNode )
15             newTree = root.right;
16         else
17         {
18             // Find the maximum in the left subtree
19             // Splay it to the root; and then attach right child
20             newTree = root.left;
21             newTree = splay( x, newTree );
22             newTree.right = root.right;
23         }
24         root = newTree;
25     }
```

Figure 12.8   Top-down deletion procedure

12.2 Red-Black Trees

A historically popular alternative to the AVL tree is the **red-black tree**. Operations on red-black trees take $O(\log N)$ time in the worst case, and, as we will see, a careful nonrecursive implementation (for insertion) can be done relatively effortlessly (compared with AVL trees).

A red-black tree is a binary search tree with the following coloring properties:

1. Every node is colored either red or black.
2. The root is black.
3. If a node is red, its children must be black.
4. Every path from a node to a `null` reference must contain the same number of black nodes.

A consequence of the coloring rules is that the height of a red-black tree is at most $2\log(N + 1)$. Consequently, searching is guaranteed to be a logarithmic operation. Figure 12.9 shows a red-black tree. Red nodes are shown with double circles.

The difficulty, as usual, is inserting a new item into the tree. The new item, as usual, is placed as a leaf in the tree. If we color this item black, then we are certain to violate condition 4, because we will create a longer path of black nodes. Thus the item must be colored red. If the parent is black, we are done. If the parent is already red, then we will violate condition 3 by having consecutive red nodes. In this case, we have to adjust the tree to ensure that condition 3 is enforced (without introducing a violation of condition 4). The basic operations that are used to do this are color changes and tree rotations.

12.2.1 Bottom-Up Insertion

As we have already mentioned, if the parent of the newly inserted item is black, we are done. Thus insertion of 25 into the tree in Figure 12.9 is trivial.

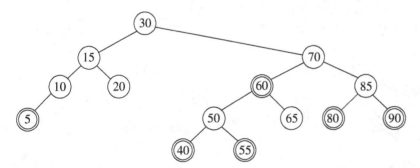

Figure 12.9 Example of a red-black tree (insertion sequence is: 10, 85, 15, 70, 20, 60, 30, 50, 65, 80, 90, 40, 5, 55)

There are several cases (each with a mirror image symmetry) to consider if the parent is red. First, suppose that the sibling of the parent is black (we adopt the convention that `null` nodes are black). This would apply for an insertion of 3 or 8, but not for the insertion of 99. Let X be the newly added leaf, P be its parent, S be the sibling of the parent (if it exists), and G be the grandparent. Only X and P are red in this case; G is black, because otherwise there would be two consecutive red nodes *prior* to the insertion, in violation of red-black rules. Adopting the splay tree terminology, X, P, and G can form either a *zig-zig* chain or a *zig-zag* chain (in either of two directions). Figure 12.10 shows how we can rotate the tree for the case where P is a left child (note there is a symmetric case). Even though X is a leaf, we have drawn a more general case that allows X to be in the middle of the tree. We will use this more general rotation later.

The first case corresponds to a single rotation between P and G, and the second case corresponds to a double rotation, first between X and P and then between X and G. When we write the code, we have to keep track of the parent, the grandparent, and, for reattachment purposes, the great-grandparent.

In both cases, the subtree's new root is colored black, and so even if the original great-grandparent was red, we removed the possibility of two consecutive red nodes. Equally important, the number of black nodes on the paths into A, B, and C has remained unchanged as a result of the rotations.

So far so good. But what happens if S is red, as is the case when we attempt to insert 79 in the tree in Figure 12.9? In that case, initially there is one black node on the path from the subtree's root to C. After the rotation, there must still be only one black node. But in both cases, there are three nodes (the new root, G, and S) on the path to C. Since only one may be black, and since we cannot have consecutive red nodes, it follows that we'd have to color both S and the subtree's new root red, and G (and our fourth node) black. That's great, but what happens if the great-grandparent is also red? In that case, we

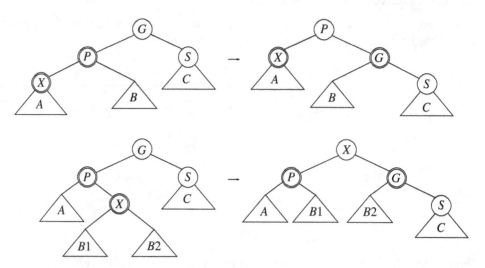

Figure 12.10   Zig rotation and *zig-zag* rotation work if S is black

could percolate this procedure up toward the root as is done for B-trees and binary heaps, until we no longer have two consecutive red nodes, or we reach the root (which will be recolored black).

12.2.2 Top-Down Red-Black Trees

Implementing the percolation would require maintaining the path using a stack or parent links. We saw that splay trees are more efficient if we use a top-down procedure, and it turns out that we can apply a top-down procedure to red-black trees that guarantees that S won't be red.

The procedure is conceptually easy. On the way down, when we see a node X that has two red children, we make X red and the two children black. (If X is the root, after the color flip it will be red but can be made black immediately to restore property 2.) Figure 12.11 shows this color flip. This will induce a red-black violation only if X's parent P is also red. But in that case, we can apply the appropriate rotations in Figure 12.10. What if X's parent's sibling is red? This possibility has been removed by our actions on the way down, and so X's parent's sibling can't be red! Specifically, if on the way down the tree we see a node Y that has two red children, we know that Y's grandchildren must be black, and that since Y's children are made black too, even after the rotation that may occur, we won't see another red node for two levels. Thus when we see X, if X's parent is red, it is not possible for X's parent's sibling to be red also.

As an example, suppose we want to insert 45 into the tree in Figure 12.9. On the way down the tree, we see node 50, which has two red children. Thus, we perform a color flip, making 50 red, and 40 and 55 black. Now 50 and 60 are both red. We perform the single rotation between 60 and 70, making 60 the black root of 30's right subtree, and 70 and 50 both red. We then continue, performing an identical action if we see other nodes on the path that contain two red children. When we get to the leaf, we insert 45 as a red node, and since the parent is black, we are done. The resulting tree is shown in Figure 12.12.

As Figure 12.12 shows, the red-black tree that results is frequently very well balanced. Experiments suggest that the average red-black tree is about as deep as an average AVL tree and that, consequently, the searching times are typically near optimal. The advantage of red-black trees is the relatively low overhead required to perform insertion, and the fact that in practice rotations occur relatively infrequently.

An actual implementation is complicated not only by the host of possible rotations, but also by the possibility that some subtrees (such as 10's right subtree) might be empty, and the special case of dealing with the root (which among other things, has no parent). Thus, we use two sentinel nodes: one for the root, and nullNode, which indicates a null reference,

Figure 12.11   Color flip: Only if X's parent is red do we continue with a rotation

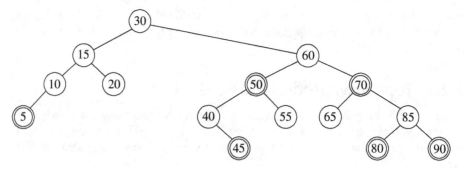

Figure 12.12   Insertion of 45 into Figure 12.9

as it did for splay trees. The root sentinel will store the key $-\infty$ and a right link to the real root. Because of this, the searching and printing procedures need to be adjusted. The recursive routines are trickiest. Figure 12.13 shows how the inorder traversal is rewritten.

     Figure 12.14 shows the `RedBlackTree` skeleton (omitting the methods), along with the constructors. Next, Figure 12.15 shows the routine to perform a single rotation. Because

```
1      /**
2       * Print the tree contents in sorted order.
3       */
4      public void printTree( )
5      {
6          if( isEmpty( ) )
7              System.out.println( "Empty tree" );
8          else
9              printTree( header.right );
10     }
11
12     /**
13      * Internal method to print a subtree in sorted order.
14      * @param t the node that roots the subtree.
15      */
16     private void printTree( RedBlackNode<AnyType> t )
17     {
18         if( t != nullNode )
19         {
20             printTree( t.left );
21             System.out.println( t.element );
22             printTree( t.right );
23         }
24     }
```

Figure 12.13   Inorder traversal for tree and two sentinels

```
 1    public class RedBlackTree<AnyType extends Comparable<? super AnyType>>
 2    {
 3        /**
 4         * Construct the tree.
 5         */
 6        public RedBlackTree( )
 7        {
 8            nullNode = new RedBlackNode<>( null );
 9            nullNode.left = nullNode.right = nullNode;
10            header        = new RedBlackNode<>( null );
11            header.left = header.right = nullNode;
12        }
13
14        private static class RedBlackNode<AnyType>
15        {
16                // Constructors
17            RedBlackNode( AnyType theElement )
18              { this( theElement, null, null ); }
19
20            RedBlackNode( AnyType theElement, RedBlackNode<AnyType> lt, RedBlackNode<AnyType> rt )
21              { element = theElement; left = lt; right = rt; color = RedBlackTree.BLACK; }
22
23            AnyType                element;    // The data in the node
24            RedBlackNode<AnyType> left;        // Left child
25            RedBlackNode<AnyType> right;       // Right child
26            int                    color;      // Color
27        }
28
29        private RedBlackNode<AnyType> header;
30        private RedBlackNode<AnyType> nullNode;
31
32        private static final int BLACK = 1;    // BLACK must be 1
33        private static final int RED   = 0;
34    }
```

Figure 12.14 Class skeleton and initialization routines

the resultant tree must be attached to a parent, rotate takes the parent node as a parameter. Rather than keeping track of the type of rotation as we descend the tree, we pass item as a parameter. Since we expect very few rotations during the insertion procedure, it turns out that it is not only simpler, but actually faster, to do it this way. rotate simply returns the result of performing an appropriate single rotation.

Finally, we provide the insertion procedure in Figure 12.16. The routine handleReorient is called when we encounter a node with two red children, and also when we insert a leaf.

```
1      /**
2       * Internal routine that performs a single or double rotation.
3       * Because the result is attached to the parent, there are four cases.
4       * Called by handleReorient.
5       * @param item the item in handleReorient.
6       * @param parent the parent of the root of the rotated subtree.
7       * @return the root of the rotated subtree.
8       */
9      private RedBlackNode<AnyType> rotate( AnyType item, RedBlackNode<AnyType> parent )
10     {
11         if( compare( item, parent ) < 0 )
12             return parent.left = compare( item, parent.left ) < 0 ?
13                 rotateWithLeftChild( parent.left )  :  // LL
14                 rotateWithRightChild( parent.left ) ;  // LR
15         else
16             return parent.right = compare( item, parent.right ) < 0 ?
17                 rotateWithLeftChild( parent.right )  :  // RL
18                 rotateWithRightChild( parent.right );  // RR
19     }
20
21     /**
22      * Compare item and t.element, using compareTo, with
23      * caveat that if t is header, then item is always larger.
24      * This routine is called if it is possible that t is header.
25      * If it is not possible for t to be header, use compareTo directly.
26      */
27     private final int compare( AnyType item, RedBlackNode<AnyType> t )
28     {
29         if( t == header )
30             return 1;
31         else
32             return item.compareTo( t.element );
33     }
```

Figure 12.15 rotate method

The most tricky part is the observation that a double rotation is really two single rotations, and is done only when branching to X (represented in the insert method by current) takes opposite directions. As we mentioned in the earlier discussion, insert must keep track of the parent, grandparent, and great-grandparent as the tree is descended. Since these are shared with handleReorient, we make these class members. Note that after a rotation, the values stored in the grandparent and great-grandparent are no longer correct. However, we are assured that they will be restored by the time they are next needed.

```
1              // Used in insert routine and its helpers
2       private RedBlackNode<AnyType> current;
3       private RedBlackNode<AnyType> parent;
4       private RedBlackNode<AnyType> grand;
5       private RedBlackNode<AnyType> great;
6
7       /**
8        * Internal routine that is called during an insertion
9        * if a node has two red children. Performs flip and rotations.
10       * @param item the item being inserted.
11       */
12      private void handleReorient( AnyType item )
13      {
14              // Do the color flip
15          current.color = RED;
16          current.left.color = BLACK;
17          current.right.color = BLACK;
18
19          if( parent.color == RED )   // Have to rotate
20          {
21              grand.color = RED;
22              if( ( compare( item, grand ) < 0 ) !=
23                  ( compare( item, parent ) < 0 ) )
24                  parent = rotate( item, grand );  // Start dbl rotate
25              current = rotate( item, great );
26              current.color = BLACK;
27          }
28          header.right.color = BLACK; // Make root black
29      }
30
31      /**
32       * Insert into the tree.
33       * @param item the item to insert.
34       */
35      public void insert( AnyType item )
36      {
37          current = parent = grand = header;
38          nullNode.element = item;
39
40          while( compare( item, current ) != 0 )
41          {
42              great = grand; grand = parent; parent = current;
43              current = compare( item, current ) < 0 ? current.left : current.right;
44
```

Figure 12.16   Insertion procedure

```
45                      // Check if two red children; fix if so
46                  if( current.left.color == RED && current.right.color == RED )
47                      handleReorient( item );
48              }
49
50              // Insertion fails if already present
51          if( current != nullNode )
52              return;
53          current = new RedBlackNode<>( item, nullNode, nullNode );
54
55              // Attach to parent
56          if( compare( item, parent ) < 0 )
57              parent.left = current;
58          else
59              parent.right = current;
60          handleReorient( item );
61      }
```

Figure 12.16   *(continued)*

12.2.3 Top-Down Deletion

Deletion in red-black trees can also be performed top-down. Everything boils down to being able to delete a leaf. This is because to delete a node that has two children, we replace it with the smallest node in the right subtree; that node, which must have at most one child, is then deleted. Nodes with only a right child can be deleted in the same manner, while nodes with only a left child can be deleted by replacement with the largest node in the left subtree, and subsequent deletion of that node. Note that for red-black trees, we don't want to use the strategy of bypassing for the case of a node with one child because that may connect two red nodes in the middle of the tree, making enforcement of the red-black condition difficult.

Deletion of a red leaf is, of course, trivial. If a leaf is black, however, the deletion is more complicated because removal of a black node will violate condition 4. The solution is to ensure during the top-down pass that the leaf is red.

Throughout this discussion, let X be the current node, T be its sibling, and P be their parent. We begin by coloring the root sentinel red. As we traverse down the tree, we attempt to ensure that X is red. When we arrive at a new node, we are certain that P is red (inductively, by the invariant we are trying to maintain), and that X and T are black (because we can't have two consecutive red nodes). There are two main cases.

First, suppose X has two black children. Then there are three subcases, which are shown in Figure 12.17. If T also has two black children, we can flip the colors of X, T, and P to maintain the invariant. Otherwise, one of T's children is red. Depending on which

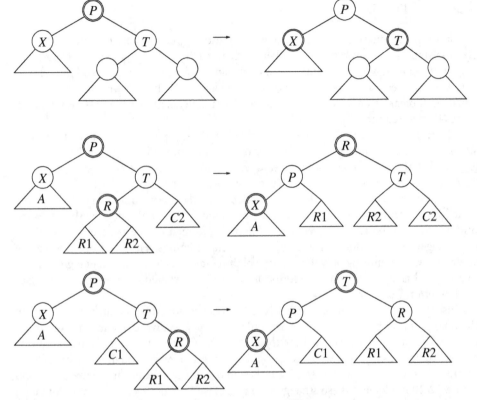

Figure 12.17 Three cases when X is a left child and has two black children

one it is,[3] we can apply the rotation shown in the second and third cases of Figure 12.17. Note carefully that this case will apply for the leaf, because nullNode is considered to be black.

Otherwise, one of X's children is red. In this case, we fall through to the next level, obtaining new X, T, and P. If we're lucky, X will land on the red child, and we can continue onward. If not, we know that T will be red, and X and P will be black. We can rotate T and P, making X's new parent red; X and its grandparent will, of course, be black. At this point we can go back to the first main case.

---

[3] If both children are red, we can apply either rotation. As usual, there are symmetric rotations for the case when X is a right child that are not shown.

12.3 Treaps

Our last type of binary search tree, known as the **treap**, is probably the simplest of all. Like the skip list, it uses random numbers and gives $O(\log N)$ expected time behavior for any input. Searching time is identical to an unbalanced binary search tree (and thus slower than balanced search trees), while insertion time is only slightly slower than a recursive unbalanced binary search tree implementation. Although deletion is much slower, it is still $O(\log N)$ expected time.

The treap is so simple that we can describe it without a picture. Each node in the tree stores an item, a left and right link, and a priority that is randomly assigned when the node is created. A treap is a binary search tree with the property that the node priorities satisfy heap order: Any node's priority must be at least as large as its parent's.

A collection of distinct items each of which has a distinct priority can only be represented by one treap. This is easily deduced by induction, since the node with the lowest priority must be the root. Consequently, the tree is formed on the basis of the $N!$ possible arrangements of priority instead of the $N!$ item orderings. The node declarations are straightforward, requiring only the addition of the `priority` field, as shown in Figure 12.18. The sentinel `nullNode` will have priority of ∞. A single, shared, `Random` object generates random priorities.

Insertion into the treap is simple: After an item is added as a leaf, we rotate it up the treap until its priority satisfies heap order. It can be shown that the expected number of rotations is less than 2. After the item to be deleted has been found, it can be deleted by increasing its priority to ∞ and rotating it down through the path of low-priority children. Once it is a leaf, it can be removed. The routines in Figure 12.19 and Figure 12.20 implement these strategies using recursion. A nonrecursive implementation is left for the reader (Exercise 12.11). For deletion, note that when the node is logically a

```
 1    private static class TreapNode<AnyType>
 2    {
 3            // Constructors
 4        TreapNode( AnyType theElement )
 5          { this( theElement, null, null ); }
 6        TreapNode( AnyType theElement, TreapNode<AnyType> lt, TreapNode<AnyType> rt )
 7          { element = theElement; left = lt; right = rt; priority = randomObj.randomInt( ); }
 8
 9        AnyType              element;      // The data in the node
10        TreapNode<AnyType> left;          // Left child
11        TreapNode<AnyType> right;         // Right child
12        int                  priority;    // Priority
13
14        private static Random randomObj = new Random( );
15    }
```

Figure 12.18 Node declaration for treaps

```
1      /**
2       * Internal method to insert into a subtree.
3       * @param x the item to insert.
4       * @param t the node that roots the subtree.
5       * @return the new root of the subtree.
6       */
7      private TreapNode<AnyType> insert( AnyType x, TreapNode<AnyType> t )
8      {
9          if( t == nullNode )
10             return new TreapNode<>( x, nullNode, nullNode );
11
12         int compareResult = x.compareTo( t.element );
13
14         if( compareResult < 0 )
15         {
16             t.left = insert( x, t.left );
17             if( t.left.priority < t.priority )
18                 t = rotateWithLeftChild( t );
19         }
20         else if( compareResult > 0 )
21         {
22             t.right = insert( x, t.right );
23             if( t.right.priority < t.priority )
24                 t = rotateWithRightChild( t );
25         }
26         // Otherwise, it's a duplicate; do nothing
27
28         return t;
29     }
```

Figure 12.19   Treaps: insertion routine

leaf, it still has nullNode as both its left and right children. Consequently, it is rotated with the right child. After the rotation, t is nullNode, and the left child stores the item that is to be deleted. Thus we change t.left to reference the nullNode sentinel. Note also that our implementation assumes that there are no duplicates; if this is not true, then the remove could fail (why?).

The treap is particularly easy to implement because we never have to worry about adjusting the priority field. One of the difficulties of the balanced tree approaches is that it is difficult to track down errors that result from failing to update balance information in the course of an operation. In terms of total lines for a reasonable insertion and deletion package, the treap, especially a nonrecursive implementation, seems like the hands-down winner.

```
 1       /**
 2        * Internal method to remove from a subtree.
 3        * @param x the item to remove.
 4        * @param t the node that roots the subtree.
 5        * @return the new root of the subtree.
 6        */
 7       private TreapNode<AnyType> remove( AnyType x, TreapNode<AnyType> t )
 8       {
 9           if( t != nullNode )
10           {
11               int compareResult = x.compareTo( t.element );
12
13               if( compareResult < 0 )
14                   t.left = remove( x, t.left );
15               else if( compareResult > 0 )
16                   t.right = remove( x, t.right );
17               else
18               {
19                   // Match found
20                   if( t.left.priority < t.right.priority )
21                       t = rotateWithLeftChild( t );
22                   else
23                       t = rotateWithRightChild( t );
24
25                   if( t != nullNode )      // Continue on down
26                       t = remove( x, t );
27                   else
28                       t.left = nullNode;  // At a leaf
29               }
30           }
31           return t;
32       }
```

Figure 12.20 Treaps: deletion procedure

12.4 Suffix Arrays and Suffix Trees

One of the most fundamental problems in data processing is to find the location of a pattern P in a text T. For instance, we may be interested in answering questions such as

- Is there a substring of T matching P?
- How many times does P appear in T?
- Where are all occurrences of P in T?

Assuming that the size of P is less than T (and usually it is significantly less), then we would reasonably expect that the time to solve this problem for a given P and T would be at least linear in the length of T, and in fact there are several $O(\,|\,T\,|\,)$ algorithms.

However, we are interested in a more common problem, in which T is fixed, and queries with different P occur frequently. For instance, T could be a huge archive of email messages, and we are interested in repeatedly searching the email messages for different patterns. In this case, we are willing to preprocess T into a nice form that would make each individual search much more efficient, taking time significantly less than linear in the size of T—either logarithmic in the size of T, or even better, independent of T and dependent only on the length of P.

One such data structure is the *suffix array* and *suffix tree* (that sounds like two data structures, but as we will see, they are basically equivalent, and trade time for space).

12.4.1 Suffix Arrays

A suffix array for a text T is simply an array of all suffixes of T arranged in sorted order. For instance, suppose our text string is banana. Then the suffix array for banana is shown in Figure 12.21:

A suffix array that stores the suffixes explicitly would seem to require quadratic space, since it stores one string of each length 1 to N (where N is the length of T). In Java, this is not exactly true, since Java strings are implemented by maintaining an array of characters and a starting and ending index. This means that when a String is created via a call to substring, the same array of characters is shared, and the additional memory requirement is only the starting and ending index for the new substring. Nonetheless, even this could be considered to be too much space: The suffix array would be constructed for the text, not the pattern, and the text could be huge. Thus it is common for a practical implementation to store only the starting indices of the suffixes in the suffix array, instead of the entire substring. Figure 12.22 shows the indices that would be stored.

The suffix array by itself is extremely powerful. For instance, if a pattern P occurs in the text, then it must be a prefix of some suffix. A binary search of the suffix array would be enough to determine if the pattern P is in the text: The binary search either lands on P, or P would be between two values, one smaller than P and one larger than P. If P is a prefix of some substring, it is a prefix of the larger value found at the end of the binary search.

0	a
1	ana
2	anana
3	banana
4	na
5	nana

Figure 12.21 Suffixes for "banana"

	Index	Substring Being Represented
0	5	a
1	3	ana
2	1	anana
3	0	banana
4	4	na
5	2	nana

Figure 12.22 Suffix array that stores only Indices (full substrings shown for reference)

Immediately, this reduces the query time to $O(\,|\,P\,|\,\log\,|\,T\,|\,)$, where the $\log|\,T\,|$ is the binary search, and the $|\,P\,|$ is the cost of the comparison at each step.

We can also use the suffix array to find the number of occurrences of P: They will be stored sequentially in the suffix array, thus two binary searches suffix to find a range of suffixes that will be guaranteed to begin with P. One way to speed this search is to compute the *longest common prefix* (LCP) for each consecutive pair of substrings; if this computation is done as the suffix array is built, then each query to find the number of occurrences of P can be sped up to $O(\,|\,P\,|\,+\,\log\,|\,T\,|\,)$ although this is not obvious. Figure 12.23 shows the LCP computed for each substring, relative to the preceding substring.

The longest common prefix also provides information about the longest pattern that occurs twice in the text: Look for the largest LCP value, and take that many characters of the corresponding substring. In Figure 12.23, this is 3, and the longest repeated pattern is ana.

Figure 12.24 shows simple code to compute the suffix array and longest common prefix information for any string. Lines 28–30 compute the suffixes, and then the suffixes are sorted at line 32. Lines 34–35 compute the suffixes' starting indices, and lines 37–39 compute the longest common prefixes for adjacent entries by calling the computeLCP routine written at lines 4–13.

The running time of the suffix array computation is dominated by the sorting step, which uses $O(N\log N)$ comparisons. In many circumstances this can be reasonably acceptable performance. For instance, a suffix array for a 3,000,000-character

	Index	LCP	Substring Being Represented
0	5	–	a
1	3	1	ana
2	1	3	anana
3	0	0	banana
4	4	0	na
5	2	2	nana

Figure 12.23 Suffix array for "banana"; includes longest common prefix (LCP)

```
 1      /*
 2       * Returns the LCP for any two strings
 3       */
 4      public static int computeLCP( String s1, String s2 )
 5      {
 6          int i = 0;
 7
 8          while( i < s1.length( ) && i < s2.length( )
 9                              && s1.charAt( i ) == s2.charAt( i ) )
10              i++;
11
12          return i;
13      }
14
15      /*
16       * Fill in the suffix array and LCP information for String str
17       * @param str the input String
18       * @param SA existing array to place the suffix array
19       * @param LCP existing array to place the LCP information
20       */
21      public static void createSuffixArray( String str, int [ ] SA, int [ ] LCP )
22      {
23          if( SA.length != str.length( ) || LCP.length != str.length( ) )
24              throw new IllegalArgumentException( );
25
26          int N = str.length( );
27
28          String [ ] suffixes = new String[ N ];
29          for( int i = 0; i < N; i++ )
30              suffixes[ i ] = str.substring( i );
31
32          Arrays.sort( suffixes );
33
34          for( int i = 0; i < N; i++ )
35              SA[ i ] = N - suffixes[ i ].length( );
36
37          LCP[ 0 ] = 0;
38          for( int i = 1; i < N; i++ )
39              LCP[ i ] = computeLCP( suffixes[ i - 1 ], suffixes[ i ] );
40      }
```

Figure 12.24 Simple algorithm to create suffix array and LCP array

English-language novel can be built in just a few seconds. However, the $O(N \log N)$ cost, based on the number of comparisons, hides the fact that a String comparison between $s1$ and $s2$ takes time that depends on $LCP(s1, s2)$, So while it is true that almost all these comparisons end quickly when run on the suffixes found in natural language processing, the comparisons will be expensive in applications where there are many long common substrings. One such example occurs in pattern searching of DNA, whose alphabet consists of four characters (A, C, G, T) and whose strings can be huge. For instance, the DNA string for a human chromosome 22 has roughly 35 million characters, with a maximum LCP of approximately 200,000, and an average LCP of nearly 2,000. And even the HTML/Java distribution for JDK 1.3 (much smaller than the current distribution) is nearly 70 million characters, with a maximum LCP of roughly 37,000 and an average LCP of roughly 14,000. In the degenerate case of a String that contains only one character, repeated N times, it is easy to see that each comparison takes $O(N)$ time, and the total cost is $O(N^2 \log N)$.

In Section 12.4.3, we will show a linear-time algorithm to construct the suffix array.

12.4.2 Suffix Trees

Suffix arrays are easily searchable by binary search, but the binary search itself automatically implies $\log T$ cost. What we would like to do is find a matching suffix even more efficiently. One idea is to store the suffixes in a *trie*. A binary trie was seen in our discussion of Huffman codes in Section 10.1.2.

The basic idea of the trie is to store the suffixes in a tree. At the root, instead of having two branches, we would have one branch for each possible first character. Then at the next level, we would have one branch for the next character, and so on. At each level we are doing multiway branching, much like radix sort, and thus we can find a match in time that would depend only on the length of the match.

In Figure 12.25, we see on the left a basic trie to store the suffixes of the string *deed*. These suffixes are *d*, *deed*, *ed*, and *eed*. In this trie, internal branching nodes are drawn in circles, and the suffixes that are reached are drawn in rectangles. Each branch is labeled with the character that is chosen, but the branch prior to a completed suffix has no label.

This representation could waste significant space if there are many nodes that have only one child. Thus in Figure 12.25, we see an equivalent representation on the right, known as a *compressed trie*. Here, single-branch nodes are collapsed into a single node. Notice that although the branches now have multicharacter labels, all the labels for the branches of any given node must have unique first characters. Thus it is still just as easy as before to choose which branch to take. Thus we can see that a search for a pattern P depends only on the length of the pattern P, as desired. (We assume that the letters of the alphabet are represented by numbers 1, 2, ... Then each node stores an array representing each possible branch and we can locate the appropriate branch in constant time. The empty edge label can be represented by 0.)

If the original string has length N, the total number of branches is less than $2N$. However, this by itself does not mean that the compressed trie uses linear space: The labels on the edges take up space. The total length of all the labels on the compressed trie in Figure 12.25 is exactly one less than the number of internal branching nodes in the original trie in Figure 12.25. And of course writing all the suffixes in the leaves could

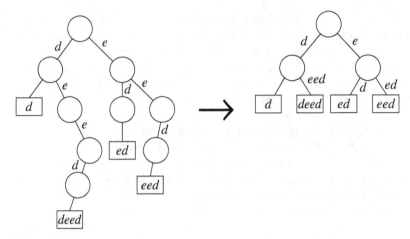

Figure 12.25 Left: trie representing the suffixes for *deed*: {*d, deed, ed, eed*}; right: compressed trie that collapses single-node branches

take quadratic space. So if the original used quadratic space, so does the compressed trie. Fortunately, we can get by with linear space as follows:

1. In the leaves, we use the index where the suffix begins (as in the suffix array).
2. In the internal nodes, we store the number of common characters matched from the root until the internal node; this number represents the *letter depth*.

Figure 12.26 shows how the compressed trie is stored for the suffixes of *banana*. The leaves are simply the indices of the starting points for each suffix. The internal node with a letter depth of 1 is representing the common string "a" in all nodes that are below it. The internal node with a letter depth of 3 is representing the common string "ana" in all nodes that are below it. And the internal node with a letter depth of 2 is representing the common string "na" in all nodes that are below it. In fact, this analysis makes clear that a suffix tree is equivalent to a suffix array plus an LCP array.

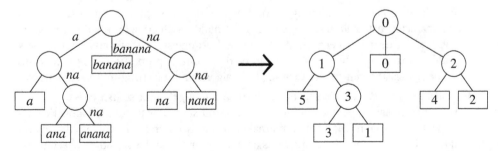

Figure 12.26 Compressed trie representing the suffixes for *banana*: {*a, ana, anana, banana, na, nana*}. Left: the explicit representation; right: the implicit representation that stores only one integer (plus branches) per node

If we have a suffix tree, we can compute the suffix array and the LCP array by performing an inorder traversal of the tree (compare Figure 12.23 with the suffix tree in Figure 12.26). At that time we can compute the LCP as follows: If the suffix node value PLUS the letter depth of the parent is equal to N, then use the letter depth of the grandparent as the LCP; otherwise use the parent's letter depth as the LCP. In Figure 12.26, if we proceed inorder, we obtain for our suffixes and LCP values

Suffix $= 5$, with LCP $= 0$ (the grandparent) because $5 + 1$ equals 6
Suffix $= 3$, with LCP $= 1$ (the grandparent) because $3 + 3$ equals 6
Suffix $= 1$, with LCP $= 3$ (the parent) because $1 + 3$ does not equal 6
Suffix $= 0$, with LCP $= 0$ (the parent) because $0 + 0$ does not equal 6
Suffix $= 4$, with LCP $= 0$ (the grandparent) because $4 + 2$ equals 6
Suffix $= 2$, with LCP $= 2$ (the parent) because $2 + 2$ does not equal 6

This transformation can clearly be done in linear time.

The suffix array and LCP array also uniquely define the suffix tree. First, create a root with letter depth 0. Then search the LCP array (ignoring position 0, for which LCP is not really defined) for all occurrences of the minimum (which at this phase will be the zeros). Once these minimums are found, they will partition the array (view the LCP as residing *between* adjacent elements). For instance, in our example, there are two zeros in the LCP array, which partitions the suffix array into three portions: one portion containing the suffixes $\{5, 3, 1\}$, another portion containing the suffix $\{0\}$, and the third portion containing the suffixes $\{4, 2\}$. The internal nodes for these portions can be built recursively, and then the suffix leaves can be attached with an inorder traversal. Although it is not obvious, with care the suffix tree can be generated in linear time from the suffix array and LCP array.

The suffix tree solves many problems efficiently, especially if we augment each internal node to also maintain the number of suffixes stored below it. A small sampling of suffix tree applications includes the following:

1. *Find the longest repeated substring in T:* Traverse the tree, finding the internal node with the largest number letter depth; this represents the maximum LCP. The running time is $O(|T|)$. This generalizes to the longest substring repeated at least k times.

2. *Find the longest common substring in two strings T_1 and T_2:* Form a string $T_1 \# T_2$ where $\#$ is a character that is not in either string. Then build a suffix tree for the resulting string and find the deepest internal node that has at least one suffix that starts prior to the $\#$, and one that starts after the $\#$. This can be done in time proportional to the total size of the strings and generalizes to an $O(kN)$ algorithm for k strings of total length N.

3. *Find the number of occurrences of the pattern P:* Assuming that the suffix tree is augmented so that each leaf keeps track of the number of suffixes below it, simply follow the path down the internal node; the first internal node that is a prefix of P provides the answer; if there is no such node, the answer is either zero or one and is found by checking the suffix at which the search terminates. This takes time proportional to the length of the pattern P and is independent of the size of $|T|$.

4. *Find the most common substring of a specified length $L > 1$:* Return the internal node with largest size amongst those with letter depth at least L. This takes time $O(|T|)$.

12.4.3 Linear-Time Construction of Suffix Arrays and Suffix Trees

In Section 12.4.1 we showed the simplest algorithm to construct a suffix array and an LCP array, but this algorithm has $O(N^2 \log N)$ worst-case running time for an N-character string and can occur if the string has suffixes with long common prefixes. In this section we describe an $O(N)$ worst-case time algorithm to compute the suffix array. This algorithm can also be enhanced to compute the LCP array in linear time, but there is also a very simple linear-time algorithm to compute the LCP array from the suffix array (see Exercise 12.9 and complete code in Figure 12.50). Either way, we can thus also build a suffix tree in linear time.

This algorithm makes use of divide and conquer. The basic idea is as follows:

1. Choose a sample A of suffixes.
2. Sort the sample A by recursion.
3. Sort the remaining suffixes, B, by using the now-sorted sample of suffixes A.
4. Merge A and B.

To get an intuition of how step 3 might work, suppose the sample A of suffixes are all suffixes that start at an odd index. Then the remaining suffixes B, are those suffixes that start at an even index. So suppose we have computed the sorted set of suffixes A. To compute the sorted set of suffixes B, we would in effect need to sort all the suffixes that start at even indices. But these suffixes each consist of a single first character in an even position, followed by a string that starts with the second character, which must be in an odd position. Thus the string that starts in the second character is exactly a string that is in A. So to sort all the suffixes B, we can do something similar to a radix sort: First sort the strings in B starting from the second character. This should take linear time, since the sorted order of A is already known. Then stably sort on the first character of the strings in B. Thus B could be sorted in linear time, after A is sorted recursively. If A and B could then be merged in linear time, we would have a linear-time algorithm. The algorithm we present uses a different sampling step, that admits a simple linear-time merging step.

As we describe the algorithm, we will also show how it computes the suffix array for the string ABRACADABRA. We adopt the following conventions:

$S[i]$	represents the ith character of string S
$S[i =>]$	represents the suffix of S starting at index i
$<>$	represents an array

Step 1: Sort the characters in the string, assigning them numbers sequentially starting at 1. Then use those numbers for the remainder of the algorithm. Note that the numbers that are assigned depend on the text. So, if the text contains DNA characters A, C, G, and T only, then there will be only four numbers. Then pad the array with three 0's to avoid boundary cases. If we assume that the alphabet is a fixed size, then the sort takes some constant amount of time.

Input String, S	A	B	R	A	C	A	D	A	B	R	A			
New Problem	1	2	5	1	3	1	4	1	2	5	1	0	0	0
Index	0	1	2	3	4	5	6	7	8	9	10	11	12	13

Figure 12.27   Mapping of character in string to an array of integers

Example:

In our example, the mapping is A $=$ 1, B $=$ 2, C $=$ 3, D $=$ 4, and R $=$ 5; the transformation can be visualized in Figure 12.27.

Step 2: Divide the text into three groups:

$$S_0 = < S[3i]S[3i+1]S[3i+2] \qquad \text{for } i = 0, 1, 2, \ldots >$$
$$S_1 = < S[3i+1]S[3i+2]S[3i+3] \quad \text{for } i = 0, 1, 2, \ldots >$$
$$S_2 = < S[3i+2]S[3i+3]S[3i+4] \quad \text{for } i = 0, 1, 2, \ldots >$$

The idea is that each of S_0, S_1, S_2 consists of roughly $N/3$ symbols, but the symbols are no longer the original alphabet, but instead each new symbol is some group of three symbols from the original alphabet. We will call these *tri-characters*. Most importantly, the suffixes of S_0, S_1, and S_2 combine to form the suffixes of S. Thus one idea would be to recursively compute the suffixes of S_0, S_1, and S_2 (which by definition implicitly represent sorted strings) and then merge the results in linear time. However, since this would be three recursive calls on problems $1/3$ the original size, that would result in an $O(N \log N)$ algorithm. So the idea is going to be to avoid one of the three recursive calls, by computing two of the suffix groups recursively and using that information to compute the third suffix group.

Example:

In our example, if we look at the original character set and use $ to represent the padded character, we get

$$S_0 = [ABR], [ACA], [DAB], [RA\$]$$
$$S_1 = [BRA], [CAD], [ABR], [A\$\$]$$
$$S_2 = [RAC], [ADA], [BRA]$$

We can see that in S_0, S_1, and S_2, each tri-character is now a trio of characters from the original alphabet. Using that alphabet, S_0 and S_1 are arrays of length four and S_2 is an array of length three. S_0, S_1, and S_2 thus have four, four, and three suffixes, respectively. S_0's suffixes are [ABR][ACA][DAB][RA\$], [ACA][DAB][RA\$], [DAB][RA\$], [RA\$], which clearly correspond to the suffixes ABRACADABRA, ACADABRA, DABRA, and RA in the original string S. In the original string S these suffixes are located at indices 0, 3, 6, and 9, respectively, so looking at all three of S_0, S_1, and S_2, we can see that each S_i represents the suffixes that are located at indices $i \bmod 3$ in S.

Step 3: Concatenate S_1 and S_2 and recursively compute the suffix array. In order to compute this suffix array, we will need to sort the new alphabet of tri-characters. This can be done in linear time by three passes of radix sort, since the old characters were already sorted in step 1. If in fact all the tri-characters in the new alphabet are unique, then we do not even need to bother with a recursive call. Making three passes of radix sort takes linear time. If $T(N)$ is the running time of the suffix array construction algorithm, then the recursive call takes $T(2N/3)$ time.

Example:

In our example

$$S_1 S_2 = [BRA], [CAD], [ABR], [A\$\$], [RAC], [ADA], [BRA]$$

The sorted suffixes that will be computed recursively will represent tri-character strings as shown in Figure 12.28.

Notice that these are not exactly the same as the corresponding suffixes in S; however, if we strip out characters starting at the first $, we do have a match of suffixes. Also note that the indices returned by the recursive call do not correspond directly to the indices in S, though it is a simple matter to map them back. So to see how the algorithm actually forms the recursive call, observe that three passes of radix sort will assign the following alphabet: $[A\$\$] = 1, [ABR] = 2, [ADA] = 3, [BRA] = 4, [CAD] = 5, [RAC] = 6$. Figure 12.29 shows the mapping of tri-characters, the resulting array that is formed for S_1, S_2, and the resulting suffix array that is computed recursively.

Step 4: Compute the suffix array for S_0. This is easy to do because

$$
\begin{aligned}
S_0[i =>] &= S[3i =>] \\
&= S[3i] \, S[3i + 1 =>] \\
&= S[3i]S_1[i =>] \\
&= S_0[i]S_1[i =>]
\end{aligned}
$$

Since our recursive call has already sorted all $S_1[i =>]$, we can do step 4 with a simple two-pass radix sort: The first pass is on $S_1[i =>]$, and the second pass is on $S_0[i]$.

	Index	Substring Being Represented
0	3	[A$$] [RAC] [ADA] [BRA]
1	2	[ABR] [A$$] [RAC] [ADA] [BRA]
2	5	[ADA] [BRA]
3	6	[BRA]
4	0	[BRA] [CAD] [ABR] [A$$] [RAC] [ADA] [BRA]
5	1	[CAD] [ABR] [A$$] [RAC] [ADA] [BRA]
6	4	[RAC] [ADA] [BRA]

Figure 12.28 Suffix array for $S_1 S_2$ in tri-character set

S_1S_2	[BRA]	[CAD]	[ABR]	[A$$]	[RAC]	[ADA]	[BRA]			
Integers	4	5	2	1	6	3	4	0	0	0
SA[S_1S_2]	3	2	5	6	0	1	4	0	0	0
Index	0	1	2	3	4	5	6	7	8	9

Figure 12.29 Mapping of tri-characters, the resulting array that is formed for S_1, S_2, and the resulting suffix array that is computed recursively

Example:

In our example

$$S_0 = [ABR], [ACA], [DAB], [RA$]$$

From the recursive call in step 3, we can rank the suffixes in S_1 and S_2. Figure 12.30, how the indices in the original string can be referenced from the recursively computed suffix array and shows how the suffix array from Figure 12.29 leads to a ranking of suffixes among $S_1 + S_2$. Entries in the next to last row are easily obtained from the prior two rows. In the last row, the ith entry is given by the location of i in the row labelled $SA[S_1, S_2]$.

The ranking established in S_1 can be used directly for the first radix sort pass on S_0. Then we do a second pass on the single characters from S, using the prior radix sort to break ties. Notice that it is convenient if S_1 has exactly as many elements as S_0. Figure 12.31 shows how we can compute the suffix array for S_0.

At this point, we now have the suffix array for S_0 and for the combined group S_1 and S_2. Since this is a two-pass radix sort, this step takes $O(N)$.

Step 5: Merge the two suffix arrays using the standard algorithm to merge two sorted lists. The only issue is that we must be able to compare each suffix pair in constant time. There are two cases.

Case 1: Comparing an S_0 element with an S_1 element: Compare the first letter; if they do not match, we are done; otherwise, compare the remainder of S_0 (which is an S_1 suffix) with the remainder of S_1 (which is an S_2 suffix); those are already ordered, so we are done.

Case 2: Comparing an S_0 element with an S_2 element: Compare at most the first two letters; if we still have a match, then at that point compare the remainder of S_0 (which

	S_1				S_2		
	[BRA]	[CAD]	[ABR]	[A$$]	[RAC]	[ADA]	[BRA]
Index in S	1	4	7	10	2	5	8
SA[S_1S_2]	3	2	5	6	0	1	4
SA using S's indices	10	7	5	8	1	4	2
Rank in group	5	6	2	1	7	3	4

Figure 12.30 Ranking of suffixes based on suffix array shown in Figure 12.29

	S_0				
	[ABR]	[ACA]	[DAB]	[RA$]	
Index	0	3	6	9	
Index of second element	1	4	7	10	*add one to above*
Radix Pass 1 ordering	5	6	2	1	*last line of Figure 12.30*
Radix Pass 2 ordering	1	2	3	4	*stably radix sort by first char*
Rank in group	1	2	3	4	*using results of previous line*
SA, using S's indices	0	3	6	9	*using results of previous line*

Figure 12.31 Computing suffix array for S_0

after skipping the two letters becomes an S_2 suffix) with the remainder of S_2 (which after skipping two letters becomes an S_1 suffix); as in case 1, those suffixes are already ordered by SA12 so we are done.

Example:

In our example, we have to merge

	A	A	D	R
SA for S_0	0	3	6	9

↑

with

	A	A	A	B	B	C	R
SA for S_1 and S_2	10	7	5	8	1	4	2

↑

The first comparison is between index 0 (an A), which is an S_0 element and index 10 (also an A) which is an S_1 element. Since that is a tie, we now have to compare index 1 with index 11. Normally this would have already been computed, since index 1 is S_1, while index 11 is in S_2. However, this is special because index 11 is past the end of the string; consequently it always represents the earlier suffix lexicographically, and the first element in the final suffix array is 10. We advance in the second group and now we have.

	A	A	D	R
SA for S_0	0	3	6	9

↑

	A	A	A	B	B	C	R
SA for S_1 and S_2	10	7	5	8	1	4	2

↑

Final SA	10										
Input S	A	B	R	A	C	A	D	A	B	R	A
Index	0	1	2	3	4	5	6	7	8	9	10

Again the first characters match, so we compare indices 1 and 8, and this is already computed, with index 8 having the smaller string. So that means that now 7 goes into the final suffix array, and we advance the second group, obtaining

SA for S_0	A	A	D	R
	0	3	6	9

↑ (under the 0)

SA for S_1 and S_2	A	A	A	B	B	C	R
	10	7	5	8	1	4	2

↑ (under the 7)

Final SA	10	7									
Input S	A	B	R	A	C	A	D	A	B	R	A
Index	0	1	2	3	4	5	6	7	8	9	10

Once again, the first characters match, so now we have to compare indices 1 and 6. Since this is a comparison between an S_1 element and an S_0 element, we cannot look up the result. Thus we have to compare characters directly. Index 1 contains a B and index 6 contains a D, so index 1 wins. Thus 0 goes into the final suffix array and we advance the first group.

SA for S_0	A	A	D	R
	0	3	6	9

↑ (under the 3)

SA for S_1 and S_2	A	A	A	B	B	C	R
	10	7	5	8	1	4	2

↑ (under the 5)

Final SA	10	7	0								
Input S	A	B	R	A	C	A	D	A	B	R	A
Index	0	1	2	3	4	5	6	7	8	9	10

The same situation occurs on the next comparison between a pair of A's; the second comparison is between index 4 (a C) and index 6 (a D), so the element from the first group advances.

SA for S_0		A	A	D	R
		0	3	6	9

↑

SA for S_1 and S_2		A	A	A	B	B	C	R
		10	7	5	8	1	4	2

↑

Final SA	10	7	0	3							
Input S	A	B	R	A	C	A	D	A	B	R	A
Index	0	1	2	3	4	5	6	7	8	9	10

At this point, there are no ties for a while, so we quickly advance to the last characters of each group:

SA for S_0		A	A	D	R
		0	3	6	9

↑

SA for S_1 and S_2		A	A	A	B	B	C	R
		10	7	5	8	1	4	2

↑

Final SA	10	7	0	3	5	8	1	4	6		
Input S	A	B	R	A	C	A	D	A	B	R	A
Index	0	1	2	3	4	5	6	7	8	9	10

Finally, we get to the end. The comparison between two R's requires that we compare the next characters, which are at indices 10 and 3. Since this comparison is between an S_1 element and an S_0 element, as we saw before, we cannot look up the result and must compare directly. But those are also the same, so now we have to compare indices 11 and 4, which is an automatic winner for index 11 (since it is past the end of the string). Thus the R in index 9 advances, and then we can finish the merge. Notice that had we not been at the end of the string, we could have used the fact that the comparison is between an S_2 element and an S_1 element, which means the ordering would have been obtainable from the suffix array for $S_1 + S_2$.

	A	A	D	R
SA for S_0	0	3	6	9

↑

	A	A	A	B	B	C	R
SA for S_1 and S_2	10	7	5	8	1	4	2

↑

Final SA	10	7	0	3	5	8	1	4	6	9	2
Input S	A	B	R	A	C	A	D	A	B	R	A
Index	0	1	2	3	4	5	6	7	8	9	10

Since this is a standard merge, with at most two comparisons per suffix pair, this step takes linear time. The entire algorithm thus satisfies $T(N) = T(2N/3) + O(N)$ and takes linear time. Although we have only computed the suffix array, the LCP information can also be

```
1      /*
2       * Fill in the suffix array and LCP information for String str
3       * @param str the input String
4       * @param sa existing array to place the suffix array
5       * @param LCP existing array to place the LCP information
6       */
7      public static void createSuffixArray( String str, int [ ] sa, int [ ] LCP )
8      {
9          if( sa.length != str.length( ) || LCP.length != str.length( ) )
10             throw new IllegalArgumentException( );
11
12         int N = str.length( );
13
14         int [ ] s = new int[ N + 3 ];
15         int [ ] SA = new int[ N + 3 ];
16
17         for( int i = 0; i < N; i++ )
18             s[ i ] = str.charAt( i );
19
20         makeSuffixArray( s, SA, N, 256 );
21
22         for( int i = 0; i < N; i++ )
23             sa[ i ] = SA[ i ];
24
25         makeLCPArray( s, sa, LCP );   // Figure 12.50 and Exercise 12.9
26     }
```

Figure 12.32 Code to set up the first call to makeSuffixArray; create appropriate size arrays, and to keep things simple; just use the 256 ASCII character codes

computed as the algorithm runs, but there are some tricky details that are involved, and often the LCP information is computed by a separate linear-time algorithm.

We close by providing a working implementation to compute suffix arrays; rather than fully implementing step 1 to sort the original characters, we'll assume only a small set of ASCII characters (residing in values 1–255) are present in the string. In Figure 12.32, we allocate the arrays that have three extra slots for padding and call makeSuffixArray, which is the basic linear-time algorithm.

Figure 12.33 shows makeSuffixArray. At lines 12–16, it allocates all the needed arrays and makes sure that S_0 and S_1 have the same number of elements (lines 17–22); it then delegates work to assignNames, computeS12, computeS0, and merge.

```
1       // find the suffix array SA of s[ 0..n-1 ] in {1..K}^n
2       // require s[ n ] = s[ n + 1 ] = s[ n + 2 ] = 0, n >= 2
3       public static void makeSuffixArray( int [ ] s, int [ ] SA,
4                                           int n, int K )
5       {
6           int n0 = ( n + 2 ) / 3;
7           int n1 = ( n + 1 ) / 3;
8           int n2 = n / 3;
9           int t = n0 - n1;  // 1 iff n%3 == 1
10          int n12 = n1 + n2 + t;
11
12          int [ ] s12  = new int[ n12 + 3 ];
13          int [ ] SA12 = new int[ n12 + 3 ];
14          int [ ] s0   = new int[ n0 ];
15          int [ ] SA0  = new int[ n0 ];
16
17          // generate positions in s for items in s12
18          // the "+t" adds a dummy S1 suffix if n%3 == 1
19          // at that point, the size of s12 is n12
20          for( int i = 0, j = 0; i < n + t; i++ )
21              if( i % 3 != 0 )
22                  s12[ j++ ] = i;
23
24          int K12 = assignNames( s, s12, SA12, n0, n12, K );
25
26          computeS12( s12, SA12, n12, K12 );
27          computeS0( s, s0, SA0, SA12, n0, n12, K );
28          merge( s, s12, SA, SA0, SA12, n, n0, n12, t );
29      }
```

Figure 12.33   The main routine for linear-time suffix array construction

```
1     // Assigns the new tri-character names.
2     // At end of routine, SA will have indices into s, in sorted order
3     // and s12 will have new character names
4     // Returns the number of names assigned; note that if
5     // this value is the same as n12, then SA is a suffix array for s12.
6     private static int assignNames( int [ ] s, int [ ] s12, int [ ] SA12,
7                                     int n0, int n12, int K )
8     {
9         // radix sort the new character trios
10        radixPass( s12 , SA12, s, 2, n12, K );
11        radixPass( SA12, s12 , s, 1, n12, K );
12        radixPass( s12 , SA12, s, 0, n12, K );
13
14        // find lexicographic names of triples
15        int name = 0;
16        int c0 = -1, c1 = -1, c2 = -1;
17
18        for( int i = 0; i < n12; i++ )
19        {
20            if( s[ SA12[ i ] ] != c0 || s[ SA12[ i ] + 1 ] != c1
21                                     || s[ SA12[ i ] + 2 ] != c2 )
22            {
23                name++;
24                c0 = s[ SA12[ i ] ];
25                c1 = s[ SA12[ i ] + 1 ];
26                c2 = s[ SA12[ i ] + 2 ];
27            }
28
29            if( SA12[ i ] % 3 == 1 )
30                s12[ SA12[ i ] / 3 ]      = name;  // S1
31            else
32                s12[ SA12[ i ] / 3 + n0 ] = name;  // S2
33        }
34
35        return name;
36    }
```

Figure 12.34 Routine to compute and assign the tri-character names

assignNames, shown in Figure 12.34, begins by performing three passes of radix sort. Then, it assigns names (i.e., numbers), sequentially using the next available number if the current item has a different trio of characters than the prior item (recall that the tri-characters have already been sorted by the three passes of radix sort, and also recall that S_0 and S_1 have the same size, so at line 32, adding n0 adds the number of elements in S_1). We can use the basic counting radix sort from Chapter 7 to obtain a linear-time sort. This

```
1      // stably sort in[0..n-1] with indices into s that has keys in 0..K
2      // into out[0..n-1]; sort is relative to offset into s
3      // uses counting radix sort
4      private static void radixPass( int [ ] in, int [ ] out, int [ ] s,
5                                    int offset, int n, int K )
6      {
7          int [ ] count = new int[ K + 2 ];
8
9          for( int i = 0; i < n; i++ )
10             count[ s[ in[ i ] + offset ] + 1 ]++;
11
12         for( int i = 1; i <= K + 1; i++ )
13             count[ i ] += count[ i - 1 ];
14
15         for( int i = 0; i < n; i++ )
16             out[ count[ s[ in[ i ] + offset ] ]++ ] = in[ i ];
17     }
18
19     // stably sort in[0..n-1] with indices into s that has keys in 0..K
20     // into out[0..n-1]
21     // uses counting radix sort
22     private static void radixPass( int [ ] in, int [ ] out, int [ ] s,
23                                    int n, int K )
24     {
25         radixPass( in, out, s, 0, n, K );
26     }
```

Figure 12.35 A counting radix sort for the suffix array

code is shown in Figure 12.35. The array in represents the indexes into s; the result of the radix sort is that the indices are sorted so that the characters in s are sorted at those indices (where the indices are offset as specified).

Figure 12.36 contains the routines to compute the suffix arrays for s12, and then s0.

Finally, the merge routine is shown in Figure 12.37, with some supporting routines in Figure 12.38. The merge routine has the same basic look and feel as the standard merging algorithm seen in Figure 7.10.

```
1        // Compute the suffix array for s12, placing result into SA12
2        private static void computeS12( int [ ] s12, int [ ] SA12,
3                                        int n12, int K12 )
4        {
5            if( K12 == n12 ) // if unique names, don't need recursion
6                for( int i = 0; i < n12; i++ )
7                    SA12[ s12[ i ] - 1 ] = i;
8            else
9            {
10               makeSuffixArray( s12, SA12, n12, K12 );
11                   // store unique names in s12 using the suffix array
12               for( int i = 0; i < n12; i++ )
13                   s12[ SA12[ i ] ] = i + 1;
14           }
15       }
16
17       private static void computeS0( int [ ] s, int [ ] s0, int [ ] SA0,
18                                      int [ ] SA12, int n0, int n12, int K )
19       {
20           for( int i = 0, j = 0; i < n12; i++ )
21               if( SA12[ i ] < n0 )
22                   s0[ j++ ] = 3 * SA12[ i ];
23
24           radixPass( s0, SA0, s, n0, K );
25       }
```

Figure 12.36 Compute the suffix array for s12 (possibly recursively) and the suffix array for s0

12.5 *k*-d Trees

Suppose that an advertising company maintains a database and needs to generate mailing labels for certain constituencies. A typical request might require sending out a mailing to people who are between the ages of 34 and 49 and whose annual income is between $100,000 and $150,000. This problem is known as a two-dimensional range query. In one dimension, the problem can be solved by a simple recursive algorithm in $O(M + \log N)$ average time, by traversing a preconstructed binary search tree. Here M is the number of matches reported by the query. We would like to obtain a similar bound for two or more dimensions.

The two-dimensional search tree has the simple property that branching on odd levels is done with respect to the first key, and branching on even levels is done with respect to the second key. The root is arbitrarily chosen to be an odd level. Figure 12.39 shows a 2-d tree. Insertion into a 2-d tree is a trivial extension of insertion into a binary search

```
1        // merge sorted SA0 suffixes and sorted SA12 suffixes
2        private static void merge( int [ ] s, int [ ] s12,
3                                   int [ ] SA, int [ ] SA0, int [ ] SA12,
4                                   int n, int n0, int n12, int t )
5        {
6            int p = 0, k = 0;
7
8            while( t != n12 && p != n0 )
9            {
10               int i = getIndexIntoS( SA12, t, n0 ); // S12 index in s
11               int j = SA0[ p ];                     // S0  index in s
12
13               if( suffix12IsSmaller( s, s12, SA12, n0, i, j, t ) )
14               {
15                   SA[ k++ ] = i;
16                   t++;
17               }
18               else
19               {
20                   SA[ k++ ] = j;
21                   p++;
22               }
23           }
24
25           while( p < n0 )
26               SA[ k++ ] = SA0[ p++ ];
27           while( t < n12 )
28               SA[ k++ ] = getIndexIntoS( SA12, t++, n0 );
29       }
```

Figure 12.37   Merge the suffix arrays SA0 and SA12

tree: As we go down the tree, we need to maintain the current level. To keep our code simple, we assume that a basic item is an array of two elements. We then need to toggle the level between 0 and 1. Figure 12.40 shows the code to perform an insertion. We use recursion in this section; a nonrecursive implementation that would be used in practice is straightforward and left as Exercise 12.17. One difficulty is duplicates, particularly since several items can agree in one key. Our code allows duplicates and always places them in right branches; clearly this can be a problem if there are too many duplicates.

A moment's thought will convince you that a randomly constructed 2-d tree has the same structural properties as a random binary search tree: The height is $O(\log N)$ on average, but $O(N)$ in the worst case.

Unlike binary search trees, for which clever $O(\log N)$ worst-case variants exist, there are no schemes that are known to guarantee a balanced 2-d tree. The problem is that such a scheme would likely be based on tree rotations, and tree rotations don't work in 2-d trees.

```
1        private static int getIndexIntoS( int [ ] SA12, int t, int n0 )
2        {
3            if( SA12[ t ] < n0 )
4                return SA12[ t ] * 3 + 1;
5            else
6                return ( SA12[ t ] - n0 ) * 3 + 2;
7        }
8
9          // True if [a1 a2] <= [b1 b2]
10        private static boolean leq( int a1, int a2, int b1, int b2 )
11          { return a1 < b1 || a1 == b1 && a2 <= b2; }
12
13          // True if [a1 a2 a3] <= [b1 b2 b3]
14        private static boolean leq( int a1, int a2, int a3, int b1, int b2, int b3 )
15          { return a1 < b1 || a1 == b1 && leq( a2, a3, b2, b3 ); }
16
17        private static boolean suffix12IsSmaller( int [ ] s, int [ ] s12,
18                                  int [ ] SA12, int n0, int i, int j, int t )
19        {
20            if( SA12[ t ] < n0 )  // s1 vs s0; can break tie after 1 char
21                return leq( s[ i ], s12[ SA12[ t ] + n0 ],
22                            s[ j ], s12[ j / 3 ] );
23            else                // s2 vs s0; can break tie after 2 chars
24                return leq( s[ i ], s[ i + 1 ], s12[ SA12[ t ] - n0 + 1 ],
25                            s[ j ], s[ j + 1 ], s12[ j / 3 + n0 ] );
26        }
```

Figure 12.38 Supporting routines for merging the suffix arrays SA0 and SA12

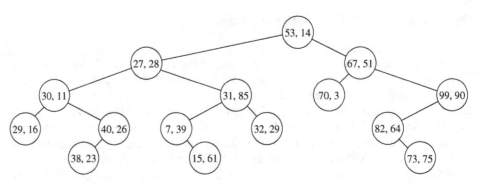

Figure 12.39 Sample 2-d tree

```
1        public void insert( AnyType [ ] x )
2        {
3            root = insert( x, root, 0 );
4        }
5
6        private KdNode<AnyType> insert( AnyType [ ] x, KdNode<AnyType> t, int level )
7        {
8            if( t == null )
9                t = new KdNode<>( x );
10           else if( x[ level ].compareTo( t.data[ level ] ) < 0 )
11               t.left = insert( x, t.left, 1 - level );
12           else
13               t.right = insert( x, t.right, 1 - level );
14           return t;
15       }
```

Figure 12.40 Insertion into 2-d trees

The best one can do is to periodically rebalance the tree by reconstructing a subtree, as described in the exercises. Similarly, there are no deletion algorithms beyond the obvious lazy deletion strategy. If all the items arrive before we need to process queries, then we can construct a perfectly balanced 2-d tree in $O(N \log N)$ time; we leave this as Exercise 12.15c.

Several kinds of queries are possible on a 2-d tree. We can ask for an exact match, or a match based on one of the two keys; the latter type of request is a **partial match query**. Both of these are special cases of an (orthogonal) **range query**.

An orthogonal range query gives all items whose first key is between a specified set of values and whose second key is between another specified set of values. This is exactly the problem that was described in the introduction to this section. A range query is easily solved by a recursive tree traversal, as shown in Figure 12.41. By testing before making a recursive call, we can avoid unnecessarily visiting all nodes.

To find a specific item, we can set low equal to high equal to the item we are searching for. To perform a partial match query, we set the range for the key not involved in the match to $-\infty$ to ∞. The other range is set with the low and high point equal to the value of the key involved in the match.

An insertion or exact match search in a 2-d tree takes time that is proportional to the depth of the tree, namely, $O(\log N)$ on average and $O(N)$ in the worst case. The running time of a range search depends on how balanced the tree is, whether or not a partial match is requested, and how many items are actually found. We mention three results that have been shown.

For a perfectly balanced tree, a range query could take $O(M + \sqrt{N})$ time in the worst case, to report M matches. At any node, we may have to visit two of the four grandchildren, leading to the equation $T(N) = 2T(N/4) + O(1)$. In practice, however, these searches tend to be very efficient, and even the worst case is not poor because for typical N, the difference between $\sqrt{N}$ and $\log N$ is compensated by the smaller constant that is hidden in the Big-Oh notation.

```
1       /**
2        * Print items satisfying
3        * low[ 0 ] <= x[ 0 ] <= high[ 0 ] and
4        * low[ 1 ] <= x[ 1 ] <= high[ 1 ].
5        */
6       public void printRange( AnyType [ ] low, AnyType [ ] high )
7       {
8           printRange( low, high, root, 0 );
9       }
10
11      private void printRange( AnyType [ ] low, AnyType [ ] high,
12                              KdNode<AnyType> t, int level )
13      {
14          if( t != null )
15          {
16              if( low[ 0 ].compareTo( t.data[ 0 ] ) <= 0 &&
17                      low[ 1 ].compareTo( t.data[ 1 ] ) <= 0 &&
18                      high[ 0 ].compareTo( t.data[ 0 ] ) >= 0 &&
19                      high[ 1 ].compareTo( t.data[ 1 ] ) >= 0 )
20                  System.out.println( "(" + t.data[ 0 ] + ","
21                          + t.data[ 1 ] + ")" );
22
23              if( low[ level ].compareTo( t.data[ level ] ) <= 0 )
24                  printRange( low, high, t.left, 1 - level );
25              if( high[ level ].compareTo( t.data[ level ] ) >= 0 )
26                  printRange( low, high, t.right, 1 - level );
27          }
28      }
```

Figure 12.41   2-d trees: range search

For a randomly constructed tree, the average running time of a partial match query is $O(M + N^\alpha)$, where $\alpha = (-3 + \sqrt{17})/2$ (see below). A recent, and somewhat surprising, result is that this essentially describes the average running time of a range search of a random 2-d tree.

For k dimensions, the same algorithm works; we just cycle through the keys at each level. However, in practice the balance starts getting worse because typically the effect of duplicates and nonrandom inputs becomes more pronounced. We leave the coding details as an exercise for the reader and mention the analytical results: For a perfectly balanced tree, the worst-case running time of a range query is $O(M + kN^{1-1/k})$. In a randomly constructed k-d tree, a partial match query that involves p of the k keys takes $O(M + N^\alpha)$, where α is the (only) positive root of

$$(2 + \alpha)^p (1 + \alpha)^{k-p} = 2^k$$

Computation of α for various p and k is left as an exercise; the value for $k = 2$ and $p = 1$ is reflected in the result stated above for partial matching in random 2-d trees.

Although there are several exotic structures that support range searching, the k-d tree is probably the simplest such structure that achieves respectable running times.

12.6 Pairing Heaps

The last data structure we examine is the **pairing heap**. The analysis of the pairing heap is still open, but when decreaseKey operations are needed, it seems to outperform other heap structures. The most likely reason for its efficiency is its simplicity. The pairing heap is represented as a heap-ordered tree. Figure 12.42 shows a sample pairing heap.

The actual pairing heap implementation uses a left child, right sibling representation as discussed in Chapter 4. The decreaseKey operation, as we will see, requires that each node contain an additional link. A node that is a leftmost child contains a link to its parent; otherwise the node is a right sibling and contains a link to its left sibling. We'll refer to this field as the prev field. Figure 12.43 shows the actual representation of the pairing heap in Figure 12.42.

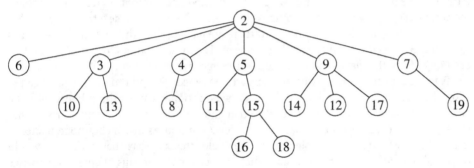

Figure 12.42 Sample pairing heap: abstract representation

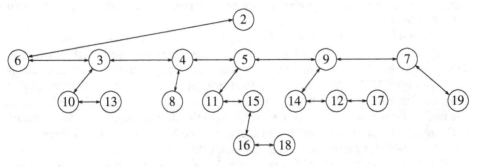

Figure 12.43 Actual representation of previous pairing heap

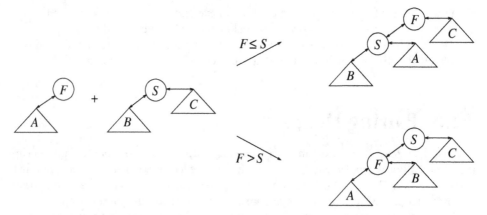

Figure 12.44   `compareAndLink` merges two subheaps

We begin by sketching the basic operations. To merge two pairing heaps, we make the heap with the larger root a left child of the heap with the smaller root. Insertion is, of course, a special case of merging. To perform a `decreaseKey`, we lower the value in the requested node. Because we are not maintaining parent links for all nodes, we don't know if this violates the heap order. Thus we cut the adjusted node from its parent and complete the `decreaseKey` by merging the two heaps that result. To perform a `deleteMin`, we remove the root, creating a collection of heaps. If there are c children of the root, then $c - 1$ calls to the merge procedure will reassemble the heap. The most important detail is the method used to perform the merge and how the $c - 1$ merges are applied.

Figure 12.44 shows how two subheaps are combined. The procedure is generalized to allow the second subheap to have siblings. As we mentioned earlier, the subheap with the larger root is made a leftmost child of the other subheap. The code is straightforward and shown in Figure 12.45. Notice that we have several instances in which a node reference is tested against `null` before assigning its `prev` field; this suggests that perhaps it would be useful to have a `nullNode` sentinel, which was customary in this chapter's search tree implementations.

`decreaseKey` requires a *position* object, which is just an interface that `PairNode` implements. Figure 12.46 shows the `PairNode` class and `Position` interface, which are both nested in the `PairingHeap` class.

The `insert` and `decreaseKey` operations are, then, simple implementations of the abstract description. Since the position of an item is determined (irrevocably) when an item is first inserted, `insert` returns the `PairNode` it creates back to the caller. The code is shown in Figure 12.47. Our routine for `decreaseKey` throws an exception if the new value is not smaller than the old; otherwise, the resulting structure might not obey heap order. The basic `deleteMin` procedure follows directly from the abstract description and is shown in Figure 12.48. The `element` field is set to `null`, so if the `Position` is used in a `decreaseKey`, it will be possible for `decreaseKey` to detect that the `Position` is no longer valid.

The devil, of course, is in the details: How is `combineSiblings` implemented? Several variants have been proposed, but none has been shown to provide the same amortized

```
1     /**
2      * Internal method that is the basic operation to maintain order.
3      * Links first and second together to satisfy heap order.
4      * @param first root of tree 1, which may not be null.
5      *    first.nextSibling MUST be null on entry.
6      * @param second root of tree 2, which may be null.
7      * @return result of the tree merge.
8      */
9     private PairNode<AnyType> compareAndLink( PairNode<AnyType> first,
10                                              PairNode<AnyType> second )
11    {
12        if( second == null )
13            return first;
14
15        if( second.element.compareTo( first.element ) < 0 )
16        {
17            // Attach first as leftmost child of second
18            second.prev = first.prev;
19            first.prev = second;
20            first.nextSibling = second.leftChild;
21            if( first.nextSibling != null )
22                first.nextSibling.prev = first;
23            second.leftChild = first;
24            return second;
25        }
26        else
27        {
28            // Attach second as leftmost child of first
29            second.prev = first;
30            first.nextSibling = second.nextSibling;
31            if( first.nextSibling != null )
32                first.nextSibling.prev = first;
33            second.nextSibling = first.leftChild;
34            if( second.nextSibling != null )
35                second.nextSibling.prev = second;
36            first.leftChild = second;
37            return first;
38        }
39    }
```

Figure 12.45 Pairing heaps: routine to merge two subheaps

```
1     public class PairingHeap<AnyType extends Comparable<? super AnyType>>
2     {
3         /**
4          * The Position interface represents a type that can
5          * be used for the decreaseKey operation.
6          */
7         public interface Position<AnyType>
8         {
9             AnyType getValue( );
10        }
11
12        private static class PairNode<AnyType> implements Position<AnyType>
13        {
14            public PairNode( AnyType theElement )
15              { element = theElement; leftChild = nextSibling = prev = null; }
16
17            public AnyType getValue( )
18              { return element; }
19
20            public AnyType               element;
21            public PairNode<AnyType>     leftChild;
22            public PairNode<AnyType>     nextSibling;
23            public PairNode<AnyType>     prev;
24        }
25
26        private PairNode<AnyType> root;
27        private int theSize;
28
29        // Rest of class follows
30    }
```

Figure 12.46 PairNode class and Position interface in PairingHeap

bounds as the Fibonacci heap. It has recently been shown that almost all the proposed methods are in fact theoretically less efficient than the Fibonacci heap. Even so, the method coded in Figure 12.49 always seems to perform as well as or better than other heap structures, including the binary heap, for the typical graph theory uses that involve a host of decreaseKey operations.

This method, known as **two-pass merging**, is the simplest and most practical of the many variants that have been suggested. We first scan left to right, merging pairs of children.[4] After the first scan, we have half as many trees to merge. A second scan is then performed, right to left. At each step we merge the rightmost tree remaining from the

---

[4] We must be careful if there is an odd number of children. When that happens, we merge the last child with the result of the rightmost merge to complete the first scan.

```
1      /**
2       * Insert into the priority queue, and return a Position
3       * that can be used by decreaseKey. Duplicates are allowed.
4       * @param x the item to insert.
5       * @return the Position (PairNode) containing the newly inserted item.
6       */
7      public Position<AnyType> insert( AnyType x )
8      {
9          PairNode<AnyType> newNode = new PairNode<>( x );
10
11         if( root == null )
12             root = newNode;
13         else
14             root = compareAndLink( root, newNode );
15
16         theSize++;
17         return newNode;
18     }
19
20     /**
21      * Change the value of the item stored in the pairing heap.
22      * @param pos any Position returned by insert.
23      * @param newVal the new value, which must be smaller than the currently stored value.
24      * @throws IllegalArgumentException if pos is null, deleteMin has
25      *         been performed on pos, or new value is larger than old.
26      */
27     public void decreaseKey( Position<AnyType> pos, AnyType newVal )
28     {
29         PairNode<AnyType> p = (PairNode<AnyType>) pos;
30
31         if( p == null || p.element == null || p.element.compareTo( newVal ) < 0 )
32             throw new IllegalArgumentException( );
33
34         p.element = newVal;
35         if( p != root )
36         {
37             if( p.nextSibling != null )
38                 p.nextSibling.prev = p.prev;
39             if( p.prev.leftChild == p )
40                 p.prev.leftChild = p.nextSibling;
41             else
42                 p.prev.nextSibling = p.nextSibling;
43
44             p.nextSibling = null;
45             root = compareAndLink( root, p );
46         }
47     }
```

Figure 12.47 Pairing heaps: insert and decreaseKey

```
1      /**
2       * Remove the smallest item from the priority queue.
3       * @return the smallest item.
4       * @throws UnderflowException if pairing heap is empty.
5       */
6      public AnyType deleteMin( )
7      {
8          if( isEmpty( ) )
9              throw new UnderflowException( );
10
11         AnyType x = findMin( );
12         root.element = null; // null it out in case used in decreaseKey
13         if( root.leftChild == null )
14             root = null;
15         else
16             root = combineSiblings( root.leftChild );
17
18         theSize--;
19         return x;
20     }
```

Figure 12.48   Pairing heap `deleteMin`

first scan with the current merged result. As an example, if we have eight children, c_1 through c_8, the first scan performs the merges c_1 and c_2, c_3 and c_4, c_5 and c_6, and c_7 and c_8. As a result we obtain $d_1, d_2, d_3,$ and d_4. We perform the second pass by merging d_3 and d_4; d_2 is then merged with that result, and then d_1 is merged with the result of the previous merge.

Our implementation requires an array to store the subtrees. In the worst case, $N - 1$ items could be children of the root, but declaring an array of size N inside of `combineSiblings` would give an $O(N)$ algorithm so we use an expanding array instead.

Other merging strategies are discussed in the exercises. The only simple merging strategy that is easily seen to be poor is a left-to-right single-pass merge (Exercise 12.29). The pairing heap is a good example of "simple is better" and seems to be the method of choice for serious applications requiring the `decreaseKey` or `merge` operation.

Summary

In this chapter, we've seen several efficient variations of the binary search tree. The top-down splay tree provides $O(\log N)$ amortized performance, the treap gives $O(\log N)$ randomized performance, and the red-black tree, gives $O(\log N)$ worst-case performance for the basic operations. The trade-offs between the various structures involve code complexity, ease of deletion, and differing searching and insertion costs. It is difficult to say

```
1      /**
2       * Internal method that implements two-pass merging.
3       * @param firstSibling the root of the conglomerate;
4       *     assumed not null.
5       */
6      private PairNode<AnyType> combineSiblings( PairNode<AnyType> firstSibling )
7      {
8          if( firstSibling.nextSibling == null )
9              return firstSibling;
10
11             // Store the subtrees in an array
12         int numSiblings = 0;
13         for( ; firstSibling != null; numSiblings++ )
14         {
15             treeArray = doubleIfFull( treeArray, numSiblings );
16             treeArray[ numSiblings ] = firstSibling;
17             firstSibling.prev.nextSibling = null;  // break links
18             firstSibling = firstSibling.nextSibling;
19         }
20         treeArray = doubleIfFull( treeArray, numSiblings );
21         treeArray[ numSiblings ] = null;
22
23             // Combine subtrees two at a time, going left to right
24         int i = 0;
25         for( ; i + 1 < numSiblings; i += 2 )
26             treeArray[ i ] = compareAndLink( treeArray[ i ], treeArray[ i + 1 ] );
27
28             // j has the result of last compareAndLink.
29             // If an odd number of trees, get the last one.
30         int j = i - 2;
31         if( j == numSiblings - 3 )
32             treeArray[ j ] = compareAndLink( treeArray[ j ], treeArray[ j + 2 ] );
33
34             // Now go right to left, merging last tree with
35             // next to last. The result becomes the new last.
36         for( ; j >= 2; j -= 2 )
37             treeArray[ j - 2 ] = compareAndLink( treeArray[ j - 2 ], treeArray[ j ] );
38
39         return (PairNode<AnyType>) treeArray[ 0 ];
40     }
```

Figure 12.49 Pairing heaps: two-pass merging

```
41        private PairNode<AnyType> [ ]
42        doubleIfFull( PairNode<AnyType> [ ] array, int index )
43        {
44            if( index == array.length )
45            {
46                PairNode<AnyType> [ ] oldArray = array;
47
48                array = new PairNode[ index * 2 ];
49                for( int i = 0; i < index; i++ )
50                    array[ i ] = oldArray[ i ];
51            }
52            return array;
53        }
54
55        // The tree array for combineSiblings
56        private PairNode<AnyType> [ ] treeArray = new PairNode[ 5 ];
```

Figure 12.49 (*continued*)

that any one structure is a clear winner. Recurring themes include tree rotations and the use of sentinel nodes to eliminate many of the annoying tests for null references that would otherwise be necessary.

The suffix tree and array are a powerful data structure that allows quick repeated searching for a fixed text. The k-d tree provides a practical method for performing range searches, even though the theoretical bounds are not optimal.

Finally, we described and coded the pairing heap, which seems to be the most practical mergeable priority queue, especially when decreaseKey operations are required, even though it is theoretically less efficient than the Fibonacci heap.

Exercises

12.1   Prove that the amortized cost of a top-down splay is $O(\log N)$.

**12.2   Prove that there exist access sequences that require $2 \log N$ rotations per access for bottom-up splaying. Show that a similar result holds for top-down splaying.

12.3   Modify the splay tree to support queries for the kth smallest item.

12.4   Compare, empirically, the simplified top-down splay with the originally described top-down splay.

12.5   Write the deletion procedure for red-black trees.

12.6   Prove that the height of a red-black tree is at most $2 \log N$, and that this bound cannot be substantially lowered.

12.7   Show that every AVL tree can be colored as a red-black tree. Are all red-black trees AVL?

12.8   Draw a suffix tree and show the suffix array and LCP array for the following input strings:
a. ABCABCABC
b. MISSISSIPPI

12.9   Once the suffix array is constructed, the short routine shown in Figure 12.50 can be invoked from Figure 12.32 to create the longest common prefix array.
a. In the code, what does rank[i] represent?
b. Suppose that LCP[rank[i]] = h. Show that LCP[rank[i+1]] $\geq h - 1$.
c. Show that the algorithm in Figure 12.50 correctly computes the LCP array.
d. Prove that the algorithm in Figure 12.50 runs in linear time.

```
1      /*
2       * Create the LCP array from the suffix array
3       * @param s the input array populated from 0..N-1, with available pos N
4       * @param sa the already-computed suffix array 0..N-1
5       * @param LCP the resulting LCP array 0..N-1
6       */
7      public static void makeLCPArray( int [ ] s, int [ ] sa, int [ ] LCP )
8      {
9          int N = sa.length;
10         int [ ] rank = new int[ N ];
11
12         s[ N ] = -1;
13         for( int i = 0; i < N; i++ )
14             rank[ sa[ i ] ] = i;
15
16         int h = 0;
17         for( int i = 0; i < N; i++ )
18             if( rank[ i ] > 0 )
19             {
20                 int j = sa[ rank[ i ] - 1 ];
21
22                 while( s[ i + h ] == s[ j + h ] )
23                     h++;
24
25                 LCP[ rank[ i ] ] = h;
26                 if( h > 0 )
27                     h--;
28             }
29     }
```

Figure 12.50 LCP array construction from suffix array

12.10   Suppose that in the linear-time suffix array construction algorithm, instead of constructing three groups, we construct seven groups, using for $k = 0, 1, 2, 3, 4, 5, 6$

$$S_k = < S[7i + k]S[7i + k + 1]S[7i + k + 2]\ldots S[7i + k + 6] \text{ for } i = 0, 1, 2, \ldots >$$

a.   Show that with a recursive call to $S_3S_5S_6$, we have enough information to sort the other four groups S_0, S_1, S_2, and S_4.
b.   Show that this partitioning leads to a linear-time algorithm.

12.11   Implement the insertion routine for treaps nonrecursively by maintaining a stack. Is it worth the effort?

12.12   We can make treaps self-adjusting by using the number of accesses as a priority and performing rotations as needed after each access. Compare this method with the randomized strategy. Alternatively, generate a random number each time an item X is accessed. If this number is smaller than X's current priority, use it as X's new priority (performing the appropriate rotation).

**12.13   Show that if the items are sorted, then a treap can be constructed in linear time, even if the priorities are not sorted.

12.14   Implement red-black trees without using the `nullNode` sentinel. How much coding effort is saved by using the sentinel?

12.15   Suppose we store, for each node, the number of `null` links in its subtree; call this the node's *weight*. Adopt the following strategy: If the left and right subtrees have weights that are not within a factor of 2 of each other, then completely rebuild the subtree rooted at the node. Show the following:
a.   We can rebuild a node in $O(S)$, where S is the weight of the node.
b.   The algorithm has amortized cost of $O(\log N)$ per insertion.
c.   We can rebuild a node in a k-d tree in $O(S \log S)$ time, where S is the weight of the node.
d.   We can apply the algorithm to k-d trees, at a cost of $O(\log^2 N)$ per insertion.

12.16   Suppose we call `rotateWithLeftChild` on an arbitrary 2-d tree. Explain in detail all the reasons that the result is no longer a usable 2-d tree.

12.17   Implement the insertion and range search for the k-d tree. Do not use recursion.

12.18   Determine the time for partial match query for values of p corresponding to $k = 3$, 4, and 5.

12.19   For a perfectly balanced k-d tree, derive the worst-case running time of a range query that is quoted in the text (see p. 581).

12.20   The **2-d heap** is a data structure that allows each item to have two individual keys. `deleteMin` can be performed with respect to either of these keys. The 2-d heap is a complete binary tree with the following order property: For any node X at even depth, the item stored at X has the smallest key #1 in its subtree, while for any node X at odd depth, the item stored at X has the smallest key #2 in its subtree.
a.   Draw a possible 2-d heap for the items (1, 10), (2, 9), (3, 8), (4, 7), (5, 6).
b.   How do we find the item with minimum key #1?

c. How do we find the item with minimum key #2?

d. Give an algorithm to insert a new item into the 2-d heap.

e. Give an algorithm to perform deleteMin with respect to either key.

f. Give an algorithm to perform buildHeap in linear time.

12.21 Generalize the preceding exercise to obtain a *k-d heap,* in which each item can have k individual keys. You should be able to obtain the following bounds: insert in $O(\log N)$, deleteMin in $O(2^k \log N)$, and buildHeap in $O(kN)$.

12.22 Show that the k-d heap can be used to implement a double-ended priority queue.

12.23 Abstractly, generalize the k-d heap so that only levels that branch on key #1 have two children (all others have one).

a. Do we need links?

b. Clearly, the basic algorithms still work; what are the new time bounds?

12.24 Use a k-d tree to implement deleteMin. What would you expect the average running time to be for a random tree?

12.25 Use a k-d heap to implement a double-ended queue that also supports deleteMin.

12.26 Implement the pairing heap with a nullNode sentinel.

****12.27** Show that the amortized cost of each operation is $O(\log N)$ for the pairing heap algorithm in the text.

12.28 An alternative method for combineSiblings is to place all of the siblings on a queue, and repeatedly dequeue and merge the first two items on the queue, placing the result at the end of the queue. Implement this variation.

12.29 Show that using a stack instead of a queue in the previous exercise is bad, by giving a sequence that leads to $\Omega(N)$ cost per operation. This is the left-to-right single-pass merge.

12.30 Without decreaseKey, we can remove parent links. How competitive is the result with the skew heap?

12.31 Assume that each of the following is represented as a tree with child and parent references. Explain how to implement a decreaseKey operation.

a. Binary heap

b. Splay tree

12.32 When viewed graphically, each node in a 2-d tree *partitions the plane* into regions. For instance, Figure 12.51 shows the first five insertions into the 2-d tree in

Figure 12.51 The plane partitioned by a 2-d tree after the insertion of $p1 = (53, 14)$, $p2 = (27, 28)$, $p3 = (30, 11)$, $p4 = (67, 51)$, $p5 = (70, 3)$

Figure 12.52 The plane partitioned by a quad tree after the insertion of $p1 = (53, 14)$, $p2 = (27, 28)$, $p3 = (30, 11)$, $p4 = (67, 51)$, $p5 = (70, 3)$

Figure 12.39. The first insertion, of $p1$, splits the plane into a left part and a right part. The second insertion, of $p2$, splits the left part into a top part and a bottom part, and so on.

  a. For a given set of N items, does the order of insertion affect the final partition?

  b. If two different insertion sequences result in the same tree, is the same partition produced?

  c. Give a formula for the number of regions that result from the partition after N insertions.

  d. Show the final partition for the 2-d tree in Figure 12.39.

12.33 An alternative to the 2-d tree is the **quad tree.** Figure 12.52 shows how a plane is partitioned by a quad tree. Initially we have a region (which is often a square, but need not be). Each region may store one point. If a second point is inserted into a region, then the region is split into four equal-sized quadrants (northeast, southeast, southwest, and northwest). If this places the points in different quadrants (as when $p2$ is inserted), we are done; otherwise, we continue splitting recursively (as is done when $p5$ is inserted).

  a. For a given set of N items, does the order of insertion affect the final partition?

  b. Show the final partition if the same elements that were in the 2-d tree in Figure 12.39 are inserted into the quad tree.

12.34 A tree data structure can store the quad tree. We maintain the bounds of the original region. The tree root represents the original region. Each node is either a leaf that stores an inserted item, or has exactly four children, representing four quadrants. To perform a search, we begin at the root and repeatedly branch to an appropriate quadrant until a leaf (or null entry) is reached.

  a. Draw the quad tree that corresponds to Figure 12.52.

  b. What factors influence how deep the (quad) tree will be?

  c. Describe an algorithm that performs an orthogonal range query in a quad tree.

References

Top-down splay trees were described in the original splay tree paper [36]. A similar strategy, but without the crucial rotation, was described in [38]. The top-down red-black tree algorithm is from [18]; a more accessible description can be found in [35]. An implementation of top-down red-black trees without sentinel nodes is given in [15]; this provides a convincing demonstration of the usefulness of nullNode. Treaps [3] are based on the **Cartesian tree** described in [40]. A related data structure is the **priority search tree** [27].

Suffix trees were first described as a **position tree** by Weiner [41], who provided a linear-time algorithm for construction that was simplified by McCreight [28], and then by Ukkonen [39], who provided the first online linear-time algorithm. Farach [13] provided an alternate algorithm that is the basis for many of the linear-time suffix array construction algorithms. Numerous applications of suffix trees can be found in the text by Gusfield [19].

Suffix arrays were described by Manber and Myers [25]. The algorithm presented in the text is due to Kärkkäinen and Sanders [21]; another linear-time algorithm is due to Ko and Aluru [23]. The linear-time algorithm for constructing the LCP array from a suffix array in Exercise 12.9 was given in [22]. A survey of suffix array construction algorithms can be found in [32].

[1] shows that any problem that is solvable via suffix trees is solvable in equivalent time with suffix arrays. Because the input sizes for practical applications are so large, space is important, and thus much recent work has centered on suffix array and LCP array construction. In particular, for many algorithms, a cache-friendly slightly nonlinear algorithm can be preferable in practice to a noncache friendly linear algorithm [33]. For truly huge input sizes, in-memory construction is not always feasible. [6] is an example of an algorithm that can generate suffix arrays for 12GB of DNA sequences in a day on a single machine with only 2GB of RAM; see also [5] for a survey of external memory suffix array construction algorithms.

The k-d tree was first presented in [7]. Other range-searching algorithms are described in [8]. The worst case for range searching in a balanced k-d tree was obtained in [24], and the average-case results cited in the text are from [14] and [10].

The pairing heap and the alternatives suggested in the exercises were described in [17]. The study [20] suggests that the splay tree is the priority queue of choice when the decreaseKey operation is not required. Another study [37] suggests that the pairing heap achieves the same asymptotic bounds as the Fibonacci heap, with better performance in practice. However, a related study [29] using priority queues to implement minimum spanning tree algorithms suggests that the amortized cost of decreaseKey is not $O(1)$. M. Fredman [16] has settled the issue of optimality by proving that there are sequences for which the amortized cost of a decreaseKey operation is suboptimal (in fact, at least $\Omega(\log \log N)$). However, he has also shown that when used to implement Prim's minimum spanning tree algorithm, the pairing heap is optimal if the graph is slightly dense (that is, the number of edges in the graph is $O(N^{(1+\varepsilon)})$ for any ε). Pettie [32] has shown an upper bound of $O(2^{2\sqrt{\log \log N}})$ for decreaseKey. However, complete analysis of the pairing heap is still open.

The solutions to most of the exercises can be found in the primary references. Exercise 12.15 represents a "lazy" balancing strategy that has become somewhat popular. [26], [4], [11], and [9] describe specific strategies; [2] shows how to implement all of these strategies in one framework. A tree that satisfies the property in Exercise 12.15 is **weight-balanced**. These trees can also be maintained by rotations [30]. Part (d) is from [31]. A solution to Exercises 12.20 to 12.22 can be found in [12]. Quad trees are described in [34].

1. M. I. Abouelhoda, S. Kurtz, and E. Ohlebush, "Replacing Suffix Trees with Suffix Arrays," *Journal of Discrete Algorithms*, 2 (2004), 53–86.

2. A. Andersson, "General Balanced Trees," *Journal of Algorithms*, 30 (1991), 1–28.

3. C. Aragon and R. Seidel, "Randomized Search Trees," *Proceedings of the Thirtieth Annual Symposium on Foundations of Computer Science* (1989), 540–545.

4. J. L. Baer and B. Schwab, "A Comparison of Tree-Balancing Algorithms," *Communications of the ACM*, 20 (1977), 322–330.

5. M. Barsky, U. Stege, and A. Thomo, "A Survey of Practical Algorithms for Suffix Tree Construction in External Memory," *Software: Practice and Experience*, 40 (2010) 965–988.

6. M. Barsky, U. Stege, A. Thomo, and C. Upton, "Suffix Trees for Very Large Genomic Sequences," *Proceedings of the 18th ACM Conference on Information and Knowledge Management* (2009), 1417–1420.

7. J. L. Bentley, "Multidimensional Binary Search Trees Used for Associative Searching," *Communications of the ACM*, 18 (1975), 509–517.

8. J. L. Bentley and J. H. Friedman, "Data Structures for Range Searching," *Computing Surveys*, 11 (1979), 397–409.

9. H. Chang and S. S. Iyengar, "Efficient Algorithms to Globally Balance a Binary Search Tree," *Communications of the ACM*, 27 (1984), 695–702.

10. P. Chanzy, "Range Search and Nearest Neighbor Search," Master's Thesis, McGill University, 1993.

11. A. C. Day, "Balancing a Binary Tree," *Computer Journal*, 19 (1976), 360–361.

12. Y. Ding and M. A. Weiss, "The k-d Heap: An Efficient Multi-Dimensional Priority Queue," *Proceedings of the Third Workshop on Algorithms and Data Structures* (1993), 302–313.

13. M. Farach, "Optimal Suffix Tree Construction with Large Alphabets," *Proceedings of the 38th Annual IEEE Symposium on Foundations of Computer Science* (1997), 137–143.

14. P. Flajolet and C. Puech, "Partial Match Retrieval of Multidimensional Data," *Journal of the ACM*, 33 (1986), 371–407.

15. B. Flamig, *Practical Data Structures in C++*, John Wiley, New York, 1994.

16. M. Fredman, "Information Theoretic Implications for Pairing Heaps," *Proceedings of the Thirtieth Annual ACM Symposium on the Theory of Computing* (1998), 319–326.

17. M. L Fredman, R. Sedgewick, D. D. Sleator, and R. E. Tarjan, "The Pairing Heap: A New Form of Self-Adjusting Heap," *Algorithmica*, 1 (1986), 111–129.

18. L. J. Guibas and R. Sedgewick, "A Dichromatic Framework for Balanced Trees," *Proceedings of the Nineteenth Annual Symposium on Foundations of Computer Science* (1978), 8–21.

19. D. Gusfield, *Algorithms on Strings, Trees and Sequences: Computer Science and Computational Biology*, Cambridge University Press, Cambridge, U.K., 1997.

20. D. W. Jones, "An Empirical Comparison of Priority-Queue and Event-Set Implementations," *Communications of the ACM*, 29 (1986), 300–311.

21. J. Kärkkäinen and P. Sanders, "Simple Linear Work Suffix Array Construction," *Proceedings of the 30th International Colloquium on Automata, Languages and Programming* (2003), 943–955.

22. T. Kasai, G. Lee, H. Arimura, S. Arikawa, and K. Park, "Linear-Time Longest Common-Prefix Computation in Suffix Arrays and Its Applications," *Proceedings of the 12th Annual Symposium on Combinatorial Pattern Matching* (2001), 181–192.

23. P. Ko and S. Aluru, "Space Efficient Linear Time Construction of Suffix Arrays," *Proceedings of the 14th Annual Symposium on Combinatorial Pattern Matching* (2003), 203–210.

24. D. T. Lee and C. K. Wong, "Worst-Case Analysis for Region and Partial Region Searches in Multidimensional Binary Search Trees and Balanced Quad Trees," *Acta Informatica,* 9 (1977), 23–29.

25. U. Manber and G. Myers, "Suffix Arrays: A New Method for On-Line String Searches," *SIAM Journal on Computing,* 22 (1993), 935–948.

26. W. A. Martin and D. N. Ness, "Optimizing Binary Trees Grown with a Sorting Algorithm," *Communications of the ACM,* 15 (1972), 88–93.

27. E. McCreight, "Priority Search Trees," *SIAM Journal of Computing,* 14 (1985), 257–276.

28. E. M. McCreight, "A Space-Economical Suffix Tree Construction Algorithm," *Journal of the ACM,* 23 (1976), 262–272.

29. B. M. E. Moret and H. D. Shapiro, "An Empirical Analysis of Algorithms for Constructing a Minimum Spanning Tree," *Proceedings of the Second Workshop on Algorithms and Data Structures* (1991), 400–411.

30. J. Nievergelt and E. M. Reingold, "Binary Search Trees of Bounded Balance," *SIAM Journal on Computing,* 2 (1973), 33–43.

31. M. H. Overmars and J. van Leeuwen, "Dynamic Multidimensional Data Structures Based on Quad and K-D Trees," *Acta Informatica,* 17 (1982), 267–285.

32. S. Pettie, "Towards a Final Analysis of Pairing Heaps," *Proceedings of the 46th Annual IEEE Symposium on Foundations of Computer Science* (2005), 174–183.

33. S. J. Puglisi, W. F. Smyth, and A. Turpin, "A Taxonomy of Suffix Array Construction Algorithms," *ACM Computing Surveys,* 39 (2007), 1–31.

34. A. H. Samet, "The Quadtree and Related Hierarchical Data Structures," *Computing Surveys,* 16 (1984), 187–260.

35. R. Sedgewick and K. Wayne, *Algorithms,* 4th ed., Addison-Wesley Professional, Boston, Mass. 2011.

36. D. D. Sleator and R. E. Tarjan, "Self Adjusting Binary Search Trees," *Journal of the ACM,* 32 (1985), 652–686.

37. J. T. Stasko and J. S. Vitter, "Pairing Heaps: Experiments and Analysis," *Communications of the ACM,* 30 (1987), 234–249.

38. C. J. Stephenson, "A Method for Constructing Binary Search Trees by Making Insertions at the Root," *International Journal of Computer and Information Science,* 9 (1980), 15–29.

39. E. Ukkonen, "On-Line Construction of Suffix Trees," *Algorithmica,* 14 (1995), 249–260.

40. J. Vuillemin, "A Unifying Look at Data Structures," *Communications of the ACM,* 23 (1980), 229–239.

41. P. Weiner, "Linear Pattern Matching Algorithm," *Proceedings of the 14th Annual IEEE Symposium on Switching and Automata Theory,* (1973), 1–11.

INDEX